Lecture Notes
in Business Information Processing 564

Series Editors

Wil van der Aalst , *RWTH Aachen University, Aachen, Germany*
Sudha Ram , *University of Arizona, Tucson, USA*
Michael Rosemann , *Queensland University of Technology, Brisbane, Australia*
Clemens Szyperski, *Microsoft Research, Redmond, USA*
Giancarlo Guizzardi , *University of Twente, Enschede, The Netherlands*

LNBIP reports state-of-the-art results in areas related to business information systems and industrial application software development – timely, at a high level, and in both printed and electronic form.

The type of material published includes

- Proceedings (published in time for the respective event)
- Postproceedings (consisting of thoroughly revised and/or extended final papers)
- Other edited monographs (such as, for example, project reports or invited volumes)
- Tutorials (coherently integrated collections of lectures given at advanced courses, seminars, schools, etc.)
- Award-winning or exceptional theses

LNBIP is abstracted/indexed in DBLP, EI and Scopus. LNBIP volumes are also submitted for the inclusion in ISI Proceedings.

Arik Senderovich · Cristina Cabanillas ·
Irene Vanderfeesten · Hajo A. Reijers
Editors

Business Process Management Forum

BPM 2025 Forum
Seville, Spain, August 31 – September 5, 2025
Proceedings

Editors
Arik Senderovich
York University
Toronto, ON, Canada

Irene Vanderfeesten
KU Leuven
Leuven, Belgium

Cristina Cabanillas
University of Seville
Seville, Spain

Hajo A. Reijers
Utrecht University
Utrecht, The Netherlands

ISSN 1865-1348   ISSN 1865-1356 (electronic)
Lecture Notes in Business Information Processing
ISBN 978-3-032-02928-7   ISBN 978-3-032-02929-4 (eBook)
https://doi.org/10.1007/978-3-032-02929-4

© The Editor(s) (if applicable) and The Author(s), under exclusive license
to Springer Nature Switzerland AG 2026

This work is subject to copyright. All rights are solely and exclusively licensed by the Publisher, whether the whole or part of the material is concerned, specifically the rights of translation, reprinting, reuse of illustrations, recitation, broadcasting, reproduction on microfilms or in any other physical way, and transmission or information storage and retrieval, electronic adaptation, computer software, or by similar or dissimilar methodology now known or hereafter developed.
The use of general descriptive names, registered names, trademarks, service marks, etc. in this publication does not imply, even in the absence of a specific statement, that such names are exempt from the relevant protective laws and regulations and therefore free for general use.
The publisher, the authors and the editors are safe to assume that the advice and information in this book are believed to be true and accurate at the date of publication. Neither the publisher nor the authors or the editors give a warranty, expressed or implied, with respect to the material contained herein or for any errors or omissions that may have been made. The publisher remains neutral with regard to jurisdictional claims in published maps and institutional affiliations.

This Springer imprint is published by the registered company Springer Nature Switzerland AG
The registered company address is: Gewerbestrasse 11, 6330 Cham, Switzerland

If disposing of this product, please recycle the paper.

Preface

This volume comprises all papers presented at the BPM Forum of the 23rd International Conference on Business Process Management (BPM), held during September 2025 in Seville, Spain. The BPM Forum provides an opportunity for papers that, while not fully meeting the quality criteria for the main conference track, make meaningful and promising contributions to the BPM community. These contributions may include novel applications, early-stage results with strong potential, or creative perspectives on known problems. Forum papers were evaluated through the same rigorous review process as main track submissions and are presented in full length in this separate post-proceedings volume. All Forum papers were presented in a special session during the main conference, offering a platform for constructive feedback and engagement from the broader BPM community.

This year, the conference received a total of 142 submissions, of which 132 were selected for full review. After a rigorous double-blind review process involving at least three Program Committee members and one Senior Program Committee member per submission, and a discussion phase followed by meta-reviews, 30 papers were accepted for presentation at the main conference. In parallel, 24 papers were accepted for inclusion in the BPM Forum.

The Forum papers cover a diverse and timely set of topics, reflecting the evolving socio-technical and AI-enhanced landscape of BPM. Of the 24 accepted Forum papers, 23 are included in this volume. These span the three thematic tracks: 8 in Foundations, 10 in Engineering, and 5 in Management. The 2025 Forum contributions explore themes such as the use of large language models in process monitoring and predictive analytics, RPA-induced technostress, blockchain-based compliance and documentation systems, process similarity and fairness in decision making, as well as new methods for model orchestration and simulation.

Open Science aiming at reproducibility and replicability of research results remained a major principle for the BPM Forum. Authors were explicitly encouraged to provide supplementary materials, such as datasets, source code, and implemented prototypes. Papers that made such artifacts available are recognized with badges in this volume.

We also promoted diversity, equity, and inclusion (DEI) across all conference activities. We extend our sincere thanks to the DEI Chairs and the entire BPM community for fostering a welcoming and representative environment.

We would like to thank all authors, both regular and senior members of the Program Committees, and all external reviewers. Their diligent reviews and effective collaboration made a rigorous and timely review process possible and enabled the high-quality research output reflected in the papers of this volume. We also gratefully acknowledge the sponsors of BPM 2025 for their valuable support, which significantly contributed to making this event possible. Special thanks to Celonis as our platinum sponsor, and to SAP Signavio and Process Science as gold sponsors, for their commitment to the BPM community. Finally, we express our heartfelt appreciation to the BPM 2025 General

Chairs and the local organizers for creating a memorable event in the vibrant setting of Seville.

We hope that the papers presented at the BPM Forum 2025 and captured in this volume will provide fertile stimulation for BPM academics and professionals.

September 2025

Arik Senderovich
Cristina Cabanillas
Irene Vanderfeesten
Hajo A. Reijers

Chun Ouyang	Queensland University of Technology, Australia
Daniel Ritter	SAP, Germany
Dirk Fahland	Eindhoven University of Technology, the Netherlands
Eric Verbeek	Eindhoven University of Technology, the Netherlands
Ekkart Kindler	Technical University of Denmark, Denmark
Ernest Teniente	UPC, Spain
Fabiana Fournier	IBM Research, Israel
Felix Mannhardt	Eindhoven University of Technology, the Netherlands
Francesco Folino	ICAR-CNR, Italy
Hugo A. López	Technical University of Denmark, Denmark
Irina Lomazova	National Research University Higher School of Economics, Russia
Ivan Donadello	Free University of Bozen-Bolzano, Italy
Izack Cohen	Bar-Ilan University, Israel
Jana-Rebecca Rehse	University of Mannheim, Germany
Johannes De Smedt	KU Leuven, Belgium
Jörg Desel	Fernuniversität in Hagen, Germany
Lorenzo Rossi	University of Camerino, Italy
Maria Teresa Gómez López	University of Seville, Spain
Mario Luca Bernardi	Università degli studi del Sannio, Italy
Massimiliano Ronzani	Fondazione Bruno Kessler (FBK), Italy
Matteo Zavatteri	University of Padova, Italy
Monique Snoeck	KU Leuven, Belgium
Patrick Delfmann	University of Koblenz-Landau, Germany
Peter Fettke	DFKI and Saarland University, Germany
Rik Eshuis	Eindhoven University of Technology, the Netherlands
Sarah M. Winkler	Free University of Bozen-Bolzano, Italy
Simone Agostinelli	Universitas Mercatorum of Rome, Italy
Stephan Fahrenkrog-Petersen	Humboldt-Universität zu Berlin, Germany
Tijs Slaats	University of Copenhagen, Denmark
Valeria Fionda	University of Calabria, Italy

Track II: Engineering

Senior Program Committee

Agnes Koschmider	University of Bayreuth, Germany
Andrea Burattin	Technical University of Denmark, Denmark
Avigdor Gal	Technion, Israel
Barbara Weber	University of St. Gallen, Switzerland
Benoît Depaire	Hasselt University, Belgium
Boualem Benatallah	University of New South Wales, Australia
Boudewijn F. van Dongen	Eindhoven University of Technology, the Netherlands
Han van der Aa	University of Vienna, Austria
Henrik Leopold	Kühne Logistics University, Germany
Ingo Weber	Technical University of Munich, Germany
Jan Mendling	Humboldt-Universität zu Berlin, Germany
Jochen De Weerdt	Katholieke Universiteit Leuven, Belgium
Jorge Munoz-Gama	Pontificia Universidad Católica de Chile, Chile
Luise Pufahl	Technical University of Munich, Germany
Manuel Resinas	University of Seville, Spain
Marcos Sepúlveda	Pontificia Universidad Católica de Chile, Chile
Marlon Dumas	University of Tartu, Estonia
Massimiliano de Leoni	University of Padua, Italy
Massimo Mecella	Sapienza University of Rome, Italy
Minseok Song	Pohang University of Science and Technology, South Korea
Moe Thandar Wynn	Queensland University of Technology, Australia
Niels Martin	Hasselt University, Belgium
Oscar Pastor	Universidad Politécnica de Valencia, Spain
Pnina Soffer	University of Haifa, Israel
Remco M. Dijkman	Eindhoven University of Technology, the Netherlands
Shazia W. Sadiq	University of Queensland, Australia
Stefanie Rinderle-Ma	Technical University of Munich, Germany

Program Committee

Abel Armas-Cervantes	University of Melbourne, Australia
Alfonso Márquez-Chamorro	Universidad de Sevilla, Spain
Andrea Delgado	Universidad de la República, Uruguay
Andrea Morichetta	TU Vienna, Austria

Andrés Jiménez Ramírez	Universidad de Sevilla, Spain
Anna Kalenkova	University of Adelaide, Australia
Barbara Re	Università di Camerino, Italy
Bernhard Axmann	Technische Hochschule Ingolstadt, Germany
Carlos Fernandez-Llatas	Universidad Politecnica de Valencia, Spain
Cesare Pautasso	University of Lugano, Switzerland
Claudia Diamantini	Università Politecnica delle Marche, Italy
Daniel Amyot	University of Ottawa, Canada
Daniela Grigori	Laboratoire LAMSADE, University Paris-Dauphine, France
Elisa Marengo	Free University of Bozen-Bolzano, Italy
Fabio Casati	Servicenow, USA
Francesco Leotta	Sapienza Università di Roma, Italy
Georg Grossmann	University of South Australia, Australia
Gert Janssenswillen	Hasselt University, Belgium
Giovanni Meroni	Technical University of Denmark, Denmark
Hye-young Paik	University of New South Wales, Australia
Jari Peeperkorn	KU Leuven, Belgium
Joerg Evermann	Memorial University of Newfoundland, Canada
Joscha Grüger	University of Trier, Germany
José González Enríquez	University of Seville, Spain
Julius Köpke	University of Klagenfurt, Austria
Karolin Winter	Technical University of Munich, Germany
Kate Revoredo	Humboldt Universität zu Berlin, Germany
Krzysztof Kluza	AGH University of Krakow, Poland
Lars Ackermann	University of Bayreuth, Germany
Laura Genga	Eindhoven University of Technology, the Netherlands
Laura Maruster	University of Groningen, the Netherlands
Luciano García-Bañuelos	Tecnológico de Monterrey, Mexico
Lucinéia Heloisa Thom	Federal University of Rio Grande do Sul, Brazil
Manuel Lama	Universidad de Santiago de Compostela, Spain
Marcelo Fantinato	University of São Paulo, Brazil
Maxim Vidgof	WU Vienna, Austria
Michael Arias	University of Costa Rica, Costa Rica
Natalia Sidorova	Eindhoven University of Technology, the Netherlands
Nick van Beest	Data61, Australia
Nicola Zannone	Eindhoven University of Technology, the Netherlands
Niek Tax	Meta, UK
Orlenys López-Pintado	University of Tartu, Estonia

Oscar González-Rojas	Universidad de los Andes, Colombia
Pascal Poizat	Université Paris Nanterre, France
Pierluigi Plebani	Politecnico di Milano, Italy
Renuka Sindhgatta	IBM, India
Robert Andrews	Queensland University of Technology, Australia
Roy J. Yang	Queensland University of Technology, Australia
Saimir Bala	Humboldt Universität zu Berlin, Germany
Sarajane Marques Peres	University of São Paulo, Brazil
Sareh Sadeghianasl	Queensland University of Technology, Australia
Sebastiaan J. van Zelst	Celonis, Germany
Simon K. Poon	University of Sydney, Australia
Stefan Schönig	Universität Regensburg, Germany
Victoria Torres	Universidad Politécnica de Valencia, Spain
Walid Fdhila	University of Vienna, Austria

Track III: Management

Senior Program Committee

Adela Del Río Ortega	University of Seville, Spain
Amy Van Looy	Ghent University, Belgium
Christian Janiesch	TU Dortmund University, Germany
Daniel Beverungen	Paderborn University, Germany
Flavia Santoro	UERJ, Brazil
Marta Indulska	University of Queensland, Australia
Maximilian Röglinger	FIM Research Center Finance & Information Management, Germany
Michael Rosemann	Queensland University of Technology, Australia
Mieke Jans	Hasselt University, Belgium
Mojca Indihar Štemberger	University of Ljubljana, Slovenia
Paul Grefen	Eindhoven University of Technology, the Netherlands
Peter Loos	IWi at DFKI, Saarland University, Germany
Peter Trkman	University of Ljubljana, Slovenia
Ralf Plattfaut	University of Duisburg-Essen, Germany
Thomas Grisold	University of St. Gallen, Switzerland

Contents

Foundations

From Sound Workflow Nets to LTL$_f$ Declarative Specifications by Casting
Three Spells 3
   Luca Barbaro, Giovanni Varricchione, Marco Montali,
   and Claudio Di Ciccio

High-Level Requirements-Driven Business Process Compliance 23
   Juanita Caballero-Villalobos, Andrea Burattin, and Hugo A. López

The WHY in Business Processes: Unification of Causal Process Models 40
   Yuval David, Fabiana Fournier, Lior Limonad, and Inna Skarbovsky

A Self-orchestration Model for Business Collaborations with Verifiable
Process History Credentials 58
   Martin Farkas, Bertalan Zoltán Péter, and Imre Kocsis

Comparing Apples with Oranges: An Assessment Framework
for Model-System Similarity 75
   Martin Kabierski, Jana-Rebecca Rehse,
   and Jan Martijn E. M. van der Werf

Stochastic BPMN and Their Conformance 92
   Aleksandar Kuzmanoski, Jan Niklas van Detten,
   and Sander J. J. Leemans

Instance Configuration and Scheduling Based on the Resource-Augmented
Process Structure Tree 110
   Felix Schumann, G. Wessel van der Heijden, and Stefanie Rinderle-Ma

Rethinking Business Process Simulation: A Utility-Based Evaluation
Framework 128
   Konrad Özdemir, Lukas Kirchdorfer, Keyvan Amiri Elyasi,
   Han van der Aa, and Heiner Stuckenschmidt

Engineering

Discovering Comprehensive Branched Declarative Process Constraints 147
   Christos Balaktsis, Ioannis Mavroudopoulos, Marco Comuzzi,
   Anastasios Gounaris, and Fabrizio Maria Maggi

SimBank: From Simulation to Solution in Prescriptive Process Monitoring 165
   *Jakob De Moor, Hans Weytjens, Johannes De Smedt,
and Jochen De Weerdt*

Detecting Undesired Process Behavior by Means of Retrieval Augmented
Generation 183
   Michael Grohs, Adrian Rebmann, and Jana-Rebecca Rehse

Layouting Object-Centric Directly Follows Graphs 204
   Deoksang Lee, Minseok Song, and Wil M. P. van der Aalst

Leveraging the Diamond Pattern for Scalable and Upgradeable
Blockchain-Based Business Process Management Applications 221
   *Victor Lemaire, Tiphaine Henry, Álvaro García-Pérez, Walid Gaaloul,
and Sara Tucci-Piergiovanni*

Balancing Confidentiality and Transparency for Blockchain-Based
Process-Aware Information Systems 238
   *Alessandro Marcelletti, Edoardo Marangone, Michele Kryston,
and Claudio Di Ciccio*

A Rollout-Based Algorithm and Reward Function for Resource Allocation
in Business Processes 256
   Jeroen Middelhuis, Zaharah Bukhsh, Ivo Adan, and Remco Dijkman

Enhancing Predictive Process Monitoring on Small-Scale Event Logs
Using LLMs 274
   *Alessandro Padella, Paolo Frazzetto, Nicolò Navarin,
and Massimiliano de Leoni*

Progression: A Lightweight BPMN Engine Simplifying the Execution
and Monitoring of Process Models 291
   Thomas M. Prinz, Yongsun Choi, and Anja Vetterlein

Predicting Newcomer Capabilities and Performance in Process Execution 308
   Roy Jing Yang, Chun Ouyang, and Remco Dijkman

Management

A Case for Public Process Documentation: Robodebt an Automated
Decision Making System 327
   Adam Banham, Azumah Mamudu, and Rehan Syed

Affective Business Process Design 345
   Thomas Grisold and Michael Rosemann

Process Autonomization: Rethinking Business Process Management 361
  Christian Janiesch, Marek Kowalkiewicz, and Michael Rosemann

Automation to Agitation: Unveiling RPA-Induced Technostress 378
  Ishadi Mirispelakotuwa, Rehan Syed, and Moe T. Wynn

FairPM: A Taxonomy of Bias and Interventions in Process Mining 395
  Kate Revoredo, Saimir Bala, and Flavia Santoro

Author Index 413

Foundations

From Sound Workflow Nets to LTL$_f$ Declarative Specifications by Casting Three Spells

Luca Barbaro[1]($\boxtimes$), Giovanni Varricchione[2], Marco Montali[3], and Claudio Di Ciccio[2]

[1] Sapienza University of Rome, Rome, Italy
luca.barbaro@uniroma1.it
[2] Utrecht University, Utrecht, The Netherlands
{g.varricchione,c.diciccio}@uu.nl
[3] Free University of Bozen-Bolzano, Bolzano, Italy
montali@inf.unibz.it

Abstract. In process management, effective behavior modeling is essential for understanding execution dynamics and identifying potential issues. Two complementary paradigms have emerged in the pursuit of this objective: the imperative approach, representing all allowed runs of a system in a graph-based model, and the declarative one, specifying the rules that a run must not violate in a constraint-based specification. Extensive studies have been conducted on the synergy and comparisons of the two paradigms. To date, though, whether a declarative specification could be systematically derived from an imperative model such that the original behavior was fully preserved (and if so, how) remained an unanswered question. In this paper, we propose a three-fold contribution. (1) We introduce a systematic approach to synthesize declarative process specifications from safe and sound Workflow nets. (2) We prove behavioral equivalence of the input net with the output specification, alongside related guarantees. (3) We experimentally demonstrate the scalability and compactness of our encoding through tests conducted with synthetic and real-world testbeds.

Keywords: Process modeling · Petri nets · Linear-time Temporal Logic on finite traces · Declare

1 Introduction

The act of modeling a process is a key element in a multitude of domains, including business process management [22], and is specifically tailored to meet the specific requirements and objectives of the individual application scenarios. Two fundamental, complementary paradigms cover the spectrum of modeling: the imperative (e.g., Worflow nets [3] and BPMN models [22]) and the declarative (e.g., DECLARE maps [29] and DCR Graphs [24]). Generally, the former class offers the opportunity to explicitly capture the set of actions available at each reachable state of the process, from start to end. However, such models often show limitations when it comes to capture flexibility in execution, since the possible runs highly vary and their graph-based structure gets cluttered. To compactly represent that variability, declarative specifications depict the rules that govern the behavior of every instance, leaving the allowed sequences implicit as long as none of those rules is violated.

Research has acknowledged that none of the available representations would be superior in all cases, as imperative and declarative approaches are apt to different comprehension tasks [30]. The ability to translate one representation to the other while preserving behavioral equivalence would allow the comparison and selection of the most suitable one. The first work in this direction is [32], where a systematic procedure is proposed to turn a declarative specification into an imperative model. Other endeavors followed to close the circle by providing an approximate solution to the inverse path (i.e., from an imperative model to a declarative specification), resorting on re-discovery over simulations [33], state space exploration [34], or behavioral comparison [7].

The goal of this work is to close the existing gap of this procedural-to-declarative direction. To this end, we show how to encode a safe and sound Workflow net [3] into a behaviorally equivalent DECLARE specification [19]. The three spells mentioned in the title of this paper refer to the fact that we only employ three parametric constraint types (*templates*) in the DECLARE repertoire. Importantly, the encoding is obtained in one pass and modularly over the net, preserving runs and choice points without incurring the state space explosion caused by concurrency unfolding. A byproduct of the encoding is that a safe and sound Workflow net induces a star-free regular language when considering transitions of the former as the alphabet of the latter. This strengthens the well-known fact that languages induced by sound Workflow nets are regular. Then, we evaluate the scalability of our approach by experimentally testing our proof-of-concept implementation against synthetic and real-world testbeds. Also, we show a downstream reasoning task on process diagnostics with public benchmarks.

The remainder of the paper is organized as follows. Section 2 provides an overview of the background knowledge our research is built upon. We describe our algorithm and formally discuss its correctness and complexity in Sect. 3. Section 4 evaluates our implementation to demonstrate the feasibility of our approach. Finally, we conclude by discussing related works in Sect. 5 and outlining future research directions in Sect. 6.

2 Background

In this section, we formally describe the foundational pillars our work is built upon.

2.1 Linear Temporal Logic on Finite Traces

LTL_f has the same syntax of LTL [31], but is interpreted on finite traces. Here, we consider the LTL dialect including past modalities [27] for declarative process specifications as in [10]. From now on, we fix a finite set Σ representing an alphabet of propositional symbols. A (finite) *trace* $\pi = \langle a_1, \ldots, a_n \rangle \in \Sigma$ is a finite sequence of symbols of length $|\pi| = n$ (with $n \in \mathbb{N}$), where the occurrence of symbol a_i at instant i of the trace represents an *event* that witnesses a_i at instant i —we write $\pi(i) = a_i$. Notice that *at each instant we assume that one and only one symbol occurs*. Using standard notation from regular expressions, Σ^* denotes the overall set of finite traces derived from events belonging to Σ.

Definition 1 (LTL_f). *(Syntax)* Well-formed LTL_f formulae are built from a finite non-empty alphabet of symbols $\Sigma \ni a$, the unary temporal operators **X** ("next") and **Y** ("yesterday"), and the binary temporal operators **U** ("until") and **S** ("since") as follows:

$$\varphi ::= a \mid (\neg \varphi) \mid (\varphi_1 \wedge \varphi_2) \mid (\mathbf{X}\, \varphi) \mid (\varphi_1\, \mathbf{U}\, \varphi_2) \mid (\mathbf{Y}\, \varphi) \mid (\varphi_1\, \mathbf{S}\, \varphi_2).$$

(Semantics) An LTL_f formula φ is inductively satisfied *in some instant i (with $1 \leq i \leq n$) of a trace π of length $n \in \mathbb{N}$, written $\pi, i \models \varphi$, if the following holds:*

$\pi, i \models a$ iff $\pi(i)$ is assigned with a; $\quad \pi, i \models \neg \varphi$ iff $\pi, i \not\models \varphi$;
$\pi, i \models \varphi_1 \wedge \varphi_2$ iff $\pi, i \models \varphi_1$ and $\pi, i \models \varphi_2$;
$\pi, i \models \mathbf{X}\, \varphi$ iff $i < n$ and $\pi, i+1 \models \varphi$; $\quad \pi, i \models \mathbf{Y}\, \varphi$ iff $i > 1$ and $\pi, i-1 \models \varphi$;
$\pi, i \models \varphi_1\, \mathbf{U}\, \varphi_2$ iff there exists $i \leq j \leq n$ s.t. $\pi, j \models \varphi_2$, and $\pi, k \models \varphi_1$ for all k s.t. $i \leq k < j$;
$\pi, i \models \varphi_1\, \mathbf{S}\, \varphi_2$ iff there exists $1 \leq j \leq i$ such that $\pi, j \models \varphi_2$, and $\pi, k \models \varphi_1$ for all k s.t. $j < k \leq i$.

A formula φ is satisfied by a trace π, written $\pi \models \varphi$, iff $\pi, 1 \models \varphi$. ◁

From the basic operators above, the following can be derived: Classical boolean abbreviations **true, false**, $\vee$, $\rightarrow$; **F** $\varphi \equiv$ **true U** φ indicating that φ eventually holds true in the trace ("*eventually*"); **G** $\varphi \equiv \neg$ **F** $\neg \varphi$ indicating that φ holds true from the current on ("*always*").

As an example, let $\pi = \langle a, b, c, e, f, g, u, v \rangle$ be a trace and φ_1, φ_2 and φ_3 three LTL_f formulae defined as follows: $\varphi_1 \doteq f$; $\varphi_2 \doteq \mathbf{Y}\, c$; $\varphi_3 \doteq \mathbf{G}\, (f \rightarrow \mathbf{Y}\, e)$. We have that $\pi, 1 \not\models \varphi_1$ whereas $\pi, 5 \models \varphi_1$; $\pi, 1 \not\models \varphi_2$ while $\pi, 4 \models \varphi_2$; $\pi, 1 \models \varphi_3$ (hence, $\pi \models \varphi_3$); in fact, $\pi, i \models \varphi_3$ for any instant $1 \leq i \leq |\pi|$.

2.2 Finite State Automata

Every LTL_f formula can be encoded into a *deterministic finite state automaton* [12].

Definition 2 (Finite State Automaton). *A (deterministic) finite state automaton (FSA) is a tuple* $\mathcal{A} = (S, s_0, s_F, \Sigma, \delta)$, *where S is a finite set of states, $s_0 \in S$ is the initial state, $s_F \subseteq S$ is the set of accepting states, Σ is the input alphabet of the automaton, and $\delta : S \times \Sigma \rightarrow S$ is the state transition function.* ◁

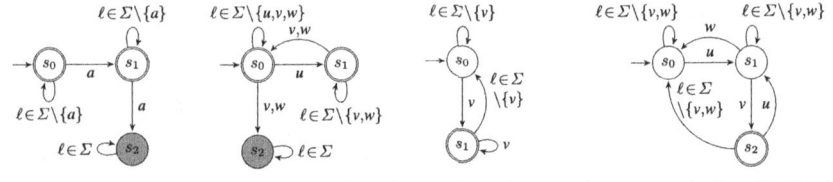

(a) ATMOSTONE(a) (b) ALT.PREC.($u,\{v,w\}$) (c) END(v) (d) {ALT.PREC.($u,\{v,w\}$),END(v)}

Fig. 1. Example FSAs of DECLARE constraints. The FSA in Fig. 1(d) is trimmed.

Table 1. Semantics of some DECLARE constraint templates

Template	LTL_f expression [10,13]	Description
ATMOSTONE (x)	$\mathbf{G}\,(x \rightarrow \neg\,\mathbf{X}\,\mathbf{F}\,x)$	x occurs at most once in the trace
END (x)	$\mathbf{G}\,\mathbf{F}\,x$	The last event of any trace is x
ALTERNATEPRECEDENCE (y,x)	$\mathbf{G}\,(x \rightarrow \mathbf{Y}\,(\neg x\,\mathbf{S}\,y))$	Every occurrence of x requires that y occurred before, with no recurrence of x in between

Figure 1 depicts three finite state automata (FSAs). An FSA reads in input sequences of symbols (*"string"*) of its input alphabet. It starts in its initial state s_0 and updates the state after having read each symbol via the state transition function δ. We say that a FSA *accepts* a string if after reading it is in one of its accepting states (i.e., a state in s_F), and otherwise we say that it *rejects* that string. The set of strings accepted by an FSA $\mathcal{A}$ is called the *language* of $\mathcal{A}$.

Definition 3 (Bisimilarity). *Two FSAs $\mathcal{A} = (S, s_0, s_F, \Sigma, \delta)$ and $\mathcal{A}' = (S', s'_0, s'_F, \Sigma, \delta')$ are bisimilar if and only if there exists a relation $\sim \subset S \times S'$ such that the following hold: $(s_0, s'_0) \in \sim$; if $(s,s') \in \sim$, then $(\delta(s, \ell), \delta'(s', \ell)) \in \sim$ for any $\ell \in \Sigma$; if $(s, s') \in \sim$, then $s \in s_F$ if and only if $s' \in s'_F$.* ◁

Observation 1. *In the case of FSAs, bisimilarity coincides with language equivalence, i.e., two FSAs are bisimilar if and only if the sets of strings that they accept are equal [25].* ◁

A direct approach that builds a non-deterministic FSA $\mathcal{A}_\varphi$ accepting all and only the traces that satisfy a given LTL_f formula φ is presented in [12]. We make two further observations from [25]: *(i)* the so-obtained FSAs can be determinized, minimized, and trimmed using standard techniques without modifying the accepted language, and *(ii)* given any two FSAs $\mathcal{A}_\varphi$ and $\mathcal{A}_{\varphi'}$, their *product* $\mathcal{A}_\varphi \times \mathcal{A}_{\varphi'}$ recognizes all and only the traces of $\varphi \wedge \varphi'$.

2.3 LTL_f-Based Declarative Specifications

The semantics of a DECLARE template is given as an LTL_f formula. Given the free variables (*"parameters"*) x and y, e.g., ALTERNATEPRECEDENCE (y, x) corresponds to $\mathbf{G}\,(x \rightarrow \mathbf{Y}\,(\neg x\,\mathbf{S}\,y))$, witnessing that for every instant in which x is verified, then a previous instant must verify y without any occurrences of x in between. Hitherto, we will occasionally use an abbreviation for the template name—ALT.PREC. (y, x). Table 1 shows the LTL_f formulae of some templates of the DECLARE repertoire. Standard DECLARE imposes that template parameters be interpreted as single symbols of Σ to build *constraints*. For example, ALT.PREC. (b, c) interprets x as c and y as b. Branched DECLARE [29] comprises the same set of templates of standard DECLARE, yet allowing the interpretation of parameters as elements of a join-semilattice $(\Sigma, \vee)$, i.e., an idempotent commutative semigroup, where $\vee$ is the join-operation [16]. We shall use a clausal set-notation whenever a parameter is interpreted as a disjunction of literals. For example, ALT.PREC. ($\{a, w\}, b$) interprets x as b and y as a $\vee$ w: for every b occurring in a trace, a previous instant must have verified a $\vee$ w, without b recurring between that instant and the following occurrence of b. The conjunction of a finite set of constraints forms a DECLARE specification. In the following, we formalize the above notions.

Definition 4. *A* Declare *specification* $\mathcal{DS} = (\text{Rep}, \Sigma, K)$ *is a tuple wherein:*

Rep *is a finite non-empty set of* templates, *or "repertoire", where each template* k$(x_1, \ldots, x_m)$ *is an* LTL$_f$ *formula parameterized on free variables* $x_1, \ldots, x_m$;
$\Sigma \ni a_i$ *is a finite non-empty alphabet of symbols* a_i *with* $1 \leq i \leq |\Sigma|$, $|\Sigma| \in \mathbb{N}$;
K *is a finite set of* constraints, *namely pairs* (k$(x_1, \ldots, x_m), \kappa$) *where* k$(x_1, \ldots, x_m)$ *is a template from* Rep, *and* $\kappa : \{x_1, \ldots, x_m\} \to 2^\Sigma \setminus \{\}$ *is a mapping from every variable* x_i *to a non-empty, finite set of symbols* $A_i = \{a_{i,1}, \ldots, a_{i,v_i}\} \subseteq \Sigma$, *with* $1 \leq i \leq m$ *and* $1 \leq v_i \leq |\Sigma|$; *we denote such a constraint with* k$(A_1, \ldots, A_m)$ *or equivalently* k$(\{a_{1,1}, \ldots, a_{1,v_1}\}, \ldots, \{a_{m,1}, \ldots, a_{m,v_m}\})$, *omitting curly brackets from the latter form whenever a variable is mapped to a singleton.* ◁

As an example, consider the following specification: REP = {ATMOSTONE (x), END (x), ALT.PREC. (y,x)}, Σ = {a, b, c, d, e, f, g, u, v, w}, and K = {ATMOSTONE (a), END (v), ALT.PREC. (e, f), ALT.PREC. ($\{a,w\}, b$), ALT.PREC. $(u, \{v, w\})$}, where, e.g., ALT.PREC. ($\{a, w\}, b$)} is derived from the template ALT.PREC. (y,x) by mapping $y \mapsto_\kappa \{a, w\}$ and $x \mapsto_\kappa b$.

Definition 5 (Constraint formula, satisfying trace). *Let* k$(A_1, \ldots, A_m)$ *be a constraint, whereby* $A_i = \{a_{i,1}, \ldots, a_{i,v_i}\}$ *for each* $1 \leq i \leq m$. *Its constraint formula, written* $\varphi_{k(A_1,\ldots,A_m)}$, *is the* LTL$_f$ *formula obtained from the template* k$(x_1, \ldots, x_m)$ *by interpreting* x_i *as* $(a_{i,1} \vee \cdots \vee a_{i,v_i})$ *for each* $1 \leq i \leq m$. *A trace* π *satisfies* k$(A_1, \ldots, A_m)$ *iff* $\pi \models \varphi_{k(A_1,\ldots,A_m)}$; *otherwise, we say that* π *violates* k$(A_1, \ldots, A_m)$. ◁

Considering Table 1 and the above example specification, we have that $\varphi_{\text{ALT.PREC.}(\{a,w\},b)} = \mathbf{G}\,(b \to \mathbf{Y}\,(\neg b\,\mathbf{S}\,(a \vee w)))$, and $\varphi_{\text{END}(v)} = \mathbf{G}\,\mathbf{F}\,v$. Traces $\langle a, b, c\rangle$, $\langle a, b, c, f, u, w, b\rangle$, and $\langle a, b, c, e, f, g, u, v\rangle$ satisfy ALT.PREC. ($\{a, w\}, b$), while only the third one satisfies END (v).

Definition 6 (Specification formula, model trace). *A given Declare specification* $\mathcal{DS} = (\text{Rep}, \Sigma, K)$ *is logically represented by conjoining its constraint formulae* $\varphi_{\mathcal{DS}} \doteq \bigwedge_{k(A_1,\ldots,A_m) \in K} \left(\varphi_{k(A_1,\ldots,A_m)}\right)$. *A trace is a* model trace *for the specification,* $\pi \models \mathcal{DS}$, *iff* $\pi \models \varphi_{\mathcal{DS}}$, *i.e., it satisfies the conjunction of all the constraint formulae,* $\pi \models \varphi_{k(A_1,\ldots,A_m)}$ *for each* k$(A_1, \ldots, A_m) \in K$. ◁

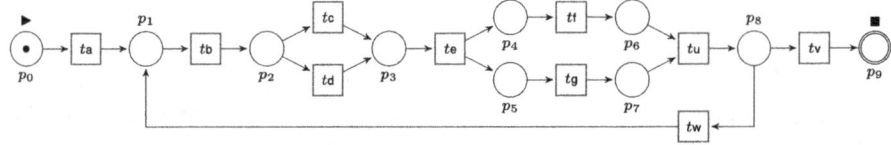

Fig. 2. A Workflow net

The specification formula of the above example is (**G** (a → ¬ **X F** a)) ∧ (**G F** v) ∧ (**G** (f → **Y** (¬f **S** e))) ∧ (**G** (b → **Y** (¬b **S** (a ∨ w)))) ∧ (**G** ((v ∨ w) → **Y** (¬ (v ∨ w) **S** u))). ⟨a, b, c, e, f, g, u, v⟩ is a model trace for it, unlike ⟨a, b, c⟩ or ⟨a, b, c, f, u, w, b⟩.

Leveraging the techniques mentioned at the end of Sect. 2.2 and the above definition, we can create an FSA that accepts all and only the traces of a single DECLARE formula φ and of a whole specification $\mathcal{DS}$.

Definition 7 (Constraint and specification FSA). *Let $\varphi_1, \ldots, \varphi_{|K|}$ be the constraint formulae of a process specification $\mathcal{DS}$. A constraint automaton $\mathcal{A}_{\varphi_i}$ is an FSA that accepts all and only those traces that satisfy φ_i [17] with $1 \leq i \leq |K|$. The product automaton $\mathcal{A}_{\varphi_1} \times \cdots \times \mathcal{A}_{\varphi_{|K|}}$ is the specification FSA, recognizing all and only the traces satisfying $\mathcal{DS}$.* ◁

Figure 1(a) and 1(c) show the automata of constraints ATMOSTONE (a), ALT.PREC. (u, {v, w}), and END (v), respectively. Figure 1(d) depicts the FSA of a specification consisting of END (v) and ALT.PREC. (u, {v, w}). Notice that the accepting state cannot be reached from s_2 in Figs. 1(a) and 1(b). Instead, the FSA in Fig. 1(d) has no such trap states due to trimming.

Aside from keeping the FSA's language unchanged, trimming caters for structural compatibility with the state space representation of Workflow nets, which we discuss next.

2.4 Workflow Nets

A Workflow net (see, e.g., Fig. 2) is a renowned subclass of Petri nets suitable for the formal representation of imperative process models [3].

Definition 8 (Petri net). *A place/transition net [15] (henceforth, Petri net) is a bipartite graph (P, T, F), where P (the finite set of "places") and T (the finite set of "transitions") constitute the nodes ($P \cap T = \emptyset$), and the flow relation $F \subseteq (P \times T \uplus T \times P)$ defines the edges.* ◁

Given a place $p \in P$, we shall denote the sets $\{t \mid (t, p) \in F\}$ and $\{t \mid (p, t) \in F\}$ with ▸p ("*preset*") and p▸ ("*postset*"), respectively. For example, in Fig. 2, ▸p_2 = {tb} and p_2▸ = {tc, td}.

Definition 9 (Workflow net). *A Workflow net $\mathcal{WN} = (P, T, F)$ is a Petri net such that:*

1. There is a unique place ("initial place", ▶ ∈ P) such that its preset is empty;
2. There is a unique place ("output place", ■ ∈ P) such that its postset is empty;
3. Every place $p \in P$ and transition $t \in T$ is on a path of the underlying graph from ▶ to ■.

◁

We remark that we operate with full knowledge of the imperative model's structure, treated as a white box. Therefore, we directly focus on transitions rather than on their labels here.

In Petri and Workflow nets, places can be *marked* with tokens, intuitively representing resources that are processed by the transitions succeeding them in the net. In Fig. 2, a token is graphically depicted as a solid circle (see p_0 in the figure). The state of a net is defined by the distribution of tokens over places. This is formalized with the notion of *marking*, a function mapping each place to the number of tokens in it. The net's state changes with the consumption and production of tokens caused by the execution ("*firing*") of transitions.

Definition 10 (Marking and firing). *Let $\mathcal{WN} = (P, T, F)$ be a Workflow net. A* marking *is a function $M : P \to \mathbb{N} \cup \{0\}$. The* initial marking M_0 *of $\mathcal{WN}$ maps ▶ to 1 and any other $p \in P \setminus \{▶\}$ to 0. A marking M of $\mathcal{WN}$ is* final *if $M(■) > 0$. A marking* enables *a transition $t \in T$ iff $M(p) > 0$ for all places p such that $t \in p▸$. An enabled transition can* fire, *i.e., turn a marking M into M' (in symbols, $M[t\rangle M'$), according to the following rule: For each place $p \in P$, $M'(p) = M(p) + 1$ if $t \in ▸p$; $M'(p) = M(p) - 1$ if $t \in ▸p$; otherwise, $M'(p) = M(p)$.*

◁

In Fig. 2, e.g., the initial marking enables t_a. Denoting markings with a multi-set notation, $\{p_0\}\,[t\mathsf{a}\rangle\,\{p_1\}$. Subsequently, tb gets enabled. After firing tb, and tc get enabled. With Petri and Workflow nets, interleaving semantics are adopted, thus only one transition can fire per timestep, thus the firing of tb and tc are mutually exclusive in that state.

Definition 11 (Firing sequence and run). *Given a Workflow net $\mathcal{WN} = (P, T, F)$, a (finite)* firing sequence *σ is $\langle\rangle$ or a sequence of transitions $\langle t_1, \ldots, t_n\rangle$ such that, for any index $1 \leq i \leq n$ with $n \in \mathbb{N}$, $t_i \in T$: (a) The i-th marking enables the i-th transition; (b) The $i+1$-th marking M_{i+1} is such that $M_i[t_i\rangle M_{i+1}$. A marking M' is* reachable *in $\mathcal{WN}$ if there exists a firing sequence σ leading from the initial marking M_0 to M' (in symbols, $M_0[\sigma\rangle M'$). A firing sequence leading from M_0 to a final marking is a* run.

◁

Given a workflow net $\mathcal{WN}$, we will use $\mathcal{M}_{\mathcal{WN}}$ to denote the set of markings that can be reached from its initial marking M_0.

Runs of the Workflow net in Fig. 2 include $\langle t\mathsf{a}, t\mathsf{b}, t\mathsf{c}, t\mathsf{e}, t\mathsf{f}, t\mathsf{g}, t\mathsf{u}, t\mathsf{v}\rangle$ and $\langle t\mathsf{a}, t\mathsf{b}, t\mathsf{d}, t\mathsf{e}, t\mathsf{f}, t\mathsf{g}, t\mathsf{u}, t\mathsf{w}, t\mathsf{c}, t\mathsf{e}, t\mathsf{f}, t\mathsf{g}, t\mathsf{u}, t\mathsf{v}\rangle$. Any prefix of the first run of length 7 or less is a firing sequence from the initial marking $\{p_0\}$ but not a run.

In this paper, we assume that Workflow nets enjoy the following properties.

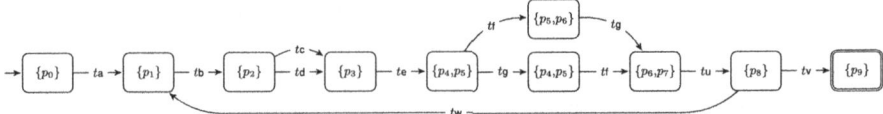

Fig. 3. Reachability graph derived from the Workflow net in Fig. 2.

Definition 12 (Soundness and safety of a Workflow net). *Let $\mathcal{WN} = (P, T, F)$ be a Workflow net. $\mathcal{WN}$ is k-bounded if the number of tokens assigned by any reachable marking M' to any place $p \in P$ is such that $M'(p) \leq k$. A 1-bounded Workflow net is safe. $\mathcal{WN}$ is sound iff it enjoys the following properties:* **Option to complete:** *from any marking M it is possible to reach the final marking;* **Proper completion:** *if a reachable marking M is such that $M(\blacksquare) > 0$, then M is the final marking;* **No dead transitions:** *for any transition $t \in T$, there exists a reachable marking M such that t is enabled by M.* ◁

Safe and sound Workflow nets (like the one depicted in Fig. 2) are a superclass of sound S-coverable nets, which in turn subsume safe and sound free-choice and well-structured nets [1]. These structural characteristics are widely recognized as recommendable in process management [3] and underpin well-formed business process diagrams [28]. Notice that, given a *safe* net, all markings that are reachable from the initial one are such that each place can be marked with at most one token. Also, the final marking of a sound workflow net is $\{\blacksquare\}$.

The state space of k-bounded Petri nets can be represented in the form of a deterministic labeled transition system that go under the name of *reachability graph* [15]. Safe and sound Workflow nets have a given initial marking and one final marking. We can thus endow the state representation with these characteristics and reinterpret the known concept of reachability graph as a finite state automaton.

Definition 13 (Reachability FSA). *Given a sound and safe Workflow net $\mathcal{WN}$, the reachability FSA $\mathcal{A}^{\mathcal{WN}} = \left(S^{\mathcal{WN}}, s_0^{\mathcal{WN}}, s_F^{\mathcal{WN}}, \Sigma^{\mathcal{WN}}, \delta^{\mathcal{WN}}\right)$ is a finite state automaton where:*

$S^{\mathcal{WN}} = \mathcal{M}_{\mathcal{WN}}$, *i.e., the set of states is the set of reachable markings in $\mathcal{WN}$;*
$s_0^{\mathcal{WN}} = \{\blacktriangleright\}$, *i.e., the initial state is the initial marking of $\mathcal{WN}$;*
$s_F^{\mathcal{WN}} = \{\{\blacksquare\}\}$, *i.e., the accepting state set is a singleton with the final marking of $\mathcal{WN}$;*
$\Sigma^{\mathcal{WN}} = T$, *i.e., the alphabet is the set of transitions of $\mathcal{WN}$;*
$\delta^{\mathcal{WN}}$ *is s.t. $\delta(M, t) = M'$ iff $M[t\rangle M'$ for every transition t and reachable M, M' in $\mathcal{WN}$.*

◁

Figure 3 depicts the reachability FSA of the Workflow net in Fig. 2.

When dealing with accepting traces and languages, working with trimmed or non-trimmed FSAs is equivalent. This is not the case if we structurally relate

Algorithm 1: Wizard's guide to synthesize DECLARE specifications from Workflow nets

Input: $\mathcal{WN} = (P, T, F)$, a workflow net;
Output: $\mathcal{DS} = (\text{Rep}, \text{Act}, K)$, a declarative process specification

1 $\Sigma \leftarrow T; \quad K \leftarrow \{\};$ # Assign the alphabet with the transition set, and initialize the constraint set
2 $\mathcal{DS} \leftarrow (\{\text{AtMostOne}(x), \text{End}(x), \text{Alt.Prec.}(x, y)\}, \Sigma, K)$ # Initialize $\mathcal{DS}$ including templates
3 **foreach** $p \in P$ **do**   # Visit all places in $\mathcal{WN}$
4   **if** $\blacktriangleright p \neq \emptyset$ and $p \blacktriangleright \neq \emptyset$ **then** $K \leftarrow K \cup \{\text{Alt.Prec.}(\blacktriangleright p, p \blacktriangleright)\};$ # Add the Alt.Prec. ($\blacktriangleright p, p \blacktriangleright$) constraint ☆
5   **else if** $\blacktriangleright p = \emptyset$ **then** $K \leftarrow K \cup \{\text{AtMostOne}(p \blacktriangleright)\};$ # Add the AtMostOne ($p \blacktriangleright$) constraint ☆
6   **else if** $p \blacktriangleright = \emptyset$ **then** $K \leftarrow K \cup \{\text{End}(\blacktriangleright p)\};$ # Add the End ($\blacktriangleright p$) constraint ☆

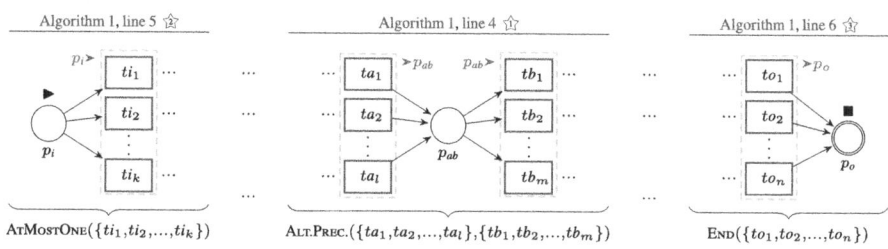

Fig. 4. A graphical sketch of the execution of Algorithm 1.

the FSA of a DECLARE specification with the reachability FSA of a Workflow net. That FSA is indeed constructed, state-by-state, considering only enabled transitions, which globally yields that it is trimmed by design. Therefore, we operate with trimmed FSAs for this comparison.

3 Synthesis of LTL$_f$ Specifications from Workflow Nets

In this section, we outline the algorithm (including the three spells to cast: ☆, ☆, and ☆) to synthesize a DECLARE specification $\mathcal{DS}$ from a given input safe and sound Workflow net $\mathcal{WN}$ ensuring behavioral equivalence between them. Algorithm 1 illustrates the transformation process. The algorithm initializes $\mathcal{DS}$ by assigning its alphabet with the transition set of $\mathcal{WN}$ (Algorithm 1, ln. 1). Given the Workflow net in Fig. 2, e.g., Σ gets $\{t_\mathsf{a}, \ldots, t_\mathsf{g}, t_\mathsf{u}, t_\mathsf{v}, t_\mathsf{w}\}$. Then, it sets the three (necessary) templates that will be used (ln. 2): $\text{AtMostOne}(x)$, $\text{End}(x)$, and $\text{Alt.Prec.}(y, x)$. A cycle begins to visit all places in $\mathcal{WN}$ and update $\mathcal{DS}$ by including a new constraint per place. Figure 4 graphically sketches this passage, which casts the three spells as follows: ☆ If p is the output place as $p \blacktriangleright$ is empty, $\text{End}(\blacktriangleright p)$ is included (ln. 6); ☆ If p is the input place as $\blacktriangleright p$ is empty, $\text{AtMostOne}(p \blacktriangleright)$ is added (ln. 5); ☆ Otherwise, $\text{Alt.Prec.}(\blacktriangleright p, p \blacktriangleright)$ becomes one of the constraints in $\mathcal{DS}$ (ln. 4). Intuitively, the rationale is that: ☆ Every time a transition in the postset of p fires, it is necessary that at least one of the transitions in the preset of p fired before and that no transition in the postset of

Table 2. DECLAREspecification generated from the Workflow net in Fig. 2

AtMostOne (ta)	End (tv)	Alt.Prec. $(\{ta, tw\}, tb)$	Alt.Prec. $(tb, \{td, tc\})$	Alt.Prec. $(\{td, tc\}, te)$
Alt.Prec. (te, tf)	Alt.Prec. (te, tg)	Alt.Prec. (tf, tu)	Alt.Prec. (tg, tu)	Alt.Prec. $(tu, \{tv, tw\})$

p has fired since then; ☆ Any of the transitions in the postset of ▶ will start the run and will not repeat afterwards (because no firing can assign ▶ with a token again by definition); ☆ Every run must terminate with one of the transitions in the preset of ■.

Table 2 shows the constraints that are generated by our algorithm if the Workflow net in Fig. 2 is fed as input. It is noteworthy to analyze in particular the non-trivial behavior entailed by constraints that stem from the parsing of places that begin or end cycles like p_8 and p_1 in Fig. 2. From the former we derive Alt.Prec. $(tu, \{tv, tw\})$. It states that before tv or tw, tu must occur. Also, *neither* tv nor tw can recur until tu is repeated. As a consequence, an *exclusive* choice between tv and tw is enforced cyclically for each recurrence of tu. Dually, with Alt.Prec. $(\{ta, tw\}, tb)$ (generated by fetching the pre- and post-sets of p_1) we demand that *each* occurrence of tb follows ta or tw. From Table 2 we notice that ta can occur only once (AtMostOne (ta)), thus the subsequent recurrences of tb are bound to tw.

Given the construction in Algorithm 1, it is clear that the semantics of the resulting DECLARE specification $\mathcal{DS}$ is an LTL_f formula, traces of which are finite sequences of transitions of the input Workflow net $\mathcal{WN}$. Notice that the mapping of the transitions of $\mathcal{WN}$ to labels (as usual in a process modeling context) can be treated as a post-hoc refinement of $\mathcal{WN}$ and equivalently of $\mathcal{DS}$: Assuming that t_1 maps to label z, e.g., the occurrence of transition t_1 will emit z regardless of the underlying behavioral representation.

As established in the beginning of this section, our goal is to now show the behavioral equivalence between the DECLARE specification given as output by Algorithm 1 and the input safe and sound Workflow net. To this end, we use the following notion of bisimilarity.

Definition 14 (Bisimilarity of Workflow nets and Declare specifications). *A safe and sound Workflow net $\mathcal{WN}$ is bisimilar to a Declare specification $\mathcal{DS}$ if and only if the reachability FSA of $\mathcal{WN}$ (as per Definition 13) is bisimilar to the specification FSA of $\mathcal{DS}$ (as per Definition 7).* ◁

Given *(i)* this notion of bisimilarity, and *(ii)* Observation 1, it suffices to show that the two automata accept the same language to prove our claim. This, in turn, means that the DECLARE specification returned by Algorithm 1 accepts all and only the runs of the input safe and sound Workflow net. According to the language-encodability criteria in [23], the obtained bisimilarity ensures operational correspondence between the source and target languages. We now proceed to formally express our claim.

Theorem 1. *Given a safe and sound Workflow net $\mathcal{WN}$, Algorithm 1 returns a Declare specification $\mathcal{DS}$ such that: (i) any run of $\mathcal{WN}$ satisfies $\mathcal{DS}$, and (ii) any trace satisfying $\mathcal{DS}$ is a run of $\mathcal{WN}$.*

Proof. We prove that $\mathcal{DS}$ and $\mathcal{WN}$ satisfy the two conditions stated in the claim.

(i) Let σ be a run of $\mathcal{WN}$. We show that $\sigma \models \varphi_{\mathcal{DS}}$. As $\mathcal{WN}$ is a Workflow net, it has a unique input place and a unique output place. Let p_i be ▶ and p_o be ■. In $\varphi_{\mathcal{DS}}$, we have only one constraint for the templates END(x) and ATMOSTONE(x), namely END$(\blacktriangleright p_o)$ and ATMOSTONE$(p_i\blacktriangleright)$. Let $\{to_1, \ldots to_n\}$ be the preset of p_o (with $n \in \mathbb{N}$). Any run of $\mathcal{WN}$ must satisfy **G F** $(to_1 \vee \ldots \vee to_n)$, i.e., $\varphi_{\text{END}(\blacktriangleright p_o)}$, as one of the transitions in the preset of p_o must fire last. Let $\{ti_1, \ldots ti_k\}$ be the postset of p_i (with $k \in \mathbb{N}$). No other place $p' \neq p_i$ can be such that $ti \in p'\blacktriangleright$ for any $ti \in p_i\blacktriangleright$, otherwise ti would be a dead transition (thus contradicting soundness): The initial marking assigns no token to p', and no marking except the initial one assigns a token to p_i. As a consequence, any run of $\mathcal{WN}$ must satisfy **G** $((ti_1 \vee \ldots \vee ti_k) \rightarrow \neg \mathbf{X} \mathbf{F} (ti_1 \vee \ldots \vee ti_k))$, i.e., $\varphi_{\text{ATMOSTONE}(p_i\blacktriangleright)}$.

It remains to show that $\sigma \models \varphi_{\text{ALT.PREC.}(\blacktriangleright p, p\blacktriangleright)}$ for any arbitrary place $p \in P \setminus \{p_i, p_o\}$. Assume by contradiction that $\sigma \not\models \mathbf{G} (p\blacktriangleright \rightarrow \mathbf{Y} (\neg p\blacktriangleright \mathbf{S} \blacktriangleright p))$. Then, there must be a timestep $\ell < |\sigma|$ such that $\sigma, \ell \models p\blacktriangleright$, i.e., a transition in $p\blacktriangleright$ was fired, but $\sigma, \ell \not\models \mathbf{Y} (\neg p\blacktriangleright \mathbf{S} \blacktriangleright p)$. Notice that, as a transition in $p\blacktriangleright$ was fired, it means that a transition in $\blacktriangleright p$ was fired at a timestep $\ell' < \ell$, as otherwise there would be no token assigned to p at timestep ℓ. Then, for $\sigma, \ell \not\models \mathbf{Y} (\neg p\blacktriangleright \mathbf{S} \blacktriangleright p)$ to be true, it must be the case that a transition in $p\blacktriangleright$ was fired at some timestep ℓ'' such that $\ell' < \ell'' < \ell$, and no transition in $\blacktriangleright p$ has been fired between timesteps ℓ'' and ℓ. However, this, in conjunction with the fact the Workflow net is safe, implies that it would not have been possible to fire a transition in $p\blacktriangleright$ at timestep ℓ: p has no token assigned at timestep ℓ as it was consumed to fire a transition in $p\blacktriangleright$ at timestep ℓ''. Therefore, $\sigma \models \varphi_{\text{ALT.PREC.}(\blacktriangleright p, p\blacktriangleright)}$ for any arbitrary place $p \in P \setminus \{p_i, p_o\}$, thus implying that $\sigma \models \varphi_{\mathcal{DS}}$.

(ii) We now show that if a trace σ is such that $\sigma \models \varphi_{\mathcal{DS}}$ then σ is a run of $\mathcal{WN}$. Let p_i be ▶ and p_o be ■ again. Since $\sigma \models \varphi_{\mathcal{DS}}$, we have that the trace correctly ends with a transition in the preset of p_o, because $\sigma \models \varphi_{\text{END}(\blacktriangleright p_o)}$. Also, in a run of $\mathcal{WN}$, the transitions in $p_i\blacktriangleright$ can only be fired once, otherwise $\blacktriangleright p_i$ would be non-empty against the definition of ▶. This holds true in σ, as $\sigma \models \varphi_{\text{ATMOSTONE}(p_i\blacktriangleright)}$. Notice that, unlike all other transitions in $\mathcal{WN}$, only those in $p_i\blacktriangleright$ do *not* map to x for ALT.PREC.(y, x) in $\mathcal{DS}$ by design of Algorithm 1. Therefore, every trace will begin with the occurrence of one of the transitions in $p_i\blacktriangleright$ as it happens with the runs of $\mathcal{WN}$.

It remains to show that every transition in the trace σ was fired in $\mathcal{WN}$ following the preceding sequence of transitions in the trace. Suppose by contradiction that this is not the case, i.e., that there is some transition t fired at a timestep ℓ which could not have been fired in $\mathcal{WN}$ given the prefix of σ from 1 to $\ell - 1$. Then, this implies that at least one of the places p such that $t \in p\blacktriangleright$ does not have a token at timestep $\ell - 1$, i.e., $M_{\ell-1}(p) = 0$. Two conditions can entail this situation: Either no transition in $\blacktriangleright p$ was fired before, or a transition in $p\blacktriangleright$ was fired since the last timestep in which a transition in $\blacktriangleright p$ was fired, consuming the only token assigned to p. Both cases contradict the fact that $\sigma \models \varphi_{\text{ALT.PREC.}(\blacktriangleright p, p\blacktriangleright)}$, thus proving that σ is a valid sequence of transitions with respect to $\mathcal{WN}$. Thus, σ is a run of $\mathcal{WN}$. ⊣

Given Observation 1, we immediately obtain the following corollary.

Corollary 1. *The* Declare *specification $\mathcal{DS}$ given as output by Alg. 1 is bisimilar to the input safe and sound Workflow net $\mathcal{WN}$.*

This result has a profound implication that transcends DECLARE and LTL$_f$ but pertains to the languages recognized by safe and sound Workflow nets.

Theorem 2. *Languages of safe and sound Workflow nets are star-free regular expressions.*

Proof. The claim follows from Theorem 1, recalling that DECLARE patterns are expressed in LTL$_f$, which is expressively equivalent to star-free regular expressions [13]. ⊣

Space and Time Complexity. Algorithm 1 outputs a DECLARE specification $\mathcal{DS} = (\text{REP}, \Sigma, K)$ which contains, for each place in the input Workflow net $\mathcal{WN} = (P, T, F)$, a constraint with the pre- and post-sets as its actual parameters. Each transition that is in relation with a place p in the flow relation F appears exactly once in the constraint stemming from p; therefore, the space complexity class of Algorithm 1 is $\mathcal{O}(|F|)$. As for the time complexity, we can assume that a pre-processing step is conducted to represent F in the form of a sequence of pairs, associating every place to its pre-set and post-set. The cost of this operation is $\Theta(F)$ and $\mathcal{O}(|P| \times |T|)$. For each place, the algorithm performs up to three if-checks and then a new constraint is created in constant time, hence $\mathcal{O}(|P|)$. Therefore, the time complexity of the algorithm is bounded by the update of K, necessitating up to $\mathcal{O}(|P| \times |T|)$ time.

Next, we experimentally validate and put the above theoretical results to the test.

4 Implementation and Evaluation

We implemented Algorithm 1 in the form of a proof-of-concept prototype encoded in Python. The tool, testbeds, and experimental results are available for public access.[1] In the following, we report on tests conducted with our algorithm's implementation to empirically confirm its soundness, assess memory efficiency, and gauge runtime performance. Finally, we showcase a process diagnostic application as a downstream task for our approach.

---

[1] Sp3llsWizard, https://doi.org/10.5281/zenodo.15528894.

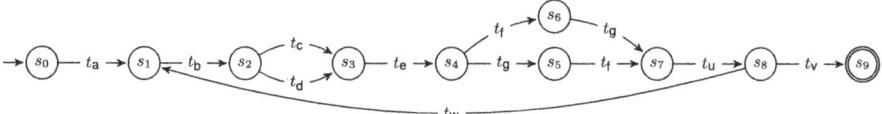

Fig. 5. FSA of the specification in Table 2

4.1 Automata Bisimulation

To experimentally validate the correctness of the implementation of Algorithm 1, we performed a preliminary comparison of the reachability FSA (Definition 13) of known Workflow nets and the specification FSA (Definition 7) consisting of the DECLARE constraints returned by our tool. Figure 5 illustrates the FSA of the specification derived from the Workflow net in Fig. 2, computed with a dedicated module presented in [17]. Also by visual inspection, we can conclude that the two FSAs are bisimilar, as expected. Owing to space constraints, we cannot portray the entire range of automata derived from the Workflow nets in our experiments. The interested reader can find the full collection (including non-free choice nets such as that of [3, Fig. 24]) in our public codebase.[1]

4.2 Performance Analysis

Here, we report on the quantitative assessment of our solution in terms of scalability given an increasing workload, and against real-world testbeds. For the former, we observe the time and space performance of our implemented prototype fed in input with Workflow nets of increasing size. We control the expansion process in two directions, so as to obtain the following separate effects: *(i)* more constraints are generated, while each is exerted on up to three literals; *(ii)* the amount of generated constraints remains fixed, while the literals mapped to their parameters increase. For the real-world testbed, we take as input processes discovered by a well-known imperative process mining algorithm from a collection of openly available event logs. We conducted the performance tests on an AMD Ryzen 9 8945HS CPU at 4.00 GHz with 32 GB RAM running Ubuntu 24.04.1. For the sake of reliability, we ran three iterations for every test configuration and averaged the outputs to derive the final result.

Increasing Constraint-Set Cardinality. To examine the effectiveness of the Algorithm 1 in handling an incremental number of constraints, we examine memory utilization and execution time through the progressive rise in the complexity of the input Workflow net. Our evaluation method relies on an expansion mechanism that iteratively applies a structured pattern of four soundness-preserving transformation rules from [2] to progressively increase the number of nodes and their configuration. This leads to a gradual increase in the number of constraints our algorithm needs to initiate. Starting from the Workflow net in Fig. 6(a), we designate transition t_1 as a fixed 'pivot', retaining the initial and final places (p_1 and p_2), and iteratively apply the expansion mechanism illustrated in Fig. 6(b)

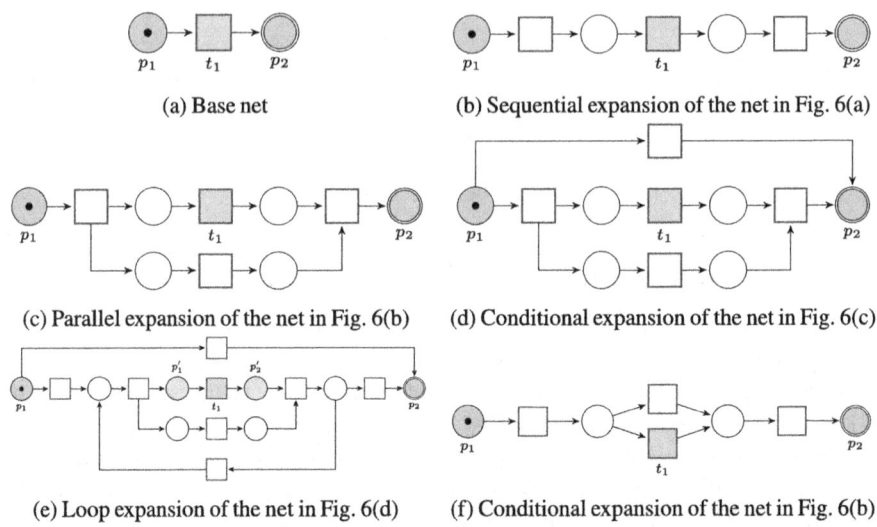

Fig. 6. Transformation rules used to iteratively expand a safe and sound Workflow net.

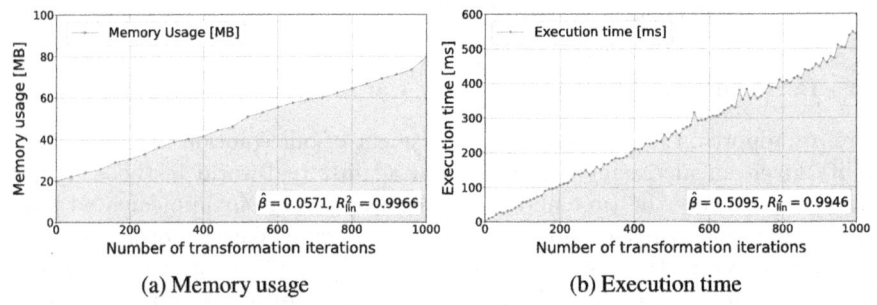

Fig. 7. Test results for the incremental number of constraints setup.

to 6(e). We apply known workflow patterns in the following order: (Fig. 6(b)) We add a transition before and after t_1; (Fig. 6(c)) We introduce a parallel execution path; (Fig. 6(d)) We insert an exclusive branch; (Fig. 6(e)) Finally, we incorporate a loop structure. Upon completion of the expansion process, we execute the algorithm, record the results, and initiate a new iteration, maintaining t_1 unchanged while reassigning p_1 and p_2 with the places that have t_1 in the preset and postset (see the places colored in blue and labeled with p'_1 and p'_2 in Fig. 6(e)). We reiterated the procedure 1000 times.

Figure 7 displays the registered memory usage and execution time. To interpret the performance trends, we employ two well-established measures: the coefficient of determination R^2_{lin}, which assesses the goodness-of-fit of the data to a linear trend, and the $\hat{\beta}$ rate, which serves as a meter for the line's slope. As depicted in Fig. 7(a), memory consumption increases linearly with the number

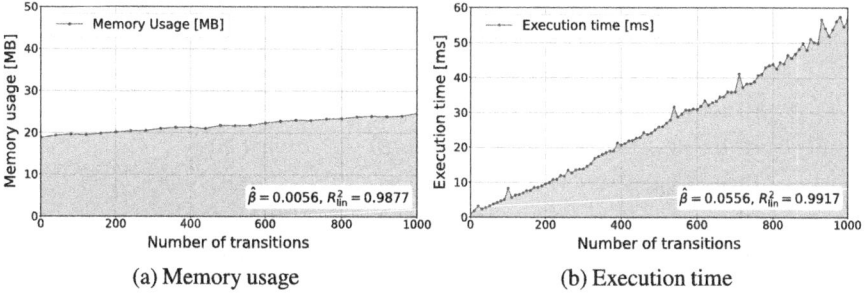

Fig. 8. Test results for the incremental constraints dimension setup.

of iterations of the expansion mechanism confirmed by $R_{\text{lin}}^2 = 0.9966$ and a low slope increase ($\hat{\beta} = 0.0571$). Figure 7(b) displays the execution time plot, with $R_{\text{lin}}^2 = 0.9946$ and $\hat{\beta} = 0.5095$, thus indicating a linear trend with a slope exhibiting a moderate incline. We remark that these results are in line with the theoretical analysis of the space and time complexity in Sect. 3.

Increasing Constraint Formula Size. Here, we configure the test on memory usage and execution time to investigate the algorithm's performance while handling an expanding constraints' formula size (i.e., with an increasing number of disjuncts). To this end, we progressively broaden the Workflow net by applying the soundness-preserving conditional expansion rule from [2] depicted in Fig. 6(f) to transition t_1 in the net of Fig. 6(b). We reiterate the process 1000 times. Figure 8 displays the results we registered. Observing Fig. 8(a), we can assert that the memory utilization increases linearly ($R_{\text{lin}}^2 = 0.9877$) with a minimal rate ($\hat{\beta} = 0.0056$). The execution time plotted in Fig. 8(b) also exhibits a linear increase ($R_{\text{lin}}^2 = 0.9917$), with a moderate slope inclination ($\hat{\beta} = 0.0556$). Once more, the results align with the theoretical complexity analysis in Sect. 3.

Real-World Process Model Testing. To evaluate the performance of our algorithm in application on real process models, we conduct the same memory usage and execution time tests employing Workflow nets directly derived from a collection of real-life event logs available at *4TU.ResearchData*.[2]

To this end, we employ the Inductive Miner algorithm version proposed in in [26], which filters out infrequent behavior while still discovering well-structured, sound models [5]. Thus, we first run the Inductive Miner on the event logs considered in [5] to generate the Workflow nets. We then apply Algorithm 1 to derive the corresponding DECLARE specification. We report the aggregate test result in Table 3, detailing the memory usage, the execution time, and all the features of the mined Workflow nets. We find that the overall differences in resource usage are negligible. These real-world test outcomes again follow the complexity assumptions outlined in Sect. 3.

---

[2] The event logs used in our experiments are publicly available at https://data.4tu.nl/.

Table 3. Performance comparison with real-world process models

Event log	Trans.	Places	Nodes	Mem.usage [MB]	Exec.time [ms]
BPIC 12	78	54	174	19.97	5.11
BPIC 13_{cp}	19	54	44	19.76	1.70
BPIC 13_{inc}	23	17	50	19.89	2.03
BPIC 14_f	46	35	102	19.90	3.31
BPIC 15_{1f}	135	89	286	20.44	8.39
BPIC 15_{2f}	200	123	422	20.91	12.30
BPIC 15_{3f}	178	122	396	20.77	11.49
BPIC 15_{4f}	168	115	368	20.55	11.38
BPIC 15_{5f}	150	99	320	20.43	9.16
BPIC 17	87	55	184	19.91	5.67
RTFMP	34	29	82	19.81	3.47
Sepsis	50	39	116	19.75	3.65

4.3 A Downstream Task: Using Constraints as Determinants for Process Diagnosis

Algorithm 1 enables the transition from an overarching imperative model to a constraint-based specification, enclosing parts of behavior into separate constraints. Herewith, we aim to demonstrate how we can single out the violated rules constituting the process model behavior, thereby spotlighting points of non-compliance with processes. In other words, we aim to use constraints as determinants for a process diagnosis. For this purpose, we created a dedicated module extending a declarative specification miner for constraint checking via the replay of runs on semi-symbolic automata like those in Fig. 1(a) to 1(c), following [18]. Without loss of generality, we build the runs from data pertaining to building permit applications in Dutch municipalities from BPIC 15_{5f} [20] and apply the preprocessing technique mentioned in [5], resulting in 975 traces. We then process the log with the α-algorithm [4] and provide the returned net as input to our implementation of Algorithm 1.

We observe that the specification consists of 129 constraints. Of those, our tool detected violations by at least a trace for 77 of those. Figure 9 illustrates the percentage of satisfying traces (henceforth, *fitness* for brevity) of the 77 constraints, which we clustered into five distinct groups to ease inspection. We first focus on violated constraints exhibiting high fitness (the blue upward triangles at the top of Fig. 9). Let us take, e.g., the constraint identified by ID 36 in the figure: ALT.PREC. ($\{t01\_HOOFD\_490\_1, t13\_CRD\_010\}, t01\_HOOFD\_490\_1a$), which exhibits a fitness of 0.997. This constraint imposes that when *"Set Decision Status"* ($t01\_HOOFD\_490\_1a$) occurs, it should be preceded by either *"Create Environmental Permit Decision"* ($t01\_HOOFD\_490\_1$) or *"Coordination of Application"* ($t13\_CRD\_010$). In the three traces violating the constraint (11369696, 9613229, 12135936), though, *"Set Decision Status"* is preceded by neither

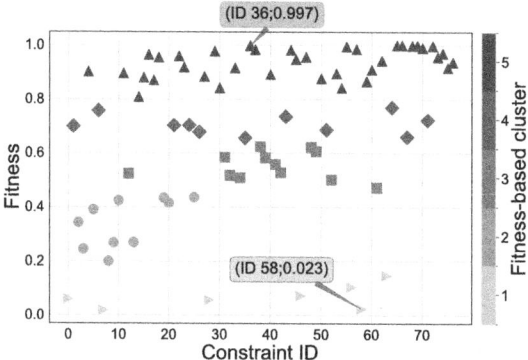

Fig. 9. Fitness-based clusters of the constraints in the descriptive model of BPIC 15^{5f}.

of the two. By further inspection, we observe "*No Permit Needed or Only Notification Needed*" ($t14$_VRIJ_010) in the trace prefix instead, suggesting that the process bypasses the standard decision-making steps defined by the reference model in favor of an alternative where a permit decision is unnecessary. On the other side of the spectrum, let us look at constraints with low trace fitness values (depicted by rightward orange triangles in Fig. 9). These constraints likely suffer from systematic defects rather than spurious alterations in the process behavior. Constraint ID 58, e.g., belongs to this group: ALT.PREC. ($t1$_HOOFD_510_2, $\{t01$_HOOFD_510_3, $t01$_HOOFD_520, tEND$\}$) (depicted in the lower section of Fig. 9). Other constraints in the same group have in common the presence of tEND in the activator's set. An explanation is that the BPIC 15 log allows a multitude of possible conclusions. The α-algorithm, though, disregards the occurrence frequency of individual transitions during model construction, resulting in a non-selective inclusion of all events. Consequently, this affects the fitness of those constraints. Our tool specifically pinpoints and isolates the effect of this tendency from the remainder of the net. Thoroughly assessing the suitability of our approach for process diagnostics transcends the scope of this paper but paves the path for future work.

5 Related Work

The relationship between imperative and declarative modeling approaches has been extensively explored in the existing literature, with a prevailing focus directed toward the development of analytics tools that effectively compare and integrate the strengths of both paradigms. Building on previous contributions aimed at establishing a formal connection between these paradigms [11,32], our research focuses on providing a systematic approach for translating safe and sound Workflow nets into their declarative counterparts. A growing research stream configures this transformation aiming to leverage the support provided by the declarative specifications for conformance checking and anomaly detection. Notably, integrating declarative constraints into event log analysis facilitates more comprehensive diagnostics than those provided by trace replaying techniques. In this regard, Rocha et al. [34] propose an automated method for generating conformance diagnostics using declarative constraints derived from an input imperative model. Their method relies on a library of templates internally maintained in the tool. Eligible constraints are generated by verifying the instantiation of those templates against the model's state space. The ones that are behaviorally compatible are then subject to redundancy removal pruning. Finally, the retained constraints are checked for conformance against log traces. In [7], the authors present a tool that derives a set of eligible constraints directly extracting relations based on a selection of BPMN models' activity patterns. The work of Rebmann et al. [33] proposes a framework for extracting best-practice declarative constraints from a collection of imperative models aiming to discover potential violations and undesired behavior. Constraints are extracted akin to [7], then refined and validated via natural-language-processing techniques to measure their relevance for a given event log. Busch et al. [8] adopt a similar technique to check constraints characterizing process model repositories against event logs. All these techniques share our aim to derive declarative constraints from imperative models given as input. However, they do so via simulation or state space exploration, with limited guarantees of behavioral equivalence. In contrast, our work proposes an algorithm that is proven to establish a formal equivalence between the given imperative model and the derived declarative specification. Being based on the sole exploration of the net's structure, it is also lightweight in terms of computational demands.

6 Conclusion and Future Work

In this paper, we presented a systematic approach to translate safe and sound Workflow nets into bisimilar DECLARE specifications. The latter are based solely on three LTL_f formula templates from the DECLARE repertoire with branching: ATMOSTONE, END, and ALTERNATEPRECEDENCE. We provide a proof-of-concept implementation, of which we evaluate scalability and showcase applications against synthetic and real-world testbeds.

We believe that the scope of this research may be expanded in a number of directions. A natural extension of our work is the inclusion of label-mappings of the Workflow net in the declarative specifications, which would turn the constraints' semi-symbolic automata into transducers that are advantageous in conformance checking contexts. To this end, investigating the handling of silent transitions and repeated labels is a path we intend to pursue. Moreover, we seek to broaden the application of our solution to detect behavioral violations, extending support to a wider range of imperative input models on one hand, and unlocking the use of quality measures for declarative specifications on imperative models [9]. Also, we aim to investigate the correspondence between specification inconsistencies and Workflow net unsafeness and unsoundness. Another promising application lies in mixed representations combining imperative and declarative paradigms [21]. In this regard, our approach could facilitate behavioral comparisons akin to [6] and enable the construction of hybrid representations tailored to diverse scenarios [14].

Acknowledgments. This work was partly funded by MUR under PRIN grant B87G22000450001 (PINPOINT) and the Latium Region under PO FSE+ grant B83C22004050009 (PPMPP). Thanks to Saba Latif for the fruitful initial discussions.

References

1. van der Aalst, W.M.P.: Structural characterizations of sound workflow nets, Technical report, 9623, Technische Universiteit Eindhoven (1996)
2. van der Aalst, W.M.P.: Verification of workflow nets. In: ICATPN, pp. 407–426 (1997)
3. van der Aalst, W.M.P.: The application of petri nets to workflow management. J. Circ. Syst. Comput. **8**(1), 21–66 (1998)
4. van der Aalst, W.M.P., Weijters, T., Maruster, L.: Workflow mining: discovering process models from event logs. IEEE Trans. Knowl. Data Eng. **16**(9), 1128–1142 (2004)
5. Augusto, A., Conforti, R., Dumas, M., La Rosa, M., Maggi, F.M., Marella, A.: Automated discovery of process models from event logs: review and benchmark. IEEE Trans. Knowl. Data Eng. **31**(4), 686–705 (2019)
6. Baumann, M.: Comparing imperative and declarative process models with flow dependencies. In: IEEESOSE 2018, pp. 63–68. IEEE Computer Society (2018)
7. Bergmann, A., Rebmann, A., Kampik, T.: BPMN2Constraints: breaking down BPMN diagrams into declarative process query constraints. In: BPM Demos, vol. 3469, pp. 137–141 (2023)
8. Busch, K., Kampik, T., Leopold, H.: xSemAD: explainable semantic anomaly detection in event logs using sequence-to-sequence models. In: BPM, vol. 14940, pp. 309–327 (2024)
9. Cecconi, A., Barbaro, L., Ciccio, C.D., Senderovich, A.: Measuring rule-based LTLf process specifications: a probabilistic data-driven approach. Inf. Syst. **120**, 102312 (2024)
10. Cecconi, A., Di Ciccio, C., De Giacomo, G., Mendling, J.: Interestingness of traces in declarative process mining: the Janus LTLpf approach. In: BPM, pp. 121–138 (2018)
11. Cosma, V.P., Hildebrandt, T.T., Slaats, T.: Transforming dynamic condition response graphs to safe petri nets. In: PETRI NETS 2023. LNCS, vol. 13929, pp. 417–439. Springer (2023)
12. De Giacomo, G., De Masellis, R., Montali, M.: Reasoning on LTL on finite traces: insensitivity to infiniteness. In: AAAI, pp. 1027–1033 (2014)
13. De Giacomo, G., Vardi, M.Y.: Linear temporal logic and linear dynamic logic on finite traces. In: IJCAI, pp. 854–860 (2013)
14. De Smedt, J., De Weerdt, J., Vanthienen, J., Poels, G.: Mixed-paradigm process modeling with intertwined state spaces. Bus. Inf. Syst. Eng. **58**(1), 19–29 (2016)
15. Desel, J., Reisig, W.: Place/transition Petri Nets, pp. 122–173 (1998)
16. Di Ciccio, C., Maggi, F.M., Mendling, J.: Efficient discovery of target-branched declare constraints. Inf. Syst. **56**, 258–283 (2016)

17. Di Ciccio, C., Maggi, F.M., Montali, M., Mendling, J.: Resolving inconsistencies and redundancies in declarative process models. Inf. Syst. **64**, 425–446 (2017)
18. Di Ciccio, C., Maggi, F.M., Montali, M., Mendling, J.: On the relevance of a business constraint to an event log. Inf. Syst. **78**, 144–161 (2018)
19. Di Ciccio, C., Montali, M.: Declarative process specifications: reasoning, discovery, monitoring. In: van der Aalst, W.M.P., Carmona, J. (eds.) Process Mining Handbook. Lecture Notes in Business Information Processing, vol. 448, pp. 108–152. Springer, Cham (2022). https://doi.org/10.1007/978-3-031-08848-3_4
20. van Dongen, B.: BPI challenge 2015 municipality 5 (2015)
21. van Dongen, B.F., De Smedt, J., Di Ciccio, C., Mendling, J.: Conformance checking of mixed-paradigm process models. Inf. Syst. **102**, 101685 (2021)
22. Dumas, M., La Rosa, M., Mendling, J., Reijers, H.: Fundamentals of Business Process Management, 2nd edn. Springer, Berlin, Heidelberg (2018)
23. Gorla, D.: Towards a unified approach to encodability and separation results for process calculi. Inf. Comput. **208**(9), 1031–1053 (2010)
24. Hildebrandt, T.T., Mukkamala, R.R.: Declarative event-based workflow as distributed dynamic condition response graphs. In: PLACES 2010, vol. 69, pp. 59–73 (2010)
25. Hopcroft, J.E., Motwani, R., Ullman, J.D.: Introduction to Automata Theory, Languages, and Computation, 3rd edn. Addison-Wesley Longman, Boston, MA, USA (2006)
26. Leemans, S.J.J., Fahland, D., van der Aalst, W.M.P.: Discovering block-structured process models from event logs containing infrequent behaviour. In: BPM Workshops 2013, vol. 171, pp. 66–78 (2013)
27. Lichtenstein, O., Pnueli, A., Zuck, L.D.: The glory of the past. In: Logics of Programs, pp. 196–218 (1985)
28. Ouyang, C., Dumas, M., van der Aalst, W.M.P., Ter Hofstede, A.H.M., Mendling, J.: From business process models to process-oriented software systems. ACM Trans. Softw. Eng. Methodol. **19**(1), 2:1-2:37 (2009)
29. Pesic, M.: Constraint-based workflow management systems: shifting control to users, Ph.D. thesis, TU Eindhoven (2008)
30. Pichler, P., Weber, B., Zugal, S., Pinggera, J., Mendling, J., Reijers, H.A.: Imperative versus declarative process modeling languages: an empirical investigation. In: BPM Workshops 2011, vol. 99, pp. 383–394 (2011)
31. Pnueli, A.: The temporal logic of programs. In: FOCS, pp. 46–57 (1977)
32. Prescher, J., Di Ciccio, C., Mendling, J.: From declarative processes to imperative models. In: SIMPDA 2014, vol. 1293, pp. 162–173 (2014)
33. Rebmann, A., Kampik, T., Corea, C., van der Aa, H.: Mining constraints from reference process models for detecting best-practice violations in event log. arXiv preprint arXiv:2407.02336 (2024)
34. Rocha, E.G., van Zelst, S.J., van der Aalst, W.M.P.: Mining behavioral patterns for conformance diagnostics. In: BPM 2024, vol. 14940, pp. 291–308 (2024)

High-Level Requirements-Driven Business Process Compliance

Juanita Caballero-Villalobos[✉][iD], Andrea Burattin[iD], and Hugo A. López[iD]

Technical University of Denmark, Kongens Lyngby 2800, Denmark
{jcavi,andbur,hulo}@dtu.dk

Abstract. Process compliance refers to the alignment between business processes and regulatory requirements. Compliance is difficult because it needs to be able to express the intent and the possible interpretations of laws into formal models, align models with traces in a process, and inspect whether these traces are generating violations. This paper focuses on a largely unexplored area within BPM: the compliance of high-level and non-functional requirements. While compliance checking has been studied through conformance checking techniques, most regulatory requirements are defined in subjective and high-level terms, limiting the application of rule-checking and alignments to specific cases. In contrast, we propose the application of requirement engineering methods for business process compliance. In particular, we raise the level of abstraction from the compliance of specific patterns to the satisfaction of high-level goals and subjective qualities. We propose a framework that connects process models with goal models, rendering explicit alternatives for the satisfaction of vague goals and subjective qualities. Compliance checking is reduced to a reachability of a state where subjective qualities are satisfied. This approach is exhibited in a data protection scenario, and we provide a prototypical implementation of the compliance checking tool.

Keywords: Compliance Checking · Business Process Compliance · Goal Modeling · Requirements Engineering

1 Introduction

Business processes are considered the heart of organizations. Through processes, companies achieve objectives, coordinate and optimize resources, and comply with regulatory requirements. Legal compliance became a substantial task for all business organizations. Yet, the majority of organizations treat law as an exogenous force and compliance is mapped and measured rather than explained [23]. The management of business processes requires traceability between high-level requirements (for instance, laws) to traces in an information system. Regulatory compliance is one of the major drivers behind the adoption of process mining in the industry [14], yet there is still a large gap between regulations and the artifacts used in PM. In particular, there is a non-trivial interpretative factor: a legal paragraph may have multiple interpretations *by-design* [10], and its disambiguation may therefore require human support. This contrasts with the intention of formalizing policies with a single, mathematical, and unequivocal semantics.

Modeling high-level business requirements is challenging, as their disambiguation must be resolved by experts in both the process and legal domains [20], and if those experts do not remain involved through later compliance assessments, the rationale behind each interpretation is lost. Multiple techniques have been applied to resolve legal compliance, including combinations of imperative models and modal logics [13], conformance checking [5], and declarative process models [21] (see [22] for a recent review). However, these works consider a low-level view of compliance, where goals in an organization can be directly mapped to activities in a process. This limits the applicability to regulatory compliance, where most requirements come in the form of high-level and subjective descriptions. For instance, consider the following excerpt: *"the company needs to send a **timely** delivery confirmation"*. Existing approaches may verify that a confirmation was sent; however, they will not be able to deal with the ambiguities generated by not being able to define how "timely" delivery confirmation was.

This work explores how goal-modelling frameworks may help in the definition of compliance checking techniques to align high-level requirements and business processes. Goal Models [29] are a well-established set of techniques used in requirements engineering to capture non-functional requirements (see [7] for a recent guide). Thanks to their graphical representation, it allows the communication of multiple stakeholders about functional and non-functional requirements. While the relation of Goal Models and Business Processes has been explored before (i.e., [12,19,24,26,28]), its use has been limited to top-down, simulation, and monitoring approaches, not considering qualities and non-functional requirements. Moreover, a typical pitfall of these works is the tight links between requirements and processes, when in reality, they evolve in different lifecycles [3].

In particular, we propose a compliance framework where 1) functional and non-functional requirements are modelled as IStar models [29], 2) any process modelling notation with an LTS semantics captures business processes, and 3) a mapping between requirements and tasks in a goal model and activities in a process model is maintained. This framework allows us to decouple the goals and processes, maintaining their independence, identify ways of satisfying high-level and non-functional goals, and reuse the goal models across multiple process notations. Moreover, we introduce a design-time compliance checking algorithm that evaluates the synchronized execution steps of process and goal models. As an example of the applicability of the framework, we use workflow nets as a process language, but the framework could be instantiated in imperative or declarative languages with an operational semantics, for instance, BPMN [9] and DCR graphs [16]. We illustrate the framework using a simple data protection guideline and provide a prototypical implementation of the framework.

Structure of this paper. Section 2 introduces our compliance framework; Sect. 3 presents the preliminaries; Sect. 4 provides a technique to check the compliance of high-level requirements; Sect. 5 details a prototypical implementation; Sect. 6 presents related work; and Sect. 7 concludes.

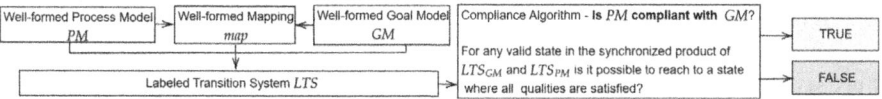

Fig. 1. Overview of the framework

2 Approach

Figure 1 illustrates our compliance checking approach, which combines three well-formed inputs: the process model, the goal model, and the mapping. The process model encodes the *how*, while the goal model captures *what* and *why*, including the high-level business requirements (HL-BR). Both are translated into labeled transition systems (LTS) and synchronized via the mapping, yielding a composite system LTS_C. Compliance holds if, from every state in LTS_C, there exists a path to a state where all HL-BR are satisfied. Formal semantics and well-formedness conditions are detailed in the following section.

2.1 Running Example

To illustrate how our framework works, consider a fictitious scenario:

"**(R) Every time a *company* releases a new feature, it must take *appropriate measures* to ensure its *data remains protected*.** The process begins with the IT department implementing a password policy by (a) *update the encryption standards* and (b) *revise access controls*. Then, the company deploys a system to monitor data security, choosing between (c) *deploy DataGrail or (d) OneTrust*. Once the monitoring system is in place, the security team (e) *conducts penetration tests* to detect potential vulnerabilities. Testing is repeated until no vulnerabilities are found. If there are anomalies, the system (f) *flags the suspicious activity*, and (g) *applies a vulnerability patch* to address the vulnerability. If no vulnerability is found, the testing is successful."

Determining the specific tasks to achieve R requires expert interpretation, and it may vary according to the understanding of *"appropriate measures"*. Most business process compliance techniques [22] focus on verifying whether the tasks carried out align with the predefined ones (i.e., a, b, c, d, e, f, g). However, they do not distinguish among the conformant traces, which are more desirable to fulfill stakeholders' goals or differences in outcome quality [4]. For instance, assume two process executions both follow tasks (a) to (g), yet in one case, the company chooses *DataGrail* and in the other *OneTrust*. Although both traces are conformant, they may differ in terms of effectiveness or cost. Existing compliance techniques would treat them as equivalent, despite these distinctions.

3 Preliminaries

This section provides the background to understand and position the contributions of the framework. Section 3.1 provides formalization of abstract models, capturing the commonalities between imperative process models and goal-oriented model languages. Sections 3.2 and 3.3 instantiate the framework using i* for goal modeling and Workflow nets for process modeling.

3.1 Abstract Models

Any model artifact from a model-driven approach with trace-based semantics can serve as a basic model, provided it uses the same notation.

Definition 1 (Abstract Model Notation (adapted from [8])). *Let A be a fixed universe of model actions, where $a \in A$ is an action. An abstract model notation $\mathcal{MN} \triangleq \langle \mathcal{M}, alph, excluded, step \rangle$ comprises a set of models $\mathcal{M}$; a labeling function $alph : \mathcal{M} \to 2^A$; an exclusion function $excluded : \mathcal{M} \to 2^A$; and a transition predicate $\textbf{step} \subseteq \mathcal{M} \times A \times \mathcal{M}$. Let M_1 and M_2 be two models in $\mathcal{M}$. We require that $\langle M_1, a, M_2 \rangle \in \textbf{step}$ implies both $a \in alph(M_1)$ and $alph(M_1) = alph(M_2)$, and if also $\langle M_1, a, M_2' \rangle \in \textbf{step}$ then $M_2 = M_2'$, then $\textbf{step}$ is action-deterministic.*

Let $\mathcal{M}$ be a set of process models. Intuitively, $alph$ bounds the actions a process may exhibit, and this bound must be preserved by $\textbf{step}$ transitions. Similarly, $exclude$ identifies actions excluded from a given process and may change over time. For non-monotonic languages, setting $excluded = \emptyset$ suffices.

We use Labeled Transition Systems (LTS) to capture configurations, actions, and transitions, where runs represent execution and traces their action sequence.

Definition 2 (Abstract Labeled Transition System). *Let $\mathcal{MN} = \langle \mathcal{M}, alph, excluded, step \rangle$ be an abstract model notation. $M \in \mathcal{M}$ is a model instance. An LTS of the model M, $LTS_M \triangleq \langle S_M, Act_M, \longrightarrow_M, s_0^M, F_M \rangle$ comprises a set of states S_M; a set of actions $Act_M = alph(M)$; a transition relation $\longrightarrow_M \subseteq S_M \times Act_M \times S_M$ is the transition relation, where $\langle M, a, M' \rangle \in \longrightarrow_M$ iff $\langle M, a, M' \rangle \in \textbf{step}$; and initial state $s_0^M = M$; and a set of final states $F_M \subseteq S_M$.*

Definition 3 (Run and traces). *Let $LTS_M = \langle S_M, Act_M, \longrightarrow_M, s_0^M, F_M \rangle$ be the LTS of a model M and s_0 be its initial state. A finite run is a sequence $s_0 \xrightarrow{a_1} s_1 \xrightarrow{a_2} \ldots \xrightarrow{a_n} s_n$ with $s_i \xrightarrow{a_{i+1}} s_{i+1}$ for all $0 \leq i < n$. An infinite run continues indefinitely. A trace σ is the sequence $\langle a_1, a_2, \ldots \rangle$ of actions in the run.*

Let $LTS_M = \langle S_M, Act_M, \to_M, s_0^M, F_M \rangle$ be the LTS of a model M with $Act_M = Act_M \cup \{\epsilon\}$, where ϵ denotes no observable action. For $s, s' \in S_M$ and $a \in Act_M$, we write $s \xrightarrow{a}_M s'$ if $(s, a, s') \in \to_M$. The abstract reachability relation $s \xrightarrow{\sigma}{}^*_M s'$ holds for a finite sequence $\sigma = \langle a_1, \ldots, a_n \rangle$ such that

$s = s_0 \xrightarrow{a_1}_M \cdots \xrightarrow{a_n}_M s_n = s'$. Reflexivity means $s \xrightarrow{\epsilon}_M s$, and transitivity means if $s \xrightarrow{\sigma_1}{}^*_M s'$ and $s' \xrightarrow{\sigma_2}{}^*_M s''$, then we write $s \xrightarrow{\sigma_1\sigma_2}{}^*_M s''$ to indicate that s'' is reachable from s via the execution of σ_1 and σ_2.

3.2 Goal Models

A goal model captures the *rationale* behind the execution of certain tasks, and *what* requirements a system should achieve, using goals, qualities, and tasks.

Definition 4 (Goal Model Elements). *Let the finite set of goal model elements be the tuple $GE \triangleq \langle IE, L \rangle$, where $IE = T \cup G \cup Q$ is a finite set of* intentional elements, *consisting of a set of tasks T, a set of goals G, and a set of qualities Q. The set $L = R \cup C$ denotes* intentional element relations *(or links), where $R \subseteq (G \cup T) \times (G \cup T) \rightharpoonup \{and, or\}$ is a set of refinement links, and $C \subseteq (G \cup T) \times Q \rightharpoonup \{Make, Break\}$ is a set of contribution links.*

We adopt naming conventions from [11] to ensure clarity and consistency in modelling: goals use passive syntax (i.e., *Data protected*), tasks use active forms (i.e., *Implement password policy*), and qualities add manner complements (i.e., *Data protected appropriately*). The model is further refined through links of the form $r(e_1, e_2)$, where element e_1 is refined by e_2. Figure 2, shows the refinement of the goal *Data Protected* (e_1) by tasks such as *Implement password policy* (e_2). Achieving e_1 contributes to fulfilling the HL-BR expressed as the quality *Appropriate measures to protect data*. Intentional elements (IE) are assigned to a truth state for reasoning over their satisfaction.

Definition 5 (Intentional Element Status). *Let $GE = \langle IE, R \rangle$ be the set of goal model elements, with $IE = G \cup T \cup Q$ (Definition 4). Let $\Delta = \{\top, \bot, ?\}$ be the set of truth-values "true," "false," and "unknown". Let $d \in \Delta$ be a status value. The status of an intentional element $e \in IE$ is defined by the function $\Phi : IE \to \Delta$, where $\Phi(e) = (d, d)$ if $e \in G \cup T$, and $\Phi(e) = d$ if $e \in Q$.*

Goals and tasks take states $(d, d) \in \Delta \times \Delta$, where $d = (\top, \top)$ marks them as achieved but pending and $d = (\top, \bot)$ means achieved and not-pending. Qualities use a single $d \in \Delta$, where $d = \top$ means satisfied and $d = \top$ indicates that it is denied. Initially, all goals and tasks are marked as unknown with state $(?, ?)$, and all qualities are set to unknown (?) (Fig. 2). A goal model represents and tracks the state of its intentional elements.

Definition 6 (Goal Model). *Let $\mathcal{MN} = \langle \mathcal{M}, alph, excluded, step \rangle$ be an abstract model notation. Let $GE = \langle IE, L \rangle$, with $IE = G \cup T \cup Q$ and $L = R \cup C$ be the set of goal model elements (Definition 4). Let $\Delta = \{\top, \bot, ?\}$ be the set of truth-values and $\Phi : IE \to \Delta$ the status function of an intentional element (Definition 5). Let $GM \in \mathcal{M}$ be a model instance. A goal model is defined as the tuple $GM \triangleq \langle GE, ID_{GM}, iden_{GM} \rangle$, where GE is a finite set of goal model elements, ID_{GM} is a finite set of identifiers, and $iden_{GM} \subseteq IE \to ID_{GM}$ is a bijection that assigns each intentional element a distinct identifier.*

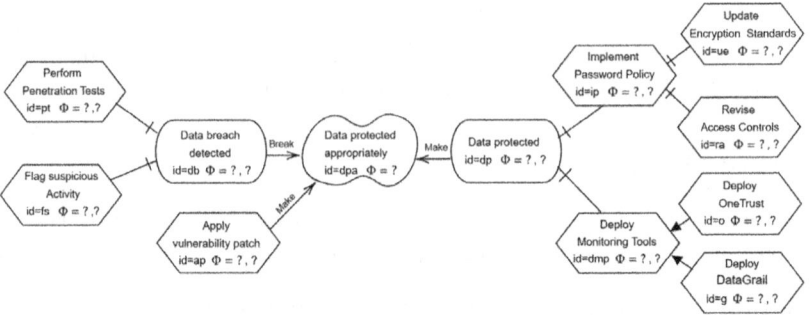

Fig. 2. A goal model GM, modeling the behaviour described in Sect. 2.1.

Figure 2 shows a goal model based on the behaviour described in Sect. 2.1, with one *quality* (⌢), two *goals* (⌢), nine *tasks* (⌢), refinements of type *and* (—+) and *or* (—→), and contribution links *Make* and *Break*. Each intentional element shows its identifier (id) and its status (Φ). A *goal model marking* μ^{GM} represents the current distribution of intentional elements' status. A *marked goal model* is a tuple of a goal model $GM = \langle GE, ID_{GM}, iden_{GM} \rangle$ and a goal model marking $\mu^{GM} \subseteq IE \to \Phi$, denoted by $\langle GM, \mu^{GM} \rangle$. For instance, the goal model marking in Fig. 2 is $\langle$dpa :?, dp : (?,?), ap : (?,?), db : (?,?), ip : (?,?), dmp : (?,?), ue : (?,?), ra : (?,?), o : (?,?), g : (?,?), fs : (?,?), pt : (?,?)$\rangle$.

Let $GE = \langle IE, L \rangle$ be the set of goal model elements, with $IE = G \cup T, \cup Q$ and $L = R \cup C$, and $e \in G \cup T$. The refinement reachability relation $\hookrightarrow \subseteq IE \times IE$ is defined as the smallest relation such that $(e, e') \in R$ implies $e' \hookrightarrow e$ (*direct*), and if $e' \hookrightarrow e$ and $e'' \hookrightarrow e'$, then $e'' \hookrightarrow^* e$ (*transitive*). Its reflexive–transitive closure $\hookrightarrow^*$ expresses reachability through zero or more refinement steps. For example, in Fig. 2, the goal "*Data Protected*" (e) is refined through "*Deploy monitoring tools*" (e'), which includes tasks such as "*Deploy OneTrust*" (e''); thus, $e'' \hookrightarrow^* e$. To ensure analyzability for HL-BR satisfaction, we define well-formedness criteria constraining the structure and labeling of goal models.

Definition 7 (Well-formed Goal Model). *Let $GM = \langle GE, ID_{GM}, iden_{GM} \rangle$ be a goal model (Definition 6). Let $GE = \langle IE, L \rangle$, with $IE = G \cup T \cup Q$ and $L = R \cup C$ be the set of goal model elements (Definition 4). GM is well-formed iff: (1) GE is acyclic; (2) R and C are asymmetric; (3) There exist a minimum amount of goal model elements $|G \cup T| \geq 1$, $|Q| \geq 1$, and $\exists e \in G \cup T, q \in Q : C(e, q) = $ Make; (4) All refinements of e are the same type $\forall e \in G \cup T, \forall e_1, e_2 : (e, e_1), (e, e_2) \in R \Rightarrow R(e, e_1) = R(e, e_2)$; (5) Between any two intentional elements, at most one refinement link exists $\forall e_1, e_2 \in G \cup T : |\{(e_1, e_2) \in R\}| \leq 1$; (6) elements that contribute conflicting values to a quality must stem from disjoint refinement branches $\forall q \in Q, (e_1, q), (e_2, q) \in C, C(e_1, q) \neq C(e_2, q) : \{x \mid x \hookrightarrow^* e_1\} \cap \{y \mid y \hookrightarrow^* e_2\} = \emptyset$.*

Figure 2 depicts a well-formed goal model. We assume that every GM provided as input to our algorithm is well-formed. Based on this structure, we define its LTS, where states (S_{GM}) correspond to configurations reachable from the initial marking μ_0^{GM} via the transition relation $\rightarrow_{GM}$. We use the shorthand $GM \models \mu_1^{GM} \xrightarrow{x} \mu_2^{GM}$ to denote the transition relation where GM does not change, that is $(\langle GM, \mu_1^{GM}\rangle, x, \langle GM, \mu_2^{GM}\rangle) \in \rightarrow_M$.

Definition 8 (Labeled Transition System Goal Model). Let $\langle GM, \mu^{GM}\rangle$ be a marked goal model. Let $GM = \langle GE, ID_{GM}, iden_{GM}\rangle$ be a goal model (Definition 6). Let $GE = \langle IE, L\rangle$, with $IE = G \cup T \cup Q$ and $L = R \cup C$ be the set of goal model elements (Definition 4) Let $LTS_M = \langle S_M, Act_M, \rightarrow_M, s_0^M, F_M\rangle$ be a labeled transition system of a model M (Definition 2). Let GM be the model M. A labeled transition system for a goal model is the tuple: $LTS_{GM} \triangleq \langle S_{GM}, Act_{GM}, \rightarrow_{GM}, s_0^{GM}, F_{GM}\rangle$, where:

1. S_{GM} is the set of states,
2. $Act_{GM} = G \cup T$ is the set of actions,
3. $\rightarrow_{GM} \subseteq S_{GM} \times Act_{GM} \times S_{GM}$, is the transition relation governed by the rules described in Fig. 3,
4. $s_0^{GM} = \mu_0^{GM}$ is the initial state, s.t. $\mu_0^{GM}(e) = (?, ?) \; \forall e \mid e \in G \cup T$ and $\mu_0^{GM}(e) = ? \; \forall e \mid e \in Q$,
5. $F_{GM} \subseteq S_{GM}$ are the final states s.t. $\forall q \in Q, \mu^{GM}(q) = \top$.

Let $LTS_{GM} = \langle S_{GM}, Act_{GM}, \rightarrow_{GM}, s_0^{GM}, F_{GM}\rangle$ be a labeled transition system of a goal model G, (Definition 8). The rules (Fig. 3) that govern the transition relation $\rightarrow_{GM}$ can be grouped in four categories: (1) *Activation*: A leaf node may fire if it has no refinements, (2) *Refinement Propagation*: If the conditions for the refinement type (i.e., *and*, *or*) are satisfied, then the element e in a refinement link (e, e') is marked as satisfied, (3) *Contribution Propagation*: If an element e with a contribution link to a quality q is satisfied, the status value of q will change according with the contribution type, (4) *Backpropagation*: As the satisfaction of the qualities evolve overtime, its effect is propagated backwards, indicating which executions are pending (i.e., need to be redone).

Let $GE = \langle IE, L\rangle$, where $IE = G \cup T \cup Q$ and $L = R \cup C$ be the set of goal model elements (Definition 4). Let $GM = \langle GE, ID_{GM}, iden_{GM}\rangle$ be the goal model (Definition 6). We write $GM = \langle G, T, Q, R, C, ID_{GM}, iden_{GM}\rangle$. Given a marking μ^{GM}, Fig. 4 shows some of the reachable markings of the goal model shown in Fig. 2. Each arrow is labeled with the identifier of the executed element and the transition rule applied for the resulting marking.

Lemma 1 (Goal Model as Abstract Model Notation). Let $\mathcal{MN} = \langle \mathcal{M}, alph, excluded, step\rangle$ be an abstract model notation (Definition 1). Let $GE = \langle IE, L\rangle$, where $IE = G \cup T \cup Q$ and $L = R \cup C$ be the set of goal model elements (Definition 4). Let $GM = \langle GE, ID_{GM}, iden_{GM}\rangle$ be the goal model (Definition 6), and $\Phi(e)$ the status of $e \in G \cup T \cup Q$ (Definition 5). Let $\mathcal{M}$ be the set of all such GM where $\mu^{GM} : G \cup T \cup Q \rightarrow \Phi$, $\lambda = iden$ the labeling function, $excluded(GM) = \emptyset$, and $step \subseteq \mathcal{M} \times (G \cup T) \times \mathcal{M}$ where $(GM, e, GM') \in step$ iff e is executed according to Fig. 3. Then $\mathcal{A} = \langle \mathcal{M}, \lambda, excluded, step\rangle$ is an abstract model notation.

$(\mathbf{P_{ie}})$ $\dfrac{e \in G \cup T \quad e \vdash \text{leafnode}}{\langle G,T,Q,R,C,ID_{GM},iden_{GM}\rangle \models \mu^{GM} \xrightarrow{e} \mu^{GM}[e \mapsto (\top,\bot)]}$

$(\mathbf{P_{and}})$ $\dfrac{(e,e') \in R \quad R(e,e')=\text{AND} \quad \forall e': \mu^{GM}(e')=(\top,\bot)}{\langle G,T,Q,R,C,ID_{GM},iden_{GM}\rangle \models \mu^{GM} \xrightarrow{e} \mu^{GM}[e \mapsto (\top,\bot)]}$

$(\mathbf{P_{or}})$ $\dfrac{(e,e') \in R \quad R(e,e')=\text{OR} \quad \exists e': \mu^{GM}(e')=(\top,\bot)}{\langle G,T,Q,R,C,ID_{GM},iden_{GM}\rangle \models \mu^{GM} \xrightarrow{e} \mu^{GM}[e \mapsto (\top,\bot)]}$

$(\mathbf{P_{Make}})$ $\dfrac{(e,e') \in C \quad e \in Q \quad \mu^{GM}(e) \in \{\top,?\} \quad \exists e': \mu^{GM}(e')=\top \quad C(e,e')=\text{Make}}{\langle G,T,Q,R,C,ID_{GM},iden_{GM}\rangle \models \mu^{GM} \xrightarrow{e} \mu^{GM}[e \mapsto (\top,\bot)]}$

$(\mathbf{P_{Break}})$ $\dfrac{(e,e') \in C \quad e \in Q \quad \mu^{GM}(e) \in \{\bot,?\} \quad \exists e': \mu^{GM}(e')=(\top,\bot) \quad C(e,e')=\text{Break}}{\langle G,T,Q,R,C,ID_{GM},iden_{GM}\rangle \models \mu^{GM} \xrightarrow{e} \mu^{GM}[e \mapsto \bot]}$

$(\mathbf{BP_{fulfill}})$ $\dfrac{(e,e') \in C \quad e \in Q \quad \mu^{GM}(e)=\bot \quad \mu^{GM}(e')=(\top,\bot) \quad C(e,e')=\text{Make}}{\langle G,T,Q,R,C,ID_{GM},iden_{GM}\rangle \models \mu^{GM} \xrightarrow{e} \mu^{GM'}}$ where $\mu^{GM'} =$ $\mu^{GM}[e \mapsto \top,\; e' \mapsto (\top,\top) \text{ for all } C(e,e')=\text{Break},\; e'' \mapsto (\top,\top) \text{ for all } e'' \hookrightarrow_R^* e']$

$(\mathbf{BP_{deny}})$ $\dfrac{(e,e') \in C \quad e \in Q \quad \mu^{GM}(e)=\top \quad \mu^{GM}(e')=(\top,\bot) \quad C(e,e')=\text{Break}}{\langle G,T,Q,R,C,ID_{GM},iden_{GM}\rangle \models \mu^{GM} \xrightarrow{e} \mu^{GM'}}$ where $\mu^{GM'} =$ $\mu^{GM}[e \mapsto \bot,\; e' \mapsto (\top,\top) \text{ for all } C(e,e')=\text{Make},\; e'' \mapsto (\top,\top) \text{ for all } e'' \hookrightarrow_R^* e']$

Fig. 3. Operational semantics of a goal model by extension $GM = \langle G,T,Q,R,C,ID_{GM},iden_{GM}\rangle$ under marking μ^{GM}, denoted as $GM \models \mu^{GM}$. Transition rules update the marking based on activation, refinement, and contribution.

3.3 Process Models

A process model captures executable workflow steps; they can be modeled using imperative or declarative languages. Here, we use a subclass of Petri nets [25], called Workflow nets [1]. Let $\mathcal{MN} = \langle \mathcal{M}, alph, excluded, step \rangle$ be an abstract model notation (Definition 1). Let a Petri net N be a model instance $N \in \mathcal{M}$.

Definition 9 (Petri net). *(taken from [2]) A Petri net is a tuple* $N \triangleq \langle P, Tr, F, ID_{PM}, iden_{PM}\rangle$ *where (1) P is a finite set of places, (2) Tr is a finite set of transitions such that $P \cap Tr = \emptyset$, (3) $F \subseteq (P \times Tr) \cup (Tr \times P)$ is a set of directed arcs, called the flow relation. A Petri net node is an element of $P \cup Tr$. (4) ID_{PM} is a finite set of unique transition identifiers, and (5) $iden_{PM} : Tr \to ID_{PM}$ is a bijection that assigns a distinct identifier to each transition.*

Figure 5a shows a Petri net, with ten places and eleven transitions. Token resides in places; p_0 has one token. A marking (μ^{PM}) represents token distribution. A *marked Petri net* is a tuple of a net $N = \langle P, Tr, F, ID_{PM}, iden_{PM}\rangle$ and a marking $\mu^{PM} \in \mathbb{B}(P)$. The initial marking in Fig. 5a is $[p_0]$; the set of all marked Petri nets is $\mathcal{N}$. A node x is called an preset of node y if $(x,y) \in F$, and an postset of y if $(y,x) \in F$. For example, $^\bullet t_4 = \{p_3, p_4\}$ and $t_4^\bullet = \{p_5\}$. A transition is enabled if all input places have tokens. Firing removes tokens from input places and adds tokens to output places.

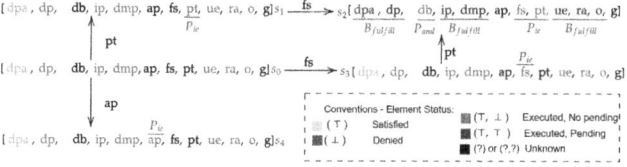

Fig. 4. Execution fragment of the LTS associated with the goal model in Fig. 6.

Definition 10 (Firing Rule). *(taken from [2])* Let $\langle N, \mu^{PM} \rangle$ be a marked Petri net, with $N = \langle P, Tr, F, ID_{PM}, iden_{PM} \rangle$ (Definition 9). A transition $t \in Tr$ is enabled, denoted $\langle N, \mu^{PM} \rangle[t\rangle$, if and only if $\bullet t \leq \mu^{PM}$. The firing rule $[-\rangle \subseteq \mathcal{N} \times Tr \times \mathcal{N}$ is the smallest relation satisfying: for any $\langle N, \mu^{PM} \rangle \in \mathcal{N}$ and any $t \in Tr$, $\langle N, \mu^{PM} \rangle[t\rangle \Rightarrow \langle N, \mu^{PM} \rangle[t\rangle \, \langle N, \langle \mu^{PM} \setminus \bullet t \rangle \cup t \bullet \rangle$

In the marking of Fig. 5a, transition t_1 is enabled. Firing t_1 yields $[p_1, p_2]$. Transition sequences determine which markings are reachable and describe the dynamic behavior of the net.

Definition 11 (Firing Sequence). *(taken from [2])* Let $\langle N, \mu^{PM} \rangle$ be a marked petri net, with $N = \langle P, Tr, F, ID_{PM}, iden_{PM} \rangle$ (Definition 9). A sequence $\sigma \in Tr^*$ is a firing sequence of $\langle N, \mu_0^{PM} \rangle$ if, there exist markings $\mu_1^{PM}, \ldots, \mu_n^{PM}$ and transitions $t_1, \ldots, t_n \in Tr \mid n \in \mathbb{N}$ such that $\sigma = \langle t_1 \ldots t_n \rangle$, and for all i with $1 \leq i < n$, it holds that $\langle N, \mu_i^{PM} \rangle[t_{i+1}\rangle$ and $\langle N, \mu_i^{PM} \rangle[t_{i+1}\rangle \rightarrow \langle N, \mu_{i+1}^{PM} \rangle$.

A marking μ^{PM} is *reachable* from μ_0^{PM} if a sequence of enabled transitions leads from μ_0^{PM} to μ^{PM}. Given an initial marking μ_0^{PM}, the set of reachable markings of $\langle N, \mu_0^{PM} \rangle$ is denoted S_{PM}, and can be computed using a *reachability graph* which is a specific type of labeled transition system.

Definition 12 (Labeled Transition System - Petri Net). Let $\langle N, \mu^{PM} \rangle$ be the marked Petri net, with $N = \langle P, Tr, F, ID_{PM}, iden_{PM} \rangle$ (Definition 9. Let N be a model instance M. Let $LTS_M \triangleq \langle S_M, Act_M, \longrightarrow_M, s_0^M, F_M \rangle$ be a labeled transition system of a model M (Definition 2). A labeled transition system for a Petri net is the tuple: $LTS_{PM} \triangleq \langle S_{PM}, Act_{PM}, \rightarrow_{PM}, s_0^{PM}, F_{PM} \rangle$ where:

1. S_{PM} is the set of states,
2. $Act_{PM} = Tr$ is the set of actions,
3. $\rightarrow_{PM} \subseteq S_{PM} \times Tr \times S_{PM}$ is the transition relation, it follows the firing rule (Definition 10), i.e., $(\mu^{PM}, t, \mu^{PM'}) \in \rightarrow_{PM}$ iff $\langle N, \mu^{PM} \rangle[t\rangle \rightarrow \langle N, \mu^{PM'} \rangle$,
4. $s_0^{PM} = \mu_0^{PM}$, is the initial state.
5. $F_{PM} \subseteq S_{PM}$ are the final states.

Figure 5b shows the LTS of the Petri net in Fig. 5a. We use Workflow nets to model processes with clear start, end, and control flow, with a unique entry point, a unique exit point, and well-defined execution semantics.

Definition 13 (Workflow nets). *(Taken from [2])* Let $N = \langle P, Tr, F, ID_{PM}, iden_{PM}\rangle$ be a Petri net (Definition 9) and $\bar{t}$ a fresh identifier not in $P \cup Tr$. N is a workflow net (WF-net) iff: (1) P contains an input place i s.t. $\bullet i = \emptyset$, (2) P contains an output place o s.t. $o\bullet = \emptyset$, and (3) $\bar{N} = (P, Tr \cup \{\bar{t}\}, F \cup \{(o, \bar{t}), (\bar{t}, i)\})$ is strongly connected.

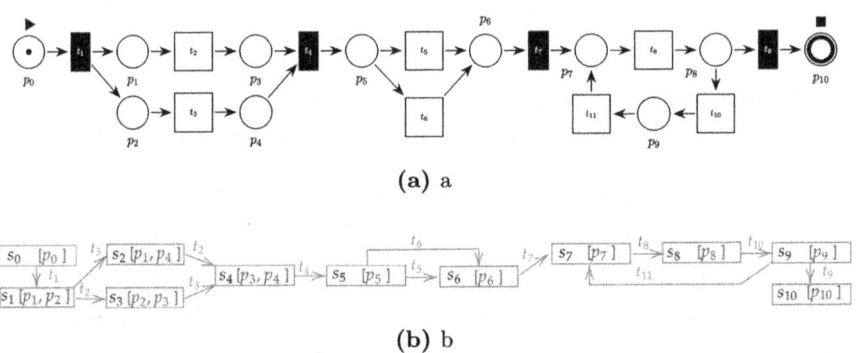

(a) a

(b) b

Fig. 5. A WF-net and its behaviour (a) A WF-net representing the example in Sect. 2.1; (b) visualizes its LTS. For example, t_2 represents *Update Encryption Standards*, t_3 *Revise the Access Controls*, t_5 and t_6 correspond to Deploy *DataGrail* or *OneTrust*, respectively, while t_8, t_{10} and t_{11} denote *Perform Penetration Tests*, *Flag suspicious activity* and *Apply Vulnerabilities Patches*.

Definition 14 (Safeness). Let $WF = \langle N, \mu^{PM}\rangle$ be a marked Workflow net, with $N = \langle P, Tr, F, ID_{PM}, iden_{PM}\rangle$ (Definition 9). We say WF is safe if, for every reachable marking μ^{PM} from the initial marking μ_0^{PM}, it holds that $\mu^{PM}(p) \leq 1$ for all $p \in P$.

Let $LTS_{PM} = \langle S_{PM}, Act_{PM}, \rightarrow_{PM}, s_0^{PM}, F_{PM}\rangle$ be the labeled transition system of a safe Workflow net, here $s_0^{PM} = \mu_0^{PM}$ with $\mu_0^{PM}(\blacktriangle) = 1$ and $\mu_0^{PM}(p) = 0$ for all $p \neq \blacktriangle$, and F_{PM} includes all markings with $\mu^{PM}(\blacksquare) = 1$ and $\mu^{PM}(p) = 0$ for all $p \neq \blacksquare$. Figure 5 depicts a safe Workflow net and its LTS. We assume that every process model used as input to our algorithm is a safe WF-net.

Lemma 2 (A Marked Workflow Net as Abstract Model Notation). Let $MWN = \langle P, Tr, F, ID_{PM}, iden_{PM}, \mu_0^{PM}\rangle$ be a marked Workflow net (Definition 13). Define $\mathcal{M} = \left\{\mu^{PM} : P \rightarrow \mathbb{N} \mid \exists \text{ firing sequence } \mu_0^{PM} \xrightarrow{t_1} \cdots \xrightarrow{t_n} \mu^{PM}\right\}$ as the set of reachable markings. Let $\lambda : \mathcal{M} \rightarrow 2^{ID_{PM}}$ be the labeling function defined by $\lambda(\mu^{PM}) = \{iden_{PM}(t) \mid t \in Tr \text{ and } \forall p \in \bullet t : \mu^{PM}(p) \geq 1\}$. Let $excluded : \mathcal{M} \rightarrow 2^{\emptyset}$ denote the exclusion function, and let $step \subseteq \mathcal{M} \times Tr \times \mathcal{M}$ be the transition relation such that $(\mu^{PM}, t, \mu'^{PM}) \in step$ if and only if $\mu^{PM} \xrightarrow{t} \mu'^{PM}$ holds according to the firing rule (Definition 10). Then $\mathcal{A} = \langle \mathcal{M}, \lambda, excluded, step\rangle$ is an abstract model notation (Definition 1).

3.4 Mapping

To enable synchronization, we map process actions (transitions) to goal model elements (goals and tasks).

Definition 15 (Mapping between Process transitions and Intentional Elements). *Let $GM = \langle GE, ID_{GM}, iden_{GM} \rangle$ be a well-formed goal model. Let $N = \langle P, Tr, F, ID_{PM}, iden_{PM} \rangle$ be a safe Workflow net (Definition 13). The process transition to intentional element mapping is a total function defined as: $map: ID_{PM}^{Tr} \to ID_{GM}^{G \cup T} \cup \{\epsilon\}$ where $ID_{PM}^{Tr} \subseteq ID_{PM}$ is the set of identifiers assigned to process transitions, $ID_{GM}^{G \cup T} \subseteq ID_{GM}$ is the set of identifiers assigned to goals and tasks only, and ϵ denotes that a transition is left unmapped.*

For example, consider the well-formed goal model in Fig. 2 and the safe Workflow net shown in Fig. 5a. Transitions $t_2, t_3, t_5, t_6, t_7, t_{10}, t_{11}$ are mapped to intentional elements with identifiers ue (*Update Encryption Standards*), ra (*Revise Access Controls*), o (*Deploy OneTrust*), g (*Deploy DataGrail*), pt (*Perform Penetration Tests*), fs (*Flag Suspicious Activity*), and ap (*Apply vulnerability patch*), respectively; all others (i.e., t_1, t_4, t_9) remain unmapped (ϵ). To ensure consistency, we introduce well-formedness criteria.

Definition 16 (Well-formed Mapping). *Let $map : ID_{PM}^{Tr} \to ID_{GM}^{G \cup T} \cup \{\epsilon\}$ be the process transition to intentional element mapping (Definition 15). Let $GE = \langle IE, L \rangle$, with $IE = \langle G \cup T \cup Q \rangle$ and $L = \langle R \cup C \rangle$ be the set of goal model elements (Definition 4). We say that map is well-formed if the following conditions holds:*

1. *Consistent contribution: Let $id \in ID_{PM}^{Tr}$ such that $map(id) = \{e_i, e_j, \dots\}$ and $|map(id)| \geq 2$. Let $e_i, e_j \in ID_{GM}^{G \cup T}$. For all intentional elements e_1, e_2 with identifier e_i, e_j and the same quality $q \in Q$:*
   (a) *if e_1 or one of its descendants ($\forall e'$ s.t. $e' \hookrightarrow^* e_1$) contributes to q,*
   (b) *if e_2 or one of its descendants ($\forall e'$ s.t. $e' \hookrightarrow^* e_2$) also contributes to q,*
   (c) *then e_1 and e_2 must contribute the same type (either Make or Break). Formally, if there exist $e'_1, e'_2 \in G \cup T$ s.t. $e_1 = e'_1$ or $e_1 \hookrightarrow^* e'_1, e_2 = e'_2$ or $e_2 \hookrightarrow^* e'_2$ and $(e'_1, q), (e'_2, q) \in C$ then $C(e'_1, q) = C(e'_2, q)$*
2. *No and-refinement co-mapping: For any process transition identifier $id \in ID_{PM}^{Tr}$, let $map(id) = \{e_i, e_j, \dots\} \subseteq ID_{GM}^{G \cup T}$. For all $e_i, e_j \in map(id)$, if there exist $e_1, e_2 \in G \cup T$ such that $iden_{GM}(e_1) = e_i, iden_{GM}(e_2) = e_j$, and there exists $R(e_1, e'_2) = and$, and $e_2 \hookrightarrow^* e_1$ or $e_1 \hookrightarrow^* e_2$, then e_1 and e_2 can not be mapped by the same transition identifier.*

For example, in the goal model shown in Fig. 2, a single process transition must not map to both the identifier db (*Data breach detected*) and ap (*Apply vulnerability patch*), since they contribute differently, one as Make, the other as Break, to the same quality. Similarly, a transition must not be mapped to both ue (*Update encryption standard*) and ip (*Implement password policy*), because there is an and-type refinement relation between them. We assume that every mapping used as input of our algorithm is well-formed.

4 Compliance Assessment of High-Level Business Requirements

We define compliance as the satisfaction of all high-level business requirements (HL-BR). HL-BR are represented as qualities in the goal model (Fig. 2). This section introduces the method to assess compliance at **design time** by evaluating whether the execution of process actions leads to a system state in which all the HL-BR are fulfilled. We introduce the synchronous product of the labeled transition system of each model and the compliance criterion.

Definition 17 (Composed Labeled Transition System). *Let* $LTS_{GM} = \langle S_{GM}, Act_{GM}, \rightarrow_{GM}, s_0^{GM}, F_{GM}\rangle$ *be the labeled transition system of the goal model (Definition 8). Let* $LTS_{PM} = \langle S_{PM}, Act_{PM}, \rightarrow_{PM}, s_0^{PM}, F_{PM}\rangle$ *be the labeled transition system of the process model (Definition 12). Their synchronous product is defined as:* $LTS_C \triangleq \langle S_C, Act_C, \rightarrow_C, s_0^C, F_C\rangle$ *where:*

1. $S_C = S_{GM} \times S_{PM}$ *is the set of composed states,*
2. $Act_C = Act_{GM} \cup Act_{PM} \cup \{\epsilon\}$ *is the set of actions, where ϵ represent a no observable action,*
3. $\rightarrow_C$ *is the transition relation, preserving synchronization of shared actions,*
4. $s_0^C = \langle s_0^{GM}, s_0^{PM}\rangle$ *is the initial state,*
5. $F_C = \{\langle s_{GM}, s_{PM}\rangle \in S_C \mid \forall q \in Q : \mu^{GM}(q) = \top\}$ *is the set of final states where all qualities are satisfied.*

Definition 18 (Compliance Criterion). *Let* $LTS_{GM} = \langle S_{GM}, Act_{GM}, \rightarrow_{GM}, s_0^{GM}, F_{GM}\rangle$ *be the LTS of a well-formed goal model (Definition 8). Let* $LTS_{PM} = \langle S_{PM}, Act_{PM}, \rightarrow_{PM}, s_0^{PM}, F_{PM}\rangle$ *be the LTS of a safe Workflow net (Definition 12). Let* $LTS_C = \langle S_C, Act_C, \rightarrow_C, s_0^C, F_C\rangle$ *be the synchronous product of* LTS_{PM} *and* LTS_{GM} *(Definition 17). The compliance satisfaction is defined as true if* $\forall s \in S_C, \exists s' \in F_C$ *such that* $s \rightarrow^* s'$, *and false otherwise.*

Let $GM = \langle GE, ID_{GM}, iden_{GM}\rangle$ be a well-formed goal model (Definition 6) Let $GE = \langle IE, L\rangle$, with $IE = G \cup T \cup Q$ and $L = R \cup C$ be the set of goal model elements (Definition 4). Let $N = \langle P, Tr, F, ID_{PM}, iden_{PM}\rangle$ be a safe Workflow net (Definition 13). Let $LTS_C = \langle S_C, Act_C, \rightarrow_C, s_0^C, F_C\rangle$ be the composed labeled transition system. Given an goal model marking μ^{GM} and process model marking μ^{PM} such that $GM \models \mu^{GM}$, and $N \models \mu^{PM}$. The operational semantics of the transition relation $\rightarrow_C$, is defined by the following rules:

The composed LTS runs the process and goal models in a unified step, so every process transition either triggers the matching goal model update or leads to no changes. For instance, Fig. 7 depicts two exemplary execution fragments of the labeled transition system composed for the goal model shown in Fig. 2 and the Workflow net in Fig. 5a. In the left side (*PM is compliant with GM*). States s_9, s_{10}, and s_{11} satisfy all the qualities (*dpa*) defined in the goal model, and therefore belong to the set F_C. If every state in LTS_C that is not shown in the exemplary execution eventually leads to at least one of these states, then the

(P$_{\text{sync}}$)
$$\frac{\begin{array}{c}map(iden_{PM}(t)) = iden_{GM}(e)\\ \langle G,T,Q,R,C,ID_{GM},iden_{GM}\rangle \models \mu^{GM} \xrightarrow{e}_{GM} \mu^{GM'}\\ \langle P,Tr,F,ID_{PM},iden_{PM}\rangle \models \mu^{PM} \xrightarrow{t}_{PM} \mu^{PM'}\end{array}}{\left\langle \begin{array}{c}G,T,Q,R,C,ID_{GM},iden_{GM},\\ P,Tr,F,ID_{PM},iden_{PM}\end{array}\right\rangle \models (\mu^{GM},\mu^{PM}) \xrightarrow{t}_C (\mu^{GM'},\mu^{PM'})}$$

(P$_{\text{local}}$)
$$\frac{\begin{array}{c}map(iden_{PM}(t)) = \epsilon\\ \langle P,Tr,F,ID_{PM},iden_{PM}\rangle \models \mu^{PM} \xrightarrow{t}_{PM} \mu^{PM'}\end{array}}{\left\langle \begin{array}{c}G,T,Q,R,C,ID_{GM},iden_{GM},\\ P,Tr,F,ID_{PM},iden_{PM}\end{array}\right\rangle \models (\mu^{GM},\mu^{PM}) \xrightarrow{t}_C (\mu^{GM},\mu^{PM'})}$$

Fig. 6. Transition rules LTS_C

process model depicted in Fig. 5a is considered compliant with the goal model presented in Fig. 2, which means that the data was protected appropriately.

In contrast, in the right side scenario (*PM is non-compliant with GM*) consider a variation in the transition t_5 labels shown in Fig. 5, instead of *"Deploy OneTrust"*, now the company will *"Deploy Grafana"* (t_5^*). Previously, the mapping of that transition was associated with the intentional element identified as *"o"* in the goal model shown in Fig. 2, that is $map(t_5) = o$, but now with the mapping would be $map(t_5^*) = \epsilon$, meaning that t_5^* is not mapped to any intentional element. Since at least one state in LTS_C does not eventually reach a state where all qualities are satisfied, the process model including t_5^* is not compliant with the goal model shown in Fig. 2.

Having established the formal rules for process execution and goal satisfaction, we now present the complete compliance-checking algorithm. The algorithm first constructs the state space of the composed Labeled Transition System LTS_C (Definition17) by iteratively applying the transition rules. Its complexity is $O(n)$ where $n = M \times K$, with $M = |S_{GM}|$ denoting the number of states in the goal model and $K = |S_{PM}|$ denoting the number of states in the process model. In the second part, compliance is verified by ensuring that for every state $s \in S_C$ there exists at least one reachable final state $s' \in F_C$ ($s \rightarrow^* s'$) such that $\forall q \in Q$, $\mu_{s'}^{GM}(q) = \top$. Termination is guaranteed by the finiteness of LTS_C. Note that while LTS_{GM} is always finite, the finiteness of LTS_{PM} holds only under specific properties of the process model (Definition 14).

The algorithm is modular and extensible, allowing improvements or adaptations to be incorporated according to the specific application domain or process analysis need. For instance, it can be easily modified based on a set of traces that determine which of these are compliant with the goal model, and the mapping can be redesigned and validated by legal and process experts.

5 Prototypical Implementation

We implemented our algorithm tasking as input a process model pattern derived from Article 17 of the GDPR, which defines the data subject's right to erasure,

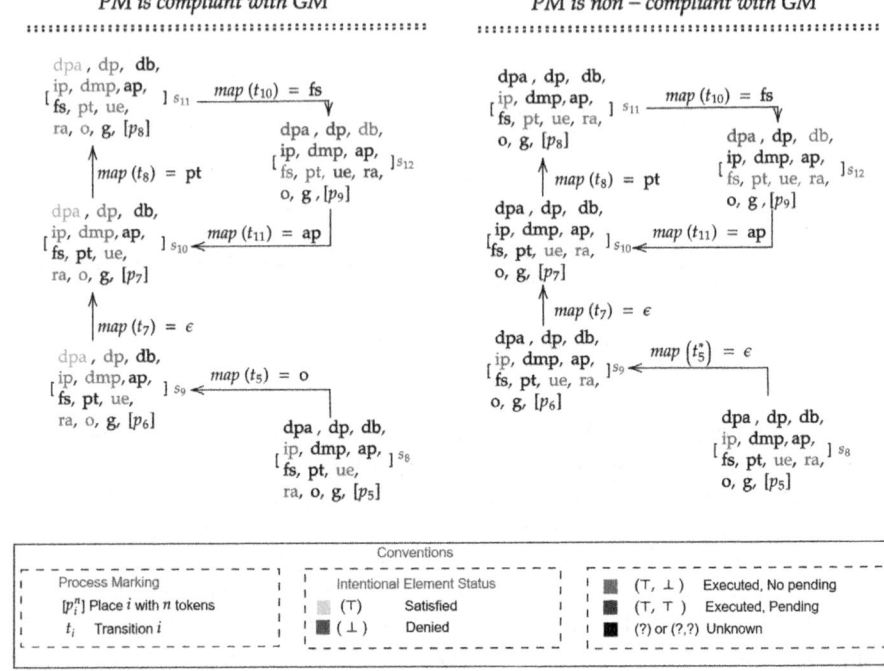

Fig. 7. Execution of the composed LTS of the Marked Workflow net in Fig. 5a, and the goal model in Fig. 6. The left side shows an example of a compliant execution, and the right side shows the counterexample of a non-compliant case.

also known as the right to be forgotten. This article contains high-level business requirements (HL-BR) such as *Data deleted without undue delay*, *when no longer necessary*, *after consent withdrawal*, and *Data retained* when *overriding legitimate grounds*, and *public interest*. The requirements were modeled in a goal model and refined into tasks. Each process activity was associated with some intentional element. Compliance was evaluated at design time by verifying the alignment between goal satisfaction and process execution behavior. We compared the capabilities of our framework against the requirements and interactions described in Art. 17 of the GDPR, such as the request for data erasure, verification of the data subject identity, evaluation of legal exceptions, and the actual deletion of personal data. These interactions serve as a representative example of the type of multi-step, conditional processes that are legally regulated. The evaluation was used to reason about the expressiveness of the framework. Our case study showed enough expressiveness to capture most requirements; however, requirements describing interprocess communication and relationships among multiple agents and their goals are still missing. Moreover, the current version does not yet support certain expressive constructs of process models, such as data conditions and timed events. The prototype implementing our approach is available in Python 3[1]. It requires three inputs to perform design-time compliance checking: a process model in PNML format, a goal model in JSON format, and a mapping file in CSV format.

Algorithm 1: Compliance Checking Algorithm

Input: Goal model $GM = \langle GE, ID_{GM}, iden_{GM} \rangle$,
  Process model $PM = \langle P, Tr, F, ID_{PM}, iden_{PM} \rangle$,
  Mapping function $map : ID_{PM}^{Tr} \to ID_{GM}^{GUT} \cup \{\epsilon\}$
Output: true if every state leads to one where all $q \in Q$ are satisfied; false otherwise
1 Initialize $S_C := \{\langle \mu_0^{GM}, \mu_0^{PM} \rangle\}$;
2 Construct state space by exploring all enabled transitions $t \in Tr$; for each, update μ^{PM} and apply goal model semantics to μ^{GM} if $map(iden_{PM}(t)) \neq \epsilon$; add resulting states to S_C;
3 **foreach** $s \in S_C$ **do**
4   **if** no reachable $s' \in S_C$ from s with $\mu^{GM'}(q) = \top$ for all $q \in Q$ **then**
5 | return false // Not compliant
6   **end**
7 **end**
8 return true // Compliant
---

6 Related Work

Recent advances in process compliance have emphasized the alignment of operational processes with regulatory and business constraints through formal methods such as conformance checking and logic-based validation. Foundational techniques focus on aligning observed executions with procedural models (i.e., Petri nets) or multi-perspective specifications that include data and resources [5,6]. However, most approaches presuppose a complete, procedural specification and struggle with high-level requirements expressed in non-operational terms, such as stakeholder goals or soft constraints. While extensions using preferences [18] or alignment of goals and processes [15] offer expressive power, they lack runtime verification semantics or cannot systematically integrate with trace-level abstractions. As such, the space between high-level intent modeling and low-level event traces remains fragmented, limiting formal reasoning about compliance.

To address this, recent efforts have explored mapping business requirements to executable process behaviors, but largely from the perspective of conformance, not enforcement or design-time guarantee. For example, works like [4] investigate the conformance space by comparing different behavioral equivalences, yet stop short of offering a mechanism to ensure satisfaction of non-functional constraints during execution. Predictive approaches [27] further extend the vision by forecasting compliance outcomes, but do not explicitly link these to stakeholder-defined goals. Moreover, alignment-based conformance checking [17] focuses on decomposing data-aware models for efficiency, not for semantic integration with intentional structures. This gap shows an opportunity for frameworks that unify abstract goal models and operational models through shared transition semantics, enabling formal guarantees over business objectives at runtime.

---
[1] https://github.com/jc4v1/HLBRBPM25.

7 Conclusion and Future Work

We proposed a design-time compliance framework that integrates Workflow Petri nets with i goal models using synchronized labeled transition systems to assess whether all high-level business requirements (HL-BR) are satisfied. The key contributions include: (1) a formal operational semantics for goal models with support for conflict resolution; (2) a compliance-checking algorithm based on synchronous composition of process and goal models; and (3) a prototype implementation that evaluates compliance from model inputs. Current limitations include a lack of support for timed events and capturing full model-driven language expressiveness. Future work will address this by extending the framework to support time, data, and multi-agent behavior, and adding a runtime monitor for detecting violations during execution.

Acknowledgement. This work was supported by the research grant "Center for Digital CompliancE (DICE)" (VIL57420) from VILLUM FONDEN, and by the Innovation Foundation project "Explainable Hybrid-AI for Computational Law and Accurate Legal Chatbots" 4355-00018B XHAILe.

References

1. van der Aalst, W.M.P.: The application of petri nets to workflow management. J. Circ. Syst. Comput. **8**(1), 21–66 (1998)
2. Aalst, W.M.V.D., Dongen, B.F.V.: Discovering petri nets from event logs. In: LNCS (including subseries Lecture Notes in Artificial Intelligence and Lecture Notes in Bioinformatics), vol. 7480 (2013)
3. Amyot, D., et al.: Combining goal modelling with business process modelling: two decades of experience with the user requirements notation standard. Enterp. Model. Inf. Syst. Archit. **17** (2022)
4. Burattin, A., Gianola, A., López, H.A., Montali, M.: Exploring the conformance space. In: CEUR Workshop Proceedings, vol. 2952 (2021)
5. Burattin, A., Maggi, F.M., Sperduti, A.: Conformance checking based on multi-perspective declarative process models. Expert Syst. Appl. **65** (2016)
6. Carmona, J., van Dongen, B., Solti, A., Weidlich, M.: Conformance Checking: Relating processes and Models. Springer (2018)
7. Dalpiaz, F., Franch, X., Horkoff, J.: ISTAR 2.0 language guide (2016)
8. Debois, S., López, H.A., Slaats, T., Andaloussi, A.A., Hildebrandt, T.T.: Chain of events: modular process models for the law. In: LNCS (including subseries Lecture Notes in Artificial Intelligence and Lecture Notes in Bioinformatics), vol. 12546 (2020)
9. Dijkman, R.M., Dumas, M., Ouyang, C.: Semantics and analysis of business process models in BPMN. Inf. Softw. Technol. **50** (2008)
10. Franceschetti, M., Seiger, R., López, H.A., Burattin, A., García-Bañuelos, L., Weber, B.: A characterisation of ambiguity in BPM. In: LNCS, vol. 14320 (2023)
11. Franch, X., López, L., Cares, C., Colomer, D.: The i* Framework for Goal-Oriented Modeling, pp. 485–506. Springer, Cham (2016)

12. Ghasemi, M., Amyot, D.: From event logs to goals: a systematic literature review of goal-oriented process mining. Requirements Eng. **25** (2020)
13. Governatori, G.: The regorous approach to process compliance. In: Proceedings of the 2015 IEEE 19th International Enterprise Distributed Object Computing Conference Workshops and Demonstrations, EDOCW 2015 (2015)
14. Grisold, T., Mendling, J., Otto, M., vom Brocke, J.: Adoption, use and management of process mining in practice. Bus. Process Manag. J. **27** (2021)
15. Guizzardi, R., Reis, A.N.: A method to align goals and business processes. In: LNCS, vol. 9381 (2015)
16. Hildebrandt, T.T., Mukkamala, R.R.: Declarative event-based workflow as distributed dynamic condition response graphs. Electron. Proc. Theor. Comput. Sci. **69** (2011)
17. de Leoni, M., Munoz-Gama, J., Carmona, J., van der Aalst, W.M.: Decomposing alignment-based conformance checking of data-aware process models. In: LNCS, vol. 8841 (2014)
18. Liaskos, S., McIlraith, S.A., Sohrabi, S., Mylopoulos, J.: Representing and reasoning about preferences in requirements engineering. Requirements Eng. **16** (2011)
19. López, H.A., Massacci, F., Zannone, N.: Goal-equivalent secure business process re-engineering. In: International Conference on Service-Oriented Computing, pp. 212–223. Springer (2007)
20. López, H.A.: Challenges in legal process discovery. In: CEUR Workshop Proceedings, vol. 2952 (2021)
21. López, H.A., Debois, S., Slaats, T., Hildebrandt, T.T.: Business process compliance using reference models of law. In: LNCS (including subseries Lecture Notes in Artificial Intelligence and Lecture Notes in Bioinformatics), vol. 12076 (2020)
22. López, H.A., Hildebrandt, T.T.: Three decades of formal methods in business process compliance: a systematic literature review (2024)
23. Monciardini, D., Bernaz, N., Andhov, A.: The organizational dynamics of compliance with the UK modern slavery act in the food and tobacco sector. Bus. Soc. **60** (2021)
24. Morandini, M., Penserini, L., Perini, A.: Operational semantics of goal models in adaptive agents. In: Proceedings of the International Joint Conference on Autonomous Agents and Multiagent Systems, AAMAS, vol. 1 (2009)
25. Reisig, W., Rozenberg, G. (eds.): Lectures on Petri Nets I: Basic Models. LNCS, vol. 1491. Springer, Berlin (1998)
26. Riemsdijk, M.B.V., Dastani, M., Winikoff, M.: Goals in agent systems: a unifying framework. In: Proceedings of the International Joint Conference on Autonomous Agents and Multiagent Systems, AAMAS, vol. 2 (2008)
27. Teinemaa, I., Dumas, M., Rosa, M.L., Maggi, F.M.: Outcome-oriented predictive process monitoring: review and benchmark. ACM Trans. Knowl. Discov. Data **13**(2) (2019)
28. Varela-Vaca, A.J., Gómez-López, M.T., Morales Zamora, Y., Gasca, M.R.: Business process models and simulation to enable GDPR compliance. Int. J. Inf. Secur. (2025)
29. Yu, E., et al.: Modeling strategic relationships for process reengineering. Soc. Model. Requirements Eng. **11**(2011), 66–87 (2011)

The WHY in Business Processes: Unification of Causal Process Models

Yuval David[iD], Fabiana Fournier[iD], Lior Limonad[✉][iD], and Inna Skarbovsky[iD]

IBM Research, Haifa, Israel
{yuval.david,fabiana,liorli,inna}@il.ibm.com

Abstract. Causal reasoning is essential for business process interventions and improvement, requiring a clear understanding of causal relationships among activity execution times in an event log. Recent work introduced a method for discovering causal process models but lacked the ability to capture alternating causal conditions across multiple variants. This raises the challenges of handling missing values and expressing the alternating conditions among log splits when blending traces with varying activities.

We propose a novel method to unify multiple causal process variants into a consistent model that preserves the correctness of the original causal models, while explicitly representing their causal-flow alternations. The method is formally defined, proved, evaluated on three open and two proprietary datasets, and released as an open-source implementation.

Keywords: Business Processes · Causal Discovery · Unification · Intervention

1 Introduction

"A rooster's crow does not cause the sun to rise, even though it always precedes the sun." [21] But will a rooster's crowing pattern remain the same if placed at the North Pole? While research by [27] has shown that a rooster's crow is influenced by its biological circadian cycle and may continue even when placed in a dark room, causal diagrams allow us to predict the effects of interventions without actually conducting an experiment [21]. Our aim is to perform the same type of analysis in the context of business processes, eliminating the need to intervene directly in the process, which is highly costly. Our developed causal execution business process (CBP) model [10] links activity executions, enabling the assessment of how omitting or modifying an activity's execution time (e.g., delaying or expediting it) impacts other activities. This eliminates the need for empirical experimentation while informing resource allocation decisions.

Causal inference and causal discovery (CD) are the main pillars of causal analysis. While causal discovery focuses on analyzing and creating models that

---

Supplementary Information The online version contains supplementary material available at https://doi.org/10.1007/978-3-032-02929-4_3.

illustrate the relationships inherent in the data [20,22,29], causal inference is the process of drawing a conclusion about a causal connection based on the conditions of the occurrence of an effect [6,12,30]. More concretely, CD aims at constructing causal graphs from data by exploring hypotheses about the causal structure [26]. Additional assumptions, such as functional forms and distributions, are often required to identify the causal graph from the data. In typical CD settings, the causal graph is assumed to be a Directed Acyclic Graph (DAG), with all common causes of observed variables also being observed (i.e., present in the event log).

The work in [10] laid the foundation for discovering CBP models using process execution times. Given multiple CBP models, each capturing a causal perspective of activity execution, unifying them into a single cohesive representation remains a challenge, referred to as the 'CBP model unification' problem. Process variability often leads to differing or even conflicting causal conditions. For example, in a loan application process, one variant may follow a sequential flow: after preliminary screening (A), a credit check (B) determines the loan amount (C). In high-demand periods, the process may be expedited, where screening (A) directly determines (C). Symbolically representing the two variants, the first states 'A' causes 'B', and 'B' causes 'C', while the second states 'A' causes 'C' directly.

In [10], a simple union of graph elements (i.e., nodes and edges) was proposed to merge multiple CBP models into a more holistic view. While valid, this approach does not account for the alternating causal execution sequences across variants. A naïve union of CBP models would state that 'A' causes both 'B' and 'C', without distinguishing between cases where sometimes 'A' causes 'B' and sometimes 'A' causes 'C'. Similarly, it fails to accurately represent the causes of 'C'. Even in this simple example, it is clear that the current CBP model lacks the semantics to capture the alternating causal conditions that unfold across different variants.

Another approach could merge the two variants by bundling their traces, where each trace records a sequence of activity executions in the event log. However, this results in an inconsistent data structure, with missing timestamp values across different execution sequences. In the earlier example, traces from the expedited variant would lack a timestamp for activity 'B'. This poses a challenge for causal discovery, which struggles with *missing data* [5]. Traditional causal discovery algorithms typically assume complete data—an assumption often violated in practice [18]. Missing data can bias estimations, distort causal structures, and lead to flawed decision-making [16]. We note that this problem is unique to CBP model discovery, due to the inability to run core causal discovery algorithms on event logs containing null values, unlike conventional process mining techniques, which can still generate models such as the Directly Follows Graphs (DFG). This limitation necessitates unification at the model level, rather than through trace-level integration. Consequently, the contribution presented here also enhances the robustness of the CBP model discovery approach by addressing its vulnerability to missing data.

Our work addresses the CBP model unification problem, substantially extending [10] by introducing a framework for unifying multiple CBP models, each corresponding to a process variant, into a holistic graph representation. This framework preserves the causal knowledge in each CBP model, explicitly captures the logic of alternating execution conditions, and accounts for missing data.

We introduce a novel method that integrates multiple CBP models into a cohesive causal execution model while preserving the individual causal dependencies in the underlying variants, and articulating the conditions under which different execution paths occur. To achieve this, we augment the causal process model with a gating mechanism that encodes the alternating logic among variants. The method is formalized and assessed. We disclose its computational properties, prove its correctness and empirically evaluate its performance and scalability using five benchmark datasets, including three open and two proprietary ones.

2 Preliminaries

This section outlines key concepts essential for understanding the proposed framework. Let L be an input process event log, where L is defined as follows:

Definition 1. *A process event log L is defined as a finite set of traces, where each trace represents an ordered sequence of activities. Formally, $L = \{\tau_1, \tau_2, \ldots, \tau_n\}$, where τ_i is a trace, and n is the total number of traces in the log. Each trace τ is associated with a unique identifier, $case_{id}$, and is an ordered sequence of activities, $\tau = \langle a_1, a_2, \ldots, a_m \rangle$, where a_j is an activity, and m is the total number of activities in the trace. An activity a is a tuple $a = (case_{id}, t, n_{id}, V)$, where $case_{id}$ is a unique identifier linking the activity to its corresponding trace, t is the timestamp indicating when the activity occurred (could be start, end, or both), n_{id} is the unique name of the activity, and V is a (possibly empty) set of additional 'payload' attributes $\{v_1, v_2, \ldots, v_k\}$. Each attribute v_k is a key-value pair, defined as $v_k = (k, v)$, where k is the key representing the attribute name, and v is the value associated with the key.*

The timestamps t in a trace τ must be non-decreasing ($t_1 \leq t_2 \leq \cdots \leq t_m$), and activity names n_{id} must belong to a predefined set of activity labels N ($n_{id} \in N$). The attributes V may vary between activities and traces.

Assume the following event log L:

$$L = (\tau_1 = \{A^1, B^3, C^6\}, \tau_2 = \{A^2, F^5\}, \tau_3 = \{F^4, G^8, H^{12}\}, \tau_4 = \{A^{10}, F^{15}\},$$

$\tau_5 = \{A^{13}, C^{14}, B^{17}\})$, where superscript numbers correspond to timestamps.

For the given log L, it may be further partitioned into a set of variants, where a variant v is defined as follows:

Definition 2. *A variant v in an event log L is a subset of traces in L, where all traces in v share the same ordered sequence of activities. Formally, for an event log $L = \{\tau_1, \tau_2, \ldots, \tau_n\}$, a variant $v \subseteq L$ is defined as $v = \{\tau \in L \mid \tau = \langle a_1, a_2, \ldots, a_m \rangle \text{ and } \tau' = \langle a_1, a_2, \ldots, a_m \rangle, \forall \tau, \tau' \in v\}$, where τ and τ' are traces,* and $\langle a_1, a_2, \ldots, a_m \rangle$ denotes the ordered sequence of activities in the trace. The set of all unique variants in L is denoted by V_L, where $V_L = \{v_1, v_2, \ldots, v_k\}$, such that $\bigcup_{i=1}^{k} v_i = L$ and $v_i \cap v_j = \emptyset$ for $i \neq j$. That is, each variant v groups traces from L that represent identical sequences of activities.

For the above example log L, the variants in the log are the following:

$$V_L = (v_1 = \{A, B, C\}_1, v_2 = \{A, F\}_{2,4}, v_3 = \{F, G, H\}_3, v_4 = \{A, C, B\}_5),$$

where subscript numbers correspond to case IDs.

A less dense partitioning of the log can be derived by combining all variants sharing the same set of activities, independent of activity ordering. Thus, we define a partition p as follows:

Definition 3. *A partition p in an event log L is a subset of traces in L, where all traces in p share the same set of activities, independent of their ordering. Formally, for an event log $L = \{\tau_1, \tau_2, \ldots, \tau_n\}$, a partition $p \subseteq L$ is defined as $p = \{\tau \in L \mid set(\tau) = set(\tau'), \forall \tau, \tau' \in p\}$, where $set(\tau)$ represents the set of activities in the trace τ and ordering is disregarded. The set of all partitions in L is denoted by P_L, where $P_L = \{p_1, p_2, \ldots, p_m\}$, such that $\bigcup_{i=1}^{m} p_i = L$ and $p_i \cap p_j = \emptyset$ for $i \neq j$.*

For the above example log L and variants V_L, log partitions are the following:

$$P_L = (p_1 = \{1, 5\}, p_2 = \{2, 4\}, p_3 = \{3\}),$$ where numbers correspond to case IDs.

In our work in [10], an algorithm for causal business process model discovery was presented that identifies all inter-activity relations implying causal execution dependencies between the activities. The core algorithm presented adapts the LiNGAM causal discovery method to timestamped event logs, treating activity execution times as observed variables to infer a directed acyclic graph (DAG) of causal dependencies between activities. Given a process event log as an input, the developed algorithm could be applied to any process variant or a set of variants. The result can be represented as a causal execution (CX) graph G defined as follows:

Definition 4. *A causal execution graph G, resulting from applying a causal discovery algorithm to a process event log L, is a tuple $G = (V, E)$, where V is a finite set of nodes, and each node represents an activity name occurring in L. The set of edges $E \subseteq V \times V$ denotes causal execution relationships, where each edge $(n_i, n_j) \in E$ signifies that the execution of the activity corresponding to n_i causes the execution of the activity corresponding to n_j.*

We assume that the name of an activity is unique in the event log (or the activity can be uniquely identified by some other attribute). The graph G can be derived from a process variant (a single sequence of traces), a partition (a set of traces with the same set of activities, independent of order), or the entire event log L, and it provides a representation of the causal structure of the process with nodes as activities and directed edges as causal execution dependencies.

Corresponding to the above example partitions, Fig. 1 depicts possible causal execution graphs corresponding to each partition.

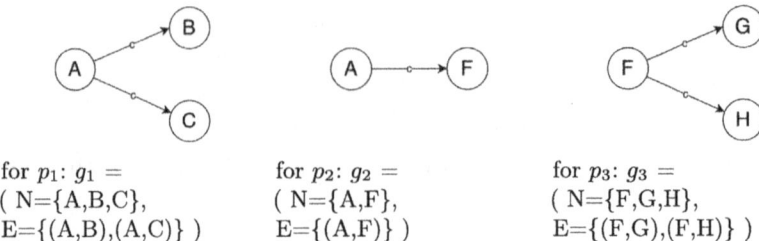

Fig. 1. Three causal execution graphs for the above example log L and partitions P_L.

3 Framework

This section presents the developed method, its underlying approach and assumptions, and concludes with a proof of its correctness.

3.1 Our Approach

While the work in [10] presented graph unification as a means to combine multiple causal execution graphs that correspond to process variants, it introduced a naïve method employing a simple logical union over the input CXs. This yields a causal graph that is semantically ambiguous concerning the alternating causal execution sequences among the variants. To generate a unified causal execution (U-CX) graph it is essential to address two challenges:

1. The need to accommodate causal process discovery employment to missing values in the data resulting from blending traces with different activities.
2. The need to express the alternating causal execution conditions among the different log splits.

To tackle the first challenge, we identify a plausible log split that relies on as many observations as possible. This is accomplished by adhering to the assumption that it is possible to combine traces of different variants as long as no known confounder may account for any differences among them. Consequently, we assume:

Assumption 1 Within each log split, there are no alternating causal execution conditions between any subset of variants that correspond to the same set of activities in the given event log.

Splitting the log into *partitions* allows the application of the CD algorithm developed in [10], yielding a corresponding CX graph for each. However, in cases where there are some known confounders to induce alternating causal conditions within a partition, we let the user override the default split into more fine-grain partitioning, yielding a split that adheres to the aforementioned assumption. For example, in a loan approval process, the *credit assessment* activity typically determines the *loan amount approved*. Given only these two activities and no additional context, it is reasonable to assume a causal relationship between them.

However, an exception may arise in scenarios where the typical adjacency is disrupted—for instance, in organizations where there is a habitual break between 2 p.m. and 4 p.m., the approval activity might consistently occur after the break, regardless of when the assessment finishes. In such cases, the temporal ordering no longer implies causality. To account for these situations, we allow the user to override the default assumption.

Addressing the first challenge enables applying the CD algorithm within each partition but entails adhering to its assumptions, as required by the LiNGAM adaptation [10]. These include treating activity execution times as continuous, non-Gaussian, linearly dependent variables with no unobserved confounders.

Given that some degree of noisy measurement is also captured in real-world datasets, some mechanism to mitigate the effect of noise must be embedded to correctly identify the conditions within each partition. For this, we also include a thresholding component in the method to cope with the presence of noise.

To address the second challenge, after identifying splits, we require a formal symbolic representation and an algorithmic method to capture the alternating causal execution conditions across partitions. For this, we extend the basic CX graph with a notation to express logical alternations in causal relations and develop the method detailed in Sect. 3.2.

Regarding the extension of the symbolic representation, we incorporated four new types of relationships that connect three or more activities (see example in Fig. 2). We introduce these new relationship types as follows. A formal definition is available in [7].

- **Causal "And" Gateway (AND_C)**: This gateway represents a situation where an activity execution always causes multiple other activities. Whenever the causing activity occurs, all connected activities must also occur.
- **Causal "Or" Gateway (OR_C)**: This gateway represents a situation where an activity execution can lead to one or more different activities, but not necessarily all of them. At least one of the connected activities must occur, but it is not required that all do.
- **Exhaustive Causal "Or" Gateway (OR_C^E)**: This is a special case of the "Or" gateway, where the causing activity can trigger any possible combination of the connected activities. That means it could cause just one, several, or even all of them to occur. We introduce this type of gateway for notational clarity, serving as syntactic sugar.
- **Causal "Xor" Gateway (XOR_C)**: This gateway represents a situation where an activity execution leads to exactly one of several possible activities, but never more than one. When the initial activity occurs, only one of the connected activities will take place.

Mirroring the above (causal-split) gateways by inverting their source and target edges, we also extend the notation with causal-join gateway types: $AND_{C>}$, $OR_{C>}$, $OR_{C>}^E$, and $XOR_{C>}$. Based on these definitions, we thus define:

Definition 5. *An Extended Causal Execution (CX) Graph is a causal execution graph $G = (V, E)$, where $V = V_A \cup V_G$ is the set of nodes and $E \subseteq V \times V$ is*

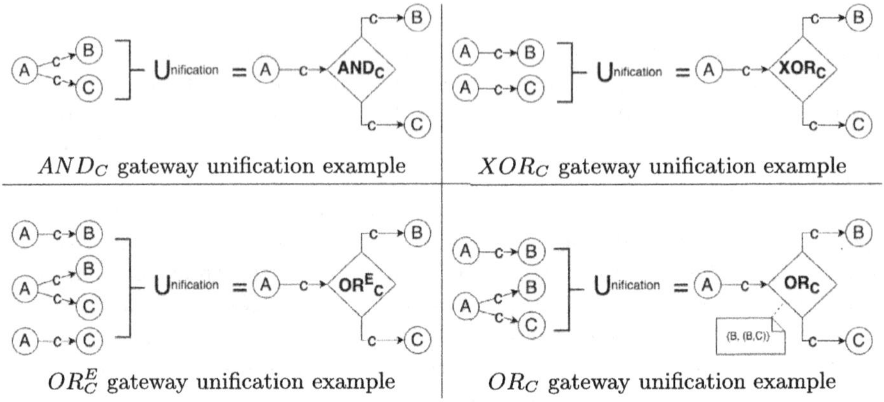

| AND_C gateway unification example | XOR_C gateway unification example |
| OR_C^E gateway unification example | OR_C gateway unification example |

Fig. 2. Four types of causal split gateways.

the set of edges. The set V_A represents activity nodes corresponding to activities in the process, and the set V_G represents gateway nodes, which include AND_C, OR_C, XOR_C, and OR_C^E, of both split and join gateway types.

3.2 Causal Graph Unification Method

The method for causal graph unification takes as input an event log L and a set of variants $V_s \subseteq V_L$, where V_L is the set of all variants in L. The output is a U-CX graph $G_U = (V_U, E_U)$, which results from the combination of the causal execution graphs corresponding to the individual partitions in the log, exclusively preserving the alternative causal relationships across the selected variants.

Thus, we define:

Definition 6. *Causal Graph Unification is a function f_C that takes as input an event log L and a set of variants $V_s \subseteq V_L$, where V_L is the set of all variants in L, and produces as output a unified extended (U-CX) form of a causal execution graph $G_U = (V_U, E_U)$. Formally, $f_C : (L, V_s) \rightarrow G_U$, where $G_U = (V_U, E_U)$ is the result of combining the set of causal execution graphs $\{G_v = (V_v, E_v) \mid v \in V_s\}$ corresponding to the individual partitions in L.*

The unified graph G_U integrates the causal relationships and employs the multilateral dependencies ($AND_C, OR_C, OR_C^E, XOR_C$) to construct a representation that maintains coherence concerning the input selected variants. Such a graph is deemed valid if it adheres to the following requirements:
Soundness. An output unified graph is sound if it has no traversal path in it that has no consistent path (i.e., denoting the same causal sequence) in any of the input causal execution graphs corresponding to the log partitions.
Completeness. An output unified graph is complete if every traversal path in the input causal execution graphs corresponding to the log partitions has a consistent path in the output graph.

The method consists of the following three high-level steps: (1) Log partitioning, (2) Unification, and (3) Simplification, elaborated next.

Log Partitioning. Careful attention should be given to the log partitioning strategy, ensuring alignment with the assumption that a genuine factor (e.g., an unknown confounder) may account for differences between variants. If no such factor exists, breaking down an identified partition into variants can lead to Berkson's paradox [21], potentially introducing spurious causal execution dependencies in the resulting CX graph.

By default, for any given set of variants in the input, we split the combined set of traces into partitions to ensure the largest possible set of observations is considered for all activities in the given variants. However, we also allow the user to explicitly select a partitioning modality that retains coherence with the original set of variants. In addition, if not explicitly specified in the input, the algorithm will run over the entire event log.

Because of the way we split the data, each partition will not include any NaN values, which allows us to apply the causal process discovery algorithm (CPD) presented in [10] to each partition, yielding a corresponding CX graph. We note that it may also be possible for the user to allow the algorithm to combine partition traces with other partitions that contain the same activities, but only when there is no prior knowledge of expected differences in the causal dependencies between the partitions. This can help mitigating the possibility of having causal inconsistencies between a partition and the overall event-log. Due to space limitation, a formal specification of the partitioning is described in [7].

A thresholding component for noise elimination. In practice, noise or data errors may cause B to occur before A, even if this is rare in a partition. To handle this, we use a thresholding mechanism that tolerates minor noise but blocks causal influences opposing the majority trend. This is achieved by blacklisting edges when violations exceed a set threshold.

We define a user-configurable threshold θ, as the maximum proportion of order violations allowed before they become significant. For each activity pair (A, B), we compute the proportion $p_{B \to A}$ of cases where B precedes A. If $p_{B \to A} \leq \theta$, we blacklist the edge from B to A, filtering minor noise while preserving meaningful temporal constraints. This mechanism ensures the causal discovery algorithm follows dominant temporal patterns without being misled by anomalies. We then apply the CPD algorithm, incorporating thresholding, as $G_i \leftarrow g_i = \text{causalDiscovery}(p_i)$, producing a CX graph for each partition.

Unification. In this step, we aim to unify the CX graphs generated for all partitions. The unification step consists of three main stages:

1. Processing the input causal execution graphs into a matrix representation.
2. Processing the matrix with a unification algorithm.
3. Forming the output U-CX graph from the eventual matrix.

The processing of the input CX graphs into a matrix representation considers an input G_i, where each graph $g_i = (N_i, E_i) \in G_i$ is defined by N_i, the set of activity nodes, and E_i, the set of directed edges (u, v) where $u, v \in N_i$. To represent these graphs as a matrix **for the case of split gateways**, we first compute the union of all activity nodes across the graphs: $N = \bigcup_{i=1}^{n} N_i$. A matrix M is initialized such that rows correspond to nodes in N, columns correspond to the graphs $g_1, g_2, \ldots, g_n$, and each cell $M[u][g_i]$ is initialized as an empty set. The matrix is then populated by iterating over the edges of each graph g_i: for each edge $(u, v) \in E_i$, v is added to $M[u][g_i]$. That is, all 'children' nodes of u are added to the column corresponding to each graph g_i. **For join gateways**, a second matrix having the same structure is created, having each row populated with the family of 'parents' nodes. That is, iterating over the edges of each graph g_i: for each edge $(u, v) \in E_i$, u is added to $M[v][g_i]$. Without loss of generality, in the remainder of the paper, we refer to the split case, considering the join case as its mirroring equivalent. The algorithm for matrix construction can be found in [7].

For example, given an input set G_i including the graphs in Fig. 1. The union of all activity nodes across the graphs is $N = \{a, b, c, f, g, h\}$, and the corresponding matrix representation M is shown in Table 1(a), where each cell contains the set of child activity nodes for the given node and graph.

Table 1. (a) Matrix representation input. (b) Matrix representation output.

Node	g_1	g_2	g_3
a	{b, c}	{f}	∅
f	∅	∅	{g, h}
b	∅	∅	∅
c	∅	∅	∅
g	∅	∅	∅
h	∅	∅	∅

(a)

Node	g_1	g_2	g_3
}a{	{(b, c)}	{f}	∅
b	∅	∅	∅
c	∅	∅	∅
d	∅	∅	∅
e	∅	∅	∅
f	∅	∅	{(g, h)}

(b)

For each row in the matrix, representing a *family of child sets* for a given node across all graphs, apply the *Unification Algorithm* [7] to classify the relationships among the child sets and mark causal gateways as follows:

1. AND_C **Identification:** For any child set that does not partially intersect with others, promote it to a single element. This is denoted by enclosing the set in round brackets, e.g., (a, b). Once identified, the promotion is propagated across all child sets in the row to ensure consistency.
2. XOR_C **Check:**
   - Invoked only if the row contains **two or more child sets**.
   - If all child sets in the row are mutually exclusive (i.e., no intersection between any two sets), the row is classified as XOR_C. This is denoted by having the node (row's name) surrounded by "}" and "{", e.g., }a{.
3. OR_C^E **(Exhaustive OR_C) Check:**
   - Invoked only if the row contains **two or more child sets**.
   - If the family of child sets is a **powerset** of the union of its child sets, the row is classified as OR_C^E. This is denoted by having the node (row's name) surrounded by "[" and "]", e.g., $[a]$.

   Otherwise, default to OR_C:

- If the row is **not classified** as OR_C^E, it is classified as OR_C. In such a case, the node (row's name) is surrounded by asterisks, e.g., *a*.
- To ensure the correctness of the result, along with any OR_C gateway that is concluded, we also store its actual set of alternatives (i.e., the Family set) in a designated map, to allow for the visual annotation of each OR_C gateway with an explicit list of its viable alternatives. This is required to ensure the *soundness* of the result.

Once the marking is determined for a row, no further steps are needed for that row. This process is repeated for all rows in the matrix. Revisiting the above example matrix, and applying the unification algorithm to it concludes with the matrix shown in Table 1(b).

As a last third step, the output U-CX graph is constructed from the annotated matrix (see Fig. 3). A formal specification of the construction is available in [7].

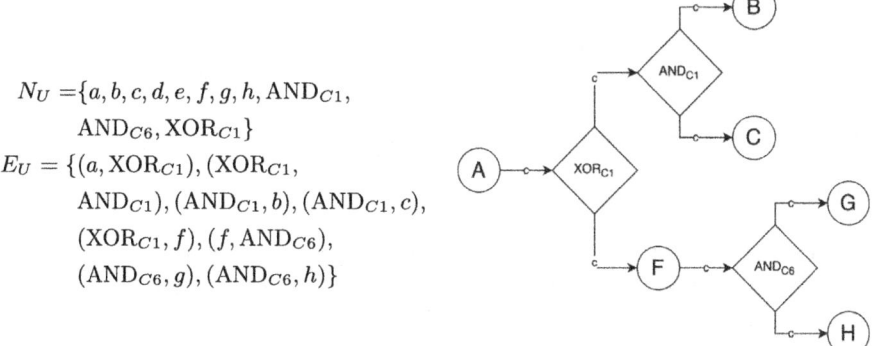

Fig. 3. U-CX graph G_U for the above example log L and graphs: g_1, g_2, g_3.

Simplification. This step is optional in the method. Various heuristics may be employed for further simplification of the non-exhaustive OR_C gateways in the result formed by the unification algorithm. For example, assume the unification of the following two graphs:

$$g_1 = (\{f, a, b\}, \{(f, a), (f, b)\}) \text{ and } g_2 = (\{f, a, c\}, \{(f, a), (f, c)\})$$

Applying our unification algorithm yields the following:

$$G_U = (N_U = \{f, a, b, c, OR_{C1}\}, \ E_U = \{(f, OR_{C1}), (OR_{C1}, a), (OR_{C1}, b), (OR_{C1}, c)\})$$
where the OR_{C1} alternatives are: $(a, b), (a, c)$

In the illustrated example, when an OR-type gateway is reached, its set of viable execution alternatives, resulting from the execution of f, is restricted exclusively to one of two possibilities, hence being non-exhaustive. Specifically, f either causes the execution of both a and b, as implied by g_1, or it causes the execution of a and c, as implied by g_2. Without delving into overly complex formalism, the consequences of executing f can be expressed as: $(a \wedge b) \oplus (a \wedge c)$. Such formula is formed as exclusive conjunction of the child nodes of the OR-type gateway.

A variety of logic simplification algorithms can be applied as a final step to simplify such expressions. For example, the same expression can be rewritten as $a \wedge (b \oplus c)$, preserving logical equivalence. This simplified expression can then be translated back into a graph representation—one that employs a combination of XOR and AND gateways to depict the same causal execution dependencies. However, note that the simplified representation does not necessarily result in a more compact visual graph compared to the original. Therefore, we leave the choice of simplification to the user's discretion.

Proof of the unification algorithms. Given a set of input causal execution graphs $G_i = [G_1, \ldots, G_n]$ and a unified, extended, causal execution output graph G_U, a proof of the algorithms' correctness was pursued. This includes the properties of *soundness* and *completeness*. Given a set of input CX graphs and an output U-CX graph, for soundness, we showed that the unification process does not introduce any causal dependencies that are not present in one of the input graphs. This excludes the case of a non-exhaustive OR_C that is sound, if and only if it is annotated with its invocation set. For correctness, we showed that none of the dependencies in the input graphs are lost during the unification process.

More formally, we prove the following:

Theorem 1. *(Soundness of the unified model) For each causal execution dependency that the unified model expresses, the same causal execution dependency is expressed by one of the underlying partition models.*

Theorem 2. *(Completeness of the unified model) For each causal execution dependency expressed by any of the partition models, the same causal execution dependency is expressed by the unified model.*

Due to space constraints, the proof of the algorithm's correctness is available in [7]. The proof confirms adherence to the requirements in Sect. 3.2.

3.3 Properties of the Unification Algorithms

The processing of the log to derive a U-CX graph involves multiple steps, with a total combined computational complexity of:

$$O(TA + VA \log A + V + TA^3 + PA^2 + A^2P^2 + A^2),$$

where T is the number of traces in the log, A is the number of activities per trace, V represents the number of variants, and P is the number of partitions. Each term corresponds to a step in the pipeline, mostly dominated by a cubic complexity in the maximal number of activities (i.e., $O(TA^3)$) in the employment of the CPD algorithm over the partitions. A detailed complexity breakdown is available in [7].

As concluded in the evaluation Sect. 4, pragmatically, the computation time is not a major concern as in most realistic processes the number of activities in partitions is typically within a range of a few dozen.

4 Evaluation

We implemented and open sourced the method at https://github.com/IBM/sax4bpm[1]. We tested the scalability and performance of our algorithm against a handful of real datasets, three open and two from industrial applications:

- Road Traffic Fines (RTF): An event log for the process of managing road traffic fines by a local police force in Italy[2].
- Sepsis: An event log obtained from a regional hospital in The Netherlands[3].
- BPIC12: BPI Challenge 2012 event log of a loan application process from a Dutch financial institution[4].
- Helpdesk: A proprietary event log for a process specifically designed for managing support tickets, tracking issues, and assigning resolutions, issued by an IBM Process Mining (IPM) client company in Italy[5].
- CROMA: A proprietary event log for a medical equipment sterilization process captured by Croma Gio.Batta company in Spain[6].

For each dataset, we measured partition-wise computation times and the total runtime for computing U-CXs across all partitions, as shown in Table 2. We also assessed performance consistency with the algorithm properties outlined in Sect. 3.3.

Table 2. Benchmark results with three open datasets and two obtained from industry.

Dataset	# Traces	# Events	# Variants	Avg event / trace	Min event / trace	Max event / trace	# Partitions	Avg time / partition [sec]	Max time / partition [sec]	Min time / partition [sec]	Total runtime [sec]
Road Traffic Fines	150370	561470	231	4	2	20	35	0.121	0.147	0.003	22.04
Sepsis	1050	15214	846	14	3	185	16	0.022	0.032	0.005	10.16
BPIC12	13087	164506	4336	12	3	96	103	0.022	0.080	0.004	118.89
HELPDESK	3664	31581	972	8	1	47	51	0.017	0.028	0.001	14.13
CROMA	595	7041	4	11	1	12	2	0.034	0.043	0.024	0.26

---

[1] https://doi.org/10.5281/zenodo.15539153.
[2] https://doi.org/10.4121/uuid:270fd440-1057-4fb9-89a9-b699b47990f5.
[3] https://doi.org/10.4121/uuid:915d2bfb-7e84-49ad-a286-dc35f063a460.
[4] https://doi.org/10.4121/uuid:3926db30-f712-4394-aebc-75976070e91f.
[5] https://www.ibm.com/products/process-mining.
[6] https://www.cromagiobatta.it/en/home/.

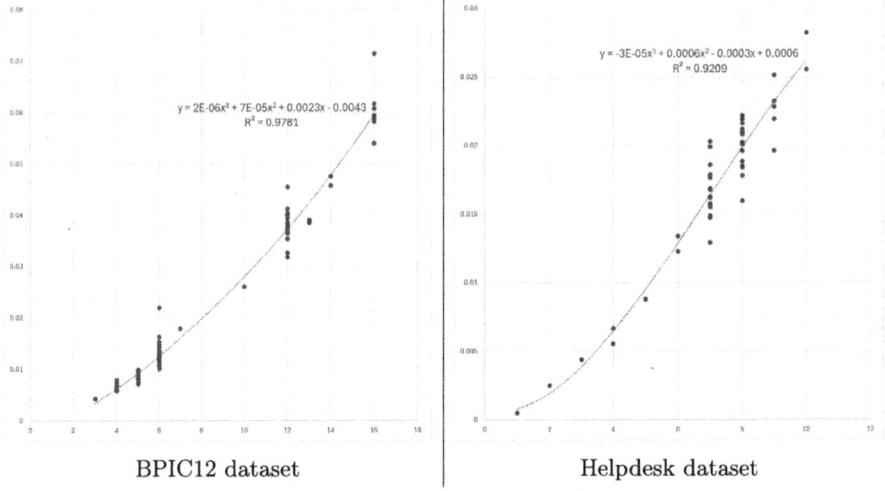

Fig. 4. Partition compute times.

Figure 4 presents scatter plots for the BPIC12 and Helpdesk datasets, depicting the relationship between the number of activities and computation time across partitions. A cubic polynomial curve fits well, with $R^2 = .9781$ for BPIC12 and $R^2 = .9209$ for Helpdesk. Linear trend lines show similar fit levels ($R^2 = .9708$ and $R^2 = .9123$, respectively). Thus, scalability remains manageable in most practical cases, even with high trace counts, as seen in the Road Traffic Fines dataset. Overall, results align with expected performance properties.

Figure 5 shows one example of two process variants from the BPIC12 event log, demonstrating the formation of a non-exhaustive OR_C gateway and two AND_C gateways. Note that the explicit invocation set associated with the newly formed gateway "or_0" is disclosed in the caption for the unified result.

Figure 6 previews a new process mining feature under development in the IBM process mining product, leveraging the developed model for prescriptive process analytics. This dashboard analyzes bottlenecks related to key performance indicators (KPIs) like process lead time, presenting historical KPI values alongside insights into activities contributing to delays while filtering out irrelevant ones. The U-CX graph enhances XAI by identifying critical activity execution times affecting the target KPI [11], ranking activities along the causal chain, and excluding those with no impact.

With regards to the simplification step in our method, we attempted integration with an available solver[7], an implementation of the QuineâĂŞMcCluskey algorithm (QMC) [23]. As noted above, this realization helped in further compacting the logical expression. However, in certain cases, it also yielded negation and "1"s as part of the concluded formulae, which is not expressible with our current graph notation.

---

[7] https://github.com/schuyler/boolgen?tab=readme-ov-file.

5 Related Work

Our work lies at the intersection of process discovery (PD) and causal discovery (CD). Addressing the multi-perspective paradigm in AI-Augmented BPMs [9], the model in [10] is the first to systematically reveal the primary causal process structure from execution timestamps. Other causal perspectives, such as quality and cost, naturally derive from it [17]. This work further enhances the original model by incorporating "causal-flow" logic, ensuring it fully captures the altering causal conditions across process variants.

Most PD approaches apply some threshold on time precedence, counting 'directly follows' or 'eventually follows' relationships. "Causal" relations are sometimes used to express the frequency of these dependencies, as in [14], which reduces representational bias in the hybrid miner algorithm by leveraging causal graph metrics for long-term dependencies. Instead, we take a bidirectional, asymmetric approach to analyzing timestamp relationships between activities.

Causal discovery aims to uncover causal relationships from observational data, distinguishing cause-effect directionality from mere correlation [20,29]. Methods divide into assessing intervention impact and identifying qualitative causal links. The Conditional Average Treatment Effect (CATE) technique gives a measurement to assess the magnitude of an intervention. It is used in [4] for decision-making and in [28] for prescriptive process monitoring. While these studies assume a given causal model, we focus on discovering causal models.

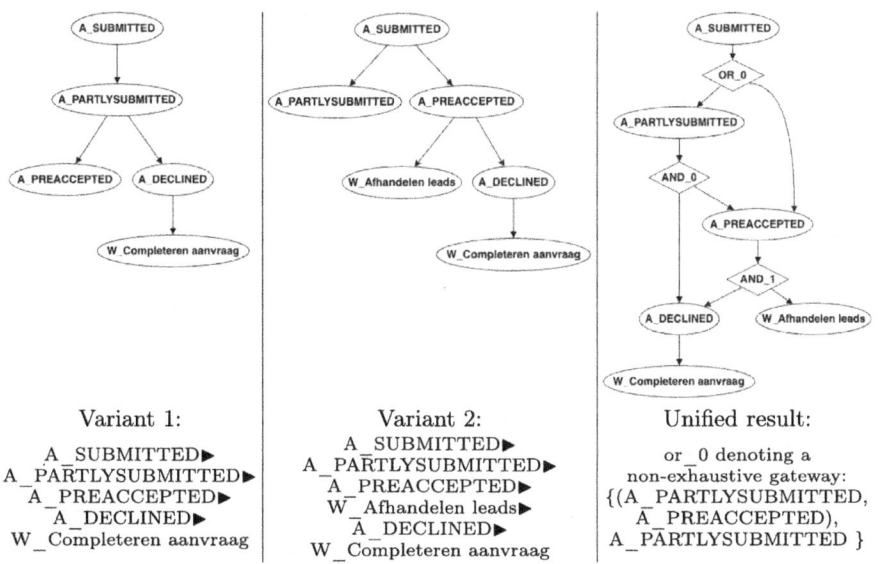

Fig. 5. BPIC12: unification of two example variants.

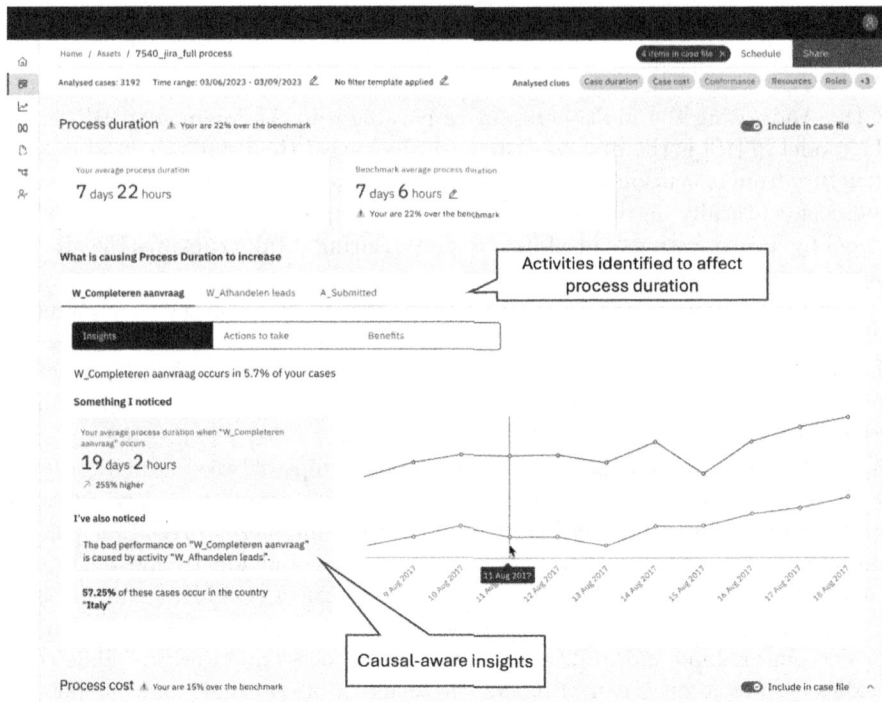

Fig. 6. Screenshot of a prescriptive process analytics dashboard.

In [13], a graph of causal factors is generated using Granger causality, a statistical test for time-series analysis. However, relations are based on timestamped KPI values, whereas we analyze activity timestamps.

Prior work has explored causal relationships among decision points [3,15,19]. Our method, however, uncovers how decisions, represented by causal gateways, influence subsequent activity executions. Causal discovery algorithms face challenges like handling latent variables, feedback loops, and missing data. Methods such as multiple imputation [24], expectation-maximization (EM) [8], and inverse probability weighting [25] address missing data but have limitations in process mining. Multiple imputations risk bias if misspecified, and EM algorithms are computationally demanding for large datasets [18]. These approaches also rely on strong assumptions about data distribution and missingness. Instead, by splitting data into subsets, we avoid the need for imputation or complex modeling. The gating mechanism integrates individual causal graphs, ensuring the unified model captures the causal relationships across subsets.

Causal discovery also employs some modeling language for result articulation. C-Nets [1,2], a rich process notation compared to others (e.g., Petri-nets, BPMN, and EPC), could serve as a viable alternative for U-CX graphs. By using input/output bindings, C-Nets capture diverse execution sequences. However, as a notation, they lack a mechanism to infer causal structures, requiring expert input or a unification method like the one proposed in our work.

6 Conclusions and Future Work

We introduce a method for integrating multiple causal graphs into a unified model, preserving individual causal dependencies while explicitly capturing alternating execution conditions via a gating mechanism. Unlike conventional *always causes* approaches, our method adopts a *sometimes causes* perspective, adapting arrow notation accordingly and extending it with graphical symbols for temporal alternations. We formalized and implemented the method, analyzed its complexity, proved its correctness, and evaluated its scalability using five benchmark datasets—three open and two proprietary.

The method offers key advantages. It scales well for large process mining datasets by operating on subsets. While extending a prior causal process discovery technique, it remains agnostic to the specific causal discovery algorithm used. By handling missing data through splitting and unification, our method mitigates the risk of introducing bias due to incorrect assumptions about missing values.

Our diamond-based gateway notation enhances visualization in a BPMN-like manner, distinguishing logical alternations (AND, OR, XOR). However, it does not explicitly convey edge coefficient values derived from the causal discovery algorithm. Users seeking these values must refer to the input graphs or rely on annotations in the extended output. While numeric figures can be added next to alternatives (e.g., the OR_C gateway in Fig. 2), such annotations can become cumbersome as the number of alternatives per gateway increases.

Our model could serve as a semantic foundation for C-Nets, where each node has inbound and outbound sets represented by dots connected by arcs. Extending this notation with numeric indicators for coefficient values could enhance clarity but may also lead to dense visualizations with multiple arcs, affecting usability. Furthermore, incorporating additional literals from the simplification step (e.g., negation, "1"s) remains an open question. Future work could explore leveraging large language models for simplification using few-shot learning and user-rated outputs to balance compactness and completeness.

Acknowledgments. This project has received funding from the European Union's Horizon research and innovation programme under grant agreements no 101094905 (AI4GOV), 101092021 (AutoTwin), and 101092639 (FAME). We thank Croma Gio.Batta and IBM Process Mining for sharing their datasets.

References

1. van der Aalst, W.: Process Mining. Springer, Berlin, Heidelberg (2016)
2. van der Aalst, W., Adriansyah, A., van Dongen, B.: Causal nets: a modeling language tailored towards process discovery. In: CONCUR 2011 - Concurrency Theory, pp. 28–42. Springer (2011)
3. Alaee, A.J., Weidlich, M., Senderovich, A.: Data-driven decision support for business processes: causal reasoning and discovery (2024)
4. Bozorgi, Z.D., Teinemaa, I., Dumas, M., La Rosa, M., Polyvyanyy, A.: Process mining meets causal machine learning: discovering causal rules from event logs. In: 2020 2nd International Conference on Process Mining (ICPM), pp. 129–136. IEEE (2020)
5. Conforti, R., La Rosa, M., Ter Hofstede, A.H.: Filtering out infrequent behavior from business process event logs. IEEE Trans. Knowl. Data Eng. **29**(2) (2017)
6. Cunningham, S.: Causal Inference: The Mixtape. Yale University Press (2021)
7. David, Y., Fournier, F., Limonad, L., Skarbovsky, I.: Supplementary material for the WHY in business processes: unification of causal process models. https://github.com/IBM/SAX/tree/main/BPM2025F
8. Dempster, A.P., Laird, N.M., Rubin, D.B.: Maximum likelihood from incomplete data via the EM algorithm. J. Roy. Stat. Soc. Ser. B Stat. Methodol. **39**(1) (1977)
9. Dumas, M., Fournier, F., Limonad, L., Marrella, A., et al.: AI-augmented business process management systems: a research manifesto. ACM Trans. Manag. Inf. Syst. **14**(1) (2023)
10. Fournier, F., Limonad, L., Skarbovsky, I., David, Y.: The WHY in business processes: discovery of causal execution dependencies. Künstliche Intelligenz (2025)
11. Galanti, R., et al.: An explainable decision support system for predictive process analytics. Eng. Appl. Artif. Intell. **120**, 105904 (2023)
12. Hernán, M.A., Robins, J.M.: Causal Inference: What If. CRC Press (2020)
13. Hompes, B.F.A., et al.: Discovering causal factors explaining business process performance variation. In: Advanced Information Systems Engineering, pp. 177–192. Springer (2017)
14. Kourani, H., Di Francescomarino, C., Ghidini, C., van der Aalst, W., van Zelst, S.: Mining for long-term dependencies in causal graphs. In: Business Process Management Workshops, pp. 117–131. Springer, Cham (2023)
15. Leemans, S.J.J., Tax, N.: Causal reasoning over control-flow decisions in process models. In: CAiSE 2022, Leuven, Belgium, June 6-10, 2022, Proceedings. LNCS, vol. 13295, pp. 183–200. Springer (2022)
16. Little, R.J., Rubin, D.B.: Statistical Analysis with Missing Data. Wiley (2014)
17. Dobson, M.S.: The Triple Constraints in Project Management. Berrett-Koehler Publishers (2004)
18. Mohan, K., Pearl, J., Tian, J.: Missing data as a causal inference problem. In: Proceedings of the 31st Conference on Uncertainty in Artificial Intelligence (UAI) (2013)
19. Narendra, T., Agarwal, P., Gupta, M., Dechu, S.: Counterfactual reasoning for process optimization using structural causal models. In: Lecture Notes in Business Information Processing, vol. 360 (2019)
20. Pearl, J.: Causality: Models, Reasoning, and Inference, Second Edition, pp. 1–464 (2011)
21. Pearl, J., Mackenzie, D.: The Book of Why: The New Science of Cause and Effect, 1st edn. Basic Books (2018)

22. Peters, J., Janzing, D., Schlkopf, B.: Elements of Causal Inference: Foundations and Learning Algorithms. The MIT Press (2017)
23. Quine, W.V.: A way to simplify truth functions. Am. Math. Month. **62**(9), 627–631 (1955)
24. Rubin, D.B.: Multiple Imputation for Nonresponse in Surveys. Wiley (1987)
25. Seaman, S.R., White, I.R.: Review of inverse probability weighting for dealing with missing data (2013)
26. Shimizu, S.: Statistical Causal Discovery: LiNGAM Approach. SpringerBriefs in Statistics, Springer, Japan, Tokyo (2022). https://doi.org/10.1007/978-4-431-55784-5
27. Shimmura, T., Yoshimura, T.: Circadian clock determines the timing of rooster crowing (2013)
28. Shoush, M., Dumas, M.: When to intervene? Prescriptive process monitoring under uncertainty and resource constraints. In: Business Process Management Forum, pp. 207–223. Springer, Cham (2022)
29. Spirtes, P., Glymour, C., Scheines, R.: Causation, Prediction, and Search. The MIT Press (2001)
30. Yao, L., Chu, Z., Li, S., Li, Y., Gao, J., Zhang, A.: A survey on causal inference. ACM Trans. Knowl. Discov. Data **15**(5), 1–46 (2021)

A Self-orchestration Model for Business Collaborations with Verifiable Process History Credentials

Martin Farkas[✉][ID], Bertalan Zoltán Péter[ID], and Imre Kocsis[ID]

Faculty of Electrical Engineering and Informatics, Department of Artificial Intelligence and Systems Engineering, Critical Systems Research Group, Budapest University of Technology and Economics, Budapest, Hungary
{martin.farkas,bpeter}@edu.bme.hu, kocsis.imre@vik.bme.hu

Abstract. In our increasingly digitalized business ecosystems, ensuring the integrity, traceability, and nonrepudiability of digital interactions in multiparty collaborations is becoming ever more critical, and Verifiable Credentials (VCs) are becoming an important element in the host of emerging cryptographic and socio-technical solutions to this challenge. The existing and envisioned applications of VCs dominantly employ them to support comparatively simple, challenge-response style claim verification scenarios. In contrast, in this paper, we propose a new usage modality: employing VCs as a vehicle for securing complex multiparty interactions. Specifically, we propose a peer-to-peer protocol model for the process "self-orchestration" of BPMN collaborations, as a novel alternative to centralized workflow engines as well as smart contract based decentralized orchestration approaches. We utilize revocable, blockchain-secured VCs to produce cryptographically verifiable proofs for the authorization of cross-organizational collaboration interactions, enabling their execution without a central authority, while maintaining trust in the shared state and the model-compliant and model-carrying, VC-based process traces, which we call "Verifiable process history Credentials" – or VphCs. Our approach prevents malicious participants from "double spending" a state by distributing divergent, contradictory state updates to different parties. Similar to stateless blockchain approaches, and in contrast with smart contract-based decentralized orchestration, our protocol model maintains a constant-sized state on-chain, with minimal data exposure to collaboration-external parties.

Keywords: Verifiable Credentials · Cross-Organizational Process Management · Blockchain

1 Introduction

Conducting business interactions between different parties, especially when performed over electronic platforms, tends to face significant challenges stemming from a lack of trust and transparency. Historically, trust and transparency issues of cross-organizational interactions have been addressed either by non-technical (e.g., *legal*) means, by the introduction of trusted third parties to mediate and govern interactions, or by (usually rather rudimentary, basic trust service-focused) trust federation solutions.

For almost a decade now, distributed ledger technology, typically implemented as a "blockchain," has been increasingly emerging as an alternative solution: instead of trusting third parties, it can be a far superior approach to trust the minority-attack tolerant integrity of a ledger-like, (usually, at least to some extent) programmable database – a distributed ledger. Distributed ledgers are maintained by sets of parties who are ideally incentivized so that - for a given use case - it is unlikely enough that a sufficient majority of them would collude to attack the integrity of the ledger; thus, trust is "decentralized," and from the point of view of ledger-using applications, it is "disintermediated."

Blockchain technology can facilitate the trustworthy *distributed execution* of cross-organizational interactions so that a decentralized platform takes over the role of the third parties we are accustomed to in hub-and-spoke style integrations. However, blockchain and distributed ledger-based collaboration orchestration have significant challenges, especially regarding privacy and confidentiality.

In this paper, we propose a novel, alternative model of distributed collaboration executions – relying on an overlapping contemporary ecosystem of trust-enabling technologies: Verifiable Credentials. To the best of our knowledge, a *systematic* use of Verifiable Credentials (VCs) for securing the decentralized execution of collaborations at the application level is an entirely new proposition. Accordingly, we first argue for their applicability (Sect. 2), define our notion of collaboration self-orchestration, and propose two core concepts for constructing self-orchestration protocols for BPMN-specified collaborations: Verifiable Process History Credentials and Non-Invoked Transition Credentials (Sect. 3). In Sect. 4, we describe a specific self-orchestration protocol and in Sect. 5, technologies that can be readily applied to implement it.

2 Business Interaction Trust with Verifiable Credentials

Business process management, as a discipline, began very early to incorporate the distributed trust models enabled by distributed ledgers, with the aim of increasing trust and transparency in cross-organizational interactions. The *leitmotif* of the corresponding body of research and industrial solutions is that blockchain-deployed smart contracts can enforce the agreed-on rules of interaction as well as act as an immutable, irrepudiable, and auditable log of the activities performed by the involved parties. This is a model of "algorithmic orchestration" in a similar way as Bitcoin implements "algorithmic money issuance:" a function performed by the ledger-maintaining peer-to-peer network, predictably, and independently from the parties relying on the trustworthiness of the ledger. [7]

The key, and still only partially solved challenges flow from the basic properties of this model of interaction facilitation, too: hiding the on-chain state from interaction-external parties (or even some interaction-internal ones), handling changes in the agreed-on rules of interactions, and accommodating the performance, data storage, and transaction cost constraints of distributed ledgers. These concerns *are* much less critical for blockchains with *permissioned* consensus, where, instead of the anonymous general public, a *consortium* of known organizations operates the ledger-maintaining nodes of the blockchain network; however, in addition to the challenges of consortium forming, users have to be able to place sufficient trust in the joint majority honesty of the consortium.

2.1 The Emergence of Verifiable Credentials

In lockstep with the proliferation of blockchain technology, "self-sovereign" approaches have been emerging for handling cryptographic and real identities of all kinds of "things" (including people and organizations), and making all kinds of "claims" about them cryptographically verifiable along complex trust structures. By now, the core standards of this *Self-Sovereign Identity* (SSI) [5] world are mature, such as the Decentralized Identifiers (DIDs) and the Verifiable Credentials (VCs) Data Model recommendations of the W3C [1,6]. DIDs and VCs were meant to rely on distributed ledgers – as Verifiable Data Registries, VDRs – to support the properties envisioned in SSI as a philosophy.

This is still more than a viable option, technologically as well as conceptually, with increasingly mature ledger, intermediary service, and "wallet" support on the end-user-facing side (for creating/issuing/receiving, storing, and presenting identities and credentials). However, it turned out that there is an immense and very broad need for Verifiable Credentials in all kinds of settings – for people and businesses to be able to present structured, digitally signed documents carrying some claim made by an *issuer* to *verifiers* in an *open* model, and in a way where the verifier does not need to turn to the issuer directly to verify the *validity* and *revocation status* of the claim.

The increasing importance of VCs is underpinned by the European Digital Identity Wallet [2] initiative, which draws heavily from the SSI paradigm; these wallets will handle for citizens a wide range of Electronic Attribute Attestations (EAAs) – VCs, conceptually as well as technically. These wallets are expected to provide at least partial support for decentralized VDRs; importantly, EBSI, the pan-European public-sector blockchain.

From the systems engineer's point of view, the key message of this evolution is that the world of systematic electronic claim handling (and revocation) is here – and the underlying trust models are very well adaptable to the needs of the various use cases.

2.2 Orchestrating Business Interactions with Verifiable Credentials

The way we used to engage in paper-based bureaucratic and business processes before the age of digitization serves as the motivation for our proposal. Let us look at an imaginary (but alas, not too unrealistic) citizen housing permission and allocation process in a major city in the Eastern Bloc, in the 80's.

The citizen makes an initial application; which the local housing authority processes and stamps. He takes it to his workplace committee; they, ideally, support the application and release a form certifying the citizen's employment and political standing, referencing the original application. The citizen takes both forms to his district administration; they verify the current dwelling and family status of the citizen and, ideally, release yet another stamped form in support of the application. The citizen carries around an ever-growing dossier of documents which (partially) reference each other – and are verifiable by trusting stamp-and-signature, or by making additional official inquiries.

Given a trustworthy history of the part of the process they need for verification and decision-making purposes, every party in the process knows how to act themselves – and that verifiable history is captured by the documents in the citizen's dossier. As a part of the long-running process (taking possibly many years), the departments of the state may engage in various interactions among themselves too, captured, similarly, through verifiable paper trails of their own. Still, if "everybody does their job" based on the verifiable process history presented to them, the process, despite being performed through the complex interactions of multiple parties, in a sense, "orchestrates itself" without a directly involved central authority. This leads to the core idea of the ***Verifiable Credential-based self-orchestrating collaboration model*** we propose in this paper: solving the general decentralized execution problem through cross-wallet Verifiable Process History Credential (VphC) presentations, instead of orchestrating (and tracking) interactions through a blockchain-deployed smart contract.

The key challenge that Bitcoin solved in a decentralized way was avoiding double spending – there shall be no alternative realities presented to various parties about the current and past ownership of a given sum of "electronic cash." A similar challenge arises for our approach, too: no party shall be able to convince other parties about alternative continuations of the process (e.g., a citizen applying at two different state housing projects with the same permit) so that process bifurcations can happen. We present a protocol that solves this problem through VC revocation registries. Revocation registries may be blockchain-based; however, our approach is qualitatively different from the smart contract-based ones in that a) the revocation registries are oblivious to the processes they support – this is a standard VC ecosystem function; b) minimal state information is leaked; c) the centralized/decentralized/federated nature of VC verification and revocation can be fully tailored to the needs of the application, without any impact on our execution model.

2.3 Running Example

To better illustrate our "self-orchestration" methodology, we introduce the imaginary citizen housing and allocation process in Sect. 2.2 as a running example for this paper. Figure 1 shows a BPMN collaboration diagram of the first few steps of the lengthy housing procedure. Three message flows have been highlighted in red that will later be used for the demonstration of our approach.

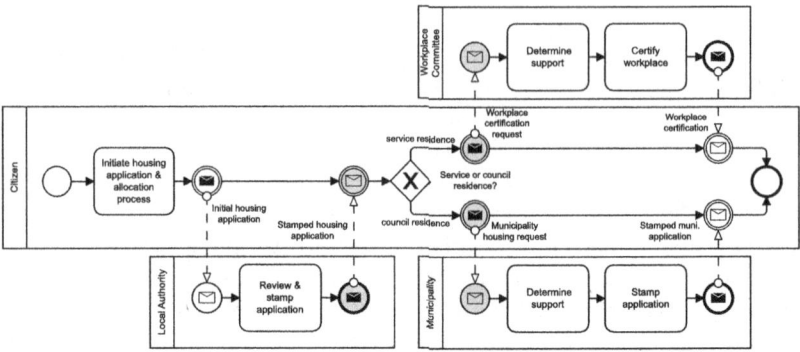

Fig. 1. Example collaboration diagram: Citizen can apply for housing either to the workplace committe or the municipality

The citizen first turns to a local authority with their initial application who review it and provide a stamp. Then, the citizen can take their stamped application document to either their workplace committee or the municipality office, depending on whether they wish to apply for a "service" or "council" housing. In either case, the relevant party reviews their request, determines whether they support it, and if so, give their own stamp, and the process continues with the next step. For simplicity, we have not modeled the rest of the process as these few interactions are already representative enough for demonstrational purposes.

2.4 Challenges of Truly Peer-to-Peer Orchestration

Blockchain-backed claim verification transforms the role of blockchains from a database to a source of trust in the classic cryptographic sense. Our key insight is that we can utilize this new capability in BPM to address the deficiencies of smart contract based orchestration by removing the smart contract as a "central party" from the point of view of interactions (not trust). The goal is to solve the trustworthy decentralized orchestration problem in a truly peer-to-peer manner, so that the process "orchestrates itself" through participants exchanging history and authorization claims as direct messages – and a VDR acting as a general-purpose source of trust.

The key challenges that an approach building on this basic idea has to solve, and which we address in this paper, are the following.

Security and Trust Without Central Enforcement. In a peer-to-peer model, each party must independently assess whether the process history presented to them is authentic and model-compliant. This requires strong cryptographic guarantees and a mechanism for verifying that the claims propagated are parts of the same shared reality – without centralized arbitration or a globally agreed ledger of state.

Preventing History Rewriting and Process Forking. A malicious participant may attempt to rewrite the process history by issuing conflicting claims to different recipients. This is analogous to a *double spend* in the context of cryptocurrencies. For collaborations, the danger is a divergence in execution, where multiple participants are convinced that mutually exclusive events have occurred. In our example in Fig. 1, this would mean applying for both a service residence and a council residence, which is forbidden by the process.

Balancing Liveness and Integrity. A key challenge arises when a participant becomes unresponsive or withholds credentials. This behavior may or may not compromise the integrity of the process state, and it may indefinitely stall progress. The core issue is enabling the advancement of the collaboration despite delays or non-cooperation, without relying on centralized coordination.

3 VC-Based Self-orchestration of Collaborations

Our discussion focuses on the trustworthy execution of BPMN *collaboration models*: models that capture both the internal processes of cooperating participants and the message exchanges between them. Issues of trust and decentralization rarely arise for single-organization *processes* (orchestrations) – the automation needs of which are well-served by centralized workflow engines, which, as a central orchestrator, all participants trust. *Choreography models* provide a much higher level of abstraction as collaboration diagrams and mostly serve as conversational "templates;" in a structured development process, collaboration diagrams are a good tool to capture the "implementations" of choreographies [11]. We believe that the important challenges we address with our revocation regimen are much easier to discuss (and are, arguably, more important) in the collaboration context; future work will extend the methodology to choreographies, to the extent it is meaningful to support them directly.

Collaborations are typically executed in a federated model, or through hub-and-spoke integration, with smart contract-based distributed execution as a relatively novel, third option. In a federated model, participants track their own state and interact through various means of message exchanges (from API calls through message queueing to email notifications); in hub-and-spoke integration, a central integration platform connects the participants. Smart contract-based approaches essentially transpose the orchestration concept of workflow engines to a decentralized platform – without a single trusted third party.

3.1 Collaboration Self-orchestration

In contrast to these approaches, we call the execution model we propose in this paper ***collaboration self-orchestration*** and define it conceptually as follows: *an execution model where each party in a BPMN collaboration, by verifying statements about the shared state and its history during message exchanges, can safely proceed with their tasks–knowing that bad actors can at most delay progress.*

The term *orchestration* underlines the similarity in *control* to classic workflow engines; and the term *self* expresses that enforced adherence to a collaboration model is an emergent property of the execution itself – depending on our philosophical stance, there is either no orchestrator at all, or, in a way, all parties act as orchestrators.

The definition is intended to be broad; it allows room for all kinds of claim-verification approaches, decentralized or otherwise, and protocols with very different properties. The protocol we propose in this paper is, to the best of our knowledge, the first of this kind – but we also acknowledge that it is only a first attempt, which is to be surpassed by further research.

3.2 BPMN Collaboration Subset

BPMN models collaborations on *collaboration diagrams*, where *participants* are represented as pools interacting via messages, and internal activity sequences. The execution semantics are defined by *message exchanges* between pools, with each pool using token-based semantics (*sequence flows, events, gateways*) for local control. *Message flows* link tasks or events across pools, allowing participants to progress independently while synchronizing through these flows.

For the purposes of this paper, without a significant loss of generality, we restrict collaborations to a heavily simplified, but still representative subset of standard BPMN. The example in Fig. 1 adheres to this subset, which is:

a) We assume only the existence of pools and not lanes; pools contain only normal start and termination events, exclusive and parallel gateways, message throw and catch events and user tasks.
b) Exclusive gateway choices are performed by the respective participants (no guard conditions).
c) We assume a directed acyclic graph model (we assume loops can be sufficiently handled by such standard model transformations as finite unrollings).
d) We consider only the simplified activity state set of Ready-Active-Completed.
e) We assume that there are no process variables.

3.3 How to Convince the Next Participant to Act

To act according to the self-orchestration model, when a message is sent, its receiver should be able to decide whether the collaboration is in the state claimed by its sender; that is, the sender has reached the claimed message send event, they have sent the previous messages, and they are *not* taking an alternative path simultaneously.

More specifically, in the context of sending the *Workplace certification request* in Fig. 1, the *Citizen* should be able to convince the *Workplace Committee* about the following facts.

- A collaboration, specifically as an instance of a BPMN model is being executed, in which the *Workplace Committee* agreed to participate or is known to be willing to join.
- That process instance is clearly distinguishable from all other instances that may run in parallel, or have terminated earlier.

– *Dear Committee, you can be sure that you are the next to act:* The *Citizen* is who they claim to be in the context of the BPMN model, and their pool is truly in a state where the *Workplace Committee* can continue execution through the message catch event, when he reaches that point in the process.

There are multiple ways to make the necessary argument; we follow an approach that builds on an explicit, recursive process history trace and commitments to not having taken alternative routes. Thus, *Committee* makes their decision based on the following set of claims presented by the *Citizen*.

a) A sequence of user activity and message throw-catch executions, as claimed by the corresponding parties in the model, which, under standard BPMN semantics, can lead to the *Citizen*'s claimed (sub)state where they have reached the message throw.
b) For each exclusive gateway where one branch is taken by the executions, a claim that the other branch was not taken, made by the excluded party.

If these claims all hold, then the *Workplace Committee* should be confident that such an evolution of the logical state of the collaboration has indeed happened – and consequently, they can proceed. The question, then, is how to represent and handle these claims so that they are *verifiable* by the *Committee*.

3.4 Verifiable Credentials as Process History Claims

Verifiable Credentials (VCs) [6] are an emerging form of cryptographically verifiable digital claims. Conceptually, they are part of the self-sovereign identity (SSI) [5] paradigm, which advocates giving people more control over their identity on the internet through advancements in cryptography and implementing decentralized trust models with distributed ledger technologies.

The key aspect of SSI for our purposes is that it establishes a new model for issuing and verifying *credentials* (digitally signed *claims*); a "trust triangle" [10], which all relevant technical approaches are able to support. In the trust triangle model, **Issuers** are the authorities that sign and issue Verifiable Credentials (VCs), which **Holders** store in their (identity) *wallets* and *present* when needed. **Verifiers** check the authenticity and revocation status of presented VCs without directly contacting the issuer; instead, they turn to a **Verifiable Data Registry** (VDR), where the (trusted) issuer can publish its cryptographic identity, manage VC schemas and revoke VCs. Although VCs can be presented directly, so-called *Verifiable Presentations* provide a way to wrap one or more VCs in a presentation container, add presentation metadata, disclose data selectively and combine VCs in a structured way. The VDR can be decentralized, but does not have to be; if it supports revocation, then it may do so in a highly privacy-preserving manner.

We propose to use VCs for representing claims about process history and commitments to branching choices. To that end, we define two types of credentials: Verifiable Process History Credentials (VphCs) and Non-Invoked Transition Credentials (NiTCs).

A **Verifiable Process History Credential** is a VC issued by a message sender to a receiver during a message passing (message throw event). It captures what *has* happened in the process up to that point – i.e., the current process history – as a chain of state transitions. A VphC represents the following claim, potentially recursively: *Here is a pool's process history, the issuer is the authorized owner of this pool and this record of their steps is compliant with the model logic.*

In contrast, a **Non-Invoked Transition Credential** captures what did *not* happen in a process instance. More precisely, by presenting an NiTC, issued by the target of a message flow, the participant, who is the source of the message flow proves: *This specific message flow in the process has not occurred.*

NiTCs are required to prevent state bifurcations on gateways in model executions. In our simplified model, in an exclusive gateway with n branches, only one branch can be taken in a process instance. To avoid malicious process bifurcation – *double-spending* –, the sender, trying to convince the recipient, would present an NiTC for each message flow that can be reached on the *non-chosen* branches of exclusive gateways, based on the presented process history.

If the recipient is honest, they will *revoke* their NiTC upon receiving and accepting a transition. This will ensure that the sender will not be able to declare other, contradictory message flows, as they will no longer be able to present all required NiTCs.

4 A Self-orchestrating Collaboration Protocol

The generic model of self-orchestration with VphCs and revokeable NiTCs lends itself to creating rather different families of self-orchestration protocols; in this paper, as an initial contribution, we propose a rather simple one, where the sender does not pass a "full dossier" (utilizing recursive VphCs) to the recipient, only non-recursive VphCs of the events (message passings) they made *her* act, and a proof that she is committed to not choosing an alternative execution route. This limits the propagation of potentially sensitive history information between the participants. We work under the following **security model**:

- There is at most one participant who may act maliciously; every other participant complies with BMPN collaboration execution semantics.
- The means of communication between the participants and the VDR, as well as the VDR itself are assumed to always behave correctly.
- The participants all have secure cryptographic identities, enabling the creation of verifiable digital signatures, and all participants are able to verify signatures created by the others (i.e., they can reliably link public keys to participant identities for asymmetric cryptography-based signing and signature verification).

We strongly believe that the protocol is actually able to tolerate two malicious parties, but, in contrast to the single malicious party assumption, a convincing argument will need to be a formal one.

Our single security goal is somewhat modest, but already provides a meaningful guarantee for practical applications: *the worst impact a malicious party may have on a collaboration execution is that they lead it to deadlock, and only so that this attack can be easily revealed.* Importantly, we explicitly defend against process bifurcations and out-of-sequence initiations of pool executions.

Allowing for deadlocks may seem to be a serious deficiency, but arguably, avoiding corrupted and bifurcating process states is more important, as process activity executions tend to have side effects that cannot be easily and systematically rolled back. This is analogous to blockchain consensus mechanisms, which always prioritize integrity over availability. Avoiding the deadlock capability (to a large extent) is also possible but requires either giving up some of the loose coupling properties of our protocol, additional VDR capabilities that are not present in the more mature solutions, or additional, somewhat involved cryptographic techniques. This aspect is the subject of ongoing research.

4.1 Verifiable Credential Abstract Data Model

The protocol relies on the VC and Verifiable Presentation conceptual data model presented in Fig. 2. It is easily translatable to specific VC data models (e.g., W3C VC Schemas [6]).

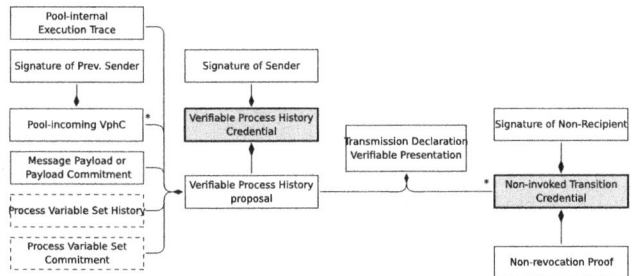

Fig. 2. Verifiable Credential data model

There are two types of VCs that both attest to the state of a specific Message Flow in a BPMN collaboration model: VphCs and NiTCs. Their differentiating factor is that the issuer of VphCs are message *senders,* while NiTCs are issued by message *recipients*. In the figure, issuers are authenticated by their signatures and we use the term *non-recipient* to signify that the claim is made by a party that has *not* received the message.

Each VphC includes a Verifiable Process History *proposal* that bundles the pool-internal execution trace of the issuing party and previous VphCs. Essentially, a VphC is the claim that a process is at that specific state, and the Verifiable Process History *proposal* is the evidence for that claim. We also envision them to include a commitment to the current process variable set, the process variable set history, and to the message payload however, these have not been incorporated into our model yet and the additional handshake mechanisms would unnecessarily complicate our discussion. These planned elements have been marked with dashed lines.

Each NiTC contains a *non-revocation proof* (borrowing from AnonCreds's [4] terminology), which is cryptographic material needed to check whether the NiTC has been revoked in the VDR, in a confidential manner. This essentially means that if somebody doesn't have this proof, they cannot deduce the existence or the revocation status of the credential based on the state of the VDR alone; thus the credential and its status are hidden from process external parties.

Finally, there are *Transmission Declaration Verifiable Presentations* (TdVP) that include all the contents of a *to-be* VphC and all relevant NiTCs (one for the current message throw event and one for each alternative execution path) as a signed document *not* a VC[1]. The TdVP is sent at the beginning of the process, when the message sender (such as the Citizen in our example) tries to persuade the recipient (e.g., the Workplace Committee) to move forward. The actual issuance of the VphC as a VC, however, only happens at the final step, when the evidence in the TdVP is accepted.

VphCs, associated with a message throw event from the issuer's point of view, and the corresponding message catch from the holder's, contain the following elements:

1. *Pool-internal Execution Trace*: This is a representation of the intra-pool sequence flows, activities, and events that led to the message throw event. This trace can be validated against the agreed-upon BPMN model and ends with the message throw event corresponding to the current VphC.
2. *Pool-incoming VphCs*: These are the VphCs linked to the message catch events that preceded the throw event in the trace. They attest that the corresponding messages have been recieved
3. *Message Payload or Commitment*: Either the actual message payload or a cryptographic commitment to it (e.g., a hash).

The dotted lines in the figure indicate that the protocol can incorporate process variables as well. However, here we exclude them from our scope, to avoid further complicating our treatment unnecessarily with discussing the additional handshake mechanisms required.

4.2 Protocol Sequence

In Fig. 3, we demonstrate the protocol sequence by illustrating key interaction patterns in the running example from Fig. 1. After an initial bootstrapping phase, decentralized self-orchestration can begin, where we differentiate between two interaction patterns: message passings where bifurcations are not a risk and interactions where they are.

---

[1] For some VC standards fitting our model and allowing for privacy-preserving revocations (e.g. AnonCreds), this is a very meaningful difference in technical complexity.

A Self-orchestration Model for Business Collaborations   69

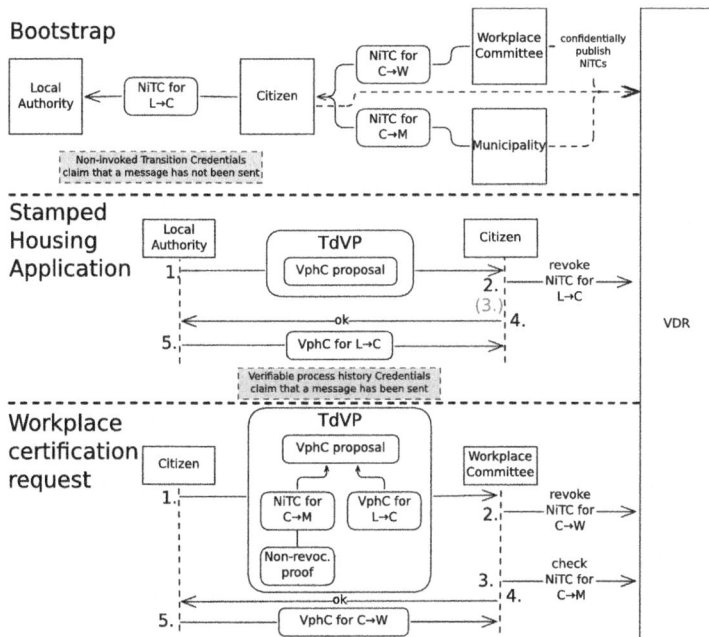

Fig. 3. Key interaction patterns of the proposed protocol demonstrated on the collaboration model in Fig. 1, around the gateway where possible process histories bifurcate.

Bootstrapping The bootstrap phase is where for each possible message flow in the agreed BPMN model, the message's receiver issues an NiTC to its senders, attesting that the message has not been sent yet. This phase can also be used to establish connections to the VDR, register public keys, establish revocation registries, set up authentication among participants, etc.

Orchestration After bootstrapping, the self-orchestration process begins. Normally, participants execute the part of the model they are responsible for – the activities, events, and gateways in their pool. They record the steps of their execution for later inclusion in a proposed Verifiable Process History Credential. Should they reach an event where they are the receiver of a cross-organizational transition (and they are willing to accept it), they store the VphC they receive.

Message throwing – declaration (1) When a participant reaches a message throw event, they send a *Transaction declaration Verifiable Presentations (TdVPs)* (TdVP) to the recipient, bundling their message passing intention, incoming VphCs, and the NiTCs for all possible, pool-internal *alternative* paths that could have been taken. For example, in our running example (Fig. 1), the citizen can choose to either apply for service housing (in which case they must request a certification from their workplace) or council housing (in which case they request a stamp from the municipality). In the execution illustrated in Fig. 3, they have chosen service housing – therefore, they must present an NiTC of the *other* execution path (i.e., where a municipality stamp is requested). For the initial stamped housing application received from the local authority, there is no need to include any NiTCs, as there are no alternative execution paths.

Verification and transmission commitment (2) The recipient of the message verifies the validity of the presented state based on the proposed and attached VphCs and the shared pool-internal execution trace. If valid, then the recipient revokes the NiTC issued by them to the sender for the currently processed message event during bootsrap.

Bifurcation check (3) At this point, the message recipient ensures that NiTCs regarding alternative paths have not been revoked. If any of them has already been revoked, then (at least this part of) the collaboration stops and (currently) protocol-external measures must be taken to rectify the situation.

VphC issuance (4–5) If all previous steps were successful, the receiver signals the sender that they accept the message passing and the sender issues a VphC to the receiver so that they may use it in *their* initiated transitions later.

For the second and third steps to work correctly (the Citizen cannot convince the Workplace and the Municipality simultaneously), we must assume that revocations in the VDR are atomic and that the VDR establishes a total order over its incoming transactions. Both are reasonable assumptions.

5 Technical System Model

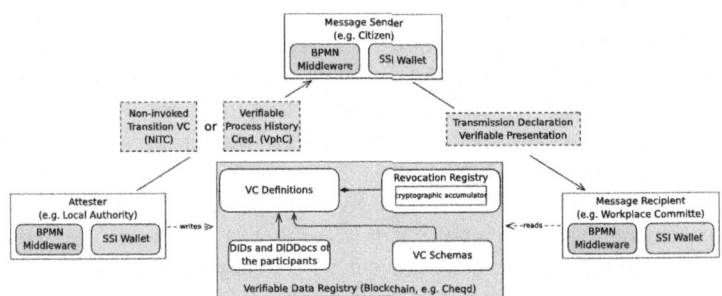

Fig. 4. System Architecture

A known set of participants wants to collaborate in the execution of a BPMN process instance, who have already verified each other's identities (e.g. via KYC/KYB processes outside our scope). Each participant runs an self-sovereign identity (SSI) [5] wallet – a secure application that handles the issuance, holding, and verification of Verifiable Credentials (VCs) over pairwise secure communication channels (e.g. DIDComm). These channels allow the participants to exchange VCs during the BPMN collaboration. We assume the SSI wallets are secure and uncompromised in terms of software, hardware, and cryptographic primitives.

All wallets connect to a verifiable data registry (VDR), which in our case is a blockchain, that holds the public keys of the participants required for the verification of the VCs, in the form of Decentralized Identifiers and DIDocs [1], as well as a revocation registry based on cryptographic accumulators. We discuss the favorable memory complexity of such registries later, in Sect. 6. In case of the AnonCreds [4] credential format, to enable better privacy characteristics, VC data schemas are also stored on the ledger, and also VC Definitions, which are essentially pairings of issuers and schemas, against which VCs can be issued and verified. We assume the integrity of the ledger (i.e., no successful consensus attacks), practical transaction finality, immutable state, and that the VDR processes revocation requests in a strictly sequential manner.

In this architecture, the wallets and blockchain are process-external: they only provide the means and cryptographic assurance for credential issuance, verification, and revocation. They do not directly execute the BPMN model.

We adopt a middleware design pattern, common in BPMN orchestration tools: each participant's BPMN engine (or local process manager) calls the wallet to request credential actions upon reaching a message throw event leading to another participant's pool. This triggers the middleware to collect or generate the necessary credentials and exchange them with relevant counterparties over secure channels, who verify their authenticity and non-revocation. This pattern ensures that trust in process transitions stems from cryptographically verifiable VCs and the decentralized VDR.

The components of the proposed system model, seen on Fig. 4, can be implemented over existing, mature technological frameworks. For BPMN execution, there is, e.g., Camunda BPM, which explicitly supports BPMN 2.0 models. For the SSI wallet, communication protocol, supporting blockchain, and VC carrying format, there are also many feasible solutions. From our evaluation, the LFDT Anoncreds [4] credential format offers the most flexible and privacy-enhancing VC data model, which can be exchanged through DIDComm 2.0, a versatile communication protocol, purposefully designed from the ground up for decentralized identity and claim management. As for the underpinning blockchain, Hyperledger Indy and cheqd both support this format. Production-ready wallet frameworks and ready-to-deploy wallet services that are pluggable into our architecture are also becoming available for enterprise as well as user wallets; key examples include credo from the OpenWallet Foundation, LFDT Identus, and ACA-Py.

6 Analysis and Discussion

In this section, we examine how the proposed *self-orchestration* model satisfies the requirements of BPMN orchestration, while preserving security and trust among participants.

Execution Trace Validity. Our approach ensures that each participant's local process state is synchronized with and validated against the overall logical state of the collaboration described by the subset of BPMN we laid out in Sect. 3.2. This is achieved by making message senders commit their state, and making sure they cannot circumvent or selectively enforce the execution semantics of BPMN.

Inherited Cryptographic Trust Guarantees. By relying on existing VC standards, we inherently benefit from well-established cryptographic signature schemes and credential exchange protocols, inheriting their cryptographic guarantees. This covers the non-tampering of credential contents, secure and private exchange of credentials, and revocation checks, as well as the indirect authentication of issuers via the verifiable data registry.

Protecting Against Rewriting and Double Spending. Once a VphC is issued, it cannot be altered or replaced without invalidating its signature or revocation status, preventing participants from rewriting process history. To ensure consistency at branching points, participants must present valid, non-revoked NiTCs for all *untaken* paths; once a path is chosen, its NiTC is revoked, making contradictory transitions impossible. This mechanism protects against both history forgery and process bifurcation, prioritizing integrity over liveness.

Effects of Partial Failure and Withholding. Two failure modes can stall execution: 1. if a participant accepts a transition but fails to revoke the corresponding NiTC, conflicting branches remain possible; 2. conversely, if the NiTC is revoked but the VphC is not issued or accepted, there is no verifiable record of the transition. In both cases, the process halts safely, ensuring that only fully consistent state changes can proceed. Any partial action can be addressed through external resolution or predefined fallback mechanisms. Overall, our model prevents unauthorized forks or modifications to the workflow state.

Memory Complexity of the Shared State. In our approach, the verifiable data registry (VDR) stores cryptographic material for verifying Verifiable Credentials and maintains per-issuer Revocation Registries, consistent with existing SSI technologies (e.g., Anoncreds [4]). These registries use cryptographic accumulators based on *quasi-commutative hash functions* [3], enabling constant-size storage. As a result, the on-chain state of a self-orchestration **does not** grow with cross-pool transitions and remains bounded per process instance.

7 Related Work

Our work addresses the challenges of orchestrating multiparty business collaborations securely and privately, placing it conceptually alongside recent advancements in smart-contract-based, confidentiality-preserving process orchestration. Specifically, the closest related contributions are the approaches of *Toldi and Kocsis* [9], and Petto et al. [8].

In the approach of Toldi and Kocsis, participants prove the validity of the communicated state by publishing its encrypted representation, and providing a zero-knowledge proof (ZKP) that this encrypted state is the continuation of the process. In contrast, our approach uses off-chain (but blockchain-backed) Verifiable Credentials for transition validation, which ideally only require constant memory space as new transitions are invoked; a solution which is likely to be more cost-effective to develop, deploy, and operate.

Petto et al. similarly utilize ZKPs, but for the interpreted confidential execution of BPMN coreographies, by validating inter-participant messaging. Their clear advantage compared to Toldi and Kocsis is that they eliminate the need to deploy smart contracts on a per-collaboration basis, requiring only a "single-setup" deployment. Our approach also focuses on cross-pool communication, but does not require the deployment of per-collaboration logic or data. During a collaboration, participants may reuse previously published generic VC Schemas, public keys, VC Definitions, and Revocation Registries.

Our approach leverages Verifiable Credentials to facilitate peer-to-peer self-orchestration of complex BPMN collaborations, differentiating itself clearly from existing centralized and blockchain-based orchestration methods, both in terms of conceptual foundation and practical implementation.

8 Conclusion and Future Work

In this paper, we introduced the concept of self-orchestration for BPMN collaborations as a decentralized, quasi-peer-to-peer approach for multiparty business interactions, eliminating the need for centralized execution engines or smart contract infrastructures. To demonstrate this concept, we proposed a model leveraging blockchain-secured Verifiable Credentials (VCs), with the key constructs of Verifiable Process History Credentials (VphCs) and Non-Invoked Transition Credentials (NiTCs). Our approach represents an advancement towards the ideal of true decentralization, and offers benefits with respect to both smart-contract-based orchestration.

Also, standard VC usage has so far focused on single-claim, challengeresponse style verification. In contrast, our model demands more expressive, semantically rich credentials that can carry structured execution traces, revocation intents, and branching commitments. This contextualization of VCs – transforming them into verifiable carriers of process control flow – is a novel usage modality.

Our model addresses core challenges in decentralized process orchestration, such as preventing process state bifurcation, minimizing data exposure, and maintaining trust while reducing coupling and computation overhead. By utilizing VCs, it provides a practical and comprehensible mechanism for managing trust and verifying compliance in multiparty collaborations.

Future work will extend the model to accommodate richer BPMN semantics – such as process variable commitments, loops, parallel gateways, and advanced event types–, to enhance its resilience against partial failures and withholding attacks, and also to increase the confidentiality of the process execution by using off-chain zero-knowledge proofs for transition validation. This research lays the foundation for advancing decentralized BPM methodologies, paving the way for secure, efficient, and autonomous cross-organizational collaborations.

Acknowledgments. The work of B. Z. P. supported by the Doctoral Excellence Fellowship Programme (DCEP) is funded by the National Research Development and Innovation Fund of the Ministry of Culture and Innovation and the Budapest University of Technology and Economics under a grant agreement with the National Research, Development and Innovation Office. The work of F. M. and I. K. was partially funded by the European Union's Digital Europe Programme (DEP) under grant agreement no. 101100768 (SME4DD).

Disclosure of Interests. The authors have no competing interests to declare.

References

1. Decentralised identifiers (2022). https://www.w3.org/TR/2022/REC-did-core-20220719/
2. Technical Specifications - EU Digital Identity Wallet (2024). https://ec.europa.eu/digital-building-blocks/sites/display/EUDIGITALIDENTITYWALLET/Technical+Specifications
3. Camenisch, J., Shoup, V.: Practical verifiable encryption and decryption of discrete logarithms. In: Boneh, D. (ed.) CRYPTO 2003. LNCS, vol. 2729, pp. 126–144. Springer, Heidelberg (2003). https://doi.org/10.1007/978-3-540-45146-4_8
4. Curran, S., Yildiz, H., Curren, S.: AnonCreds specification (2022). https://hyperledger.github.io/anoncreds-spec/
5. Drummond, R., Preukschat, A.: Self-sovereign identity: decentralized digital identity and verifiable credentials. Manning (2021)
6. Sporny, M., Chadwick, D., Longley, D.: Verifiable credentials data model v2.0 (2024). https://www.w3.org/TR/vc-data-model-2.0/
7. Mendling, J., et al.: Blockchains for business process management - challenges and opportunities. ACM Trans. Manage. Inf. Syst. **9**(1), 4:1–4:16 (2018). https://doi.org/10.1145/3183367
8. Petto, O., Preindl, T., Kjäer, M.: Interpreted and confidential execution of process choreographies on a blockchain. In: Di Ciccio, C., et al. (eds.) BPM 2024. LNBIP, vol. 527, pp. 40–54. Springer, Cham (2024). https://doi.org/10.1007/978-3-031-70445-1_3
9. Toldi, B.A., Kocsis, I.: Blockchain-based, confidentiality-preserving orchestration of collaborative workflows. Infocommunications J. **15**(3), 72–81 (2023). https://doi.org/10.36244/ICJ.2023.3.8
10. Trust Over IP Foundation: The ToIP Model (2022). https://trustoverip.org/toip-model/
11. Weske, M.: Business process management architectures. In: Weske, M. (ed.) Business Process Management: Concepts, Languages, Architectures, pp. 387–420. Springer, Heidelberg (2024). https://doi.org/10.1007/978-3-662-69518-0_8

Comparing Apples with Oranges: An Assessment Framework for Model-System Similarity

Martin Kabierski[1](✉) 📖, Jana-Rebecca Rehse[2] 📖, and Jan Martijn E.M. van der Werf[3] 📖

[1] University of Vienna, Vienna, Austria
martin.kabierski@univie.ac.at
[2] University of Mannheim, Mannheim, Germany
rehse@uni-mannheim.de
[3] Utrecht University, Utrecht, The Netherlands
j.m.e.m.vanderwerf@uu.nl

Abstract. Process models and event logs are the two most important process representations in process mining. When comparing the two, e.g., to reveal discrepancies between them, we can consider either the model-log perspective, meaning the degree to which the model represents the event log, or the model-system perspective, meaning the degree to which the model represents the underlying process. The two perspectives complement each other, but so far, most research has focused on assessing model-log similarity. In this paper, we propose a novel framework for assessing model-system similarity that addresses the two major challenges that this task poses. First, it enables a comparison between model and system that is independent from the concrete modeling formalism by measuring their similarity based on the n-grams of their respective languages. Second, it abstracts from the event log as an incomplete sample of the system by projecting the likely language of the system by means of statistical estimators. Our empirical evaluation shows that this framework provides a valid way to assess model-system similarity, which can be used in many different applications.

Keywords: Process mining · Process representations · Model-system similarity · Process discovery quality · Statistical estimators

1 Introduction

Process mining is a family of analytical techniques aimed at gaining insights into business processes for their continuous improvement [26]. It builds on (digital) representations of these processes, which can be analyzed by computational methods. The two most important process representations are event logs, which contain records of process executions [3], and process models, which describe the set of allowed process executions [26]. A fundamental process mining task is to compare an event log and a process model that represent the same process, e.g., to ensure their consistency [2] or reveal discrepancies between them [1].

Such a comparison can be conducted with two different aims [3,10]. If we assess *model-log similarity*, we want to compare the model directly with the event log, measuring how well it represents the captured process executions [29]. This is important when exploring those executions, e.g., to find patterns or identify bottlenecks. On the other hand, if we assess *model-system similarity*, we want to compare the model with the underlying true process, which log and model supposedly represent [10]. This is important when confirming assumptions about the process as a whole, e.g., with regard to control-flow patterns.

As an example, consider process discovery, which aims to construct a process model for a given event log. If we assess the model-log similarity of a discovered process model, we focus on how well it represents the traces in that log, meaning that we aim for a close correspondence between the two [3,10]. However, if we assess the model-system similarity of that same model, we focus on how well it represents the true process, i.e., the likelihood that it generated the event log [27]. This means that we also need to take unobserved behavior into account [3,10,21].

Model-log similarity and model-system similarity complement one another, but need to be measured differently [10]. Most existing research, however, focuses on measuring model-log similarity [10,11,29], which might become problematic: as process mining aims to reason about the "true" process on the basis of available data [3], properly measuring model-system similarity is crucial for assessing whether the model is likely to actually represent this true process. For process discovery, empirical research has shown that the four quality dimensions fitness, precision, generalization, and simplicity, which are state-of-the-art for model-log similarity, are not well suited to measure model-system similarity [10,12,21,27].

This means that in order to achieve a comprehensive perspective on different process representations, we need to measure model-system similarity [10]. The problem with this task, however, is that the only information we have about the true process stems from the event log, which consists of sequences that constitute a sample of all possible process behavior. This leads to two major challenges. First, we need to compare a process model with a set of sequences in the event log, without being impacted by the specifics of the modeling formalism. For example, the assessment of model-system similarity should not be impacted if one model is represented by a BPMN and another by a process tree [2]. Second, we need to account for the fact that the event log is an incomplete, imperfect sample of the true process [16]. For example, two different event logs from the same true process should achieve a similar model-system similarity [12,21].

In this paper, we propose a novel framework to assess model-system similarity. This framework addresses those two challenges by (1) measuring the similarity between this projected system and a process model based on the n-grams of their respective languages and (2) projecting the language of the system from the event log by means of statistical estimators. To elaborate on these ideas, the following Sect. 2 further discusses the two challenges of measuring model-system similarity. Section 3 introduces our framework, explaining how its components address these challenges. In Sect. 4, we conduct a controlled experiment that empirically demonstrates the framework's capabilities to measure model-system similarity. We discuss related work in Sect. 5, before concluding in Sect. 6.

2 Challenges in Assessing Model-System Similarity

For assessing model-system similarity, we assume three entities [3,10]. The *true process* is unknown, but its executions can be observed and recorded in the *event log*, which is a finite sample of observations of the true process. The *process model* aims to represent the true process. Our framework is set out to assess the achievement of this aim and measure how well the model represents the true process. In the following, we explain the challenges of this task (see Fig. 1).

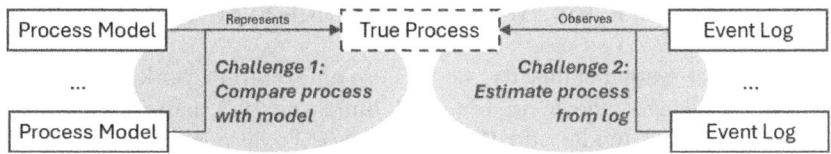

Fig. 1. Entities and challenges in assessing model-system similarity.

Challenge 1: Compare the True Process with the Process Model. To assess model-system similarity, we need to compare the process model with the true process. In this comparison, the model is represented by some modeling formalism, such as a Petri net or a process tree. The true process is represented by the observations in the event log. The challenge is that the extent to which this comparison is possible depends on the chosen modeling formalism. Currently, there are individual comparison techniques for each formalism. For example, for a Petri net, the comparison can be done by means of an alignment [1], but this technique is not directly applicable to a process tree, which needs to be converted into a Petri net first [2]. Potential differences in the results of these two comparisons can either be caused by differences in the model-system similarity or introduced by the required conversion [18]. Even for modeling formalisms that can be compared by means of an alignment, this can become computationally infeasible, particularly for large and complex event logs [1]. Hence, the comparison often approximated, e.g., by comparing only the directly-follows relationships of log and model [30]. However, this might exclude relevant information from the event log, hence biasing the assessment. For those reasons, our framework should facilitate a comparison between the process model and the true process that is independent of the chosen modeling formalism, computationally feasible, and retains as much information as possible.

Challenge 2: Estimate the True Process from the Event Log. Even when the comparison between model and true process is possible, it is not sufficient to base model-system similarity solely on the observations of the true process that can be found in the event log [10]. First, the event log is likely to be incomplete, so it does not contain all possible observations of the true process [12,21]. This is particularly relevant when the potential observations of the true process are infinite and can never be fully captured by a finite event log. Second, the event log may subject to sampling errors, so that individual observations are oversampled or undersampled compared to the true process [15,27]. We can conclude that, at least when taken at face value, the event log may not be a good representation of the true process. However, the true process is–by definition–unknown and the only information we have about it stems from the observations captured in the event log. This means that to properly assess model-system similarity, we need to take the event log as a starting point and extrapolate from it, estimating its full amount of potential observations and their distribution based on the partial amount in the event log. This allows us to establish a better representation of the true process to compare with the model.

3 A Framework for Assessing Model-System Similarity

To assess model-system similarity, we propose a novel framework that addresses these two challenges, shown in Fig. 2. It builds on two main ideas: First, it represents both the model and the true process by means of a language and compares their similarity based on their respective n-grams. Second, it projects the language of the process from the event log by means of statistic estimators.

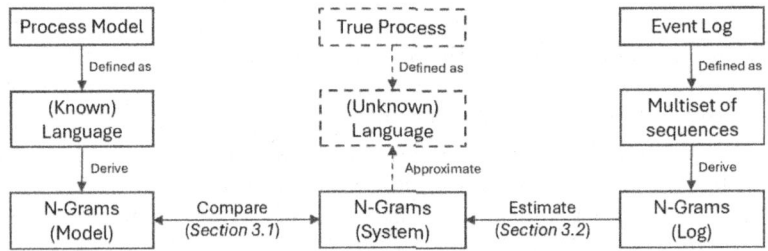

Fig. 2. Framework for Assessing Model-System Similarity.

We assume that the true process can be represented by a language $\mathcal{P}$. An event log L is a collection of sequences recorded from observing the true process, i.e., a multiset of finite sequences over some alphabet. We further assume that the process model is represented by a modeling formalism that can generate a language $\mathcal{M}$. In the following, we first explain how to compare the true process and the model independent of the model representation (Sect. 3.1) and then discuss how to estimate the language of the true process (Sect. 3.2).

3.1 Comparing the Languages of the Process and the Model

Formal methods use equivalences to describe the notions of similarity. One of the strongest notions is isomorphism, meaning that two models are identical, e.g., if a bijection exists on their underlying graph structures. A weaker notion of similarity is bisimulation, defined as a relation on the underlying semantics. Informally, two models are bisimilar if in related states, one model can take an action, the other model should be able to take that action as well, and the resulting states should be related again [28]. The problem, however, of these equivalences is that they are defined on the semantics of the modeling notation. If two modeling notations use different semantics, one needs to be translated into the other. This is not a straightforward task, as the many different and sometimes even conflicting proposals for translating BPMN into Petri nets show (e.g., [18]). In addition, not all language constructs can be formally specified, as the discussion of the semantics of the Or-operator exemplifies (e.g., [7]).

A notion that does not depend on the underlying formalism is language equivalence. Two behavioral models are language equivalent if both produce the same set of accepting sequences; any sequence one model produces, the other should be able to accept. Trace equivalence, i.e., any sequence one model can produce, the other model can produce as well, is a slightly weaker equivalence notion. These equivalence notions form a ladder: if two models are isomorphic, they are bisimilar; if they are bisimilar, they are language equivalent.

Language equivalence relies on comparing full sequences, which is known to be too strong for process mining [24], shown by the many methods for model-log comparisons, such as token replay [23] and alignments [1]. Hence, logs and models are often compared based on the directly-follows relation and its generalizations in footprints [30]. For some formalisms, such as process trees and free choice Petri nets, the directly-follows relation implies language equivalence [31]. However, this only holds if the models that generated the language are expressed in the same formalism, which is a strong assumption for which no checks exist [27].

In this paper, we introduce a new equivalence notion that compares languages based on its n-grams. Whereas footprints generalize the directly-follows relation over pairs of activities, we generalize the directly-follows relation to n-grams, i.e., the substrings of a language. We consider a language to be a set of words over some alphabet with a probability function over the words.

Definition 1 (Language). *A language is a triple $\mathcal{L} = (W, S, \pi)$ with a possibly infinite set of finite sequences (words) $W \subseteq S^*$ over some alphabet S with some probability distribution function $\pi : W \to [0, 1]$, such that $\sum_{\sigma \in W} \pi(\sigma) = 1$, where S^* denotes the set of all possible finite sequences over S. We use $\sigma \in \mathcal{L}$ as a shorthand notation for $\sigma \in W_{\mathcal{L}}$.*

The n-gram representation of a language considers all possible substrings contained in the language, with a probability distribution function that takes the occurrence frequency of each substring into account. It assumes that the n-grams are uniformly distributed within each sequence, i.e., the number of occurrences of an n-gram divided by the total number of n-grams in a sequence. In other words, the n-gram representation $\mathcal{N}(\mathcal{L})$ contains all possible substrings of $\mathcal{L}$ and the corresponding distribution function μ is updated accordingly.

Definition 2 (n-gram representation of a language). *Given a language $\mathcal{L} = (W, S, \pi)$ and some $n \in \mathbb{N}$, we define its n-gram representation as the language $\mathcal{N}_n(L) = (G, S, \mu)$ with:*

- $G = \{\tau \in S^n \mid \exists \sigma_1, \sigma_2 \in S^* : \sigma_1 \tau \sigma_2 \in W\}$ *and*
- $\mu(\tau) = \sum_{\sigma \in W} \frac{2 \cdot \#(\tau, \sigma)}{|\sigma| \cdot (|\sigma| + 1)} \pi(\sigma).$

where $\#(\tau, \sigma)$ denotes the number of times τ occurs in σ and $S^n = \{\sigma \mid \sigma \in S^* \wedge |\sigma| = n\}$ denotes the set of words of length n. We use $\mathcal{N}(L)$ for the n-gram representation over all n-grams, i.e., taking n-grams from S^* instead of S^n.

Each sequence in an event log is an observation of the true process. As the start and end of each sequence are unknown, event logs are considered a multiset of sequences, rather than a language.

Definition 3 (Event log). *An event log L is a multiset of sequences over some alphabet S, i.e., $L : S^* \to \mathbb{N}^0$, where $\mathbb{N}^0$ denotes the set of natural numbers including 0. We use $supp(L) = \{\sigma \mid L(\sigma) > 0\}$ for the support of the multiset.*

Similar to languages, event logs are defined on sequences. Consequently, event logs can also be represented using n-grams:

Definition 4 (n-gram representation of an event log). *Given an event log L over some alphabet S and some $k \in \mathbb{N}$, its n-gram representation is the language $\mathcal{N}_n(L) = (G, S, \mu)$ with:*

- $G = \{\tau \in S^n \mid \exists \sigma_1, \sigma_2 \in S^* : \sigma_1 \tau \sigma_2 \in supp(L)\}$ and
- $\mu(\tau) = \sum_{\sigma \in supp(L)} \left(\frac{2 \cdot \#(\tau, \sigma)}{|\sigma| \cdot (|\sigma|+1)} \cdot \frac{L(\sigma)}{|L|} \right)$.

As n-gram representations are defined as a set of words with a probability distribution function, two n-gram representations (of a model and a log) can be similar with regard to (1) *identity*, i.e., they contain the same n-grams, and (2) *distribution*, i.e., the n-grams have the same probability distribution. Hence, measuring these two properties gives us an instrument to establish the relation between the true process and the generated process model. We say that wo n-gram representations are *language indistinguishable* if they satisfy the identity property, and *fully indistinguishable* if they satisfy both properties. However, although any sequence in a language is finite, an n-gram representation can be unbounded. Consider, e.g., the language $\mathcal{L} = a^+ = \{\langle a \rangle, \langle a, a \rangle \ldots\}$, whose set of n-grams is infinite. Therefore, we introduce an upper bound n on the size of n-grams and define indistinguishability using this bound. Note that unboundedness does not occur for event logs, which always contain a finite number of sequences.

Definition 5 (Indistinguishable n-gram representations). *Given two n-gram representations $\mathcal{N}_1 = (G_1, S, \mu_1)$ and $\mathcal{N}_2 = (G_2, S, \mu_2)$, we say they are:*
- *language n-indistinguishable if $(G_1 \cap S^n) = (G_2 \cap S^n)$;*
- *fully n-indistinguishable if $(G_1 \cap S^n) = (G_2 \cap S^n)$ and $\mu_1(\sigma) = \mu_2(\sigma)$ for all $\sigma \in S^n$;*
- *language indistinguishable if they are language n-indistinguishable for all $n \in \mathbb{N}$;*
- *fully indistinguishable if they are fully n-indistinguishable for all $n \in \mathbb{N}$;*

As an example, consider languages $\mathcal{L}_1$ and $\mathcal{L}_2$ in Fig. 3. Both contain the same words, so their 2-grams are identical: $\{(A, B), (B, C), (C, D), (A, C), (C, B), (B, D), (D, E), (E, F)\}$. As $\mathcal{N}_n(L_1) = \mathcal{N}_n(L_2)$ for all $n \in \mathbb{N}$, $\mathcal{L}_1$ and $\mathcal{L}_2$ are language indistinguishable, but given the different occurrence probabilities of words, their n-gram representations are not fully indistinguishable.

$\langle A,B,C,D\rangle^{0.2}$ $\langle A,C,B,D\rangle^{0.2}$ $\langle A,B,C,D\rangle^{0.2}$ $\langle A,C,B,D\rangle^{0.1}$
$\langle A,B,C,D,E\rangle^{0.2}$ $\langle A,C,B,D,E\rangle^{0.2}$ $\langle A,B,C,D,E\rangle^{0.2}$ $\langle A,C,B,D,E\rangle^{0.4}$
$\langle A,B,C,D,E,F\rangle^{0.2}$ $\langle A,B,C,D,E,F\rangle^{0.1}$

(a) $\mathcal{L}_1$ (b) $\mathcal{L}_2$

Fig. 3. Languages $\mathcal{L}_1$ and $\mathcal{L}_2$ with the probability of each sequence in superscript.

In general, if two languages are equivalent, i.e., they contain the same set of accepting sequences, they are language indistinguishable. The converse does not hold. Consider, e.g., languages $\mathcal{L}_3$ with sequences $\langle A,B,C,D\rangle$ and $\langle A,B\rangle$ and $\mathcal{L}_4$ with $\langle A,B,C,D\rangle$ and $\langle B,C\rangle$. These languages are language indistinguishable, but the languages are not language equivalent. Note that if an event log L is directly-follows complete with respect to some process model $\mathcal{M}$, then the n-gram representations of L and $\mathcal{M}$ are language 2-indistinguishable.

3.2 Estimating the Language of the Process

Assuming two languages A and B are known, n-gram indistinguishability can be assessed by comparing the distribution of n-grams, one-by-one. This can be quantified using set similarity measures, such as the Jaccard coefficient [9]:

$$J(A,B) = \frac{|A \cap B|}{|A| + |B| - |A \cap B|} \quad (1)$$

If $J(\mathcal{N}(\mathcal{L}_1), \mathcal{N}(\mathcal{L}_2)) = 1$, both sets contain exactly the same elements, and thus, the two languages are n-indistinguishable. As an example, consider event logs $S_1 = \{\langle A,B,C,D\rangle^1, \langle A,B,C,D,E\rangle^1, \langle A,C,B,D,E\rangle^2\}$ and $S_2 = \{\langle A,B,C,D\rangle^2, \langle A,C,B,D,E\rangle^1, \langle A,B,C,D,E,F\rangle^1\}$, both observations from the true process $\mathcal{L}_1$. Their 2-gram representations are $\mathcal{N}_2(S_1) = \{\langle A,B\rangle^{2/12}, \langle B,C\rangle^{2/12}, \langle C,D\rangle^{2/12}, \langle D,E\rangle^{3/12}, \langle A,C\rangle^{1/12}, \langle C,B\rangle^{1/12}, \langle B,D\rangle^{1/12}\}$ and $\mathcal{N}_2(S_2) = \{\langle A,B\rangle^{2/12}, \langle B,C\rangle^{2/12}, \langle C,D\rangle^{2/12}, \langle D,E\rangle^{2/12}, \langle A,C\rangle^{1/12}, \langle C,B\rangle^{1/12}, \langle B,D\rangle^{1/12}, \langle E,F\rangle^{1/12}\}$. Then, $J(\mathcal{N}_2(S_1), \mathcal{N}_2(S_2)) = \frac{7}{7+8-7} = 0.875$, which signifies that the two logs are not language 2-indistinguishable. Both event logs are incomplete, due to the small sample size. Also, the probabilities of the 2-grams do not match.

Chao et al. [5] showed that established set similarity measures, such as the Jaccard coefficient, are negatively biased, and thus unsuitable as estimators of language equivalence in sample-based settings like process mining. Similarly, full n-indistinguishability depends on the probability distribution captured in the event log, which is subject to minor deviations due to sampling errors. As it is unknown whether the event log is a good representation of the true process, small changes in the log may have a large effect on deciding full n-indistinguishability. Therefore, measures for indistinguishability on event logs need to account for the incomplete set of observed words and be robust against sample-induced differences between observed occurrence probabilities.

As discussed in Challenge 2, the language of the true process is unknown and only represented by a finite, possibly incomplete and over- or undersampled event log. Hence, we need to estimate the likelihood that two languages $\mathcal{L}_1$ and $\mathcal{L}_2$ are language n-indistinguishable, or even fully n-indistinguishable, given only logs S_1 and S_2. Note, that such an estimate does not provide undisputable evidence, but rather a sample-based likelihood for language-indistinguishability. In this paper, we propose to estimate this likelihood by adapting the Jaccard coefficient. Using the Chao1 estimator [4] as well as an estimate [6] for the true number of shared elements in $A \cap B$, the adapted Jacquard coefficient considers the estimated true number of elements in the population of A, B, and $A \cap B$. For estimating the likelihood of full n-indistinguishability, we adopt the Morisita-Horn similarity index [8].

Estimating Language n-Indistinguishability. The Jaccard coefficient relies on the set sizes, which are only given partially, as incomplete event logs S_A and S_B. By estimating the true size of those sets as $\widehat{|A|}, \widehat{|B|}$ and $\widehat{|A \cap B|}$, we can adopt the Jaccard coefficient to estimate the similarity of A and B based on the available information in S_A and S_B as:

$$\widehat{J}(S_A, S_B) = \frac{\widehat{|A \cap B|}}{\widehat{|A|} + \widehat{|B|} - \widehat{|A \cap B|}} \quad (2)$$

Set sizes $\widehat{|A|}$ and $\widehat{|B|}$, i.e., the number of distinct elements in populations A and B, are estimated using the Chao1-estimator for species richness [4], which has also been applied in the context of sample-based process analysis [16]. For estimating $\widehat{|A \cap B|}$, i.e., the number of distinct elements in the population of elements shared by A and B, we utilize an estimator for the shared species richness of two or more populations [6].

Let $M : A \to \mathbb{N}^0$ be a multiset over some set A. We define $single(M) = \{a \in A \mid M(a) = 1\}$ as the set of all singletons of M, i.e., the set of all elements that only occur once in M. Similarly, $double(M) = \{a \in A \mid M(a) = 2\}$ defines the set of all doubletons, i.e., elements that occur twice in M. The Chao1-estimator, estimating the unknown number $|\hat{M}|$, is defined as:

$$|\hat{M}| = \begin{cases} |supp(M)| + \frac{|single(M)|^2}{2 \cdot |double(M)|} & \text{if } double(M) \neq \emptyset \\ |supp(M)| + |single(M)| \cdot \frac{1}{2}(|single(M)| - 1) & \text{if } double(M) = \emptyset. \end{cases} \quad (3)$$

This estimator is a lower bound on the true number of elements in the corresponding population, and an unbiased point estimator, when occurrence probabilities of unobserved elements and singleton elements in $single(M)$ are the same. Note that it only considers elements of M that occur once or twice. For the 2-gram representation of event log S_1 (set A), there are 3 singletons and 3 doubletons, so $\widehat{A} = \lceil 7+1.5 \rceil = 9$. This means that, after rounding, two additional elements are expected. Event log S_2 contains 4 singletons and 4 doubletons, thus $\widehat{B} = \lceil 8 + 2 \rceil = 10$, also expecting two additional elements.

For estimating $\widehat{|A \cap B|}$, we rely on an estimator for shared species richness of multiple samples [6], which constructs a lower bound on the number of shared elements of multiple samples. It decomposes the set of distinct elements in A and B, into four subsets and estimates *species richness* for each one. Let $M_A : A \to \mathbb{N}^0$ and $M_B : B \to \mathbb{N}^0$ be multisets over sets A and B. Given a shared element $e \in A \cap B$, then e is either (1) observed in both M_A and M_B, i.e., $M_A(e) > 0$ and $M_B(e) > 0$, (2) only observed in M_A or in M_B, i.e., $M_A(e) > 0$ or $M_B(e) > 0$ but not both, or (3) neither observed in M_A or in M_B, i.e., $M_A(e) = M_B(e) = 0$. For each of these four options, the estimator constructs the estimated species richness, so that the final estimate is the sum of these values. For this, we introduce the following functions: $f_{x,y} = |\{e \in A \cap B \mid M_A(e) = x \wedge M_B(e) = y\}|$, counting the number of elements in $A \cap B$ that occur x times in M_A and y times in M_B. We replace x, respectively y, by a + if the element should be at least one, e.g. $f_{+,y} = |\{e \in A \cap B \mid M_A(e) > 0 \wedge M_B(e) = y\}|$ counts the number of elements in $A \cap B$ that occurs at least once in M_A and y times in M_B. Then, the estimator for $|\widehat{A \cap B}|$ is defined as:

$$|\widehat{A \cup B}| = f_{+,+} + \frac{|supp(M_A)|}{|supp(M_A)|-1} \cdot \frac{(f_{1,+})^2}{2 \cdot f_{2,+}} + \frac{|supp(M_B)|}{|supp(M_B)|-1} \cdot \frac{(f_{+,1})^2}{2 \cdot f_{+,2}} + \frac{|supp(M_A)|}{|supp(M_A)|-1} \cdot \frac{|supp(M_B)|}{|supp(M_B)|-1} \frac{(f_{1,1})^2}{4 \cdot f_{2,2}} \quad (4)$$

The 2-gram representations for the example event logs S_1 and S_2 give $\widehat{A \cup B} \approx 7 + 1.64 + 1.23 + 0.89 = 10.76$. This lower bound may be larger than the individual estimations of set sizes for the individual sets. In such a case, we may assume both sets to be equal. Using $\widehat{A}, \widehat{B}$ and $\widehat{A \cup B}$ in Eq. 2, results in a similarity of $\widehat{J}(S_1, S_2) = \frac{11}{(9+10-11)} \approx 1.375$. As $\widehat{A \cup B}$ is larger than both $\widehat{A}$ and $\widehat{B}$, we conclude that the languages from which S_A and S_B stem are language 2-indistinguishable.

Estimating full n-Indistinguishability. The estimator in Eq. 2 only considers the existence of elements, so it does not allow for a comparison based on probability distributions. Two sets with equal elements but significantly different distributions will be considered as perfectly similar. Consider $S_3 = \{\langle A, C, B, D\rangle^1, \langle A, B, C, D, E\rangle^1, \langle A, C, B, D, E^2\rangle\}$ to be a log of $\mathcal{L}_2$ from Fig. 3. Its 2-gram representation is $\mathcal{N}_2(S_3) = \{\langle A, C\rangle^{2/10}, \langle C, B\rangle^{2/10}, \langle B, D\rangle^{2/10}, \langle A, B\rangle^{1/10}, \langle B, C\rangle^{1/10}, \langle C, D\rangle^{1/10}, \langle D, E\rangle^{1/10}\}$. As the 2-gram representations of S_3 and S_1 are identical, we have $J(\mathcal{N}_2(S_1), \mathcal{N}_2(S_3)) = 1$ and $\widehat{J}(\mathcal{N}_2(S_1), \mathcal{N}_2(S_3)) = 1$, thus they are language 2-indistinguishable. However, the occurrence distribution is completely different, and the event logs are clearly not fully 2-indistinguishable.

If two distributions are fully indistinguishable, occurrence frequencies over all elements are equal. Given two n-gram representations $\mathcal{N}_A = (G_A, S, \mu_A)$ and $\mathcal{N}_B = (G_B, S, \mu_B)$ their similarity can be expressed using the *Morisita-Horn index of similarity* [8]:

$$D(\mathcal{N}_A, \mathcal{N}_B) = \frac{2\sum_{e \in G_A \cap G_B} \mu_A(e)\mu_B(e)}{(\lambda_A + \lambda_B)|G_A||G_B|}, \text{ where}$$

$$\lambda_A = \frac{\sum_{e \in A} \mu_A(e)^2}{|G_A|^2} \text{ and } \lambda_B = \frac{\sum_{e \in B} \mu_B(e)^2}{|G_B|^2}. \quad (5)$$

The index considers both the set of shared elements in $\mathcal{N}_A$ and $\mathcal{N}_B$, as well as differences in the occurrence frequencies between elements. The 2-gram representations $\mathcal{N}_{S_1}$ and $\mathcal{N}_{S_3}$ of event logs S_1 and S_3, results in $D(\mathcal{N}_{S_1}, \mathcal{N}_{S_3}) \approx 0.843$.

The estimator quantifies the sample-based similarity of distributions and thus the likelihood of two languages being fully n-indistinguishable. If $D(\mathcal{N}_A, \mathcal{N}_B) = 1$, both distributions share the same elements, and those elements have the same probabilities. If both sets disagree on shared elements or their occurrence probabilities, $D(\mathcal{N}_A, \mathcal{N}_B)$ decreases, signaling a reduced likelihood of A and B being fully-indistinguishable, given the information in their respective samples.

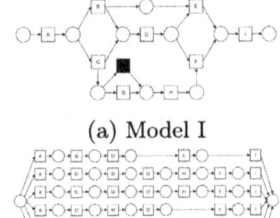

(a) Model I

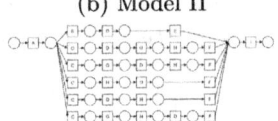

(b) Model II

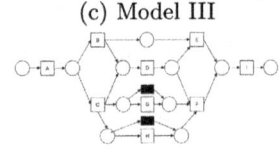

(c) Model III

4 Experimental Evaluation

To assess the adequacy and applicability of the proposed estimators for n-indistinguishability (Eq. 2, called the *I-estimator* in the following) and full n-indistinguishability (Eq. 5, called the *D-estimator* in the following), we conducted an evaluation on a set of publicly available benchmark models [21]. The selected models are varied in terms of their complexity, and their respective languages, while being defined over the same set of activities, and as such exhibit different degrees of similarity in their n-gram representations.

Here, we only compare event logs to study the ability of the estimators to provide useful estimates of the true process. We want to answer the following research questions:

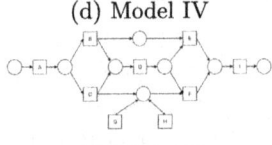

(d) Model IV

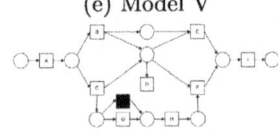

(e) Model V

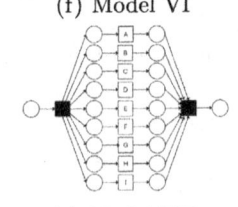

(f) Model VI

RQ1 Do both estimators properly capture (full) n-indistinguishability of logs and models?
RQ2 Do both estimators properly signify when they are (fully) n-distinguishable?
RQ3 How sensitive are the estimators to sample size and relevant parameters?

RQ1 validates the estimators' ability to detect n-indistinguishable languages. RQ2 validates the opposite, i.e., that they do not falsely signify indistinguishability when the languages are distinguishable. RQ3 investigates under which situations correct estimates can be expected, and how sensitive they are to sample size and language representations.

(g) Model VII

Fig. 4. The utilized Benchmark Models.

This evaluation is an initial demonstration of the estimators applicability. We deem the extensive evaluation of estimator accuracy and reliability on more complex synthetic and real process data a valuable avenue for future research.

Note, that we here evaluate the estimator's ability to signal n-indistinguishability between *multiple* language samples. Thus, we do not compare against estimators of species richness of a *single* sample [15–17], which do not capture the similarity between multiple samples or systems. Furthermore, other approaches for quantifying the similarity between languages either consider the given languages to be complete system representations [20], or, if they consider incomplete languages, are limited to systems, that can be represented by directly-follows graphs, i.e., 2-gram representations [19]. As such, we do compare against these approaches here.

4.1 Experimental Setup

We implemented the proposed estimators as part of the public python-based *SpeciAL4PM* package [14] for the species-based analysis of event log samples and process models. All scripts, models and results can be accessed in our repository [13].

For the evaluation, we utilized seven models, shown in Fig. 4, comparing samples of a sample sizes 1000, 2000, and 3000 obtained from each of the models using token playout with uniform selection probabilities of fireable transitions at each step. For each model and sample size, we generated 100 sample logs. As the basis for comparing the languages we retrieved n-gram representations of each sample using n ranging from 1 to 9, as well as the distribution of trace variants. All values reported in the following sections are also summarized in Table 1.

4.2 Results

RQ1: Detecting n-Indistinguishability. We compared models I, II, and III, which are language equivalent despite looking differently. Model II and III are even fully language equivalent. Estimation results for these comparisons are shown in Fig. 5. Since the largest n (8, 9) exceed the longest traces in the models' language (7), no values were obtained for those.

We can see that the I-estimator properly identifies all three languages to be equal (all estimates are 1) and the D-estimator properly identifies the fully indistinguishable models II and III, with values of or close to 1. Given that models I and II and models I and III generate traces with different probabilities, the D-estimator shows that their languages are not equal in terms of their n-gram distribution, with estimates around 0.75. For all but 7-grams, due to the differences in the respective occurrence distributions, full indistinguishability cannot be assumed. For model II and model III, which have equal trace variant distributions, the D-estimator also correctly signifies full n-indistinguishability. Regarding **RQ1**, both estimators properly quantify (full) n-indistinguishability for the selected models, given that the given languages are indeed (fully) n-indistinguishable, and the sample sizes are large enough. We expect this trend to

Table 1. Mean estimates for assessing considered models at sample size 1000 regarding equivalence in identity (I) and distribution (D)

Models	Estimator	Language Representation									
		1-gram	2-gram	3-gram	4-gram	5-gram	6-gram	7-gram	8-gram	9-gram	tv
I vs II	I	1.0	1.0	1.0	1.0	1.0	1.0	1.0	-	-	1.0
	D	0.943	0.817	0.770	0.737	0.729	0.954	0.994	-	-	0.706
I vs III	I	1.0	1.0	1.0	1.0	1.0	1.0	1.0	-	-	1.0
	D	0.942	0.815	0.767	0.735	0.727	0.955	0.994	-	-	0.704
II vs III	I	1.0	1.0	1.0	1.0	1.0	1.0	1.0	-	-	1.0
	D	1.0	0.999	0.997	0.996	0.995	0.995	0.996	-	-	0.995
IV vs V	I	1.0	0.750	0.509	0.246	0.096	0.025	0.004	-	-	0.011
	D	0.901	0.883	0.878	0.850	0.746	0.161	0.049	-	-	0.983
IV vs VI	I	1.0	0.666	0.400	0.250	0.147	0.063	0.017	-	-	0.032
	D	0.961	0.736	0.539	0.321	0.268	0.157	0.085	-	-	0.361
V vs VI	I	1.0	0.727	0.317	0.121	0.041	0.011	0.002	0.000	0.000	0.005
	D	0.883	0.683	0.528	0.355	0.359	0.060	0.031	0.000	0.000	0.368
VI vs VI	I	1.0	1.0	1.0	0.992	0.927	0.213	0.007	0.000	0.000	0.000
	D	1.0	0.992	0.933	0.664	0.248	0.062	0.016	0.005	0.003	0.003

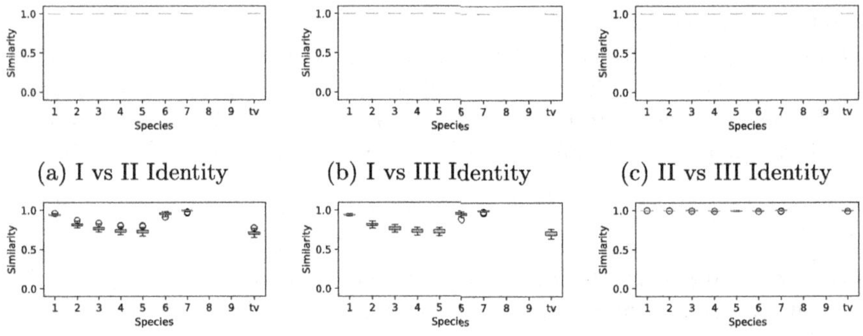

(a) I vs II Identity   (b) I vs III Identity   (c) II vs III Identity

(d) I vs II Distribution   (e) I vs III Distribution   (f) II vs III Distribution

Fig. 5. Comparison of Models I, II, and III for sample size 1000.

hold for more complex models as well, given that enough shared n-grams are sampled.

RQ2: Detecting n-Distinguishability. We compared models IV, V, and VI, which depict a similar process with minor differences. In model IV, activities G and H can each be executed once or skipped. In model V, they can each be executed an arbitrary number of times. In model VI, activity G can be skipped, but activity H must be executed exactly once. In addition, activity D, which must be executed exactly once in IV and V, can be repeated or skipped in VI.

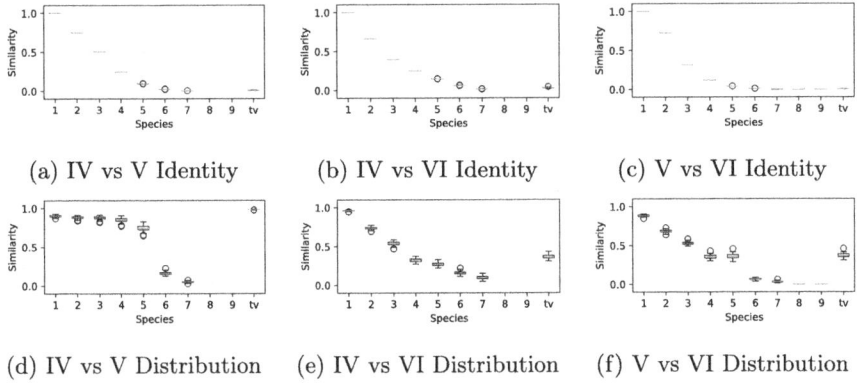

Fig. 6. Comparison of Models IV, V, and VI for sample size 1000.

Consequently, the languages of models IV, V, and VI share many n-grams, but should be found distinguishable. Figure 6 shows the results of our estimators.

Both estimators properly indicate that the models are not indistinguishable for all but 1-grams. The larger the n, the more nuanced the differences are because any difference for samples of a particular n will result in larger differences for larger n. Similarly, for none of the n-gram representations, full indistinguishability can be assumed. Nonetheless, when comparing models IV and V, we see that the likelihood of full indistinguishability (D-estimator) is larger then the likelihood of indistinguishability (I-estimator) for n. The most extreme case is the trace variant view where we have a mean likelihood of 0.011 for the I-estimator and of 0.989 for the D-estimator. Since the latter is sensitive to the most abundant elements in the sets, the impact of rarely occurring elements is small. To avoid misinterpretation, it is thus necessary to take the I-estimator into account when evaluating the D-estimator. Here, this allows us to conclude that the languages are highly different regarding their trace variants, but the trace variants they share make up the majority of the probability space. As indistinguishability is a necessary criterion for full indistinguishability, we can only assume the latter when the former is highly likely.

Answering **RQ2**, we conclude that both estimators do not signify indistinguishability when the samples stem from distinguishable languages. Nonetheless, to avoid misinterpretation when evaluating whether two samples are likely to be fully indistinguishable, both estimates should co-evaluated, given that the samples are only fully indistinguishability if they are also indistinguishable.

RQ3: Parameter Sensitivity. We investigated the impact of sample size and the considered language representation on the estimates using model VII, which represents the largest language of models. In Fig. 7, we show the estimates for comparing sample logs of different sizes of this model with each other.

We observe that, independently of the language representation, estimate accuracy increases for increasing sample sizes. For the I-estimates for the 6-

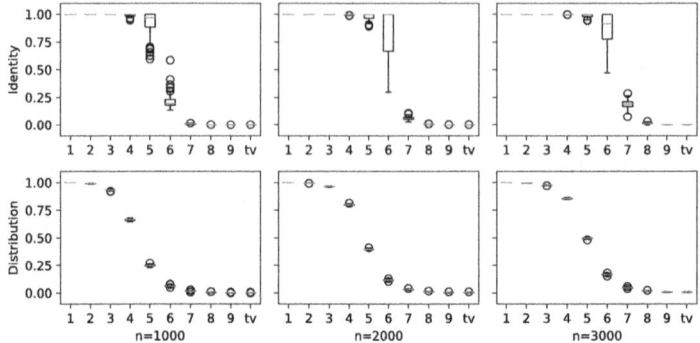

Fig. 7. Comparison of Model VII against itself for different sample sizes.

gram representation, a sample size of 1000 yields a mean likelihood of 0.213, but for a sample size of 3000, this value increases to 0.867. As estimators rely on the observation of shared elements in the samples, they can only be expected to be reasonable once a critical number of shared species have been observed. For more complex representations, this takes longer than for simpler ones. Similarly, as the D-estimator relies on occurrence frequencies, it takes significantly longer for estimates to reach the expected likelihood of 1.0. Considering 4-grams, we observe values of 0.664, 0.799, and 0.856, for sample sizes 1000, 2000, and 3000, respectively. Thus, even though the likelihood steadily increases, a sample size of 3000 does not yield enough evidence to conclude full 4-indistinguishability.

Answering **RQ3**, we observe that both estimators are sensitive to the sample size and the complexity of the considered representation. Estimation accuracy is expected to increase when increasing sample size, but the more complex the considered representation, the larger the sample size needs to be. Lastly, we note that the diversity of the models, and the coverage of the available samples directly impact the required sample size for obtaining accurate estimations [16]. While we here observe that the sample size impacts expected estimator accuracy, we leave the determination of optimal sample sizes for obtaining accurate estimated as future work. For the very complex model VII, a sample of 3000 does not allow for meaningful interpretation of estimates for more complex representations. In contrast, for models I, II and III, which are significantly less complex, a sample size of 1000 suffices.

5 Related Work

Most existing work on comparing different process representation focuses on model-log similarity, typically assessed by means of the four quality dimensions fitness, precision, generalization, and simplicity [23]. When considered together, they are assumed to balance all quality requirementsl [3]. Of these dimensions, simplicity is often omitted, as it focuses solely on the model [2,11,12,24] The most studied dimension is fitness, for which plenty of measures exist [12,29].

They mostly agree with one another, although they sometimes differ in sensitivity [11]. Existing measures for precision [12,29] were shown not to fulfill desired properties [25], questioning their validity. Finally, the few existing measures for generalization [12,29] were empirically shown not to work reliably [11].

Recognizing these divergences in existing measures and the problems that this might cause, requirements et al. [24] introduced 21 so-called conformance propositions, i.e., properties that the measures should fulfill. Based on these propositions, new measures for precision [20] and generalization [19] have been presented, which might lead to more consistent model-log similarity assessments in the future [27].

Regarding model-system similarity, most research so far has focused on empirically demonstrating the representativeness (or lack thereof) of event logs with regard to underlying systems. One notable exception is the contribution by Rogge-Solti et al. [22], who noted that neither log nor model could be good representations of the underlying process and hence proposed to assign a trust score to each of them, but it is unclear whether this allows to estimate the underlying process [10]. Besides that, research has shown that the above-mentioned measures are not good estimators for model-system similarity, as they are heavily influenced by noise and incompleteness [12] as well as unobserved behavior [21]. Building on observations by Buijs et al. [3], Janssenswillen & Depaire suggested to distinguish exploratory (model-log similarity) from confirmatory analysis (model-system similarity), but they did not provide an according assessment framework. As a first step, however, multiple authors have proposed approaches for assessing the representativeness of an event log for an (unknown) system [15–17], which provided the foundations for this paper.

6 Discussion and Conclusion

In this paper, we present a novel framework for assessing model-system similarity, i.e., a process model's ability to represent a true process only known through observations in an event log. This assessment presents two challenges: (1) comparing process and model independent of a formalism and (2) estimating the process from the log. We address them by (1) defining an n-gram representation for all three entities, on whose basis they can be compared and (2) estimating the n-gram representation of the process from that of the log by means of statistical estimators. Based on that, we define indistinguishability and full indistinguishability of two n-gram representations as equivalence in terms of the set of n-grams and their distribution. Our evaluation shows that the estimators assess both the presence and the absence of (full) indistinguishability.

At this stage, our work remains at the conceptual level. Although we see many applications, a number of aspects still have to be addressed before our framework can be used in practice. First, in our experiments, we only evaluated the estimators' ability to project the n-gram representation of the true process. They should also be evaluated with actual process models, e.g., comparing the outcome of different process discovery algorithms. For this context, we also need to adjust our framework such that it can assess whether one n-gram representation over- or underestimates another one. Third, our estimators rely on a critical mass of observed species, which means that the larger and the more unevenly distributed a language, the larger the amount of observations needs to be. One way to address this is to relate our framework to coverage measures [16] to assess whether a given log size is sufficient to properly estimate a true process.

Hence, the next step would be to move beyond the controlled experiments in this paper and apply our framework in a "real-world" setting, to test its abilities to handle actual event logs and the data quality issues that arise from them [24]. This will eventually allow us to improve our understanding of the representativeness of process mining results with respect to the true process.

References

1. Adriansyah, A., van Dongen, B., van der Aalst, W.: Conformance checking using cost-based fitness analysis. In: Enterprise Distributed Object Computing, pp. 55–64. IEEE (2011)
2. Augusto, A., et al.: Automated discovery of process models from event logs: review and benchmark. Trans. Knowl. Data Eng. **31**(4), 686–705 (2018)
3. Buijs, J., van Dongen, B., van der Aalst, W.: Quality dimensions in process discovery: the importance of fitness, precision, generalization and simplicity. Int. J. Coop. Inf. Syst. **23**(1) (2014)
4. Chao, A.: Nonparametric estimation of the number of classes in a population. Scand. J. Stat., 265–270 (1984)
5. Chao, A., Chazdon, R., Colwell, R., Shen, T.J.: Abundance-based similarity indices and their estimation when there are unseen species in samples. Biometrics **62**(2), 361–371 (2006)
6. Chao, A., Lin, C.W.: Nonparametric lower bounds for species richness and shared species richness under sampling without replacement. Biometrics **68**(3), 912–921 (2012)
7. Corradini, F., Muzi, C., Re, B., Rossi, L., Tiezzi, F.: Global vs. local semantics of BPMN 2.0 or-join. In: Theory Practice of Computer Science, pp. 321–336. Springer (2018)
8. Horn, H.S.: Measurement of overlap in comparative ecological studies. Am. Nat. **100**(914), 419–424 (1966)
9. Jaccard, P.: Étude comparative de la distribution florale dans une portion des alpes et des jura. Bull. Soc. Vaudoise Sci. Nat. **37**, 547–579 (1901)
10. Janssenswillen, G., Depaire, B.: Towards confirmatory process discovery: making assertions about the underlying system. Bus. Inf. Syst. Eng. **61**(6), 713–728 (2019)
11. Janssenswillen, G., Donders, N., Jouck, T., Depaire, B.: A comparative study of existing quality measures for process discovery. Inf. Syst. **71**, 1–15 (2017)
12. Janssenswillen, G., Jouck, T., Creemers, M., Depaire, B.: Measuring the quality of models with respect to the underlying system: an empirical study. In: Business Process Management, pp. 73–89 (2016)
13. Kabierski, M.: Evaluation scripts and result files for the paper comparing apples with oranges: an assessment framework for model-system similarity (2025). https://doi.org/10.5281/zenodo.15652385
14. Kabierski, M., Imenkamp, C., Koschmider, A., Weidlich, M.: SpeciAL4PM: species analysis of event logs for process mining. In: BPM Demos/Resources Forum (2024)

15. Kabierski, M., Richter, M., Weidlich, M.: Addressing the log representativeness problem using species discovery. In: International Conference on Process Mining, pp. 65–72. IEEE (2023)
16. Kabierski, M., Richter, M., Weidlich, M.: Quantifying and relating the completeness and diversity of process representations using species estimation. Inf. Syst. **130**, 102512 (2025)
17. Karunaratne, A., Polyvyanyy, A., Moffat, A.: The role of log representativeness in estimating generalization in process mining. In: International Conference on Process Mining, pp. 33–40. IEEE (2024)
18. Lohmann, N., Verbeek, E., Dijkman, R.M.: Petri net transformations for business processes - a survey. Trans. Petri Nets Other Model. Concurr. **2**, 46–63 (2009)
19. Polyvyanyy, A., Moffat, A., García-Bañuelos, L.: Bootstrapping generalization of process models discovered from event data. In: Advanced Information Systems Engineering (2022)
20. Polyvyanyy, A., Solti, A., Weidlich, M., Di Ciccio, C., Mendling, J.: Monotone precision and recall measures for comparing executions and specifications of dynamic systems. Trans. Softw. Eng. Methodol. **29**(3), 1–41 (2020)
21. Rehse, J.R., Fettke, P., Loos, P.: Process mining and the black swan: an empirical analysis of the influence of unobserved behavior on the quality of mined process models. In: BPM Workshops, pp. 256–268 (2018)
22. Rogge-Solti, A., Senderovich, A., Weidlich, M., Mendling, J., Gal, A.: In log and model we trust? A generalized conformance checking framework. In: Business Process Management, pp. 179–196. Springer (2016)
23. Rozinat, A., Alves de Medeiros, A.K., Günther, C., Weijters, T., van der Aalst, W.: The need for a process mining evaluation framework in research and practice. In: BPM Workshops, pp. 84–89 (2008)
24. Syring, A., Tax, N., van der Aalst, W.: Evaluating conformance measures in process mining using conformance propositions. Trans. Petri Nets Other Model. Concurr., 192–221 (2019)
25. Tax, N., Lu, X., Sidorova, N., Fahland, D., van der Aalst, W.: The imprecisions of precision measures in process mining. Inf. Proc. Lett. **135**, 1–8 (2018)
26. van der Aalst, W.: Process mining: a 360 degree overview. In: Process Mining Handbook, pp. 3–34. Springer, Cham (2022)
27. van der Werf, J., Polyvyanyy, A., van Wensveen, B., Brinkhuis, M., Reijers, H.: All that glitters is not gold: four maturity stages of process discovery algorithms. Inf. Syst. **114**, 102155 (2023)
28. van Glabbeek, R.: The linear time - branching time spectrum II: the semantics of sequential systems with silent moves. In: CONCUR, vol. 715, pp. 66–81. Springer (1993)
29. vanden Broucke, S., De Weerdt, J., Vanthienen, J., Baesens, B.: A comprehensive benchmarking framework (CoBeFra) for conformance analysis between procedural process models and event logs in ProM. In: Computational Intelligence and Data Mining, pp. 254–261. IEEE (2013)
30. Weidlich, M., Polyvyanyy, A., Mendling, J., Weske, M.: Causal behavioural profiles - efficient computation, applications, and evaluation. Fundam. Inf. **113**(3–4), 399–435 (2011)
31. Weidlich, M., van der Werf, J.M.E.M.: On profiles and footprints – relational semantics for petri nets. In: Petri Nets, pp. 148–167. Springer (2012)

Stochastic BPMN and Their Conformance

Aleksandar Kuzmanoski[1](✉), Jan Niklas van Detten[1,2],
and Sander J. J. Leemans[1]

[1] BPM (Informatik 9), RWTH Aachen University, Aachen, Germany
aleksandar.kuzmanoski@rwth-aachen.de,
{n.vandetten,s.leemans}@bpm.rwth-aachen.de
[2] Celonis, Munich, Germany

Abstract. Organizations need to continuously improve their processes to stay competitive and relevant. Process mining provides data-driven insights into business processes, typically conveyed to stakeholders using BPMN models. A common use case is to compare such BPMN models to real-life execution records of the process, called event logs. Since different process scenarios appear with different likelihoods, event logs are inherently stochastic. Adding this stochastic perspective to BPMN can yield more accurate conformance checking results. Furthermore, existing stochastic conformance checking techniques consider totally ordered traces, which does not align well with BPMN's explicit parallel gateways semantics. In this paper, we introduce the first stochastic conformance checking method for BPMN, a formalization of Stochastic BPMN (SBPMN), and a backward-compatible extension of the BPMN 2.0 specification. Our method accounts for partially ordered behavior and behavioral errors like livelocks and deadlocks in SBPMN models. We implemented our approach in ProM and evaluated it on three real-life event logs. Our experimental results demonstrate the computational feasibility of our method and highlight the importance of incorporating the stochastic perspective while preserving concurrency in conformance checking.

Keywords: Stochastic BPMN · Stochastic Conformance Checking · Earth Movers' Distance · Partial Orders · Stochastic Reachability Graph

1 Introduction

Process modeling and analysis enable organizations to streamline their operations and eliminate inefficiencies. Process mining leverages event logs to achieve this goal [28]. Real-life processes are collections of scenarios, including sequential and concurrent process steps (activities). Scenarios occur at different frequencies, making the process stochastic by nature, which is reflected in the event logs. This creates a demand for stochastic process mining methods.

Business Process Model and Notation (BPMN) models are widely used by the industry for communication and decision-making. To these ends, these models should be validated against reality using conformance checking techniques.

As a result, organizations need a stochastic extension of BPMN and a stochastic conformance checking technique that can be easily applied to these models. Moreover, such a technique should preserve the concurrent semantics of BPMN and not allow concurrent activities to influence the conformance. Similarly, which of a and b happened *first* should not influence the stochastic conformance, as both occurred simultaneously per the model. Finally, handling complex models and models with various behavioral issues, such as deadlocks and livelocks, is important as models may have been obtained through manual modeling or automated techniques that do not guarantee the absence of such issues [10].

Existing conformance checking methods for BPMN [12,22] do not consider the stochastic perspective. Although simulation models include stochastic information [5], they aim to represent timed behavior and are not used for conformance checking. In contrast, there is a rise in research on stochastic conformance checking for Generalized Stochastic Labeled Petri Nets (GSLPNs) [17,19,23]. However, these approaches require a total order, thus ignoring explicit concurrency. Recent work also considers the concurrency of events [14], though not on BPMN models and without resilience in the case of deadlocks and livelocks.

We propose a stochastic conformance method for BPMN that emphasizes the stochastic information of the process while preserving true concurrency. We focus on supporting BPMN models with behavioral issues like livelocks, and our only requirement is that the model is k-bounded, i.e., it has a finite state space.

We start by proposing an extension of the BPMN standard by introducing weights to the choices in the model. We call this extension Stochastic BPMN (SBPMN). Our method compares the stochastic languages from the event log and the SBPMN using Earth Movers' Stochastic Conformance (EMSC) [19]. While acquiring the stochastic language from the event log is trivial, we need to sample the SBPMN. To achieve this, we first introduce the Stochastic Partially Ordered Reachability Graph (SPORG) to represent the state space of the SBPMN. This reachability graph has probabilities associated with each transition, and to preserve concurrency, each transition represents a partially ordered execution of BPMN nodes that leads from one marking to the next. Second, we sample paths from the SPORG, resulting in SBPMN Partially Ordered Runs (PO runs). Each PO run is converted to Partially Ordered Trace (PO trace) with attached probability. We also use the SPORG to detect and avoid markings without an option to complete, e.g., livelocks. To compensate for model issues and sampling infinite languages, we provide confidence interval bounds instead of a single conformance value. We implemented our approach in ProM[1].

We show the computational feasibility of our method on three real-life event logs and 42 models acquired from various discovery techniques. We also showcase the importance of considering true concurrency in the stochastic perspective with a small artificial model and an event log.

In the remainder of this paper, we first explore related work in Sect. 2 and introduce the necessary background in Sect. 3. In Sect. 4, we introduce our SBPMN extension. We present our BPMN Partially Ordered Earth Movers'

---

[1] https://github.com/promworkbench/POEMSConformanceCheckingForBPMN.

Stochastic Conformance Checking (POEMS) approach in Sect. 5, and show evaluation results in Sect. 6. Finally, we conclude our work in Sect. 7.

2 Related Work

Conformance checking is an essential task in process mining and has been explored extensively in the past two decades [25–27]. Conformance measures have also been proposed for BPMN [12,22]. Such methods are usually based on token replay, alignments, or rule checking [6]. However, these approaches lack the stochastic perspective in the models, which is inherently present in the event logs. The methods that account for the stochastic perspective mainly refer to Stochastic Petri nets (SPNs) [14,17,19,23].

The method in [19] extracts stochastic languages from both the model and the event log and performs Earth Movers' Distance (EMD) to calculate the conformance measure. In [17], the authors build two Stochastic Deterministic Finite Automatons (SDFAs) from the event log and the SPN. They calculate the entropy of each SDFA and then produce a ratio between them to estimate precision and recall measures. The authors in [14] extend [19] with partially ordered traces to represent the event log to deal with uncertainty and the GSLPN to represent concurrent behavior. This approach is most closely related to our approach, but there are two important differences: (1) we work with SBPMNs instead of GSLPNs which introduces different semantics, (2) we support a larger subset of models, specifically models that are not safe or sound.

While an alternative to our approach for sound and safe models would be to convert SBPMN to GSLPN and use [14], to the best of our knowledge, such a conversion procedure is not defined. Moreover, GSLPNs assigns probability to every transition, including those executed in parallel, making the conversion non-trivial and potentially prone to errors. Finally, the obsoleteness of such probabilities makes GSLPNs and their conformance results harder to interpret, although only a user study could support this argument.

The work in [16] proposes several analytical solutions to problems defined for GSLPNs and relates them to existing stochastic conformance techniques. We are particularly interested in the outcome probability, which calculates the probability that a bounded GSLPN evolves from a given marking to one of the final markings. This is done on the stochastic reachability graph of a GSLPN. The standard reachability graph for BPMN is formalized in [13], while the Concurrent Automatons (CAs) introduced in [8] have partially ordered transition semantics.

Our SBPMN extends the semantics of BPMN defined in [13]. Models closest in terms of semantics to our SBPMN are simulation models [5,9].

The distance between PO traces can be computed using graph isomorphisms [31] or as the minimal distance between a pair of sequences in their total extensions. Moreover, the authors in [11] define two PO traces as equivalent when their corresponding Hasse diagrams are isomorphic. In our approach, we opt for the method presented in [14]. The authors use the normalized Levenshtein distance [30], making the problem similar to computing optimal alignments [27].

3 Preliminaries

In this section, we introduce the notations we use throughout this paper. We denote *sets* and *multisets* with uppercase letters $X = \{a, b, c\}$ and $M = [a^2, b, c]$, *tuples* and *sequences* with lowercase letters $t = (a, b, c)$ and $\sigma = \langle a, b, c \rangle$. We access tuple elements by name; for example, $t.a$ is the element a in the tuple t. A *partial function* is $f : X \nrightarrow Y$ with domain $dom(f) \subseteq X$ and range $rng(f) = \{f(x) | x \in dom(f)\} \subseteq Y$. $f : X \to Y$ is a *total function*, i.e., $dom(f) = X$.

Given a finite set of elements X, a *multiset* is $M : X \to \mathbb{N}$ where $\forall x \in X : M(x)$ denotes the number of times x appears in the multiset M. For two multiset M and M' over a set X we define the following operators: (1) sum $M \uplus M'$ where $\forall x \in X : (M \uplus M')(x) = M(x) + M'(x)$; (2) union $M \cup M'$ where $\forall x \in X : (M \cup M')(x) = \max(M(x), M'(x))$; (3) difference $M \setminus M'$ such that $\forall x \in X : (M \setminus M')(x) = \max(0, M(x) - M'(x))$; (4) subset relation $M \subseteq M'$ iff $\forall x \in X : M(x) \leq M'(x)$; (5) size of M denoted as $|M| = \sum_{x \in X} M(x)$; (6) number occurrences of a multisetset M in M' denoted as $M'(M) = \min_{x \in M} \lfloor \frac{M'(x)}{M(x)} \rfloor$.

The *powerset* of X is given by $\mathbb{P}(X)$ and the *set of all multiset* over X is $\mathbb{M}(X)$. $f(X)$ applies the function f to every element in X, i.e., $f(X) = \{f(x) \mid x \in X\}$. $\mathcal{A}$ is the universe of activities. $\langle a_1, \ldots, a_n \rangle \in \mathcal{T} = \mathcal{A}^*$ is a *trace*. Let $A \subseteq \mathcal{A}$ be a set of activities. An *event log* $EL \in \mathbb{M}(A^*)$ is a multiset of traces.

Partial Orders. A *strict partial order* is a homogeneous relation $\prec$ on a set X that is irreflexive ($a \not\prec a$), asymetric ($a \prec b \Rightarrow b \not\prec a$), and transitive ($a \prec b \wedge b \prec c \Rightarrow a \prec c$). $\prec^*$ is the transitive closure of $\prec$. A *Labeled Partial Order (LPO)* $\rho = (E, \prec, \lambda)$ over some set Σ consists of a set E of events, a strict partial order $\prec$ on E, and a labelling function $\lambda : E \to \Sigma$. $LPO(\Sigma)$ denotes the set of all LPOs over Σ. A *PO trace* is a LPO $\rho_t = (E, \prec, \lambda) \in \mathcal{P}$ where $A \subseteq \mathcal{A}$ and $\lambda : E \to A$ is a labeling function. $\mathcal{P}$ is the universe of all PO traces.

To be able to compare PO traces with traces, in Definition 1, we extract all traces from a given PO trace. For example, the PO trace $\rho_{t1} = (\{a, b, c\}, \{(a, b)\}, ID)$ maps to $\langle a, b, c \rangle$, $\langle a, c, b \rangle$, and $\langle c, a, b \rangle$.

Definition 1 (Traces of a PO trace). *Let $\rho_t = (E, \prec, \lambda) \in \mathcal{P}$ be a PO trace. Then $\sigma_t = \langle a_1, \ldots, a_{|\sigma_t|} \rangle \in \mathcal{T}$ is a trace of ρ_t iff there is a bijection $f : \{1, \ldots, |\sigma_t|\} \to E$ such that $\forall i \in \{1, \ldots, |\sigma_t|\}$ it holds that $\lambda(f(i)) = a_i$, and $\forall i, j \in 1, \ldots, |\sigma_t|$ such that $f(i) \prec f(j) \Rightarrow i < j$. Let $\mathcal{L}(\rho_t)$ denote all traces of ρ_t.*

Next, in Definition 2, we define the subset operator over PO traces, which defines omitting of elements while preserving the order. For example, $(\{b, c\}, \emptyset, ID)$ is subset of ρ_{t1}, but $(\{b, c\}, \{(b, c)\}, ID)$ is not.

Definition 2 (Partial Order Subset Operator). *Given $\rho_1 = (E_1, \prec_1, \lambda_1)$ and $\rho_2 = (E_2, \prec_2, \lambda_2)$ we say $\rho_1 \subseteq \rho_2$ iff $[\lambda_1(e_1) \mid e_1 \in E_1] \subseteq [\lambda_2(e_2) \mid e_2 \in E_2]$ and $[(\lambda_1(e_1), \lambda_1(e'_1)) \mid (e_1, e'_1) \in \prec_1] \subseteq [(\lambda_2(e_2), \lambda_2(e'_2)) \mid (e_2, e'_2) \in \prec_2]$.*

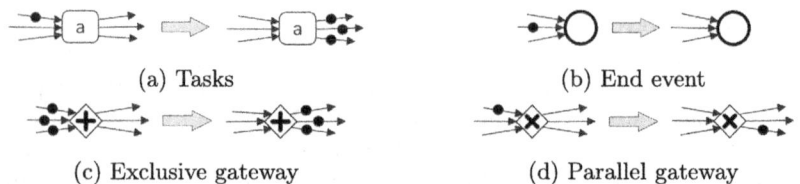

Fig. 1. BPMN elements enablement and firing semantics.

Stochastic Languages. A *stochastic language* $L\colon \mathcal{U} \to [0,1]$ assigns a probability to each element in the universe, such that $\sum_{x\in\mathcal{U}} L(x) = 1$. We write $\tilde{L} = \{x \in \mathcal{U} \mid L(x) > 0\}$. A *truncated* stochastic language L' of L assigns non-zero probability to finite number of elements such that $\forall x \in \tilde{L'} : L'(x) = L(x)$ and $\sum_{x\in\tilde{L'}} \leq 1$. We refer to the truncated probability mass as $L'(\bot) = 1 - \sum_{x\in\tilde{L'}} L'(x)$. A *Stochastic Trace Language (STL)* is a finite $L_{\mathcal{T}}\colon \mathcal{T} \to [0,1]$. A *Stochastic Partial Order Trace Language (SPOTL)* is possibly infinite $L_{\mathcal{P}}\colon \mathcal{P} \to [0,1]$.

BPMN Semantics. The BPMN formalisms here are identical to those in [13].

Definition 3 (BPMN). *A BPMN model is represented by the tuple* $\mathcal{M} = (N, e_{start}, A, G_{XOR}, G_{AND}, E_{end}, SF, \lambda)$ *where:*

- *N is a set of flow nodes;*
- *$e_{start} \in N$ is the start event, such that $\{(n, e_{start}) \mid n \in N\} = \emptyset$;*
- *$A \subseteq N$ is a set of tasks;*
- *$G_{XOR} \subseteq N$, $G_{AND} \subseteq N$ are sets of exclusive and parallel gateways;*
- *$E_{end} \subseteq N$ is a set of end events, where $\forall e_{end} \in E_{end}$: $\{(e_{end}, n) \mid n \in N\} = \emptyset$;*
- *$\{e_{start}\} \cup A \cup G_{XOR} \cup G_{AND} \cup E_{end} = N$;*
- *$SF \subseteq N \times N$ is a set of sequence flows;*
- *$\lambda\colon N \twoheadrightarrow \mathcal{A}$ is a labeling function, where $\mathcal{A}$ is the universe of activity labels;*

The *preset* $\bullet n := \{sf \in SF \mid sf = (n', n) \land n' \in N\}$ and the *postset* $n\bullet := \{sf \in SF \mid sf = (n, n') \land n' \in N\}$ are the incoming and outgoing sequence flows of a node $n \in N$. A *BPMN marking* $M \in \mathbb{M}(SF)$ is a multiset over the sequence flows. The *initial marking* M_{init} is a BPMN marking, such that $\forall sf \in SF$ $M_{init}(sf) = 1$ if $sf \in e_{start}\bullet$, and $M_{init}(sf) = 0$ otherwise.

BPMN node enablement and firing. (1) A start event $e_{start} \in E_{start}$ is never enabled. (2) A task $a \in A$ is enabled in marking M iff $\exists sf \in \bullet a\colon [sf^1] \subseteq M$. When fired produces new marking $M' = (M\backslash[sf^1]) \uplus a\bullet$ (Fig. 1a). (3) An exclusive gate $g_{XOR} \in G_{XOR}$ is enabled in marking M iff $\exists sf \in \bullet g_{XOR}\colon [sf^1] \subseteq M$. When fired produces new marking $M' = (M\backslash[sf^1]) \uplus [sf'^1]$ such that $sf' \in g_{XOR}\bullet$ (Fig. 1d). (4) A parallel gateway $g_{AND} \in G_{AND}$ is enabled in marking M iff $\bullet g_{AND} \subseteq M$. When fired produces new marking $M' = (M\backslash \bullet g_{AND}) \uplus g_{AND}\bullet$ (Fig. 1c). (5) An end event $e_{end} \in E_{end}$ is enabled in M iff $\exists sf \in \bullet e_{end}\colon [sf^1] \subseteq M$. Firing produces new marking $M' = M\backslash[sf^1]$ (Fig. 1b).

When node $n \in N$ is *enabled* in marking M, we write $M[n\rangle$. When it *fires* and produces marking M', we write $M[n\rangle M'$. We write $M[\sigma\rangle M'$ for some sequence of nodes $\sigma = \langle n_1, \ldots, n_k\rangle \in N^*$ iff there are markings $M_1, \ldots, M_k$, such that $M_0 = M$, $M_k = M'$, and for $0 \leq i < k$ holds $M_i[n_{i+1}\rangle M_{i+1}$.

We denote the set of *consume markings* of a node n, or all possible markings consumed by node n as $cm: N \to \mathbb{P}(\mathbb{M}(SF))$ such that $cm(n) = \{M_c \mid M_c[n\rangle \land \nexists M'_c: M'_c \subset M_c \land M'_c[n\rangle\}$. For example, each exclusive gateway is enabled by each incoming token. The set of *produce markings* of a node n is denoted as $pm: N \to \mathbb{P}(\mathbb{M}(SF))$ such that $pm(n) = \{M_p \mid M[n\rangle M' \land M_p = M'\backslash M\}$. Exclusive gateways produce marking for each outgoing sequence flow.

We say that a marking M' is *reachable* from M iff there is a sequence $\sigma \in N^*$, such that $M[\sigma\rangle M'$. $\mathcal{R}_\mathcal{M}(M)$ is the set of all markings reachable in $\mathcal{M}$ from the marking M. A marking M_{final} is *final* iff it is the empty marking $|M_{final}| = 0$. M_{dead} is a deadlock marking iff $M_{dead} \in \mathcal{R}_\mathcal{M}(M_{init}) \land M_{dead} \neq M_{final} \land \nexists n \in N: M_{dead}[n\rangle$. $M_{live} \in \mathcal{R}_\mathcal{M}(M_{init})$ is a livelock marking iff $\forall M \in \mathcal{R}_\mathcal{M}(M_{live}), \exists n \in N: M[n\rangle$.

We use these semantics later to define the extraction of a stochastic language from SBPMN. In some situations, we also impose restrictions on the model, such as k-boundedness. A BPMN model is *k-bounded* iff $\forall k \in \mathbb{N} \land k < \infty \land \forall M \in \mathcal{R}_\mathcal{M}(M_{init}) \land \forall sf \in SF: M(sf) \leq k$ or there is no sequence flow in any marking with more than k tokens. A BPMN model is *safe* iff it is k-bounded with $k = 1$. To represent the unfolding of the BPMN execution, we define a synchronization model in Definition 4 and use it to define PO runs in Definition 5.

Definition 4 (BPMN Synchronization Model). *A BPMN synchronization (parallel-sequential) model is a BPMN model* $\mathcal{M}_{sync} = N, e_{start}, A, G_{XOR}, G_{AND}, E_{end}, SF, \lambda$ *where* $\forall g_{XOR} \in G_{XOR} \Rightarrow |g_{XOR}\bullet| = 1$ *and the transitive closure of SF is irreflexive (i.e., no loops).*

Definition 5 (BPMN PO run). *A BPMN PO run* $r = (\mathcal{M}_{sync}, \eta)$ *of a BPMN model* $\mathcal{M} = (N, e_{start}, A, G_{XOR}, G_{AND}, E_{end}, SF, \lambda)$ *is composed of a BPMN synchronization model* $\mathcal{M}_{sync} = (N', e_{start}, A', G'_{XOR}, G_{AND}', E_{end}', SF', \lambda')$ *and a mapping* $\eta : N' \to N$ *such that* $\eta(A') \subseteq A$, $\eta(G'_{XOR}) \subseteq G_{XOR}$, $\eta(G_{AND}') \subseteq G_{AND}$, *and* $\eta(E_{end}') \subseteq E_{end}$ *(1)* $\forall n' \in N'$: η *induces a bijection from* $\bullet n'$ *to* $\bullet \eta(n')$. *(2)* $\forall n' \in N' \land n' \notin G'_{XOR}$: η *induces a bijection from* $n'\bullet$ *to* $\eta(n')\bullet$ *(3)* $\forall g_{XOR}' \in G'_{XOR}$ η *induces an injection from* $g_{XOR}'\bullet$ *to* $\eta(g_{XOR}')\bullet$ *(4)* $\forall n' \in N': \lambda'(n') = \lambda(\eta(n'))$. $crs(\mathcal{M})$ *is the set of all complete PO runs of* $\mathcal{M}$ *which we also call scenarios.*

A BPMN PO run defines a partial order $\rho_r = (E, \prec, \lambda)$, such that $\lambda: E \to N$ and $\prec = \{ (e, e') \mid (\lambda(e), \lambda(e')) \in SF\}$. Finally, in Definition 6, we transform PO runs to PO traces, by keeping only the elements representing tasks.

Definition 6 (PO run to PO trace). *Given a PO run* $\rho_r = (E, \prec, \lambda')$ *of a BPMN model* $\mathcal{M} = (N, e_{start}, A, G_{XOR}, G_{AND}, E_{end}, SF, \lambda)$, *we can convert it into a PO trace* $\rho_t = (E', \prec', \lambda'')$ *such that: (1)* $E' = \{e \in E \mid \lambda(\lambda'(e)) \neq \bot\}$ *(2)* $\prec' = \{(e, e') \mid e \in E' \land e' \in E' \land (e, e') \in \prec^*\}$ *(3)* $\forall e \in E': \lambda''(e) = \lambda(\lambda'(e))$.

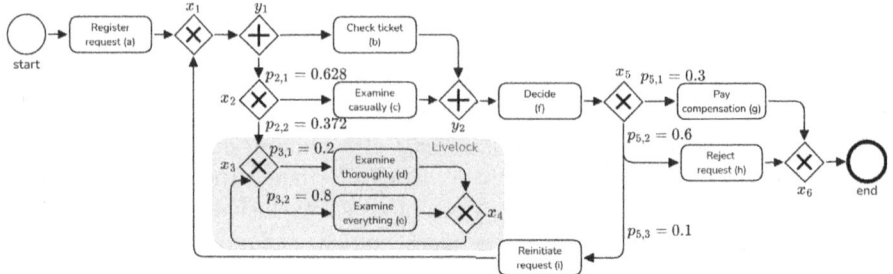

Fig. 2. SBPMN example: Handling of Compensation Requests, inspired by [28, p. 69]. Introducing stochastic weights p_* for exiting the exclusive gateways.

EMSC. We use the *normalized Levenshtein distance* [30] for the trace-trace distance measure δ. Given a trace σ_t and a partially ordered trace ρ_t, we calculate the distance between them by $\Delta(\sigma_t, \rho_t) = \min_{\sigma_{t'} \in \mathcal{L}(\rho_t)} \delta(\sigma_t, \sigma_{t'})$. Note that we use the optimizations suggested in [14].

Definition 7 (Earth Movers' Stochastic Conformance). *Let $L_\mathcal{T}$ be a STL and $L_\mathcal{P}$ be a SPOTL, and let $\Delta \colon \mathcal{T} \times \mathcal{P} \to [0,1]$ be a trace - PO trace distance function. The reallocation $R \colon \tilde{L}_\mathcal{T} \times \tilde{L}_\mathcal{P} \to [0,1]$ is a mapping such that: $cost_\Delta(R, L_\mathcal{T}, L_\mathcal{P}) = \sum_{\sigma_t \in \tilde{L}_\mathcal{T}, \rho_t \in \tilde{L}_\mathcal{P}} R(\sigma_t, \rho_t) \Delta(\sigma_t, \rho_t)$ is minimized under the constraint that $\forall \sigma_t \in \tilde{L}_\mathcal{T} \colon L_\mathcal{T}(\sigma_t) = \sum_{\rho_t \in \tilde{L}_\mathcal{P}} R(\sigma_t, \rho_t)$ and $\forall \rho_t \in \tilde{L}_\mathcal{P} \colon L_\mathcal{P}(\rho_t) = \sum_{\sigma_t \in \tilde{L}_\mathcal{T}} R(\sigma_t, \rho_t)$. Given such mapping R, the* Earth Movers' Stochastic Conformance' *is $emsc_\Delta(L_\mathcal{T}, L_\mathcal{P}) = 1 - cost_\Delta(R, L_\mathcal{T}, L_\mathcal{P})$.*

4 Stochastic BPMN

In this section, we formalize our stochastic extension of BPMN. Additionally, we provide a backward-compatible extension of the BPMN 2.0 standard. We consider processes to be collections of scenarios resulting from the choices taken in a process (Definition 5). In our stochastic extension of BPMN, we introduce weights that coincide with the probabilities of those choices. The probability of a scenario is equal to the product of the choices representing that scenario.

Definition 8 (SBPMN). *A Stochastic BPMN (SBPMN) model is a tuple $\mathcal{M}_s = (N, e_{start}, A, G_{XOR}, G_{AND}, E_{end}, SF, \lambda, w)$ where $(N, e_{start}, A, G_{XOR}, G_{AND}, E_{end}, SF, \lambda)$ is a BPMN model (Definition 3), and $w \colon C \to \mathbb{R}+$ is a weight function where $C := \{sf \in SF \mid sf \in n\bullet \land n \in G_{XOR}\}$.*

We formalize our SBPMN extension in Definition 8. Weights are assigned only to sequence flows exiting exclusive gateways. Everything else is sequential or parallel behavior, and no weights are assigned. Weights are normalized to probabilities by division with the gateway weight sum.

In our SBPMN setting, probabilities can only be assigned to scenarios represented as PO runs (Definition 5), not individual execution sequences. Each

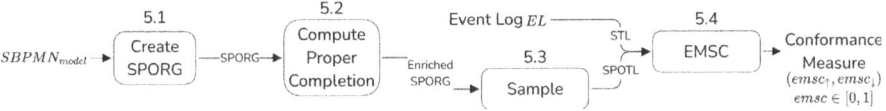

Fig. 3. Our method pipeline of four consecutive steps that computes a conformance interval between an SBPMN model and an event log.

scenario represents factorially many sequences (Definition 1) and covers their total probability. An example SBPMN model is given on Fig. 2 describing the handling of a compensation requests process. In this process, there are scenarios in which requests are examined casually, and others in which they are examined thoroughly. The probability of these scenarios, as shown by the weights $p_{2,1}$ and $p_{2,2}$ for gateway x_2 are 0.628 and 0.372 respectively. The scenario where the requests are examined casually, and the compensation is paid out has a probability of $p_{2,1} \cdot p_{5,1} = 0.628 \cdot 0.3 = 0.1884$ and is represented by the partial order $(\{a, b, c, f, g\}, \{(a, b), (a, c), (b, f), (c, f), (f, g)\})$.

We provide a backward-compatible schema extension[2] of the BPMN 2.0 standard. We use the *BPMN extension elements* to enable weight distribution over all outgoing sequence flows for any supported gateway. This schema can be freely used to distribute SBPMN models easily. Due to its backward compatibility, it can be imported into any commercially available tool.

5 BPMN Stochastic Conformance Checking

In this section, we introduce our BPMN Partially Ordered Earth Movers' Stochastic Conformance Checking (POEMS) approach. The key steps of our pipeline are illustrated in Fig. 3. First, in Sect. 5.1, we detail the construction of the SPORG. Then, in Sect. 5.2, we enrich the SPORG states with probability to reach the final state, which serves as input for sampling the SPOTL in Sect. 5.3. Finally, in Sect. 5.4, we leverage the SPOTL and the STL of the event log to apply the EMSC method for conformance calculation.

5.1 Building SPORG

For our SBPMN, the standard reachability graph [13] is not applicable. Instead, we introduce the SPORG to account for the probability distribution across partially ordered scenarios. SPORG represents a potentially infinite state-space, with one initial and multiple final states. We transition from one state to the next via a partial order over a given set Σ instead of single labels. Transitions are assigned a probability, such that all outgoing transitions from a given state sum up to 1. The deadlock states with no outgoing transition are an exception.

---

[2] https://github.com/promworkbench/StochasticBPMN/blob/main/resources/inputModel/StochasticBPMN/XMLSchema/BPMN_StochasticExtension.xsd.

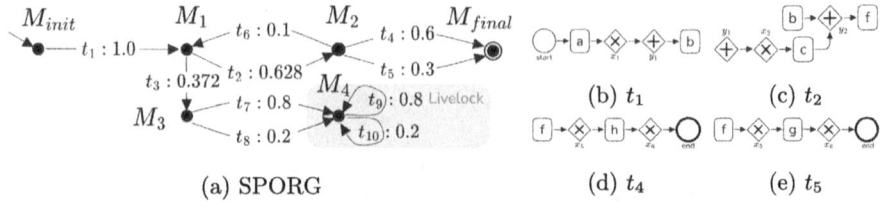

Fig. 4. SPORG of the SBPMN model shown in Fig. 2. State space of reachable markings where each transition has a probability and a PO run (subfigures).

Definition 9 (SPORG). *A Stochastic Partially Ordered Reachability Graph (SPORG) is a tuple $SPORG = (\Sigma, S, s_{in}, S_f, T, \varrho, p)$ where:*

- *Σ is a finite set of labels;*
- *S is possibly infinite set of states;*
- *$s_{in} \in S$ is the initial state;*
- *$S_f \subseteq S$ is the set of final states;*
- *$T \subseteq LPO(\Sigma)$ is possibly infinite set of partial orders over Σ;*
- *$\varrho \colon (S \times T) \nrightarrow S$ is a partial transition function, such that, given a state $s \in S$ and a partial order $\rho \in T$, is either undefined or returns a state $s' = \varrho(s, \rho)$;*
- *$p \colon \varrho \to [0,1]$ is a transition probability mapping such that for every non-deadlock state $s \in S \wedge \exists \rho \in T : \varrho(s, \rho)$ is defined, $\sum_{t=(s,\rho,s') \in \varrho} p(t) = 1$;*

SBPMN to SPORG. We represent the SBPMN markings and their reachability relations with an SPORG. Transitions between states are the node executions that lead from one marking into other markings. The initial state of the SPORG is the initial marking of the SBPMN, and every other state is a reachable marking. However, not every reachable marking is represented by a state in SPORG. The markings represented are the deadlock markings, including the final marking and markings with only exclusive gates enabled. Transitions have probabilities assigned to them, proportional to the weights of the choices in the SBPMN model. Moreover, transitions represent partially ordered executions.

To build the SPORG on Fig. 4a from the SBPMN given on Fig. 2, we start with the initial marking M_{init} as an initial state. We execute all enabled non-choice nodes recursively until only exclusive gateways are enabled. This results in a Maximal Parallel/Sequential Run (MPS run) as defined in Definition 10, which has a probability of 1. This MPS run is given on Fig. 4b and leads to state (marking) $M_1 = [(y_1, x_2), (b, y_2)]$. Note that the marking $[(x_1, y_1)]$ is not represented with a state because the parallel gateway y_1 is enabled. Then, we take a cartesian product of all enabled choices, calculate their probability, and collect the markings resulting from these choices (Definition 11). We repeat the procedure for each resulting marking. In addition to the MPS runs (Fig. 4c to 4e), transitions are assigned probability based on the choices. We formally give this procedure in Definition 12 and depict it in Fig. 5.

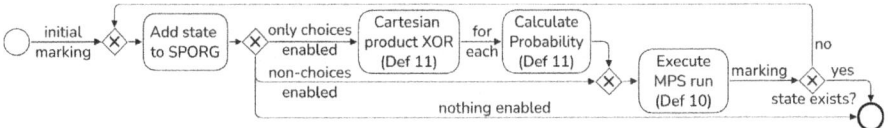

Fig. 5. SBPMN to SPORG procedure as defined in Definition 12.

Definition 10 (MPS run). *Given a SBPMN $\mathcal{M}_s = (N, e_{start}, A, G_{XOR}, G_{AND}, E_{end}, SF, \lambda, w)$ and a marking M, a PO run ρ_{pr} is a maximal choice-free run enabled in M iff $M[\rho_{pr}\rangle M'$ and $\neg \exists n \in N \setminus \{g_{XOR} \in G_{XOR} \mid |g_{XOR}\bullet| > 1\}: M'[n\rangle$. With $mpsr(M)$ we denote all maximal choice-free runs enabled in M.*

Definition 11 (Stochastic Marking Produce Options). *Given a $\mathcal{M}_s = (N, e_{start}, A, G_{XOR}, G_{AND}, E_{end}, SF, \lambda, w)$ and a marking M, we get the enabled nodes $N' = \{n \in N \mid M[n\rangle\}$ and define the stochastic produce options of a node $n \in N' \land n \in G_{XOR}$ as $spm(n) := \{(M', p) \mid M' \in pm(n) \land sf \in M' \land p = \frac{w(sf)}{\sum_{sf' \in n\bullet} w(sf')}\}$ and of the marking as $spm(M) = \{(pm, p') \mid C \in \prod_{\forall n \in N', \forall M_c \in cm(n)} \{(M_c, n, M_p, p) \mid (M_p, p) \in spm(n)\}, pm = M \setminus (\uplus_{\forall fo \in C} fo.M_c) \uplus (\uplus_{\forall fo \in C} fo.M_p), p' = \prod_{\forall fo \in C} fo.p\}$.*

Definition 12 (SBPMN to SPORG). *Given a $\mathcal{M}_s = (N, e_{start}, A, G_{XOR}, G_{AND}, E_{end}, SF, \lambda, w)$ and the initial marking M_{init} we build the SPORG inductively. (1) $\Sigma = N$, $S = \{M_{init}\}$, $s_{in} = M_{init}$, $S_f = \{M_{final}\}$, and $next(M_{init}) = \{(M_{init}, 1)\}$, where next: $\mathcal{R}_\mathcal{M}(M_{init}) \rightarrowtail \mathbb{P}(\mathcal{R}_\mathcal{M}(M_{init}) \times [0,1])$ returns the possible successor markings and their probabilities after resolving a choice in marking M. (2) For every $s \in S$ and $(s_n, p_n) \in next(s)$ let $\rho_{pr} \in mpsr(s_n)$ be a MPS run enabled in s_n, such that $s_n[\rho_{pr}\rangle s'$ Then (i) $s' \in S$ and $next(s') = spm(s')$; (ii) $\forall t \in \{\rho'_{pr} \mid s[\rho'_{pr}\rangle s' \land \rho_{pr} \subseteq \rho'_{pr}\}, t \in T$; (iii) $\varrho(s,t) = s'$; (iv) $p(s,t,s') = p_n$.*

Avoiding Choice Free Livelocks. Our approach depends on collapsing MPS runs into single transitions. Our procedure can get stuck in a livelock between nodes that don't have any choice (exclusive gateway) in between. Meaning the procedure will never finish. To avoid such a case, we prevent revisiting the same node twice $\nexists e \in E: e' \in E \land e \prec^* e' \land \lambda(e') = \lambda e$. This means we are in a livelock that can not be avoided, and we can safely prune this path and stop exploring in that direction. Note that this procedure ensures we can build the SPORG and does not detect all livelocks present in the model. Some livelocks have choices in between, so they will be represented as loops, and dealt with later.

Handling Infinite State-Space. If the BPMN model is not k-bounded, sequence flows can hold infinitely many tokens, and both the state space and the length of the transition PO run can be infinite. To address this issue, we use a rather myopic pruning strategy to stop the exploration once we encounter a superset marking of one previously visited markings $\exists M'' \in SPORG.S$: $M'' \subset M'$. While this rule ensures that we detect any place in the model where we can generate tokens indefinitely, there are cases where those tokens can be consumed, and the model can properly finish. This means we are pruning some valid paths which can increase the uncertainty in our measure. Dealing with such paths is out of the scope of this paper. Therefore we assume k-bounded models on input.

5.2 Compute Proper Completion

We calculate the probability that an SBPMN model will complete properly in a final marking M_{final}. The probability of not reaching a final marking is equal to the probability the model ends in a marking that has no option to complete. Therefore, we first detect all reachable markings from the initial marking that don't have an option to complete. Then, we calculate the probability of proper completion for all remaining markings.

Detecting Markings with No-Option to Complete. Markings with no-option to complete include markings that are deadlocks, livelocks, or markings that always lead to livelocks and/or deadlocks (Definition 13). Considering the inductive property of Definition 12, all states represent markings reachable from the initial marking, including the final marking. Therefore, the ancestors of the final marking in an SPORG represent all markings with an option to complete. The remaining states not on the ancestors list are the markings with no option to complete. In the example on Fig. 4a markings M_3 and M_4 are not ancestors of M_{final}. Therefore, they are markings with no option to complete.

Definition 13 (BPMN Marking with no Option to Complete). *Let* $\mathcal{M} = (N, e_{start}, A, G_{XOR}, G_{AND}, E_{end}, SF, \lambda)$ *be a BPMN model with initial marking* M_{init} *and a final marking* M_{final}. $M_{noc} \in \mathcal{R}_{\mathcal{M}}(M_{init})$ *is marking with no option to complete iff* $\forall M \in \mathcal{R}_{\mathcal{M}}(M_{noc})$: $M \neq M_{final}$. *With* $\mathbb{M}_{noc}$ *we denote the set of all markings without option to complete.*

Marking Probability to Complete. The final marking M_{final} has a completion probability of 1, and the markings with no option to complete $M_{noc} \in \mathbb{M}_{noc}$ have a probability of 0. We make a system of linear equations for the remaining states in the SPORG such that the probability of each state is represented as a linear function of the probabilities of the target states that this state can transition to. We transform these equations into matrix form and solve $x = A^{-1}b$ where x is the vector of probabilities of proper completion for each state. This expression always has a solution because A is never indefinite. This follows from the fact that all included states have an option to complete, meaning they always have outgoing edges with non-zero probability (no zero rows or columns). If we take our example in Fig. 4a, $M_{final} = 1$ and $M_3 = M_4 = 0$. For the remaining states, we build the following system of equations:

$$\begin{cases} M_0 = t_1 \cdot M_1, \\ M_1 = t_2 \cdot M_2 + t_3 \cdot M_3, \\ M_2 = (t_4 + t_5) \cdot M_{final} + t_6 \cdot M_1 \end{cases} = \begin{cases} M_0 - \quad M_1 \quad\quad\quad = 0, \\ \quad\quad M_1 - 0.628M_2 = 0, \\ \quad -0.1M_1 + \quad M_2 = 0.9 \end{cases}$$

writing this system in a matrix form will result in the following:

$$A = \begin{bmatrix} 1 & -1 & 0 \\ 0 & 1 & -0.628 \\ 0 & -0.1 & 1 \end{bmatrix}, b = \begin{bmatrix} 0 \\ 0 \\ 0.9 \end{bmatrix}, x = A^{-1}b = \begin{bmatrix} 0.60307 \\ 0.60307 \\ 0.96031 \end{bmatrix}$$

5.3 Sample SPOTL from SPORG

To sample an SPOTL from an SPORG, we define a path through an SPORG, which relates to a BPMN PO run. We sample PO runs and convert them to PO traces to acquire the final stochastic language. A path through SPORG, as given in Definition 16, starts in the initial state s_{in} and ends in the final state s_f. Such a path is assigned a probability equal to the product of the probabilities of all the transitions along the way. The BPMN PO run is constructed by concatenating the partial orders from the transitions and formally defined in Definition 15.

Definition 14 (Occurrence Partial Order). *An occurrence partial order $\rho_\chi = (E, \prec, \lambda, \chi)$ is a labeled partial order $(E, \prec, \lambda)$ where $\chi : E \to \mathbb{Z}$ assigns an execution counter for each label such that $\chi(e) < \chi(e')$ if $e \prec e' \wedge \lambda(e) = \lambda(e')$.*

Definition 15 (Concatenation of Partial Orders). *Given two occurrence partial orders $\rho_{\chi 1} = (E_1, \prec_1, \lambda_1, \chi_1)$ and $\rho_{\chi 2} = (E_2, \prec_2, \lambda_2, \chi_2)$ we denote the last event with label $a \in \Sigma$ as $last_1(a) = 0$ if no event is labeled with a or $last_1(a) = \max\{\chi_1(e) \mid e \in E_1 \wedge \lambda_1(e) = a\}$, the connection points in $\rho_{\chi 2}$ with $C_2 = \{e \in E_2 \mid \chi_2(e) \leq 0\}$, and a mapping function $f : C_2 \to E_1, f(e) = e'$ where $\lambda_1(e') = \lambda_2(e) \wedge \chi_1(e') = last_1(\lambda_2(e)) + \chi_2(e)$. Then $\rho_\chi = \rho_{\chi 1} \circ \rho_{\chi 2}$ represents the concatenation where: (1) $E = E_1 \cup E_2 \backslash C_2$ (2) $\prec = \prec_1 \cup \{(e, e') \mid (e \notin C_2 \wedge (e, e') \in \prec_2) \vee (e'' \in C_2 \wedge (e'', e') \in \prec_2 \wedge e = f(e''))\}$ (3) $\lambda(e) = \lambda_1(e)$ if $e \in E_1$ and $\lambda_2(e)$ if $e \in E_2$ (4) $\chi(e) = \chi_1(e)$ if $e \in E_1$ and $last_1(\lambda(e)) + \chi_2(e)$ if $e \in E_2 \backslash C_2$.*

We concatenate occurrence partial orders Definition 14, which are indexed executions of events based on their labels, to preserve the event dependencies in case multiple events with the same label have to be connected. Such cases can appear if we are dealing with models that are not safe.

Definition 16 (Path through SPORG). *Given a $SPORG = (\Sigma, S, s_{in}, S_f, T, \varrho, p)$ a path through the SPORG is given by $\pi = s_{in} \xrightarrow{t_1} \ldots \xrightarrow{t_n} s_f$ and the SBPMN PO run represented by this path is given by $\rho_\pi = \rho_{t_1} \circ \ldots \circ \rho_{t_n}$. Finally, we calculate the probability of this run with $p_\pi = \prod_{i=1}^n p(s_i, t_i, s_{i+1})$.*

We sample the PO runs from SPORG by performing a random walk. The PO runs are transformed into PO traces using Definition 6 by dropping all non-labeled SBPMN nodes. One PO trace can be represented by multiple PO runs. To acquire the probability of the PO trace, we sum the probabilities of all matching PO runs, i.e., accuracy increases by increasing the sample size. We stop our sampling procedure once the desired probability mass is reached, which can not exceed the probability of proper completion from the initial marking.

5.4 EMSC

Since we obtained the SPOTL, obtaining the STL from the event log is trivial. We do this by dividing the cardinality of each trace variant by the total number of traces. Each pair of traces from the SPOTL and the STL are compared, resulting in a distance matrix used in the EMSC method as defined in Definition 7.

Obtaining the true value might not be always possible, as two sources of error might occur: (1) the model might have markings with no option to complete and (2) the model might represent infinite stochastic languages as discussed in Sect. 5.3. The language is truncated for both scenarios. Thus, not all behaviors can be considered in the computation.

If we consider $\perp$ to represent the truncated behavior from the $L_\mathcal{P}$, as $\perp$ is not present in the objective function $cost_\Delta$, any probability mass of the log mapped to it in R has a distance of zero. This yields a lower bound for the $cost_\Delta$ function, and upper bound in $emsc_\Delta$ as the truncated behaviour is not penalized. Therefore, if we add the probability of $\perp$ to the obtained lower bound, then all the truncated behavior will not be penalized, thus leading to an upper bound to the cost and a lower bound to $emsc_\Delta$.

Definition 17 ($emsc_\Delta$ Bounds). *Let $L_\mathcal{T}$ be a STL representing the event log, $L_\mathcal{P}$ be a truncated SPOTL representing the $\mathcal{M}_s$, and let $L_\mathcal{P}(\perp)$ denote the truncated probability mass from $L_\mathcal{P}$. Take $cost_\Delta(R, L_\mathcal{T}, L_\mathcal{P}) = \sum_{\sigma_t \in L_\mathcal{T}, \rho_t \in L_\mathcal{P}} R(\sigma_t, \rho_t) \Delta(\sigma_t, \rho_t)$ with $R : \bar{L}_\mathcal{T} \times (\bar{L}_\mathcal{P} \cup \perp) \to [0, 1]$ such that $\forall \sigma_t \in \bar{L}_\mathcal{T} : L_\mathcal{T}(\sigma_t) = \sum_{\rho_t \in \bar{L}_\mathcal{P} \cup \perp} R(\sigma_t, \rho_t)$ and $\forall \rho_t \in \bar{L}_\mathcal{P} : L_\mathcal{P}(\rho_t) = \sum_{\sigma_t \in \bar{L}_\mathcal{T}} R(\sigma_t, \rho_t)$ and $cost_\Delta(R, L_\mathcal{T}, L_\mathcal{P})$ is minimal. Then the lower bound is $1 - (L_\mathcal{P}(\perp) + cost_\Delta(R, L_\mathcal{T}, L_\mathcal{P}))$ and the upper bound $1 - cost_\Delta(R, L_\mathcal{T}, L_\mathcal{P})$.*

6 Evaluation

We evaluate our method POEMS using various real-life event logs and process discovery algorithms. We assess POEMS's computational feasibility. The experiments were conducted on a 10-core $i7-1355U$ CPU, 1.7 GHz, with 32 GB RAM.

Setup: We consider three real-life event logs: Road Traffic Fine Management Process (RTFM) [7], Sepsis Cases - Event Log (Sepsis) [20], and BPI Challenge 2018 Reference Log (BPIC18r) [29]. Since there are no stochastic discovery techniques for BPMN at the moment of writing this paper, we convert SPN to

Table 1. POEMS results: p_{oc} is the maximal probability the language can reach (due to locks). p_{L_P} is the sample probability, and # is the sample size. The conformance has a lower ↑ and upper bound ↓ and execution time in seconds.

Log	Alg.	M_{noc}	ABE or Stochastic Discovery						FDE						MSAPE					
			p_{oc}	p_{L_P}	#	↑	↓	sec	p_{oc}	p_{L_P}	#	↑	↓	sec	p_{oc}	p_{L_P}	#	↑	↓	sec
RTFM	DFM	0	1	1	3	.949	.949	<1	1	1	3	.877	.877	<1	1	1	3	.861	.861	<1
	HM	0	1	1	3	.949	.949	<1	1	1	5	.867	.867	<1	1	1	5	.867	.867	<1
	IMF02	0	1	1	59	.796	.796	<1	1	1	13	.518	.518	<1	1	1	56	.621	.621	<1
	SM	0	1	1	40	.976	.976	<1	1	.999	144	.525	.525	<1	1	1	144	.645	.645	<1
	RDS	0	1	1	674	.57	.57	18	stochastic discovery											
	TP02	0	1	1	10	.881	.881	<1	stochastic discovery											
Sepsis	DFM	0	out of memory while sampling																	
	HM	8	.999	.363	19k	.316	.953	6k	.999	.349	12k	.284	.935	729	.999	.417	17k	.349	.932	2k
	IMF02	0	1	1	46k	.771	.771	11k	1	1	382	.57	.57	17	1	1	11k	.678	.678	434
	SM	2	out of memory while sampling																	
	TP02	0	1	1	85	.722	.722	<1	stochastic discovery											
BPIC18r	DFM	0	1	1	30	.984	.984	<1		.001	61	.001	.999	<1		1	28	.696	.696	<1
	HM	3	.973	.973	227	.964	.99	2	.(6)	.(6)	61	.578	.912	<1	.(6)	.501	2k	.5	.999	80
	IMF02	0	1	1	73	.985	.985	<1	1	1	22	.822	.822	<1	1	1	49	.883	.883	<1
	SM	0	1	1	215	.986	.986	<1	1	1	2k	.718	.718	3	1	1	3k	.763	.763	6
	RDS	0	1	1	2k	.957	.957	26	stochastic discovery											
	TP02	0	1	1	20	.903	.903	<1	stochastic discovery											

SBPMN models[3]. The SPN models are the ones used in the evaluation of [2], which are publicly available. These models are obtained by combining four non-stochastic discovery techniques with three weight estimators and two additional stochastic discovery techniques, resulting in 14 models per event log (42 in total). The non-stochastic discovery algorithms are: Directly-Follows Miner (DFM) [18], Heuristic Miner (HM) [21], Inductive Miner with noise threshold 0.2 (IMf02) [15], and Split Miner (SM) [1]. The weight estimators are: Alignment-Based Estimator (ABE), Fork Distribution Estimator (FDE), and Mean-Scaled Activity Pair Estimator (MSAPE) from [4]. Finally, the stochastic discovery techniques are: Rogge Solti SLPN (RDS) [24] and Toothpaste Miner with threshold 0.2 (TP02) [3].

We run our experiments using the random sampling strategy and a stopping criterion approximating the SBPMN probability of proper completion as discussed in Sect. 5.2 with a precision of six decimal points.

Results: In Table 1, we show the results of our experiments. The rows are grouped by event log, and each row represents one of the discovery algorithms used to obtain the SBPMN models. The columns are grouped by weight estimators, where the first group is for the ABE estimator or stochastic discovery algorithms, the second is for FDE, and the last is for MSAPE. The first numeric column shows the number of markings with no option to complete. For each weight estimator, we give the total achievable probability of the model, the sample probability, the sample size, the lower and upper conformance measure bounds, and the execution time in seconds.

---

[3] We adapted the conversion technique from [13] to incorporate probabilities.

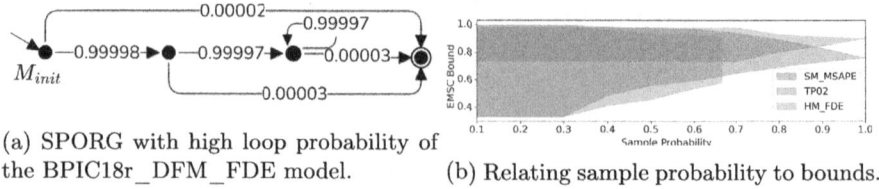

(a) SPORG with high loop probability of the BPIC18r_DFM_FDE model.

(b) Relating sample probability to bounds.

Fig. 6. BPIC18r language size analysis.

In most cases, our approach finishes within 1 s. However, there are some cases where it takes considerably longer. This is especially true for the Sepsis event log, which, although smaller, has more trace variants and longer traces. Our method has quadratic complexity in the size of stochastic languages and factorial worst-case in the length of traces. Therefore, models that generate longer traces and larger stochastic languages require longer processing time. For example, the SPORG on Fig. 6a has a high probability of looping, generating traces that are very long and have a small probability, thus resulting in bigger stochastic languages. Models like this arise from conversion errors in SPNs, where some estimators assign a fixed weight to all silent transitions. Moreover, looping with such probability rarely reflects reality. Addressing long-term event dependencies could resolve this issue, but it is outside the scope of this paper.

As shown in Table 1, some models have markings with no option to complete, and therefore, they have a lower probability of proper completion. We cannot sample the whole language from such models. Thus, the uncertainty of the missing data produces wider confidence bounds. Figure 6b relates the sample probability to the bounds discovered by our method. We can see that as the sample probability increases, the bounds shrink. The HM_FDE model has three markings with no option to complete and achieves a max probability of 0.(6) with a confidence interval from 0.578 to 0.912. When we can sample the whole language the confidence interval collapses to a single value.

Showcase. With this showcase demonstrate the importance of considering concurrency in stochastic conformance checking methods. In BPMN, concurrent tasks are modeled with a parallel gateway. According to the BPMN semantics, those tasks can start and finish independently from each other at any given time. However, if we assign probabilities to totally ordered traces we assign a probability to one of the events happening before the others. Such probabilities won't be accurate, because they will depend on the sample, and not on real dependencies.

The SBPMN model on Fig. 7a generates the SPOTL $L_\mathcal{P}$ given in Fig. 7b. From the same model, according to [19], and assuming uniform distribution over all unweighted sequence flows, we get the STL $L_\mathcal{T} = [\langle a, b, c, e\rangle^{0.45}, \langle a, c, b, e\rangle^{0.45}, \langle d, e\rangle^{0.1}]$. Now imagine we obtain the following event log $EL = [\langle a, b, c, e\rangle^{30}, \langle a, c, b, e\rangle^{60}, \langle d, e\rangle^{10}]$. If we compare $L_\mathcal{P}$ and EL with our POEMS approach, we will get a perfect conformance. However, when comparing the $L_\mathcal{T}$ with the EL we don't get perfect conformance. This difference results from the fact that we don't penalize different orderings of concurrent events. Our approach assumes that all concurrently modeled events in the SBPMN are time-independent and can occur in any order. Thus, getting a particular order in the collected event log is based only on chance and not real dependency.

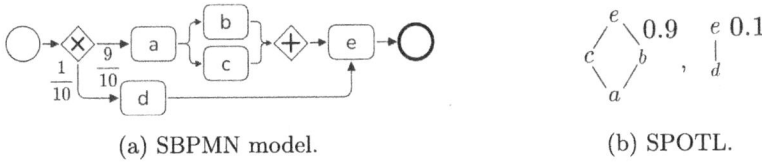

(a) SBPMN model. (b) SPOTL.

Fig. 7. Showcase of concurrency in stochastic settings.

7 Conclusion

We introduced POEMS as the first stochastic conformance checking technique for SBPMN. We adapted the existing technique introduced in [19] to BPMN and generalized it in three ways: (1) focus exclusively on the stochastic perspective of choice without forcing concurrent behavior to be described in a total order; (2) support models that may contain deadlocks and livelocks; (3) provide bounds to represent uncertainty caused by infinite behavior and behavioral model issues. Our evaluation showed that our method can be applied to real-life logs.

However, a remaining key limitation is the scalability of our method concerning the language and trace sizes. Additionally, our method assumes k-bounded models, meaning that all unbounded branches are not properly explored. Future work should investigate if the influence of these limitations can be diminished.

References

1. Augusto, A., Conforti, R., Dumas, M., La Rosa, M., Polyvyanyy, A.: Split miner: automated discovery of accurate and simple business process models from event logs. Knowl. Inf. Syst. **59**(2), 251–284 (2019)
2. Brockhoff, T., Uysal, M.S., van der Aalst, W.M.P.: Wasserstein weight estimation for stochastic Petri nets. In: ICPM, pp. 81–88 (2024)
3. Burke, A., Leemans, S.J.J., Wynn, M.T.: Discovering stochastic process models by reduction and abstraction. In: Petri Nets, pp. 312–336 (2021)
4. Burke, A., Leemans, S.J.J., Wynn, M.T.: Stochastic process discovery by weight estimation. In: ICPM Workshops, pp. 260–272 (2021)
5. Camargo, M., Dumas, M., González-Rojas, O.: Automated discovery of business process simulation models from event logs. Decis. Support Syst. **134**, 113284 (2020)
6. Carmona, J., Van Dongen, B., Solti, A., Weidlich, M.: Conformance Checking: Relating Processes and Models. Springer, Cham (2018). https://doi.org/10.1007/978-3-319-99414-7
7. de Leoni, M., Mannhardt, F.: Road Traffic Fine Management Process (2015)

8. Deussen, P.: Concurrent Automata. Tehcnical 1-05/1998,, Brandenburg Technical University Cottbus Computer Science Institute (1998)
9. Dumas, M., La Rosa, M., Mendling, J., Reijers, H.A.: Fundamentals of Business Process Management, 2nd edn. Springer, Heidelberg (2018). https://doi.org/10.1007/978-3-662-56509-4
10. Fahland, D., et al.: Instantaneous soundness checking of industrial business process models. In: BPM, pp. 278–293 (2009)
11. Fattore, M., Grassi, R., Arcagni, A.: Measuring structural dissimilarity between finite partial orders. In: Brüggemann, R., Carlsen, L., Wittmann, J. (eds.) Multi-indicator Systems and Modelling in Partial Order, pp. 69–84. Springer, New York (2014). https://doi.org/10.1007/978-1-4614-8223-9_4
12. García-Bañuelos, L., van Beest, N.R.T.P., Dumas, M., La Rosa, M., Mertens, W.: Complete and interpretable conformance checking of business processes. IEEE Trans. Software Eng. **44**(3), 262–290 (2018)
13. Kalenkova, A.A., van der Aalst, W.M.P., Lomazova, I.A., Rubin, V.A.: Process mining using BPMN: relating event logs and process models. Softw. Syst. Model. **16**(4), 1019–1048 (2017)
14. Leemans, S.J.J., Brockhoff, T., van der Aalst, W.M.P., Polyvyanyy, A.: Partially ordered stochastic conformance checking. Knowl. Inf. Syst. (2024)
15. Leemans, S.J.J., Fahland, D., van der Aalst, W.M.P.: Discovering block-structured process models from event logs containing infrequent behaviour. In: BPM Workshops, pp. 66–78 (2014)
16. Leemans, S.J.J., Maggi, F.M., Montali, M.: Enjoy the silence: analysis of stochastic Petri nets with silent transitions. Inf. Syst. **124**, 102383 (2024)
17. Leemans, S.J.J., Polyvyanyy, A.: Stochastic-aware conformance checking: an entropy-based approach. In: CAiSE, pp. 217–233 (2020)
18. Leemans, S.J.J., Poppe, E., Wynn, M.T.: Directly follows-based process mining: exploration & a case study. In: ICPM, pp. 25–32 (2019)
19. Leemans, S.J.J., Syring, A.F., van der Aalst, W.M.P.: Earth movers' stochastic conformance checking. In: BPM Forum, pp. 127–143 (2019)
20. Mannhardt, F.: Sepsis Cases - Event Log (2016)
21. Mannhardt, F., de Leoni, M., Reijers, H.A., van der Aalst, W.M.P.: Data-driven process discovery - revealing conditional infrequent behavior from event logs. In: CAiSE, pp. 545–560 (2017)
22. Molka, T., Redlich, D., Drobek, M., Caetano, A., Zeng, X.J., Gilani, W.: Conformance checking for BPMN-based process models. In: SAC, pp. 1406–1413 (2014)
23. Polyvyanyy, A., Moffat, A., García-Bañuelos, L.: An entropic relevance measure for stochastic conformance checking in process mining. In: ICPM, pp. 97–104 (2020)
24. Rogge-Solti, A., van der Aalst, W.M.P., Weske, M.: Discovering stochastic Petri nets with arbitrary delay distributions from event logs. In: BPM Workshops, pp. 15–27 (2014)
25. Rozinat, A., van der Aalst, W.M.P.: Conformance checking of processes based on monitoring real behavior. Inf. Syst. **33**(1), 64–95 (2008)
26. Tax, N., Lu, X., Sidorova, N., Fahland, D., van der Aalst, W.M.P.: The imprecisions of precision measures in process mining. Inf. Process. Lett. **135**, 1–8 (2018)
27. van der Aalst, W.M.P., Adriansyah, A., van Dongen, B.: Replaying history on process models for conformance checking and performance analysis. WIREs Data Mining Knowl. Discov. **2**(2), 182–192 (2012)
28. van der Aalst, W.M.P.: Process Mining: Data Science in Action, 2nd edn. Springer, Heidelberg (2016). https://doi.org/10.1007/978-3-662-49851-4

29. van Dongen, B., Borchert, F.: BPI Challenge 2018 (2018)
30. Yujian, L., Bo, L.: A normalized Levenshtein distance metric. IEEE Trans. Pattern Anal. Mach. Intell. **29**(6), 1091–1095 (2007)
31. Zelinka, B.: Distances between partially ordered sets. Math. Bohem. **118**(2), 167–170 (1993)

Instance Configuration and Scheduling Based on the Resource-Augmented Process Structure Tree

Felix Schumann, G. Wessel van der Heijden, and Stefanie Rinderle-Ma

Technical University of Munich, TUM School of Computation, Information, and Technology, Garching, Germany
{felix.schumann,wessel.heijden,stefanie.rinderle-ma}@tum.de

Abstract. Real-world applications often require the resource-specific configuration of process instances based on system states. In clinical processes, for example, the number and order of documentation tasks depend on whether they are conducted by an intern or the head physician. One approach supporting resource-specific instance configuration is the Resource-Augmented Process Structure Tree (RA-PST). It also offers the flexibility to drive the configuration through optimized resource allocation. In particular, combinatorial optimization approaches can be explored to plan optimal combinations of multiple configurable process instances into a system. We use Mixed Integer Programming (MIP) to find optimal process configurations and Constraint Programming (CP) to schedule the process instances. To combine configuration and scheduling of process instances, we propose an integrated CP formulation as well as a Logic-Based Benders Decomposition (LBBD) inspired by the integrated process planning and scheduling problem (IPPS). The approaches are applied and evaluated in an offline and an online scheduling setting. We show the suitability of the CP and LBBD formulation for the configuration and scheduling problem. The integrated CP can find optimal solutions for small problems but struggles to prove optimality in a timely manner when the problem size grows. Here, the LBBD can identify better lower bounds. For the online setting, the integrated CP finds optimal solutions for most process instances while considering the planned resource availabilities in the system.

Keywords: Process Configuration · Resource Allocation · Scheduling · Constraint Programming · Logic-Based Benders Cuts

1 Introduction

Process-Aware Information Systems (PAIS) and the underlying real-world processes in applications such as manufacturing or clinical services move towards flexible processes [10,14]. Configurable process models enable process modelers to incorporate flexibility at design time and allow them to better reflect reality.

When executing the process models, multiple process instances are in competition over the human and non-human resources needed to execute their tasks. Configurable process models, such as the RA-PST formalized in [17], can be utilized to bypass resource bottlenecks by adapting their control-flow configuration to the current resource availability. Approaches from combinatorial optimization can leverage this flexibility to optimize process performance and resource utilization in a system.

The RA-PST intertwines the flexibility required by real-world processes with a representation of possible resource allocations for each task and ensures the structural soundness of the resulting process models. At design time, the RA-PST allows for separate modeling of the "core" Process Structure Tree (PST), which represents the business logic and the separate modeling of resources that later execute the tasks (resource augmentation). Based on the PST, an RA-PST is executable by a process engine [17,21]. So far, no approach addresses the integrated configuration and scheduling of process instances to optimize process performance. Hence, in this paper, we propose and compare three configuration and scheduling approaches shown in Fig. 1 based on the RA-PST to find a combination of process configurations that optimizes an objective function. While approach a) separates configuration and scheduling for each instance, approaches b) and c) utilize the provided flexibility by configuring and scheduling multiple process instances simultaneously. The proposed mathematical programming models are inspired by optimal solution approaches for the Integrated Process-Planning and Scheduling Problem (IPPS) from Operations Research (OR). The proposed optimization models enable the assessment of the quality of a found solution w.r.t. optimality by specifying an optimality gap through an upper and a lower bound.

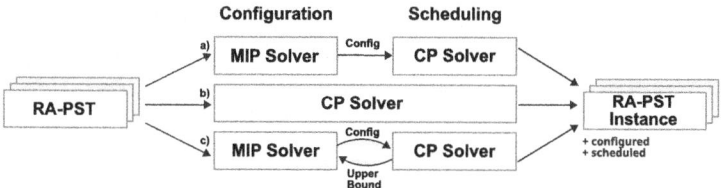

Fig. 1. Different solution approaches for the combined configuration and scheduling problem with RA-PSTs. a) separated approach, b) integrated CP approach, c) Logic-Based Benders Decomposition.

In Fig. 1b) we employ Constraint Programming (CP) to solve an integrated version of the problem, which configures and schedules the instances simultaneously. CP has grown in attention in recent years to solve scheduling problems in Operations Research (OR) [11] and Business Process Management (BPM) [18]. It performs specifically well on sequencing problems. In pure assignment decisions, Mixed Integer Programming (MIP) models perform better [11]. The approach in Fig. 1c) is based on the IPPS, e.g. [1,10]. By separating configuration and scheduling, the strengths of both approaches can be fostered for the problem at hand. We present a novel Logic-Based Benders Decomposition (LBBD) of the problem, where the configuration is solved through a MIP, and the scheduling through a CP. In the LBBD, information is exchanged between both optimization steps. Solving both problems in a sequential manner as in [9] is presented in Fig. 1a). As described in [11], the integration of configuration and scheduling is an open challenge and an active field of research for optimal solution approaches. In coherence with this research field, we work with purely sequential processes. We propose three solution approaches for the integrated configuration and scheduling problem. The proposed approaches can find optimal solutions to the problem and report an optimality gap for any found solution. By basing the approach on the RA-PST process model, we generate configured, executable process instances with allocated resources. Each process instance is configured in context with all other process instances to optimize makespan.

The remainder of this paper is structured as follows: Sect. 2 defines preliminaries for RA-PSTs. Section 3 analyzes the complexity of the configuration problem, and formalizes the optimization methods. These methods are applied in an offline and an online setting in Sect. 4. Section 5 discusses the findings in the context of related work, and 6 concludes the paper.

2 Preliminaries and Example

To combine a structural process model with resource allocation, we utilize the *RA-PST* [17] that augments a process structure tree [21] with the flexible allocation of resources. Figure 2a) shows an example taken from the diagnostic process of a radiology department. The PST reflects a sequence of two tasks, i.e., report and approve where report can be processed by a resource with roles L_1, L_2, L_3. Let R be the set of resources (cf. Figure 2b). The RA-PST creates flexibility by enabling resources to apply changes to the control flow in case they are allocated to a task, e.g., inserting an additional read task if the allocated resource is an intern. To this end, a resource $r \in R$ can have one or several *resource profiles* $rp := (r, role, t, Attr, cp)$ where r has role $role$ to specify access rights and is assigned to task t in the PST. r can be equipped with attributes $Attr$ and change patterns $cp \in$ {insert, delete, replace}, which modify the PST if r is allocated to t. In the example, resource p has two resource profiles rp_2 and rp_3. i has resource profile rp_1 referring to task report and specifying an insert of task read after report. $Attr$ typically contains the costs for allocating a resource.

The RA-PST is constructed by appending branches to each leaf task t in the PST (cf. Figure 2c). A branch consists of the resource node r allocated to t, the corresponding resource profile node, and potentially the associated change pattern node referring to a task t'. If the change pattern inserts t', the branch might be expanded. For the example, three branches are appended to task report, one for each resource i, p, h having the associated roles L1, L2, L3. Take the branch for i: below node i, the associated resource profile rp_1 is appended and below that, the change pattern node for insertion of read. For read, resource p is authorized and appended, followed by the resource profile rp_3. Here, the branch ends as rp_3 does not specify a change pattern.

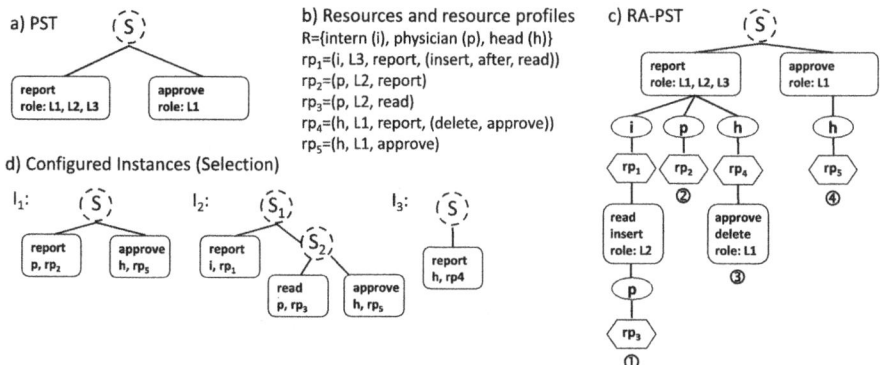

Fig. 2. RA-PST representation of the diagnostics process in a radiology department. a) Process as PST, b) Resource description, c) full RA-PST with branches, d) Possible configured process instances from the RA-PST.

As shown in [17], a branch is *valid*, i.e., can be employed for *process instance configuration*, if it ends in a resource profile or a delete pattern. A process instance configuration is achieved by choosing a branch assignment for each PST task and then resolving all change patterns in the selected branches. Figure 2d) shows three possible process instance configurations I_1, I_2, and I_3 for the RA-PST in Fig. 2c). For I_2, as i is allocated to report, task read is inserted after report. For I_3, as h is allocated to report, task approve is deleted. These configurations are sound due to the PST soundness, and the resource allocation is possible due to the validity property. For the example, several configurations are possible. In the real-world application behind this example, a traditional process model containing all variants would become complex. The RA-PST declutters the model and pushes the flexibility to the resources which are the root cause for variants. Hence, a separation of process perspectives is achieved.

The choice of branches can depend on different aspects, such as costs modeled as attributes in the resource profiles, and can be solved as an optimization problem. So far, the scope of the allocation in [17] has been limited to single process instances. In the following, we will tackle the combined configuration and scheduling of multiple process instances $\mathcal{I}$ as an optimization problem to decide on the optimal resource allocation among all tasks and process instances.

For the remainder of this paper, the term "instance" is used in its BPM context, where it denotes an instance of a process. We refer to each resource allocated to a task as a *job* $j \in J$ that has an individual cost c_j where J is the set of all jobs. A branch $b \in B$ represents each unique path from the root node to a leaf, where B is the set of all branches. Each branch contains a set of jobs $J_b \subseteq J$. The overall cost c_b of a branch $b \in B$ can be calculated as the sum of all allocated resources in this branch, i.e. $c_b = \sum_{j \in J_b} c_j$. From here on, the costs represent the processing time of a job and for each job, this cost is predetermined.

For the RA-PST we rely on the validity notion given in [17], which only allows for the deletion of top-level tasks, and exclude invalid branches in the optimization problem. To ensure a sequential process model as in the IPPS problem [1], we limit the change patterns to deletion and insertion of tasks and do not allow for parallel or exclusive choices.

3 Configuration and Scheduling of RA-PST Instances

The optimization problem consists of finding a valid configuration for each process instance (configuration problem) and fitting each process instance into a system schedule (scheduling). We leverage the variability of RA-PSTs to improve the makespan of the system's schedule through combination of configurations. As an input, we are given a set of process instances $\mathcal{I}$. Each process instance $I \in \mathcal{I}$ contains a set of tasks $\mathcal{T}(I)$, a set of jobs $J(I)$, and a set of branches $B(I)$.

3.1 Configuration Problem

For the configuration problem alone, we consider a single instance I, and show that configuring this single instance alone is NP-hard. A branch can delete another task in the same process instance if it contains a delete change pattern for that task. In the configuration problem, for each task that is not deleted exactly one branch must be selected. If e.g., branch ③ is selected for report in Fig. 2, ①, ② and ④ cannot be selected. Definition 1 formally defines the minimum cost RA-PST configuration problem. Figure 3 gives an illustrative example of the configuration problem.

Definition 1 (Minimum cost RA-PST configuration problem). *Given are n branches $B = \{b_1, \ldots, b_n\}$ partitioned over k tasks $\mathcal{T} = \{T_1, \ldots T_k\}$, i.e. $T_i \subseteq B$. Each branch $b \in B$ has a cost c_b and a set of tasks $\tau(b) \subseteq \mathcal{T}$ deleted by branch b. Find the minimum cost set of branches $X \subseteq B$ such that for each task $T_i \in \mathcal{T}$ either*

1. *there is exactly one branch $b \in X$ such that $b \in T_i$*
2. *there is a $b \in X$ such that $T_i \in \tau(b)$ and for all $b' \in T_i$ it holds that $b' \notin X$.*

NP-Hardness. We reduce from the minimum weight maximal independent set problem (WMMIS), which was shown to be NP-hard by Demange [3].

Definition 2 (Minimum weight maximal independent set problem (WMMIS), [3]). *Given a graph $G = (V, E)$, an independent set is a subset $V' \subseteq V$ such that no two vertices in V' are linked by an edge E; an independent set V' is maximal if for every vertex $v \in V \setminus V'$, there exists $v' \in V'$ such that $(v, v') \in E$. Let w_v be the weight of vertex v for every vertex $v \in V$. For a set of vertices $V' \subseteq V$, its weight $w(V')$ is the sum of weights of the vertices in V'. Find a maximal independent set $V' \subseteq V$ with minimum weight $w(V')$.*

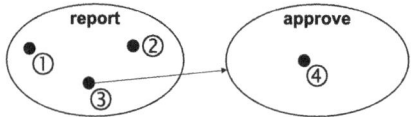

Fig. 3. A graph representation of an RA-PST instance of the example. Elements represent branches, sets represent PST tasks (Fig. 2a)), arrows indicate a delete change pattern. If branch ③ is chosen, no element from the set "approve" may be chosen.

Lemma 1. *If a polynomial time algorithm exists for Minimum Cost RA-PST configuration problem, then the minimum weight maximal independent set can be polynomially decided.*

Proof. Given is some WMMIS instance $G = (V, E)$. Index each vertex in V such that $V = \{v_1, \ldots, v_n\}$. Define branch b_i with $c_i = w_{v_i}$ for every vertex $v_i \in V$. Let task T_i be the set of branches only containing b_i for every $v_i \in V$. For every edge $(v_i, v_j) \in E$, add task T_j to $\tau(b_i)$. In the minimum cost RA-PST configuration problem, for any branch b_i that is selected, all tasks $\tau(b_i)$ are deleted. This means no branch that are part of any task in $\tau(b_i)$ can be selected. Therefore, for any b_i that is selected, no neighboring b_j with $(v_i, v_j) \in E$ can be selected, since T_j was deleted. This means that any solution $X \subseteq B$ to the RA-PST configuration problem is an independent set. By the definition of the minimum cost RA-PST, there must be a branch in each task T_i that is in X, or T_i is deleted. Therefore, X is a maximal independent set. Since the minimum cost RA-PST problem minimizes the cost of all selected branches, X is a minimum weight maximal independent set.

Demange showed that the WMMIS problem can not be approximated independently of the node weights [3]. This gives the following corollary for the minimum cost RA-PST configuration problem.

Corollary 1. *The minimum cost of the RA-PST configuration problem can not be approximated in polynomial time independently of branch weights, unless P=NP.*

We define the configuration problem as an ILP. Define the closed neighborhood of b as $N[b] = \bigcup_{T_i \in \tau(b) \cup \{T_j : b \in T_j\}} T_i$, which is the set of branches that can not be selected if branch b is selected, including branch b. The open neighborhood is defined as $N(b) = N[b] \setminus \{b\}$. If branch $b \in X$, then $b' \notin X$ for any $b' \in N(b)$. Since we have to select a maximal independent set, exactly one branch $b' \in N[b]$ must be selected. This gives the following ILP formulation of the RA-PST configuration problem.

$$\min \sum_{b \in B} c_b x_b \qquad \text{(Config)}$$

$$\text{s.t.} \sum_{b' \in N[b]} x_{b'} = 1 \qquad \forall b \in B \qquad (1)$$

$$x_b \in \{0, 1\} \qquad \forall b \in B$$

In this ILP, the variable x_b indicates whether branch b is selected, c_b is the cost of branch b, and constraint (1) ensures that exactly one branch of the neighborhood is selected.

3.2 Scheduling Problem

In the scheduling problem, a set of process instances $\mathcal{I}$ is given with selected branches X. All jobs $j \in J_b$ of each branch $b \in X$ must be scheduled on resource $r_j \in R$ adhering to the precedence constraints of each process instance. Any pair of jobs $i, j \in J(I)$ for all $I \in \mathcal{I}$ have a sorted order, which gives precedence constraints $i \prec j$. Furthermore, each resource $r \in R$ may process at most one job at any time. Each job j must be processed non-preemptively for processing length c_j. The objective is to find a schedule that schedules all selected jobs that adheres to the precedence constraints while minimizing the total makespan. The scheduling problem is a generalized job-shop scheduling problem, which is known to be NP-hard ([4], problem SS18). In the job-shop scheduling problem all individual jobs in a process instance must be scheduled on different resources [13], which is not a requirement for our scheduling problem. Since our scheduling problem is more general than job-shop scheduling, our scheduling problem is also NP-hard.

Consider the example in Fig. 2 for which two process instances i and j arrive simultaneously. Both instances are configured as variant I_3 (Fig. 2 d)). This configuration results in two tasks report that must be executed by resource h. Since one resource can only process one task at a time, resource h first processes task report for i before processing report for j. If a second resource h was available, both report tasks could be processed in parallel.

3.3 Integrated Constraint Programming Formulation

We combine the configuration and scheduling problem in a single formulation. The CP formulation is given in (ICP). The configuration constraint is formulated similarly as in the configuration problem in (Config). In the integrated version, the goal is to minimize the makespan instead of the sum of the branch costs.

Consider the goal of processing two process instances in the example over the given resources with minimum makespan. Configuring one process instance to I_1 and the other to I_3 can lead to a shorter overall makespan if the approval by h takes less time than writing a full report.

$$\min \max_{j \in \cup_{I \in \mathcal{I}} J(I)} (\text{EndOf}(Task_j)) \qquad \text{(ICP)}$$

s.t. $\quad Task_j = \text{IntervalVar}(c_j, optional) \qquad \forall I \in \mathcal{I}, j \in J(I)$ (2)

$\quad \text{NoOverlap}(Task_i, Task_j) \qquad \forall r \in R, i,j \in J_r$ (3)

$\quad \text{EndBeforeStart}(Task_i, Task_j) \qquad \forall I \in \mathcal{I}, i,j \in J(I), i \prec j$ (4)

$\quad \text{PresenceOf}(Task_i) = \text{PresenceOf}(Task_j) \quad \forall I \in \mathcal{I}, b \in B(I), i,j \in J_b$ (5)

$\quad \sum_{j_1 \in b': b' \in N[b]} \text{PresenceOf}(Task_{j_1}) = 1 \qquad \forall I \in \mathcal{I}, b \in B(I)$ (6)

The objective of (ICP) is to minimize the makespan. The function "EndOf(x)" returns the end of interval variable x. Equation (2) defines the interval variables for each job in all process instances. "IntervalVar(a, $optional$)" returns an interval variable of length a for which it is optional whether it is present. Equation (3) ensures that each resource processes at most one task at any time. The constraint "NoOverlap(x,y)" ensures that no overlap between interval variables x, y exist. Equation (4) enforces the precedence constraints of the jobs. The constraint "EndBeforeStart(x,y)" ensures that interval variable x must have its end before the start of interval variable y. Equation (5) ensures that all jobs in the same branch are scheduled when one job in the same branch is selected. The boolean function "PresenceOf(x)" returns a 0 if interval variable x is not present, and a 1 if it is. Equation (6) is the configuration constraint, where the first job j_1 in a branch $b' \in N[b]$ has to be present. In combination with (5), this ensures exactly one branch is selected.

3.4 Logic Based Benders Decomposition

CP is especially effective for the scheduling part of the problem. It can, however, struggle with finding good configurations for the process instances, which is a strength of MIP formulations [11]. In order to leverage the strength of MIP formulations for the configuration problem and CP for the scheduling problem, we use LBBD, which is the state-of-the-art approach to find solutions for IPPS [1, 10]. We adapt the approach presented in [1] for our problem with the expectation to find better lower bounds than the (ICP). To this end, we decompose the problem into a master problem and a subproblem. The master problem solves the configuration problem. Additional constraints to the master problem provide a lower bound on the makespan for the selected combination of configurations. The jobs that are selected by the master problem are then scheduled by the CP. The CP returns a makespan for the specific configuration combination, for which benders cuts are added to the master problem. If the makespan is not equal to the lower bound, the master problem selects a new configuration combination with the minimum lower bound until the termination requirement has been reached.

The objective for the combined problem is to minimize the makespan $c_{\max}$. We pick a lower bound on the makespan $c_{\max}$ in order to aid picking the master problem when selecting an assignment. This lower bound is based on the lower bounds of the IPPS, and the job-shop scheduling problem [1,2]. For each resource

$r \in R$, we find the sum of processing times of all jobs that must be processed by resource r and add the minimum *head* and *tail* of the jobs $j \in J_r$. This gives a lower bound on the makespan of jobs scheduled on resource $r \in R$. The minimum *head time* of resource $r \in R$ is defined as the minimum starting time of any job $j \in J_r$. The minimum *tail time* of resource $r \in R$ is defined as the minimum remaining processing time at all jobs processed on r. We define variable E_r^1 (resp. E_r^2) to be the earliest head (resp. tail) time on resource $r \in R$. This gives the following lower bound on the makespan where $c_b(r)$ is the minimum starting time of the first job of branch b on resource $r \in R$.

$$c_{\max} \geq E_r^1 + E_r^2 + \sum_{I \in \mathcal{I}} \sum_{b \in B(I)} c_b(r) x_b \qquad \forall r \in R \qquad (7)$$

Finding the minimum head/tail time on resource $r \in R$ involves additional constraints and variables. In this formulation, Y_r is a binary variable indicating if any job is assigned to resource $r \in R$. Binary variable $Q_{r,b}^h$ indicates whether the branch b on resource $r \in R$ is picked as the earliest head or tail branch of resource r ($h \in \{1, 2\}$). For a large number, we use $M = \sum_{I \in \mathcal{I}} \sum_{b \in B(I)} c_b$ as the sum of all branch cost as an upper bound on the makespan. The complete master problem is shown in (MP). For ease of notation, we let B_r be the set of all branches that contain a job that must be processed by resource $r \in R$.

$$\min\ c_{\max} \qquad \text{(MP)}$$

$$\text{s.t.} \sum_{b' \in N[b]} x_{b'} = 1 \qquad \forall I \in \mathcal{I}, b \in B(I) \qquad (8)$$

$$c_{\max} \geq E_r^1 + \sum_{I \in \mathcal{I}} \sum_{b \in B} c_b(r) x_b + E_r^2 \qquad \forall r \in R \qquad (9)$$

$$Y_r \geq x_b \qquad \forall r \in R, b \in B_r \qquad (10)$$

$$\sum_{b \in B_r} Q_{r,b}^h = Y_r \qquad \forall r \in R, h \in \{1, 2\} \qquad (11)$$

$$Q_{r,b}^h \leq x_b \qquad \forall r \in R, b \in r, h \in \{1, 2\} \qquad (12)$$

$$E_r^1 \geq \sum_{\beta \prec b} c_\beta x_\beta + c_b(r) x_b - M(1 - Q_{r,b}^1) \qquad \forall r \in R, b \in B_r \qquad (13)$$

$$E_r^2 \geq \sum_{\beta \succ b} c_\beta x_\beta - c_b(r) x_b - M(1 - Q_{r,b}^2) \qquad \forall r \in R, b \in B_r \qquad (14)$$

$$Y_r, Q_{r,b}^h \in \{0, 1\} \qquad \forall r \in R, b \in r, h \in \{1, 2\} \qquad (15)$$

$$E_r^h \geq 0 \qquad \forall r \in R, h \in \{1, 2\} \qquad (16)$$

$$x_b \in \{0, 1\} \qquad \forall I \in \mathcal{I}, b \in B(I) \qquad (17)$$

$$c_{\max} \geq 0 \qquad (18)$$

The variable Y_r is only equal to 1 if there exists a job on the resource by Constraint (10). Constraint (11) ensures that at most one branch from each resource is selected and Constraint (12) ensures that only selected branches can be considered for the earliest head/tail time. The selected branch is then used to find the earliest head/tail time E_r^h on each resource $r \in R$ in Constraint (13) and Constraint (14). E_r^h is minimized since it is positively correlated with makespan $c_{\max}$.

We add benders cuts to the master problem. Since the master problem selects the branches, different to the ICP, the CP subproblem does not have any optional jobs and we can use the pure scheduling formulation (SP). We use standard monotone cuts on the selected branches as described by Hooker [8]. Here $\overline{x}$ represents all variables x_j for which $\overline{x}_j = 1$, $v^*(\overline{x})$ is the optimal makespan value for the schedule induced by $\overline{x}$, and $\underline{v}$ is the global lower bound on the makespan. The branches selected by $\overline{x}$ is indicated as $B(\overline{x})$.

$$c_{\max} \geq v^*(\overline{x}) - (v^*(\overline{x}) - \underline{v}) \sum_{b \in B(\overline{x})} (1 - x_b) \qquad (19)$$

In this monotone cut, only one specific configuration is eliminated. Ideally we would like to eliminate multiple configurations through a single cut. To do this, we find a lower bound on the makespan for a subset of process instances that is greater than the global lower bound. We select a subset $\mathcal{I}' \subseteq \mathcal{I}$ with branch selection $B(\overline{x}') = (\cup_{I \in \mathcal{I}'} I) \cap B(\overline{x})$. If the makespan lower bound for the branches $B(\overline{x}')$ is greater than the global lower bound, we can add this as a strengthened cut. This means all possible configurations can be eliminated for any $I \notin \mathcal{I}'$. Solving a CP for all subsets $\mathcal{I}'$ is computationally intensive, so we consider a lower bound that can be found efficiently. For this lower bound, we compute the minimum resource usage starting after every integer time t. Let $c_r^I(t)$ indicate the cost of all jobs that must schedule at or after t on resource $r \in R$ for process instance $I \in \mathcal{I}'$. This gives the lower bound in (20) for the branches $B(\overline{x}')$.

$$v^*(\overline{x}') \geq \max_{r \in R}(\max_{t \in T}(t + \sum_{I \in \mathcal{I}'} c_r^I(t))) \qquad \forall \mathcal{I}' \subseteq \mathcal{I} \qquad (20)$$

The subproblem only schedules the jobs J that have been selected by the master problem. Different to the integrated CP, the subproblem no longer has to enforce the presence of individual jobs, since all must be present.

$$\min \max_{j \in \cup_{I \in \mathcal{I}} J(I)} (\text{EndOf}(Task_j)) \qquad \text{(SP)}$$

$$\text{s.t. } Task_j = \text{IntervalVar}(c_j) \qquad \forall I \in \mathcal{I}, j \in J(I) \qquad (21)$$
$$\text{NoOverlap}(Task_i, Task_j) \qquad \forall r \in R, i, j \in J_r \qquad (22)$$
$$\text{EndBeforeStart}(Task_i, Task_j) \qquad \forall I \in \mathcal{I}, i, j \in J(I), i \prec j \qquad (23)$$

Constraint (21) defines the interval variables with length c_j. Constraint (22) ensures that each resource processes at most one job at any time. Constraint (23) enforces the precedence constraints of the selected jobs. (SP) has as objective the makespan of the schedule.

4 Application in Offline and Online Scenarios

We test three optimization approaches for the configuration and scheduling problem of RA-PSTs. Following Sect. 3 we use (ICP) as integrated configuration and scheduling CP. LBBD uses (MP) and (SP). As comparison, we combine (Config) and (SP), which solves configuration and scheduling sequentially (Config+CP).

We test the approaches in an offline and an online setting as shown in Fig. 4. In the offline application, all process instances to be scheduled are known beforehand and are configured and scheduled simultaneously. In the BPM-driven online setting, process instances arrive over time. Each process instance is configured and scheduled independently. The solver uses current schedule information to find a configuration for the arriving instance.

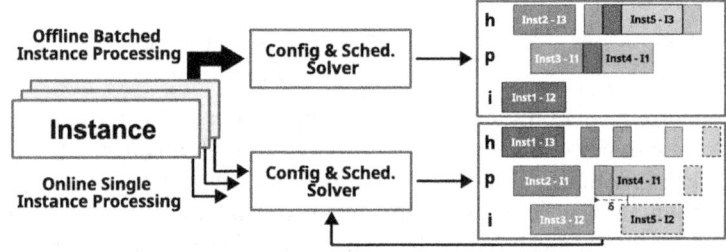

Fig. 4. Visualization of the offline and online scheduling problem. Parameter δ represents possible right-shift of already scheduled tasks. "Inst5-I2" is the process instance to be configured & scheduled next in the online setting.

All approaches are implemented in Python using IBM CPLEX 22.1 for CP. The *(optional) interval* and *sequencing* variables available in the CP Optimizer allow for intuitive modeling of the scheduling problem [11]. The MIP is implemented in Gurobi 12.0 by Gurobi[1]. Experiments were run on a computer with Core^{TM} i7-1165G7 CPU @ 2.80 GHz. The quality of a solution obtained by a solver is defined by a lower and an upper bound. The upper bound is the objective value of the best feasible solution found. The lower bound is a value that is no bigger than the objective value of any feasible solution and can be found through a relaxation of the problem. When the upper and lower bounds are equal, a solution is optimal.

4.1 Offline Scheduling:

We test ICP, LBBD, and Config+CP on an offline test set. Table 1 gives an overview of the used test sets. We design RA-PSTs with growing task numbers for the process model, namely 2, 5, and 10 Tasks. The 2-task process reflects the example presented in Sect. 2. Additionally, the problem size is increased by

---

[1] Code: https://github.com/Schlixmann/Instance_config_and_schedule/.

using process models with more branches, i.e., more flexibility on which resource can execute a task. For each test set, 8 process instances are scheduled, and all instances are released at the same time, i.e., at 0 time units. In test sets 7 and 8, different RA-PSTs are chosen randomly, i.e., for the first process instance, a different RA-PST is used than for the second instance. In this setting, the number of branches and also possible configurations per instance vary. Each test set is run for each approach for 7200s. For the LBBD we run the (SP) for 5 s.

Table 1. Description of test sets, since multiple RA-PSTs are used in 7, 8, branch and configuration per RA-PST is not given

id	unique RA-PST	tasks	resources	branches	configs p. Instance
1	1	2	4	5	4
2	1	5	5	7	4
3	1	5	5	24	32
4	1	10	5	20	1.024
5	1	10	5	37	442.368
6	1	10	5	29	39.366
7	5	10	5	-	-
8	5	10	5	-	-

Table 2 presents the computational results for the offline use case. We can see that when configuration and scheduling are separated (Config+CP), an optimal schedule can be found for all tests (LB=UB). Since this approach schedules the same configuration for each instance, as expected, Config+CP does not find a better upper bound than ICP and LBBD. For the small test sets 1, 2, and 3, ICP finds the optimal solution in all cases. LBBD finds the best upper bound for test sets 1, 2, and 3 and the optimal lower bound for test sets 1 and 3. For test set 2 LBBD fails to find a better lower bound than ICP. For larger instances in test sets 4–8, ICP always finds the best upper bounds but struggles to find meaningful lower bounds. Here, the LBBD approach finds much better lower bounds but cannot always find upper bounds as good as the ICP. ICP and LBBD perform best for test sets 7 and 8. Specifically, the LBBD approach benefits from the differentiation of RA-PSTs. This is promising for integrating the optimization approaches into scenarios with a multitude of simultaneously running processes and process variants.

4.2 Online Scheduling and Rescheduling

The online setting refers to process instances executed by a process engine that are typically deployed based on their arrival [7]. In the following, we apply CP, LBBD, and Config+CP in this "online" fashion in order to test their readiness

Table 2. Lower Bound (LB), Upper Bound (UB), Gap to best LB among all solution approaches and computing time, Config+CP serves as UB reference

	ICP				LBBD				Config+CP			
	LB	UB	Gap best LB	Time s	LB	UB	Gap best LB	Time s	LB	UB	Gap best LB	Time s
1	115	115	0.0%	0.41	115	115	0.0%	2032.9	165	165	30.3%	0.02
2	63	63	0.0%	0.16	57	63	0.0%	7200	84	84	25.0%	0.02
3	102	102	0.0%	24.27	102	102	0.0%	0.05	144	144	29.2%	0.01
4	21	94	23.4%	7200	72	103	30.1%	7200	108	108	33.3%	1.04
5	14	81	11.1%	7200	72	94	23.4%	7200	107	107	32.7%	0.02
6	21	101	23.8%	7200	77	112	31.2%	7200	163	163	52.8%	0.06
7	29	94	6.4%	7200	88	94	6.4%	7200	139	139	36.7%	0.02
8	29	94	4.3%	7200	90	95	5.3%	7200	145	145	37.9%	0.01

for the online setting. Instances arriving over time add uncertainty to the problem at hand. The flexibility in the RA-PST enables a reaction to counter this uncertainty. To show the effect of this flexibility, we also use a direct allocation approach that operates directly on the RA-PST as a comparison. It resembles the ad-hoc approaches often used in BPM by assigning tasks to resources at task release. While being the most reactive allocation approach, it offers no stable execution plan for a process instance, and tasks lying in the past can not be deleted. The greedy allocation allocates one task at a time. Once the execution of a task is finished, the next task is allocated by choosing the branch with the earliest finish time.

To harden scheduling approaches against uncertainty at runtime, rescheduling is a common approach [15]. We balance complexity and enable a reaction to newly arriving instances for the optimization approaches by allowing to right-shift already scheduled tasks by factor δ as a rescheduling measure. A re-configuration of planned instances is not possible.

To test the online set-up, multiple RA-PSTs have been designed. Due to the single process instance usage, the approach can handle bigger process models. We test processes with 10, 20, and 30 tasks and have five resources. From an RA-PST perspective, up to 4 branches per task are allowed. For each process size, 10 test sets are used. Again, 2 test sets consist of different RA-PSTs. Factor δ is set to the length of one average task length in the RA-PST, and arrival times follow a Poisson distribution; the time between each instance arrival is an exponential distribution with a scaling factor of one average task length [12]. The time needed to find a solution is important in the online setting. Integrated into a BPM System, the process engine keeps running processes in parallel to the scheduling algorithm. We allow for 100s of runtime for the configuration and scheduling of a single instance. As a comparison oracle, we let the ICP solve the problem globally for all process instances simultaneously. We use the objective found after 100s of runtime as a comparison.

Table 3 shows the median relative deviation for each allocation type from the oracle. For processes with 10 Tasks, the integrated ICP with rescheduling finds an optimal solution for each process instance in each test set. The optimum for the bigger test sets cannot be found in all cases within 100s. Through more detailed data analysis, we found that in many cases, the ICP struggles to find the best schedule for the first instance in a test set but configures the following process instances optimally. This could easily be overcome by using the (Config) formulation to schedule the first instance. The impact of rescheduling is substantial. The deviation from the oracle rises if rescheduling is prohibited. As in the offline setting, for the Config+CP approach the found solution is worse than the other approaches due to the lack of configuration combination. For the overall makespan within each test set, the greedy allocation can outperform the ICP for the processes with more tasks where the higher level of reactivity proves helpful. The LBBD approach does not perform well in the online setting. In the majority of cases, the optimum cannot be found.

Table 3. Performance of the different approaches in the online setting

	10 Tasks			20 Tasks			30 Tasks		
	Deviation	Optimal	Percentage	Deviation	Optimal	Percentage	Deviation	Optimal	Percentage
Oracle	0.0	-	-	0.0	-	-	0.0	-	-
Resched. ICP	**22.1**	80/80	100.0%	36.9	77/80	96.2%	75.1	63/80	78.8%
Resched. LBBD	26.2	43/80	53.8%	59.4	12/80	15.0%	102.3	10/80	12.5%
Greedy	30.0	-	-	**26.4**	-	-	**58.4**	-	-
No. Resched. ICP	38.3	80/80	100.0%	73.1	78/80	97.5%	135.4	74/80	92.5%
No. Resched. LBBD	38.7	34/80	42.5%	77.6	11/80	13.8%	127.6	10/80	12.5%
Resched. Config+CP	58.5	80/80	100.0%	95.0	80/80	100.0%	130.6	80/80	100.0%

The core idea of RA-PSTs is to provide the option of flexibility for resource allocation. To better analyze the different solution approaches, we compare the performances based on the flexibility provided by the RA-PST at design time. We adopt the entropy measure proposed in [16], which, in our proactive case, represents the flexibility of configurable PSTs. A higher entropy represents a higher flexibility since more similar branches are available for a decision. The entropy of an RA-PST is calculated as the mean entropy over all PST tasks. The entropy for a single task t with a set of branches B and the corresponding costs of a branch c_b can be calculated by using the Sum of all inverted costs: $\sigma = \sum_{b' \in B} \frac{1}{c_{b'}}$ The entropy for one task is then defined as:

$$H_t = -\sum_{b \in B} \frac{1}{c_b \cdot \sigma} \log(\frac{1}{c_b \cdot \sigma}) \qquad \forall b \in B \qquad (24)$$

Figure 5 shows the relative makespan deviation for each test over the entropy normalized over all test sets. We can see that the greedy allocation performs best when the entropy, and therefore flexibility, is high. The high reactivity of the ad-hoc approach facilitates the flexibility the best. For RA-PSTs with lower entropy, the greedy allocation cannot find as good objective values as the rescheduling ICP. The degree of flexibility given in a test set barely impacts the objective value if no rescheduling is allowed. We suspect that rescheduling ICP could perform better if the δ parameter is increased.

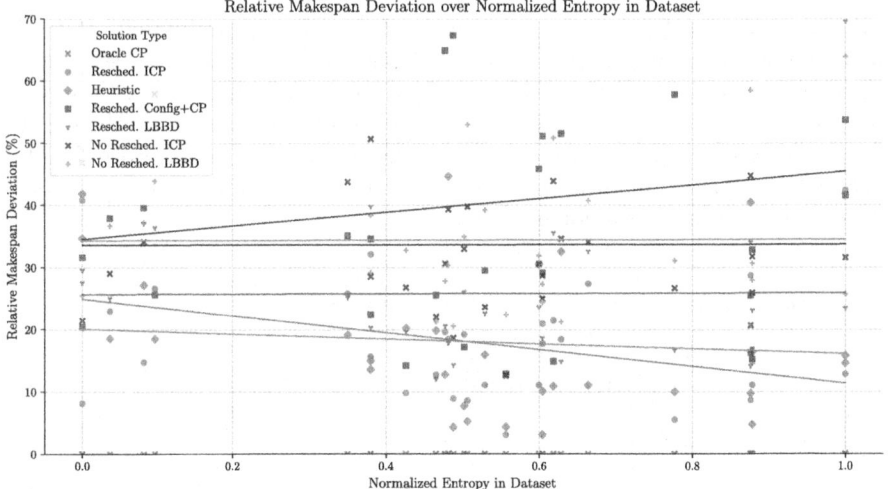

Fig. 5. Relative makespan deviation by normalized entropy over all RA-PSTs. Lines represent the trend over all data points for each solution approach.

5 Related Work and Discussion

The problem of (optimal) business process scheduling has been studied in the BPM community with different focus areas on the combination of BPM and OR methods. [6] proposes process fragmentation to deal with uncertainties in process models and uses Answer Set Programming to solve the scheduling problem. The approach can be used offline and in a quasi-online approach. In [7] different Answer-Set-Programming solvers are used to solve a scheduling problem and compared. The problem complexity is defined by the parallelity of the business process and the number of tasks. The maximum problem size contains 64 tasks and 32 resources and uses only a single process instance with parallelities. [18] proposes a method to learn scheduling models from timed petri nets and automatically solves these scheduling models with the help of CP. As stated in [18], most approaches for scheduling of timed petri nets focus on basic scheduling elements like precedence constraints and resources with limited capacities. When looking at scheduling of different alternatives for a process from the OR perspective, [19,20] propose metaheuristics to deal with the Resource Constrained

Project Scheduling Problem with arbitrary subgraphs. These problems allow for different execution alternatives. In distinction to the aforementioned RA-PST problem, alternatives are not linked to the resources executing them.

Discussion: Besides precedence constraints, process models can also contain exclusive choices, loops, and parallels as part of the control flow. In accordance with [1,10], we deal with purely sequential processes in which the variability is created through the branches of the RA-PST. Adding exclusive choices requires online process execution data to deal with the added uncertainty and is therefore not feasible for this study [6]. Dealing with uncertainty and online schedules is also a separate research field in OR [5,15]. By scheduling multiple process instances, we already deal with parallelities in terms of the scheduling problem itself. Comparable approaches in BPM create parallelity by scheduling a single instance with parallel gateways [7]. While the presented solution models can be easily adapted to allow for parallel gateways within a process instance, it would further increase the computational complexity of the problem.

Literature on the IPPS limits the number of alternative process plans (here, configurations of an instance) that are given to the optimization algorithm. In [9], a preselection of only four possible process plans for a part type (process instance) is done by choosing the cheapest options. This lowers the complexity of the problem drastically. For the BPM community, larger instances, such as those used for our evaluation, are typical, and the RA-PST naturally leads to bigger problem spaces. This might be an explanation of why the LBBD approach did not always find the optimum. For future work, better strengthened cuts can improve the performance of LBBD. [1] proposes additional cuts, which could lead to the deletion of optimal solutions. We refrain from using these cuts on our problem.

6 Conclusion

We presented three scheduling methods for the configuration and scheduling problem on RA-PSTs. The ICP formulation, which handles configuration and scheduling combined, finds good upper bounds for any problem in our experiments. Yet the approach struggles to find meaningful lower bounds, once the problem size grows. In contrast, the LBBD finds better lower bounds for larger problem sizes but struggles to find the good upper bounds the ICP can deliver in a timely manner. In the online setting, the ICP finds the optimal solution for single instances in most cases. Due to the reduced problem size in this setting also bigger process models with many tasks and branches become solvable. The LBBD approach is struggles in the online setting. For the BPM community, this work opens a path to using combinatorial optimization methods for configurable, executable process models. We show that the ICP formulation can be used in offline and online settings. In future work, introducing runtime data from an execution engine to online decision-making could be used to further improve both configuration and scheduling at process instantiation. Tying scheduling closer to business process execution will help to make BPM more proactive not only from a control flow but also from a resource perspective.

Acknowledgments. This work was funded by the Deutsche Forschungsgemeinschaft (DFG, German Research Foundation) – GRK2201 - Projektnummer – 277991500.

References

1. Barzanji, R., Naderi, B., Begen, M.A.: Decomposition algorithms for the integrated process planning and scheduling problem. Omega **93** (2020)
2. Carlier, J., Pinson, E.: An algorithm for solving the job-shop problem. Manage. Sci. **35**(2), 164–176 (1989)
3. Demange, M.: A note on the approximation of a minimum-weight maximal independent set. Comput. Optim. Appl. **14**(1), 157–169 (1999)
4. Garey, M.R., Johnson, D.S.: Computers and Intractability: A Guide to the Theory of NP-Completeness. W. H. Freeman (1979)
5. Gupta, D., Maravelias, C.T., Wassick, J.M.: From rescheduling to online scheduling. Chem. Eng. Res. Des. **116**, 83–97 (2016)
6. Havur, G., Cabanillas, C.: History-aware dynamic process fragmentation for risk-aware resource allocation. In: OTM, pp. 533–551 (2019)
7. Havur, G., Cabanillas, C., Polleres, A.: Benchmarking answer set programming systems for resource allocation in business processes. Expert Syst. Appl. **205** (2022)
8. Hooker, J.: Logic-Based Benders Decomposition: Theory and Applications. Springer (2023)
9. Jain, A., Jain, P., Singh, I.: An integrated scheme for process planning and scheduling in FMS. Int. J. Adv. Manuf. Technol. **30**, 1111–1118 (2006)
10. Naderi, B., Begen, M.A., Zaric, G.S.: Type-2 integrated process-planning and scheduling problem: reformulation and solution algorithms. Comput. Oper. Res. **142**, 105728 (2022)
11. Naderi, B., Ruiz, R., Roshanaei, V.: Mixed-integer programming vs. constraint programming for shop scheduling problems: new results and outlook. INFORMS J. Comput. **35**(4), 817–843 (2023)
12. Nie, L., Gao, L., Li, P., Li, X.: A GEP-based reactive scheduling policies constructing approach for dynamic flexible job shop scheduling problem with job release dates. J. Intell. Manuf. **24**(4), 763–774 (2013)
13. Pinedo, M.L.: Scheduling: Theory, Algorithms, and Systems, 3rd edn. Springer (2008)
14. Reichert, M., Weber, B.: Enabling Flexibility in Process-Aware Information Systems - Challenges, Methods, Technologies. Springer, Heidelberg (2012)
15. Sabuncuoglu, I., Goren, S.: Hedging production schedules against uncertainty in manufacturing environment with a review of robustness and stability research. Int. J. Comput. Integr. Manuf. **22**(2), 138–157 (2009)
16. Saidi, M., Tissaoui, A., Benslimane, D., Faiz, S.: Uncertainty measurement of a configurable business process. Syst. Eng. **26**(2), 199–215 (2023)
17. Schumann, F., Rinderle-Ma, S.: Optimizing resource-driven process configuration through genetic algorithms. In: BPM, vol. 14940, pp. 3–20 (2024)
18. Senderovich, A., Booth, K.E.C., Beck, J.C.: Learning scheduling models from event data. In: ICAPS, pp. 401–409 (2019)

19. Servranckx, T., Coelho, J., Vanhoucke, M.: A genetic algorithm for the resource-constrained project scheduling problem with alternative subgraphs using a Boolean satisfiability solver. Eur. J. Oper. Res. **316**(3), 815–827 (2024)
20. Servranckx, T., Vanhoucke, M.: A tabu search procedure for the resource-constrained project scheduling problem with alternative subgraphs. Eur. J. Oper. Res. **273**(3), 841–860 (2019)
21. Vanhatalo, J., Völzer, H., Leymann, F.: Faster and more focused control-flow analysis for business process models through SESE decomposition. In: ICSOC, pp. 43–55 (2007)

Rethinking Business Process Simulation: A Utility-Based Evaluation Framework

Konrad Özdemir[1]([✉]), Lukas Kirchdorfer[1,2], Keyvan Amiri Elyasi[1], Han van der Aa[3], Heiner Stuckenschmidt[1]

[1] Data and Web Science Group, University of Mannheim,
Mannheim, Germany
{konrad,heiner}@informatik.uni-mannheim.de
[2] SAP Signavio, Walldorf, Germany
lukas.kirchdorfer@sap.com
[3] Faculty of Computer Science, University of Vienna,
Vienna, Austria
han.van.der.aa@univie.ac.at

Abstract. Business process simulation (BPS) is a key tool for analyzing and optimizing organizational workflows, supporting decision-making by estimating the impact of process changes. The reliability of such estimates depends on the ability of a BPS model to accurately mimic the process under analysis, making rigorous accuracy evaluation essential. However, the state-of-the-art approach to evaluating BPS models has two key limitations. First, it treats simulation as a forecasting problem, testing whether models can predict unseen future events. This fails to assess how well a model captures the as-is process, particularly when process behavior changes from train to test period. Thus, it becomes difficult to determine whether poor results stem from an inaccurate model or the inherent complexity of the data, such as unpredictable drift. Second, the evaluation approach strongly relies on Earth Mover's Distance-based metrics, which can obscure temporal patterns and thus yield misleading conclusions about simulation quality. To address these issues, we propose a novel framework that evaluates simulation quality based on its ability to generate representative process behavior. Instead of comparing simulated logs to future real-world executions, we evaluate whether predictive process monitoring models trained on simulated data perform comparably to those trained on real data for downstream analysis tasks. Empirical results show that our framework not only helps identify sources of discrepancies but also distinguishes between model accuracy and data complexity, offering a more meaningful way to assess BPS quality.

Keywords: Process simulation · Process mining · Deep learning

1 Introduction

Business process simulation (BPS) plays a crucial role in the analysis and redesign of organizational processes. By creating a digital process twin [8], sim-

---

* K. Özdemir, L. Kirchdorfer and K. Amiri Elyasi—Equal contribution.

ulation enables the estimation of the impact of a process change on key performance indicators, such as cycle time, resource utilization, and waiting time for specific activities—an approach commonly referred to as counterfactual reasoning or "what-if" analysis [9]. By providing such estimates in advance, BPS can significantly enhance decision-making by improving efficiency and reducing the risks associated with process redesign [9]. Furthermore, automated approaches that derive process simulation models from historical execution data [3,6,13,17,19] eliminate the need for manual model construction, which is both time-consuming and error-prone [1]. However, the reliability of process simulation hinges on the accuracy of the underlying model. Only models that faithfully replicate the as-is process behavior can provide meaningful and trustworthy insights into the effects of potential changes. This raises a fundamental question: how can we effectively assess the quality of a simulation model?

In this work, we argue that the current approach to evaluating BPS models has two key issues. First, it frames simulation as a forecasting problem, evaluating the quality of a BPS model by comparing simulated event logs to unseen future process executions. We argue that this does not evaluate the model's accuracy in capturing the as-is process, in particular when the process behavior in the test period differs significantly from the training period. Thus, the current approach makes it difficult to determine whether poor results stem from an inaccurate BPS model or from dynamics not captured in the training data, such as an increasing workload due to higher customer demand reflected by more frequent case arrivals during the test period. Second, the metrics commonly used to compare simulated logs against real logs are mostly based on the Earth Mover's Distance (EMD). However, we give theoretical and empirical evidence that EMD can obscure temporal patterns and has a bias toward favoring estimates close to the mean, potentially yielding misleading conclusions about simulation quality.

In this work, we critically examine the limitations of the current evaluation approach for BPS and propose a novel framework that better aligns with the fundamental purpose of simulation. Rather than treating simulation as a forecasting problem, we advocate for an evaluation paradigm that assesses whether a simulation model generates process behavior that accurately mimics the observed as-is process. Our proposed framework shifts the focus from direct log-to-log comparisons toward assessing the *utility* of simulated data in downstream tasks. Specifically, we evaluate the quality of simulated event logs by measuring their effectiveness in training predictive process monitoring (PPM) models. If a model trained on simulated data—tasked with predicting next activities, remaining time, or other process-related properties—achieves performance comparable to one trained on real event logs, it suggests that the simulation model provides a realistic representation of reality. This utility-based perspective offers a more meaningful assessment of simulation accuracy and allows to discern between the pure model quality and data complexity.

The remainder starts with discussing related work in Sect. 2, before motivating our new framework by showing the limitations of the existing approach in Sect. 3. Afterward, Sect. 4 describes our evaluation framework with an experimental evaluation in Sect. 5. Finally, Sect. 6 concludes our work.

2 Related Work

This section reviews prior work on BPS model evaluation and discusses the use of downstream tasks for assessing the quality of synthetic data in the broader Machine Learning (ML) community.

Evaluation of BPS Models. The evaluation of BPS models has long been secondary to the development of novel simulation approaches, often treated as a by-product rather than a research focus in itself. Early approaches, such as Rozinat et al. [19], relied on manual comparisons of simulated and real event logs, analyzing activity execution times and gateway probabilities. Khodyrev and Popova [12] incorporated the idea of a temporal train-test split, comparing simulated and test logs along metrics such as number of case arrivals and activity durations. With the rise of data-driven simulation, Camargo et al. [5] proposed a more structured evaluation combining control-flow and temporal aspects, using metrics like control-flow log similarity, mean absolute error (MAE) of cycle times, and EMD of activity durations. In a subsequent work [6], the same authors refined this approach, using MAE of cycle times and EMD of activity timestamps, also influencing later BPS studies [16]. Despite these advancements, evaluation in BPS remained fragmented until Chapela-Campa et al. [7] explicitly tackled the question of how to assess BPS models. Their work introduced a more holistic evaluation framework, incorporating metrics from control-flow, temporal, and congestion perspectives, primarily relying on EMD and Wasserstein-1 distance to compare simulated and test logs. This framework has quickly become the de facto standard, having been applied in several recent BPS studies [13,15,17]. However, while this framework represents a significant step forward, it has critical shortcomings, as we will demonstrate in Sect. 3.

Downstream Tasks as Means for Evaluation. Since BPS focuses on the generation of synthetic data, our work draws inspiration from the way in which the quality of synthetically generated data is assessed in other contexts. Specifically, a common practice is the assessment of data quality through downstream task performance, following the *Train on Synthetic, Test on Real* (TSTR) paradigm [11]. This evaluation method has been widely applied across domains, including time series augmentation [24], natural language processing [18], object detection [20], and crowd counting [22]. In TSTR, a predictive model is trained on generated synthetic data and evaluated on hold-out real-world test data, with its performance compared to a model trained on real data. This approach provides a meaningful assessment of the generative model, as it measures how well the generated data captures relevant patterns required for generalization and how much utility has been retained by the generation w.r.t. a specific ML task [11]. While TSTR has predominantly been used for image and text data, its adaptation to process data presents new opportunities. A key challenge lies in identifying suitable downstream tasks that account for process-specific characteristics such as time-dependent behavior and resource constraints. We close this gap by adapting the TSTR principle to the process mining domain, including appropriate downstream models and tasks.

3 Motivation

In this section, we examine the current state-of-the-art approach to evaluating BPS models proposed by Chapela-Campa et al. [7], which we refer to as the *Standard Practice*. We then identify and discuss its limitations, which can be summarized by two key issues: *Objective Mismatch* and *Metric Shortfall*. The former addresses the conflation of simulation and forecasting objectives, while the latter provides theoretical and empirical evidence of critical flaws in the evaluation metrics used.

3.1 The Standard Practice of BPS Evaluation

Preliminaries. Automated BPS approaches rely on historical process execution data, which is typically captured in *event logs* [9]. An event log $\mathcal{L}$ is a multi-set of traces, where each trace $\sigma \in \mathcal{L}$ represents a sequence of events $(e_1, \ldots, e_n)$, capturing the execution of a single process instance. Each event e_i is defined as a tuple $\langle a, r, \tau_{\text{start}}, \tau_{\text{end}} \rangle$, where a denotes the executed activity, r the responsible resource, and τ_{start} and τ_{end} the corresponding start and end timestamps.

Evaluation Procedure. The state-of-the-art approach for evaluating simulation accuracy follows a structured flow consisting of five key steps and is largely inspired by the works of Chapela-Campa et al. (cf. [7]).

1. The traces of a given event log $\mathcal{L}$ are partitioned temporally into a training $\mathcal{L}_{\text{train}}$ and a testing log $\mathcal{L}_{\text{test}}$.
2. A BPS model S is discovered using $\mathcal{L}_{\text{train}}$. For example, via *Simod* [3], *AgentSimulator* [13], or *DeepSimulator* [6].
3. S is used to simulate a new event log $\mathcal{L}_{\text{sim}}$, starting at the same time as $\mathcal{L}_{\text{test}}$ and including the same number of traces.
4. To facilitate a direct comparison between $\mathcal{L}_{\text{sim}}$ and $\mathcal{L}_{\text{test}}$, *proxies*[1] reflecting the property of interest are derived from each event log. The prevalent approach for proxy derivation across nearly all metrics involves sequence binning, where events are grouped into B one-hour intervals based on their τ_{start} and τ_{end} timestamps. Counting the events in each bin produces two *count-sequences* $\bar{x} = (x_1, \ldots, x_B)$ for $\mathcal{L}_{\text{sim}}$ and $\bar{y} = (y_1, \ldots, y_B)$ for $\mathcal{L}_{\text{test}}$.
5. The discrepancy between $\mathcal{L}_{\text{sim}}$ and $\mathcal{L}_{\text{test}}$ is quantified using the Wasserstein-1 (W_1) distribution distance applied to the binned *count-sequences*. Formally: $\text{Dist}(\mathcal{L}_{\text{sim}}, \mathcal{L}_{\text{test}}) = W_1(\bar{x}, \bar{y})$. Simply put, W_1 measures the *minimal effort* needed to redistribute probability mass from one histogram to match the

---

[1] We use the term proxy in the econometric sense, referring to a variable that serves as a representation of another variable of interest that cannot be directly quantified.

other. In this sense, a lower score indicates a better result. Following Chapela-Campa et al. (cf. [7]), we refer to the W_1 distance only, as it poses a more efficient Earth Mover's Distance implementation.[2]

Having established the *Standard Practice*, we now highlight the most critical issue inherent in this approach.

3.2 Objective Mismatch

Establishing a simulation model involves extracting meaningful properties from a reference dataset ($\mathcal{L}_{\text{train}}$) and embedding these into the model S to effectively serve as a digital process twin [8]. The overall quality of this model should then be measured by how well the model's simulated dataset ($\mathcal{L}_{\text{sim}}$) captures the reference dataset's statistical properties [10]. The *Standard Practice* violates this principle through an *Objective Mismatch*: the simulated event log is compared to a test log that may significantly diverge from the training log. In our view, this conflates simulation with forecasting and undermines a clear assessment of model quality. Consequently, we propose to fix this mismatch by using $\mathcal{L}_{\text{train}}$ instead of $\mathcal{L}_{\text{test}}$ as reference for assessing $\mathcal{L}_{\text{sim}}$. The following scenario illustrates how this mismatch can distort evaluation outcomes.

Scenario: Time-Varying Behavior in Standard Practice. In Table 1, we examine a *Loan Application* process (cf. [7]) under two scenarios: the *original* log with stable arrival rates and another where we introduced a *drift* (arrival rate surge) close to the start of the test period. First, we simulate two event logs via *Agentsimulator* [13], one for the *original*- and one for the *drift*-scenario. Second, six distribution distances were used to evaluate each simulated log against its respective test log: *NGD* (N-Gram), *AEDD* (Absolute Event), *CADD* (Case Arrival), *CEDD* (Circadian Event), *REDD* (Relative Event), and *CTDD* (Cycle Time) [7]. Now, the results obtained in the *drift* scenario might lead one to conclude that the simulation model is rather poor, as indicated by high error rates in Table 1. However, *Agentsimulator* actually provides a good representation of the process, as indicated by small error values in the *original* scenario. In fact, it is the evaluation method under the *Standard Practice* that does not adequately distinguish between the intrinsic quality of the BPS model and the complexity

Table 1. Comparison of distribution distances between the *Original* version of the Loan Application process and a version with altered arrival rate (*Drift*).

Log	NGD	AEDD	CADD	CEDD	REDD	CTDD
Original	0.07	2.97	0.00	0.27	1.66	2.71
Drift	0.20	95.03	121.87	0.49	26.32	32.85

---

[2] For more details: https://github.com/konradoezdemir/Rethinking-BPS.

introduced by concept drift. For example, *AEDD* increases from 2.97 under stable conditions to 95.03 when a drift is introduced, while *CADD* rises from 0.00 to 121.87. In conclusion, this scenario demonstrates that the perceived quality of a BPS model can appear extremely poor when evaluated using the *Standard Practice* on a process with changing behavior.

Process behavior changes, as in the above scenario, are not merely superficial. In fact, half of the processes commonly examined in BPS studies [6,13,17] exhibit substantial drifts in cycle time (cf. Table 3).

One might naturally think that this *Objective Mismatch* issue can be resolved by computing distribution distances between the simulated log $\mathcal{L}_{\text{sim}}$ and the training log $\mathcal{L}_{\text{train}}$, rather than the test log $\mathcal{L}_{\text{test}}$. However, we argue that even when this mismatch is accounted for, the *Standard Practice* remains inadequate, as we discuss next.

3.3 Metric Shortfall

From a statistical perspective, W_1-based distance metrics (e.g., *AEDD*, *CADD*) require large sample sizes to accurately estimate the underlying probability distributions. This ensures that the computed differences between these estimates are both robust and reliable. In high-dimensional settings where the sample complexity can grow significantly, this necessity becomes especially pronounced (cf. [23]). Considering this, we encounter a fundamental difficulty when applying the W_1 distance to said *count-sequences* (cf. step 4): How should these sequences be interpreted in light of the assumptions made by W_1-based metrics? One option is to view the entire sequence as a single representation of the event log. However, this leads to a drastic sampling bias because each log ($\mathcal{L}_{\text{sim}}$ and $\mathcal{L}_{\text{test}}$) is represented by only one sequence. Alternatively, and as implicitly assumed in these distance metrics, one might treat each individual count within the sequence as an *independent observation* from the underlying event log's distribution. While this approach aligns with the formulation of the W_1 distance, it neglects the temporal dependencies and structural relationships between activities that are inherent to event logs.

We formally prove that—despite the statistical appeal of treating individual counts as independent realizations—this approach undermines the proper representation of temporal properties, as detailed in the following theorem.

Theorem 1. *Let* **P** *and* **Q** *denote the probability distributions associated with* $\mathcal{L}_{\text{train}}$, $\mathcal{L}_{\text{test}}$ *with samples* $(x_i)_{i=1}^{B}$ *and* $(y_i)_{i=1}^{B}$ *in* $\mathbb{R}^1$. *Then, one can show*[3] *that the Wasserstein-1 distance admits the following form:*

$$W_1(\mathbf{P}, \mathbf{Q}) = \frac{1}{B} \sum_{i=1}^{B} |x_{(i)} - y_{(i)}|. \quad (1)$$

As established in Theorem 1, computing the W_1 distance between two *count-sequences* requires sorting each sequence in non-decreasing order; indicated via

---

[3] For a formal proof: https://github.com/konradoezdemir/Rethinking-BPS.

'(i)'. In doing so, the smallest value from the first sequence is compared with the smallest from the second, the second smallest with the second smallest, etc. This reordering removes any original temporal structure, rendering likely interdependencies between events irrelevant. A simple example may look as follows:

Example 1. Consider hourly call center dial-ins from 10am to 3pm (i.e., $B = 5$ hours), which are down-trending with true counts $y_. = (5,4,3,1,1)$ (**Q**). Consider a pattern-neglecting *bad* estimate $\tilde{x}_. = (1,3,1,5,4)$ ($\widetilde{\mathbf{P}}$), and a pattern-recognizing *good* estimate $x_. = (5,5,3,1,1)$ (**P**). Sorting non-decreasingly yields $y_{(.)} = (1,1,3,4,5)$, $\tilde{x}_{(.)} = (1,1,3,4,5)$ and $x_{(.)} = (1,2,3,5,5)$. By Theorem 1, $W_1(\widetilde{\mathbf{P}}, \mathbf{Q}) = 0$ and $W_1(\mathbf{P}, \mathbf{Q}) = \frac{1}{5}(|4-5|) = 0.2$, preferring $\tilde{x}_.$ over $x_.$ here.

This example underscores that, by neglecting temporal ordering, the W_1 distance can falsely favor a bad estimate of the true data ($\tilde{x}_.$), while a more realistic one ($x_.$) is penalized. Moreover, Theorem 1 reveals another important aspect about W_1: *susceptibility to outliers*. With similarity to the *MAE*, once a largely deviating pair of values occurs, the overall distance falls at risk to explode. With the following scenario, we illustrate how the *Standard Practice* can fail in selecting the optimal simulation model under this exact premise.

Scenario: Model Selection via Standard Practice. Fig. 1 overlays three histograms representing empirical inter-arrival time distributions for the *BPIC12W* event log. One stems from the test log $\mathcal{L}_{\text{test}}$, while the others are generated by *AgentSimulator* [13]. In the 'Simulated' case, *AgentSimulator* discovers and models arrivals (i.e., $\mathcal{L}_{\text{sim}}$), whereas in 'Simulated Mean', arrivals are generated using only the mean inter-arrival time from the train log $\mathcal{L}_{\text{train}}$. While both 'Simulated' and 'Test' exhibit similar inter-arrival distributions (resembling an exponential distribution), 'Simulated Mean' collapses into a degenerate distribution, concentrating all mass at a single value. From a practitioner's perspective, the choice is clear: the mean-based approach is unsuitable for meaningful simulation, whereas *AgentSimulator* reasonably approximates the distribution associated with $\mathcal{L}_{\text{test}}$. However, under the *Standard Practice*, i.e., using the *CADD* metric, 'Simulated Mean' paradoxically achieves a better score (42.7) than 'Simulated' (55.8). This is because, as outlined before, the W_1

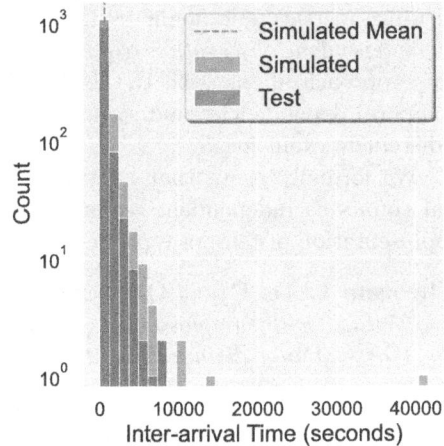

Fig. 1. Inter-arrival time histograms for the BPIC12W log.

distance is susceptible to outliers. This means that it heavily *penalizes* large, wrong estimates and instead, tends to favor observations close to the mean value (cf. [21]). This phenomenon is clearly illustrated in the histogram: although 'Simulated' effectively covers the 'Test' distribution better than 'Simulated Mean', the outlier estimate just after the 40,000 mark produces a substantial error that significantly increases the overall score. Consequently, when model selection is based on this metric, one would inadvertently favor a model that employs the simulated mean approach, even though it fails to capture the true variability of the event log.

In light of these pitfalls, we propose a new, comprehensive evaluation framework for business process simulation.

4 Utility-Based Evaluation Framework

In this section, we introduce the details of our utility-based BPS evaluation framework. The core idea is to assess the quality of BPS models by measuring how well their simulated data supports downstream prediction tasks compared to real training data. As shown in the visualization in Fig. 2, our proposed evaluation framework consists of five steps. We describe each of these next, while also highlighting key differences to the *Standard Practice* discussed in Sect. 3.1.

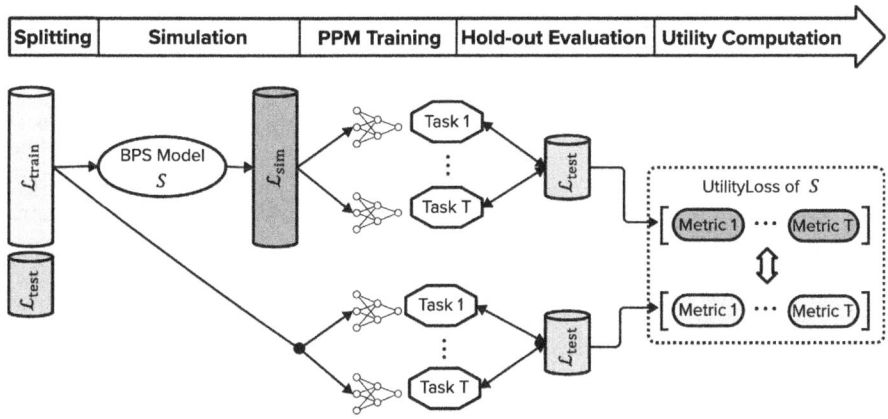

Fig. 2. Overview of our proposed BPS evaluation framework.

Splitting. The first step is to split an event log temporally into train log $\mathcal{L}_{\text{train}}$ and test log $\mathcal{L}_{\text{test}}$.

Simulation. Then, $\mathcal{L}_{\text{train}}$ is used to train a BPS model S. Unlike the *Standard Practice*, which aims to align $\mathcal{L}_{\text{sim}}$ with the characteristics of $\mathcal{L}_{\text{test}}$—implicitly treating simulation as a forecasting task and leading to an objective mismatch—we ensure that S generates a simulated log $\mathcal{L}_{\text{sim}}$ that mirrors the properties of $\mathcal{L}_{\text{train}}$. Specifically, $\mathcal{L}_{\text{sim}}$ should contain the same number of cases as $\mathcal{L}_{\text{train}}$ and start at the same point in time.

PPM Training. Next, we train predictive process monitoring (PPM) models w.r.t. T different *downstream* tasks, separately using real data $\mathcal{L}_{\text{train}}$ and simulated data $\mathcal{L}_{\text{sim}}$. This results in two distinct predictive models per task $t \in \{1, \ldots, T\}$: $\text{PM}_t(\mathcal{L}_{\text{train}})$ and $\text{PM}_t(\mathcal{L}_{\text{sim}})$. To enable a comprehensive evaluation, our framework incorporates prediction tasks spanning the four key process perspectives commonly assessed in BPS evaluation [7]:

Control-Flow. The control-flow perspective captures the order and dependency relations among activities. To capture this, we instruct our models to predict the next activity (NAP) in ongoing cases, assessing whether the BPS model preserves event order.

Resource. The resource perspective captures the workforce of the process. By predicting the next role (NRP) in ongoing cases, we assess whether the BPS model considers role assignment rules, ordering, and interaction patterns.

Temporal. The temporal perspective captures process timing, including event durations, case durations, and inter-event intervals. This is covered by predicting the next activity processing time (NPP), the next waiting time between the end of the previous activity and the start of the new activity (NWP), and the remaining time until the completion of the case (RTP).

Congestion. The congestion perspective captures the workload over time, including queuing effects and resource contention. We cover this perspective by already introduced tasks: NWP (waiting time), RTP (remaining time), and NRP (role). These tasks are suitable proxies for congestion as they capture the amount of waiting time (e.g., due to resource contention), the overall length that a process instance stays in the system, and role assignments.

Note that the downstream tasks described above may not encompass all relevant process characteristics, such as comprehensive workload or queuing dynamics. However, the modular structure of our framework makes it easy to extend the set of tasks to include additional process characteristics as needed. Also, it is the practitioner's choice if one model should be trained per task or if all tasks should be trained in a single model (Multi-Task Learning). Additionally, to account for potential model bias, each task can be trained using K different architectures—such as an LSTM and a Transformer model, as will be done in our evaluation in Sect. 5.

Hold-Out Evaluation. Having trained the PPM models on both real data $\mathcal{L}_{\text{train}}$ and simulated data $\mathcal{L}_{\text{sim}}$, we assess their performance on the hold-out test log $\mathcal{L}_{\text{test}}$. Each task $t \in \{1, \ldots, T\}$ is evaluated using a specific metric $\mathcal{M}_t$, such as MAE for RTP and accuracy for NAP. Rather than combining these metrics into a single score, we maintain a *process perspective-specific evaluation* to preserve interpretability. Thus, for an event log $\mathcal{L} \in \{\mathcal{L}_{\text{train}}, \mathcal{L}_{\text{sim}}\}$, the resulting metrics for T tasks—averaged over K model architectures—are represented as a vector:

$$\mathcal{M}(\mathcal{L}) = \left[\frac{1}{K} \sum_{k=1}^{K} \mathcal{M}_1^{\mathcal{L}_{\text{test}}}(\text{PM}_1^k(\mathcal{L})), \ldots, \frac{1}{K} \sum_{k=1}^{K} \mathcal{M}_T^{\mathcal{L}_{\text{test}}}(\text{PM}_T^k(\mathcal{L})) \right] \qquad (2)$$

Utility Computation. To evaluate how well a BPS model S preserves the predictive utility of the original process data, we introduce the concept of *utility loss*. This metric quantifies the extent to which a model trained on simulated data deviates in performance from one trained on real data across various downstream tasks. The rationale behind this choice is that task-specific performance metrics can serve as a proxy for the practical utility of the data. Computing these metrics for both the real and simulated logs yields $\mathcal{M}(\mathcal{L}_{\text{train}})$ and $\mathcal{M}(\mathcal{L}_{\text{sim}})$. These vectors can then be compared element-wise to assess how much utility is lost (or preserved) when training PPM models on simulated instead of real data. A small absolute deviation in accuracy or error suggests that the BPS model S successfully captures key patterns of $\mathcal{L}_{\text{train}}$, while a larger deviation—regardless of direction—indicates a loss in fidelity. Formally, we define the utility loss of a model S as the element-wise absolute difference between $\mathcal{M}(\mathcal{L}_{\text{train}})$ and $\mathcal{M}(\mathcal{L}_{\text{sim}})$:

$$\text{UtilityLoss}(S) = \left|\mathcal{M}(\mathcal{L}_{\text{train}}) - \mathcal{M}(\mathcal{L}_{\text{sim}})\right|. \tag{3}$$

Note that our framework is intended to be used for the evaluation of BPS models capable of supporting what-if analysis. Therefore, although we recognize that approaches that effectively copy the initial event log $\mathcal{L}_{\text{train}}$ would yield a near-optimal UtilityLoss, they fail to meet a fundamental requirement of BPS. As such, they are not suitable for evaluation within our framework.

5 Evaluation

This section outlines the experiments conducted to evaluate our utility-based framework for measuring BPS quality. We perform two experiments: The first (Sect. 5.1) tests the framework's applicability by verifying whether known modifications to a simulation model are reflected accordingly. The second (Sect. 5.2) benchmarks state-of-the-art BPS approaches, highlighting their strengths, weaknesses, and insights gained through our framework. Our Python implementation and additional results are available in our public repository (cf. [25]).

5.1 Experiment 1: Applicability of Our Framework

To validate our framework's applicability, we examine whether known modifications to a BPS model result in a corresponding UtilityLoss. An effective evaluation should (i) identify inaccurate models and (ii) reveal which perspective of the process they fail to capture appropriately. We first describe the experimental setup before discussing the results.

Experimental Setup. In this experiment, we largely follow the synthetic evaluation from Chapela-Campa et al. [7], using the *Loan Application* process as the basis (cf. [7]; see Table 3 for details). We temporally split the event log trace-wise into 80% training and 20% test data. The process comprises 12 activities, beginning with *Check application form completeness*. Its control-flow structure includes a loop, a parallel branch involving three activities, three exclusive gateways, and three possible end points: *Approve application*, *Reject application*, and *Cancel application*. In total, the process is carried out by 19 distinct resources.

Scenarios. Given the train set of this process, we use the *Simod* BPS approach [3] to discover a ground truth simulation model ($Loan_{GT}$) and create different modifications (following Chapela-Campa et al. [7]), which will allow us to assess whether our framework effectively penalizes these models in the respective downstream tasks. For each of the following models[4], we simulate 10 event logs that align with the train log in terms of start time and number of cases:

- $Loan_{SEQ}$: arranging the three parallel activities as a sequence.
- $Loan_{S\text{-}G}$: altering, on top of $Loan_{SEQ}$, the branching probabilities.
- $Loan_{RC}$: halving the available resources.
- $Loan_{EXT}$: adding extraneous waiting time to delay the start of activities.
- $Loan_{DUR}$: increasing the duration of the activities of the process.
- $Loan_{CAL}$: changing resource working schedules from 9am–5pm to 2pm–10pm.
- $Loan_{ARR}$: increasing the rate of case arrivals.

Tasks and Evaluation Metrics. For downstream utility assessment, we consider the five PPM tasks described in Sect. 4. To quantify loss, we use *accuracy* as the metric for the two classification tasks (NAP and NRP) and *MAE* for the three regression tasks (NPP, NWP, and RTP).

PPM Approach. We use *ProcessTransformer* [2] as the approach for the PPM tasks. Since this model was designed to just predict the next activity (NAP), remaining time (RTP), and the next timestamp, we adapted it so that it can distinguish between processing and waiting times, thus enabling the NPP and NWP tasks. We also extended it to predict the role responsible for the next activity, enabling NRP. We use the model's default configuration.

Results and Discussion. Table 2 presents the predictive performance obtained with the simulated data from the ground truth model $Loan_{GT}$ and the Utility-Loss for each modification, illustrating how our framework effectively detects alterations in the respective process perspectives.

Control-Flow. Significant deviations from the ground truth model ($Loan_{GT}$) in the control-flow perspective are expected only in cases where the activity order is altered, as in $Loan_{SEQ}$ and $Loan_{S\text{-}G}$. Since these changes disrupt the original sequence of events, our framework correctly assigns a UtilityLoss in NAP for these two models, while all other models align with the ground truth, as expected.

Resource. The resource-related metric NRP should reflect deviations not only for $Loan_{SEQ}$ and $Loan_{S\text{-}G}$ (where activity order changes can impact resource allocation) but also for $Loan_{RC}$, where resource assignments are altered due to halving available resources. Our framework appropriately penalizes these models, capturing the expected disruptions in resource allocation.

---

[4] The specifics of these modifications and their respective logs are in our repository.

Table 2. Average utility loss (with standard deviation) for modifications of the $Loan_{GT}$ process. NPP, NWP, and RTP errors are measured in minutes.

	NAP	NRP	NPP	NWP	RTP
$Loan_{GT}$	0.71 (0.01)	0.75 (0.00)	67.60 (2.80)	12.13 (1.48)	238.26 (8.38)
$Loan_{SEQ}$	0.26 (0.03)	0.26 (0.04)	16.12 (3.29)	12.03 (0.52)	35.88 (18.04)
$Loan_{S-G}$	0.32 (0.07)	0.35 (0.04)	12.40 (2.91)	715.84 (112.23)	4131.58 (1802.63)
$Loan_{RC}$	0.00 (0.01)	0.12 (0.14)	16.55 (3.31)	164.28 (63.69)	771.89 (368.17)
$Loan_{EXT}$	0.00 (0.01)	0.00 (0.00)	24.30 (5.19)	74.37 (6.92	508.74 (53.15)
$Loan_{DUR}$	0.00 (0.01)	0.00 (0.00)	49.63 (7.42)	7.27 (6.01)	263.29 (39.86)
$Loan_{CAL}$	0.00 (0.01)	0.00 (0.00)	0.44 (5.38)	0.45 (1.68)	5.43 (8.93)
$Loan_{ARR}$	0.00 (0.01)	0.00 (0.00)	10.60 (2.28)	35.14 (3.14)	90.1 (15.21)

Temporal. In the temporal perspective, the most substantial UtilityLoss occurs—as expected—in NPP for $Loan_{DUR}$, where activity durations are explicitly modified. Since duration changes influence process timing, this effect propagates to RTP, demonstrating our framework's ability to capture temporal deviations.

Congestion. Regarding congestion, NWP shows notable losses for $Loan_{S-G}$, $Loan_{RC}$, $Loan_{EXT}$, and $Loan_{ARR}$ aligning with the expected impact of sequential execution, resource contention, extraneous delays, and higher arrival frequency respectively. These congestion effects propagate to RTP, reflecting significantly longer cycle times for the three mentioned modifications.

Notably, $Loan_{CAL}$, modifying only absolute timestamps without affecting the relative temporal structure, correctly receives a near-zero UtilityLoss, as our framework evaluates temporal relationships rather than absolute timestamps.

In summary, our framework effectively identifies BPS models that deviate from the ground truth and precisely determines which process perspective is inadequately captured, ensuring a comprehensive evaluation of BPS accuracy.

5.2 Experiment 2: Benchmark

In this experiment, we showcase the practicality of our framework in benchmarking state-of-the-art BPS approaches. We first explain the experimental setup and then discuss the results.

Experimental Setup. For the benchmark, we rely on 8 event logs[5] (see details in Table 3) commonly used to evaluate BPS approaches [6,13,17]. As in the first experiment, we split event logs temporally, with 80% for training and 20% for testing. Each BPS model generates 10 simulated logs per process, which are used to train and evaluate the PPM models. We aggregate results across these 10 runs and compare them to models trained on real data. To account for performance variance, PPM models trained on real data use 10 random seeds. The tasks and evaluation metrics in Experiment 1 (Sect. 5.1) are also applied here.

---

[5] Datasets can be downloaded from here: https://zenodo.org/records/5734443.

Table 3. Description of event log properties. Average cycle time for train and test sets (CT-train, CT-test) are reported in days.

Log	Traces	Events	#Act/#Res	CT-train	CT-test
Loan App	1000	7492	12 / 19	0.42	0.46
P2P	608	9119	21 / 27	12.14	30.82
C. 1000	1000	38160	42 / 14	0.96	0.80
C. 2000	2000	77418	42 / 14	0.86	0.77
CVS	10000	103906	15 / 6	5.74	12.05
Production	225	4503	24 / 41	17.44	4.32
CDM	954	6870	18 / 432	10.54	3.60
BPI12W	8616	59302	6 / 52	7.92	8.16
BPI17W	30276	240854	8 / 136	12.47	11.50

BPS Approaches. We evaluate three state-of-the-art BPS approaches:

- *Simod* [3] is a traditional approach that derives a BPMN model along with a set of simulation parameters. In this work, we use an enhanced version of *Simod* that accounts for differentiated resource behavior [14].
- *DeepSimulator (DSim)* [6] is a hybrid approach that integrates a BPMN model to govern control flow while leveraging multiple deep learning models for learning temporal dynamics.
- *AgentSimulator (ASim)* [13] is an agent-based approach that represents each resource as an autonomous agent, simulating the process through agent interactions.

PPM Approaches. We use two PPM model architectures in our benchmark evaluation. The first, *ProcessTransformer* [2], is described in Sect. 5.1. The second, proposed by Camargo et al. [4], represents an *LSTM* architecture, capable of predicting the five downstream tasks. For each task, we train an LSTM with two hidden layers of size 50, each, and use fixed-size n-grams of 10.

Results and Discussion. Table 4 summarizes the results[6] of our second experiment. Each log presents results obtained using real training data, followed by the UtilityLoss of each BPS approach, separated by a dashed horizontal line. The results shown are averaged over both PPM model architectures, with the best-performing BPS approach per task boldened.

Overall, no single BPS approach consistently dominates across all logs and tasks. However, ASim consistently outperforms others in the control-flow dimension (NAP), while also most frequently leading in resource (NRP) and temporal tasks (NPP, NWP). Regarding RTP, results are rather mixed, with both *Simod* and ASim leading in three logs, respectively. In contrast, DSim only claims two.

---

[6] More detailed results on individual architecture runs can be found in our repository.

Table 4. Utility-based performance comparison of BPS models.

Log	Data	NAP	NRP	NPP(min)	NWP(hour)	RTP(day)
P2P	real	0.84 (0.02)	0.86 (0.01)	67.11 (9.38)	62.88 (0.76)	17.82 (0.63)
	Simod	0.03 (0.00)	0.36 (0.00)	65.13 (0.00)	**3.44** (0.00)	**1.44** (0.00)
	DSim	0.14 (0.02)	NA	51.46 (51.7)	6.12 (4.54)	6.65 (9.81)
	ASim	**0.00** (0.02)	**0.35** (0.02)	**0.60** (5.37)	9.74 (2.56)	3.05 (3.92)
C.1000	real	0.72 (0.01)	0.44 (0.01)	19.14 (0.15)	0.28 (0.01)	0.39 (0.03)
	Simod	0.22 (0.02)	**0.06** (0.02)	2.83 (0.40)	2.07 (0.39)	3.53 (1.48)
	DSim	0.52 (0.04)	NA	6.44 (2.84)	**0.19** (0.04)	**0.07** (0.06)
	ASim	**0.13** (0.02)	0.09 (0.01)	**0.20** (0.05)	2.83 (0.40)	0.17 (0.07)
C.2000	real	0.73 (0.01)	0.46 (0.01)	19.30 (0.29)	0.25 (0.00)	0.34 (0.03)
	Simod	0.20 (0.01)	**0.07** (0.02)	2.83 (0.73)	1.90 (0.38)	2.58 (1.38)
	DSim	0.53 (0.03)	NA	5.75 (2.19)	0.34 (0.08)	0.86 (1.37)
	ASim	**0.10** (0.02)	0.08 (0.07)	**2.40** (0.27)	**0.15** (0.04)	**0.09** (0.04)
CVS	real	0.76 (0.01)	0.80 (0.01)	3.26 (0.10)	20.31 (0.25)	3.16 (0.05)
	Simod	0.27 (0.01)	0.60 (0.04)	0.28 (0.20)	14.64 (1.79)	15.09 (1.05)
	DSim	0.14 (0.02)	NA	0.29 (0.22)	31.83 (9.75)	9.88 (2.00)
	ASim	**0.05** (0.01)	**0.02** (0.01)	**0.09** (0.01)	**14.17** (2.44)	**0.54** (0.03)
Production	real	0.57 (0.02)	0.49 (0.03)	120.83 (1.91)	12.45 (0.99)	8.18 (2.11)
	Simod	0.33 (0.02)	0.22 (0.02)	34.74 (32.17)	27.11 (10.90)	**1.83** (12.34)
	DSim	0.46 (0.06)	NA	**12.20** (6.74)	**8.25** (1.38)	91.06 (58.85)
	ASim	**0.02** (0.02)	**0.10** (0.06)	16.68 (4.02)	15.88 (1.53)	3.89 (0.74)
CDM	real	0.74 (0.01)	0.62 (0.04)	6.90 (0.48)	10.45 (0.49)	2.91 (0.07)
	Simod	0.21 (0.05)	**0.29** (0.08)	1.25 (0.96)	13.60 (2.25)	5.39 (0.40)
	DSim	0.45 (0.07)	NA	**0.13** (0.50)	5.22 (0.76)	**0.32** (0.14)
	ASim	**0.15** (0.03)	0.34 (0.11)	0.22 (1.06)	**3.03** (0.15)	0.72 (0.02)
BPI12W	real	0.67 (0.03)	0.80 (0.00)	8.81 (0.16)	28.06 (0.22)	5.62 (0.52)
	Simod	0.43 (0.00)	0.01 (0.00)	**0.26** (0.00)	8.77 (0.00)	0.42 (0.00)
	DSim	0.44 (0.02)	NA	0.90 (0.38)	2.25 (0.76)	1.13 (0.20)
	ASim	**0.01** (0.01)	**0.00** (0.00)	0.27 (0.09)	**0.19** (0.16)	**0.20** (0.91)
BPI17W	real	0.67 (0.03)	0.94 (0.00)	6.12 (0.41)	33.12 (0.29)	6.20 (0.23)
	Simod	0.43 (0.00)	0.17 (0.00)	**0.40** (0.00)	4.68 (0.00)	**0.69** (0.00)
	DSim	0.44 (0.02)	NA	1.77 (1.94)	10.93 (6.03)	4.35 (1.44)
	ASim	**0.01** (0.01)	**0.08** (0.07)	0.82 (1.29)	**4.61** (3.67)	1.67 (2.32)

ASim's robust performance in control-flow accuracy is attributed to its reliance on observed frequentist activity transition probabilities. Conversely, *Simod* and DSim employ model discovery algorithms with pruning, simplifying the discovered model but compromising accuracy in simulated activity sequences.

Unlike the aggregated temporal perspective offered by the *Standard Practice*, our evaluation separately addresses next activity processing (NPP) and waiting times (NWP). Notably, NPP and NWP exhibit no correlation, reinforcing the importance of their independent assessment to reveal distinct BPS improvement ideas. While *Simod* and ASim model processing times using parameterized distributions, and account for extraneous delays, DSim relies on LSTM models. Given DSim's poorer results for temporal metrics, the necessity of computationally expensive, black-box deep learning approaches in BPS warrants reconsideration.

Our evaluation framework's ability to separate model accuracy from data complexity is particularly evident in the P2P log, which exhibits substantial cycle time drift between the training and test sets (Table 3). PPM approaches trained on real data show significantly higher errors across all temporal tasks compared to other event logs. While the *Standard Practice* would simply suggest poor BPS model performance, our framework provides deeper insights. For example, although ASim's absolute error in the NPP task is high, its UtilityLoss is nearly zero, indicating that ASim effectively captures processing times despite the dataset's inherent complexity due to evolving process characteristics.

Finally, our framework facilitates diagnosing root causes behind poor BPS performance. In the CVS log, for instance, *Simod* and ASim yield similar results for NPP and NWP, yet *Simod* exhibits an exceptionally high RTP UtilityLoss (15.09 days), far exceeding ASim's 0.54 days. Given their comparable temporal metrics, this suggests that *Simod*'s poor RTP performance primarily stems from inaccuracies in its control-flow and resource modeling (NAP and NRP). This highlights the interdependence of process perspectives and underscores the necessity of ensuring accuracy across all simulation components, as errors in one aspect can propagate and severely degrade overall simulation utility.

In summary, the benchmark shows that our framework can effectively compare BPS approaches and identify their strengths and weaknesses across different process perspectives.

6 Conclusion

In this work, we introduced a framework for evaluating the quality of business process simulation (BPS) models by assessing the utility of simulated event logs via downstream tasks from predictive process monitoring (PPM). Unlike traditional evaluation methods that rely on direct log-to-log comparisons in a forecasting manner, our framework shifts the focus towards assessing the *utility* of simulated data in downstream tasks compared to the original data. Our results demonstrate that the proposed downstream tasks capture and differentiate modifications to a process model across control-flow, resource, temporal, and congestion dimensions. Additionally, we showed that our framework can effectively benchmark various BPS approaches to identify the respective strengths and weaknesses. Thereby, contrary to previous evaluation methods, it (1) can discern model accuracy from data complexity, (2) focuses on temporal relationships rather than absolute timestamps, and (3) can provide more fine-grained insights for potential areas of model improvement.

Limitations. Despite these advantages, our framework has some key limitations. The current set of downstream tasks, while effective, may not fully capture all relevant process characteristics, presenting an opportunity for further refinement. For instance, the resource perspective so far only considers role assignments, without explicit insights into overall occupation and work in progress. Additionally, our framework introduces significantly higher computational costs compared to traditional evaluation methods, as it requires training predictive models rather than computing log distances. Finally, despite using multiple network architectures and seeds, our evaluation framework may introduce variability due to training stochasticity, potentially limiting reproducibility and comparability. However, these limitations are expected, given that our approach introduces a fundamentally new perspective on evaluating BPS models.

Future Work. Therefore, future work will focus on expanding the set of downstream tasks, exploring alternative predictive model architectures beyond the LSTM and Transformer considered in this study, and improving the efficiency of the framework. Furthermore, the integration of BPS and PPM opens a promising research direction, where simulated data could be leveraged as an augmentation tool to enhance PPM models without requiring additional real-world data.

Acknowledgement. The authors acknowledge support by the state of Baden-Württemberg through bwHPC.

References

1. van der Aalst, W.M.P.: Business process simulation survival guide. In: vom Brocke, J., Rosemann, M. (eds.) Handbook on Business Process Management 1, Introduction, Methods, and Information Systems. International Handbooks on Information Systems, 2nd edn., pp. 337–370. Springer (2015)
2. Bukhsh, Z.A., Saeed, A., Dijkman, R.M.: Processtransformer: predictive business process monitoring with transformer network. arXiv preprint arXiv:2104.00721 (2021)
3. Camargo, M., Dumas, M., González-Rojas, O.: Automated discovery of business process simulation models from event logs. Decis. Supp. Syst. **134** (2020)
4. Camargo, M., Dumas, M., Rojas, O.G.: Learning accurate LSTM models of business processes. In: BPM. Springer (2019)
5. Camargo, M., Dumas, M., Rojas, O.G.: Discovering generative models from event logs: data-driven simulation vs deep learning. PeerJ Comput. Sci. **7**, e577 (2021)
6. Camargo, M., Dumas, M., Rojas, O.G.: Learning accurate business process simulation models from event logs via automated process discovery and deep learning. In: CAiSE. Springer (2022)
7. Chapela-Campa, D., Benchekroun, I., Baron, O., Dumas, M., Krass, D., Senderovich, A.: A framework for measuring the quality of business process simulation models. Inf. Syst. **127**, 102447 (2025)
8. Dumas, M.: Constructing digital twins for accurate and reliable what-if business process analysis. In: CEUR Workshop Proceedings, vol. 2938, pp. 23–27 (2021)

9. Dumas, M., Rosa, M.L., Mendling, J., Reijers, H.A.: Fundamentals of Business Process Management. Springer (2013)
10. Endres, M., Mannarapotta Venugopal, A., Tran, T.S.: Synthetic data generation: a comparative study. In: Proceedings of the 26th International Database Engineered Applications Symposium, pp. 94–102 (2022)
11. Esteban, C., Hyland, S.L., Rätsch, G.: Real-valued (medical) time series generation with recurrent conditional GANs. arXiv preprint arXiv:1706.02633 (2017)
12. Khodyrev, I., Popova, S.: Discrete modeling and simulation of business processes using event logs. In: ICCS. Elsevier (2014)
13. Kirchdorfer, L., Blümel, R., Kampik, T., van der Aa, H., Stuckenschmidt, H.: Agentsimulator: an agent-based approach for data-driven business process simulation. In: ICPM, pp. 97–104. IEEE (2024)
14. López-Pintado, O., Dumas, M.: Business process simulation with differentiated resources: does it make a difference? In: BPM. Springer (2022)
15. López-Pintado, O., Murashko, S., Dumas, M.: Discovery and simulation of data-aware business processes. In: ICPM. IEEE (2024)
16. Meneghello, F., Francescomarino, C.D., Ghidini, C.: Runtime integration of machine learning and simulation for business processes. In: ICPM. IEEE (2023)
17. Meneghello, F., Francescomarino, C.D., Ghidini, C., Ronzani, M.: Runtime integration of machine learning and simulation for business processes: time and decision mining predictions. Inf. Syst. **128**, 102472 (2025)
18. Okimura, I., Reid, M., Kawano, M., Matsuo, Y.: On the impact of data augmentation on downstream performance in natural language processing. In: Proceedings of the Third Workshop on Insights from Negative Results in NLP Association for Computational Linguistics (2022)
19. Rozinat, A., Mans, R.S., Song, M., van der Aalst, W.M.P.: Discovering simulation models. Inf. Syst. **34**(3), 305–327 (2009)
20. Tremblay, J., et al.: Training deep networks with synthetic data: bridging the reality gap by domain randomization. In: CVPR WS (2018)
21. Villani, C.: Topics in Optimal Transportation. Graduate Studies in Mathematics, American Mathematical Society (2003)
22. Wang, Q., Gao, J., Lin, W., Yuan, Y.: Learning from synthetic data for crowd counting in the wild (2019)
23. Weed, J., Berthet, Q.: Estimation of smooth densities in Wasserstein distance. In: Beygelzimer, A., Hsu, D. (eds.) Proceedings of the Thirty-Second Conference on Learning Theory, vol. 99, pp. 3118–3119. PMLR (2019)
24. Yoon, J., Jarrett, D., van der Schaar, M.: Time-series generative adversarial networks. In: NeurIPS, vol. 32. Curran Associates, Inc. (2019)
25. Özdemir, K., Kirchdorfer, L., Amiri Eliasi, K., Van der Aa, H., Stuckenschmidt, H.: Codebase: rethinking business process simulation: a utility-based evaluation framework (2025). https://doi.org/10.5281/zenodo.15489551

Engineering

Discovering Comprehensive Branched Declarative Process Constraints

Christos Balaktsis[1(✉)], Ioannis Mavroudopoulos[1], Marco Comuzzi[2], Anastasios Gounaris[1], and Fabrizio Maria Maggi[3]

[1] Aristotle University of Thessaloniki, Thessaloniki, Greece
{balaktsis,mavroudo,gounaria}@csd.auth.gr
[2] Ulsan National Institute of Science and Technology, Ulsan, South Korea
mcomuzzi@unist.ac.kr
[3] Free University of Bozen-Bolzano, Bolzano, Italy
maggi@inf.unibz.it

Abstract. The more nuanced the declarative process constraints discovered from an event log, the more likely they are to capture meaningful business knowledge, thus fostering the application of declarative process modeling in practice. Branching the activation (source) or the target of the constraints, that is, allowing more than one event type to appear as the source or the target, is a typical way to increase their expressivity. For the discovery of DECLARE constraints, only the case of branching the constraint target considering the inclusive disjunction policy has been considered. In this paper, we present CBDECLARE, a comprehensive approach to branched declarative process constraints, contributing in two key dimensions. First, we define a semantics of both source- and target-branched DECLARE constraints considering different branching policies. Second, we devise methods to discover the newly defined DECLARE constraint types from an event log. Our solution leverages the SIESTA framework's scalable and incremental infrastructure for event processing, achieving significant performance gains in mining these extended constraint when compared to the solutions for target-branched constraint discovery available in the literature.

Keywords: Process Mining · Big Data · Declarative Process Modeling

1 Introduction

Declarative process models describe a business process through a set of constraints on the order and occurrence of its activities, possibly associated with data conditions. They fit scenarios of high process variability, where imperative process models that capture all the admissible process behavior are likely to be too complex and not understandable by humans [11]. Among declarative modeling languages, DECLARE stands out as a language to manage the complexity of real-world processes by employing a range of templates (or constraint *types*) that capture temporal and logical constraints between activities, formally grounded in Linear Temporal Logic (LTL) over finite traces. DECLARE constraints are usually extracted (or *discovered*) from process data logged in a so-called event log [10].

The richer the constraints that can be discovered, the more likely they are to capture properties of the process execution interesting from a business standpoint. The literature on DECLARE process discovery has thus evolved from discovering basic constraint types involving one process activity in the source and the target [10], e.g., activity A (the source) is eventually followed by activity B (the target) in a trace, to adding data conditions on the constraint activation sources, that is, multi-perspective DECLARE discovery [12], e.g., A executed by resource R is eventually followed by B, or target-branched constraints [5], e.g., A is eventually followed by activity B or C. However, discovering DECLARE constraints from an event log is a computationally intensive task, and adding richness to the constraints implies additional complexity in constraint discovery.

In this work, our objective is to develop a comprehensive framework for branched DECLARE, named CBDECLARE, devising two key contributions: (i) a semantics of source- and target-branched DECLARE constraint types involving the three possible branching policies AND, OR, XOR, and (ii) a set of methods for the discovery of the newly defined branched constraints from an event log. A well-defined semantics for all types of branched DECLARE constraints is currently missing in the literature. Regarding discovery algorithms for branched constraints, the ad-hoc algorithms proposed in the literature consider only OR policy based target-branched constraints [5]. Other discovery methods based on SQL query generation allow for discovering all the branched constraints. However, as we demonstrate in our experiments, they are not efficient [17]. The proposed algorithms extend and customize an application-agnostic big data pattern analysis framework, named SIESTA [15], which has already shown to be effective in extracting basic DECLARE constraints from large event logs [16]. The proposed solution is publicly available as an extension to [16][1] and has been evaluated against state-of-the-art solutions for extracting target-branched constraints.

The paper is organized as follows. Section 2 reviews the related work, followed by Sect. 3, defining the semantics of the branched DECLARE constraints. Section 4 presents the SIESTA extension to discover the constraints, while the experimental evaluation is presented in Sect. 5. The conclusions are eventually drawn in Sect. 6.

2 Related Work

Declarative process modeling is not limited to approaches based on DECLARE, e.g. the approaches in [1,2] discover declarative similar process constraints. However, branched constraints have been explored mainly in the context of the DECLARE language.

---

[1] https://github.com/siesta-tool/declareIncremental/tree/tb-declare.

The literature on defining and discovering DECLARE constraints has flourished during the last fifteen years. The first approaches considered one source and one target activity [10]. This basic setting has been considered by other works focusing on adding conditions on the data attributes associated with the source and target event types [12]. More recently, other works have focused on ways to discriminate the interesting constraints by considering semantics vacuity of activations [13], constraint relevance [7], and interestingness based on constraint support and confidence [4], or on different discovery settings, such as event streaming [3].

Branched DECLARE constraints deal with extending the event type set that may activate a constraint (i.e., the source) or the target set. This type of constraints can be crucial in declarative process modelling because, in many real-world scenarios, a single activity may lead to multiple outcomes or involve choices between alternative behaviors. The closest work to CBDECLARE is TBDECLARE [6], which extends the original DECLARE framework by allowing constraints with disjunctions of multiple activities as targets. Our work, apart from being more comprehensive in terms of the branching types and policies covered, differs also in the manner the constraint discovery is performed, since our approach discovers the non-branched constraints first and then enriches them as needed. Finally, as already mentioned, SQL-based solutions can be used to discover branched constraints [17], but they are less efficient.

3 Defining Branched Constraints in CBDECLARE

In this section, we define the semantics of source- and target-branched DECLARE constraint types for different branching policies, namely inclusive disjunction (OR), exclusive disjunction (XOR), and conjunction (AND).

Let A be a finite set of activities (a.k.a. tasks). A log $\mathcal{L}$ is defined as $\mathcal{L} = (E, C, \gamma, \delta, \prec)$, where E is a finite set of events, C is a finite set of trace identifiers, $\gamma : E \to C$ is a function assigning events to trace identifiers, $\delta : E \to A$ is a function assigning events to activities. An event $ev \in E$ is a tuple that consists of at least a recorded timestamp ts, denoting the recording of task execution, and an event type $type \in A$ (i.e., $\delta(ev) = ev.type$). Let $\prec$ be the strict total ordering over events deriving from the event timestamps. A trace $\sigma \in \mathcal{L}$ is the ordered (established by $\prec$) sequence of events having the same trace id, that is $\forall ev_i, ev_j \in \sigma, \gamma(ev_i) = \gamma(ev_j)$. Finally, $e.pos$ denotes the position of the event in σ (i.e., $\sigma = \langle ev_1, ev_2, \ldots, ev_n \rangle$ and $ev_i.pos = i$).

While DECLARE constraint types fall into three distinct categories: existence and unordered relations, position, and ordered relations [10], CBDECLARE considers the six ordered relation constraint types already defined by TBDECLARE (Target-Branched DECLARE, proposed by [5]) and extends them (i) for source and target branching, and (ii) for different branching policies: AND-, XOR-branched policy for target-branched[2] constraints; AND-, XOR-, and OR-branched policy for the source-branched constraints. We denote the branching

---

[2] TBDECLARE [5] only considers the OR policy for the target branching type.

Table 1. Target-Branched CBDECLARE semantics

Constraint type	Semantics
Constraint type for $\vee$ policy	
response(a,[$b \vee c$])	$\forall ev_i \mid ev_i.\text{type} = a, \exists ev_j \mid ev_j.\text{type} = b \vee ev_j.\text{type} = c, i < j$
precedence([$b \vee c$], a)	$\forall ev_i \mid ev_i.\text{type} = a, \exists ev_j \mid ev_j.\text{type} = b \vee ev_j.\text{type} = c, j < i$
alternate response(a,[$b \vee c$])	$\forall ev_i \mid ev_i.\text{type} = a, \exists ev_j \mid ev_j.\text{type} = b \vee ev_j.\text{type} = c, i < j \wedge \nexists ev_k \mid ev_k.\text{type} = a, i < k < j$
alternate precedence([$b \vee c$], a)	$\forall ev_i \mid ev_i.\text{type} = a, \exists ev_j \mid ev_j.\text{type} = b \vee ev_j.\text{type} = c, j < i \wedge \nexists ev_k \mid ev_k.\text{type} = a, j < k < i$
chain response(a,[$b \vee c$])	$\forall ev_i \mid ev_i.\text{type} = a, ev_{i+1}.\text{type} = b \vee ev_{i+1}.\text{type} = c$
chain precedence([$b \vee c$], a)	$\forall ev_i \mid ev_i.\text{type} = a, ev_{i-1}.\text{type} = b \vee ev_{i-1}.\text{type} = c$
Constraint type for $\oplus$ policy	
response(a,[$b \oplus c$])	$\forall ev_i \mid ev_i.\text{type} = a, (\exists ev_j \mid ev_j.\text{type} = b, i < j \wedge \nexists ev_k \mid ev_k.\text{type} = c, i < k) \vee$ $(\exists ev_j \mid ev_j.\text{type} = c, i < j \wedge \nexists ev_k \mid ev_k.\text{type} = b, i < k)$
precedence([$b \oplus c$], a)	$\forall ev_i \mid ev_i.\text{type} = a, (\exists ev_j \mid ev_j.\text{type} = b, j < i \wedge \nexists ev_k \mid ev_k.\text{type} = c, k < i) \vee$ $(\exists ev_j \mid ev_j.\text{type} = c, j < i \wedge \nexists ev_k \mid ev_k.\text{type} = b, k < i)$
alternate response(a,[$b \oplus c$])	$\forall ev_i \mid ev_i.\text{type} = a, (\exists ev_j \mid ev_j.\text{type} = b, i < j \wedge \nexists ev_k \mid ev_k.\text{type} = a \vee ev_k.\text{type} = c, i < k < j \wedge$ $(\exists ev_h \mid ev_h.\text{type} = c, j < h) \rightarrow (\exists ev_l \mid ev_l.\text{type} = a, j < l < h)) \vee$ $(\exists ev_j \mid ev_j.\text{type} = c, i < j \wedge \nexists ev_k \mid ev_k.\text{type} = a \vee ev_k.\text{type} = b, i < k < j \wedge$ $(\exists ev_h \mid ev_h.\text{type} = b, j < h) \rightarrow (\exists ev_l \mid ev_l.\text{type} = a, j < l < h))$
alternate precedence([$b \oplus c$], a)	$\forall ev_i \mid ev_i.\text{type} = a, (\exists ev_j \mid ev_j.\text{type} = b, j < i \wedge$ $\nexists ev_k \mid ev_k.\text{type} = a \vee ev_k.\text{type} = c, j < k < i \wedge (\exists ev_h \mid ev_h.\text{type} = c, h < j) \rightarrow$ $(\exists ev_l \mid ev_l.\text{type} = a, h < l < j)) \vee (\exists ev_j \mid ev_j.\text{type} = c, j < i \wedge \nexists ev_k \mid ev_k.\text{type} = a \vee$ $ev_k.\text{type} = b, j < k < i \wedge (\exists ev_h \mid ev_h.\text{type} = b, h < j) \rightarrow (\exists ev_l \mid ev_l.\text{type} = a, h < l < j))$
chain response(a,[$b \oplus c$])	$\forall ev_i \mid ev_i.\text{type} = a, (ev_{i+1}.\text{type} = b \wedge \nexists ev_k \mid ev_k.\text{type} = c, i+1 < k) \vee (ev_{i+1}.\text{type} = c \wedge$ $\nexists ev_k \mid ev_k.\text{type} = b, i+1 < k)$
chain precedence([$b \oplus c$], a)	$\forall ev_i \mid ev_i.\text{type} = a, (ev_{i-1}.\text{type} = b \wedge \nexists ev_k \mid ev_k.\text{type} = c, k < i-1) \vee (ev_{i-1}.\text{type} = c \wedge$ $\nexists ev_k \mid ev_k.\text{type} = b, k < i-1)$
Constraint type for $\wedge$ policy	
response(a,[$b \wedge c$])	$\forall ev_i \mid ev_i.\text{type} = a, \exists ev_j \mid ev_j.\text{type} = b, i < j \wedge \exists ev_k \mid ev_k.\text{type} = c, i < k$
precedence([$b \wedge c$], a)	$\forall ev_i \mid ev_i.\text{type} = a, \exists ev_j \mid ev_j.\text{type} = b, j < i \wedge \exists ev_k \mid ev_k.\text{type} = c, k < i$
alternate response(a,[$b \wedge c$])	$\forall ev_i \mid ev_i.\text{type} = a, (\exists ev_j \mid ev_j.\text{type} = b, i < j \wedge \exists ev_k \mid ev_k.\text{type} = c, i < k) \wedge$ $(\nexists ev_h \mid ev_h.\text{type} = a, i < h < j \wedge \nexists ev_l \mid ev_l.\text{type} = a, i < l < k)$
alternate precedence([$b \wedge c$], a)	$\forall ev_i \mid ev_i.\text{type} = a, (\exists ev_j \mid ev_j.\text{type} = b, j < i \wedge \exists ev_k \mid ev_k.\text{type} = c, k < i) \wedge$ $(\nexists ev_h \mid ev_h.\text{type} = a, j < h < i \wedge \nexists ev_l \mid ev_l.\text{type} = a, k < l < i)$
chain response(a,[$b \wedge c$])	$\forall ev_i \mid ev_i.\text{type} = a, (ev_{i+1}.\text{type} = b \wedge \exists ev_k \mid ev_k.\text{type} = c, i+1 < k) \vee$ $(ev_{i+1}.\text{type} = c \wedge \exists ev_k \mid ev_k.\text{type} = b, i+1 < k)$
chain precedence([$b \wedge c$], a)	$\forall ev_i \mid ev_i.\text{type} = a, (ev_{i-1}.\text{type} = b \wedge \exists ev_k \mid ev_k.\text{type} = c, k < i-1) \vee$ $(ev_{i-1}.\text{type} = c \wedge \exists ev_k \mid ev_k.\text{type} = b, k < i-1)$

policies using $\vee$, $\wedge$, and $\oplus$ for the OR-, AND-, and XOR branching policy, respectively.

Let $\mathcal{CB}$ be the universe of branched constraints. A constraint $c \in \mathcal{CB}$ is a tuple $c = \langle t, b, p, S, T \rangle$, where t identifies one of the six constraint types considered in this work, $b \in \{source, target\}$ identifies to which element of c the branching is applied (hereinafter referred to as branching *type*), $p \in \{\vee, \wedge, \oplus\}$ is the branching policy, $S \subseteq A$ is the set of source activities of c, and $T \subseteq A$ is the target set of activities of c. For example, $c = \langle response, target, \oplus, \{a\}, \{b, c\} \rangle$ is the response constraint activated by event type a and with target the exclusive disjunction of event types b and c, which for conciseness we can also write as $response(a, [b \oplus c])$. Note that if $b = source$ ($b = target$), then $|T| = 1$ ($|S| = 1$). The branching factor bf [5] is equal to $|S|$ ($|T|$) if $b = source$ ($b = target$). When necessary, we refer to the elements of a constraint using the dotted notation, e.g., $c.S$, $c.T$ to

Table 2. Source-Branched CBDECLARE semantics

Constraint type	Semantics
Constraint type for $\vee$ policy	
response($[a \vee b]$, c)	$(\forall ev_i \mid ev_i.type = a, \exists ev_j \mid ev_j.type = c, i < j) \wedge (\forall ev_i \mid ev_i.type = b, \exists ev_j \mid ev_j.type = c, i < j)$
precedence(c, $[a \vee b]$)	$(\forall ev_i \mid ev_i.type = a, \exists ev_j \mid ev_j.type = c, j < i) \wedge (\forall ev_i \mid ev_i.type = b, \exists ev_j \mid ev_j.type = c, j < i)$
alternate response($[a \vee b]$, c)	$(\forall ev_i \mid ev_i.type = a, (\exists ev_j \mid ev_j.type = c, i < j \wedge \nexists ev_k \mid ev_k.type = a, i < k < j)) \wedge$
	$(\forall ev_i \mid ev_i.type = b, (\exists ev_j \mid ev_j.type = c, i < j \wedge \nexists ev_k \mid ev_k.type = b, i < k < j))$
alternate precedence(c, $[a \vee b]$)	$(\forall ev_i \mid ev_i.type = a, (\exists ev_j \mid ev_j.type = c, j < i \wedge \nexists ev_k \mid ev_k.type = a, j < k < i)) \wedge$
	$(\forall ev_i \mid ev_i.type = b, (\exists ev_j \mid ev_j.type = c, j < i \wedge \nexists ev_k \mid ev_k.type = b, j < k < i))$
chain response($[a \vee b]$, c)	$(\forall ev_i \mid ev_i.type = a, ev_{i+1}.type = c) \wedge (\forall ev_i \mid ev_i.type = b, ev_{i+1}.type = c)$
chain precedence(c, $[a \oplus b]$)	$(\forall ev_i \mid ev_i.type = a, ev_{i-1}.type = c) \wedge (\forall ev_i \mid ev_i.type = b, ev_{i-1}.type = c)$
Constraint type for $\oplus$ policy	
response($[a \oplus b]$, c)	$(\forall ev_i \mid ev_i.type = a, (\nexists ev_j \mid ev_j.type = b) \rightarrow (\exists ev_k \mid ev_k.type = c, i < k)) \wedge$
	$(\forall ev_i \mid ev_i.type = b, (\nexists ev_j \mid ev_j.type = a) \rightarrow (\exists ev_k \mid ev_k.type = c, i < k))$
precedence(c, $[a \oplus b]$)	$(\forall ev_i \mid ev_i.type = a, (\nexists ev_j \mid ev_j.type = b) \rightarrow (\exists ev_k \mid ev_k.type = c, k < i)) \wedge$
	$(\forall ev_i \mid ev_i.type = b, (\nexists ev_j \mid ev_j.type = a) \rightarrow (\exists ev_k \mid ev_k.type = c, k < i))$
alternate response($[a \oplus b]$, c)	$(\forall ev_i \mid ev_i.type = a, (\exists ev_j \mid ev_j.type = c, i < j \wedge \nexists ev_k \mid ev_k.type = a, i < k < j) \vee$
	$(\exists ev_h \mid ev_h.type = b, i < h \wedge \nexists ev_l \mid ev_l.type = c, i < l < h)) \wedge$
	$(\forall ev_i \mid ev_i.type = b, (\exists ev_j \mid ev_j.type = c, i < j \wedge \nexists ev_k \mid ev_k.type = b, i < k < j) \vee$
	$(\exists ev_h \mid ev_h.type = a, i < h \wedge \nexists ev_l \mid ev_l.type = c, i < l < h))$
alternate precedence(c, $[a \oplus b]$)	$(\forall ev_i \mid ev_i.type = a, (\exists ev_j \mid ev_j.type = c, j < i \wedge \nexists ev_k \mid ev_k.type = a, j < k < i) \vee$
	$(\exists ev_h \mid ev_h.type = b, h < i \wedge \nexists ev_l \mid ev_l.type = c, h < l < i)) \wedge$
	$(\forall ev_i \mid ev_i.type = b, (\exists ev_j \mid ev_j.type = c, j < i \wedge \nexists ev_k \mid ev_k.type = b, j < k < i) \vee$
	$(\exists ev_h \mid ev_h.type = a, h < i \wedge \nexists ev_l \mid ev_l.type = c, h < l < i))$
chain response($[a \oplus b]$, c)	$(\forall ev_i \mid ev_i.type = a, ev_{i+1}.type = c \vee (\exists ev_h \mid ev_h.type = b, i < h \wedge \nexists ev_l \mid ev_l.type = c, i < l < h)) \wedge$
	$(\forall ev_i \mid ev_i.type = b, ev_{i+1}.type = c \vee (\exists ev_h \mid ev_h.type = a, i < h \wedge \nexists ev_l \mid ev_l.type = c, i < l < h))$
chain precedence(c, $[a \oplus b]$)	$(\forall ev_i \mid ev_i.type = a, ev_{i-1}.type = c \vee (\exists ev_h \mid ev_h.type = b, h < i \wedge \nexists ev_l \mid ev_l.type = c, h < l < i)) \wedge$
	$(\forall ev_i \mid ev_i.type = b, ev_{i-1}.type = c \vee (\exists ev_h \mid ev_h.type = a, h < i \wedge \nexists ev_l \mid ev_l.type = c, h < l < i))$
Constraint type for $\wedge$ policy	
response($[a \wedge b]$, c)	$(\forall ev_i \mid ev_i.type = a, (\exists ev_j \mid ev_j.type = b, i < j) \rightarrow (\exists ev_k \mid ev_k.type = c, j < k)) \wedge$
	$(\forall ev_i \mid ev_i.type = b, (\exists ev_j \mid ev_j.type = a, i < j) \rightarrow (\exists ev_k \mid ev_k.type = c, j < k))$
precedence(c, $[a \wedge b]$)	$(\forall ev_i \mid ev_i.type = a, (\exists ev_j \mid ev_j.type = b, j < i) \rightarrow (\exists ev_k \mid ev_k.type = c, k < j)) \wedge$
	$(\forall ev_i \mid ev_i.type = b, (\exists ev_j \mid ev_j.type = a, j < i) \rightarrow (\exists ev_k \mid ev_k.type = c, k < j))$
alternate response($[a \wedge b]$, c)	$(\forall ev_i \mid ev_i.type = a, (\exists ev_j \mid ev_j.type = b, i < j) \rightarrow (\exists ev_k \mid ev_k.type = c, j < k \wedge$
	$\nexists ev_h, ev_l \mid ev_h.type = a, ev_l.type = b, j < h < k, j < l < k)) \wedge$
	$(\forall ev_i \mid ev_i.type = b, (\exists ev_j \mid ev_j.type = a, i < j) \rightarrow (\exists ev_k \mid ev_k.type = c, j < k \wedge$
	$\nexists ev_h, ev_l \mid ev_h.type = a, ev_l.type = b, j < h < k, j < l < k))$
alternate precedence(c, $[a \wedge b]$)	$(\forall ev_i \mid ev_i.type = a, (\exists ev_j \mid ev_j.type = b, j < i) \rightarrow (\exists ev_k \mid ev_k.type = c, k < j \wedge \nexists ev_h, ev_l \mid ev_h.type = a,$
	$ev_l.type = b, k < h < j, k < l < j)) \wedge (\forall ev_i \mid ev_i.type = b, (\exists ev_j \mid ev_j.type = a, j < i) \rightarrow$
	$(\exists ev_k \mid ev_k.type = c, k < j \wedge \nexists ev_h, ev_l \mid ev_h.type = a, ev_l.type = b, k < h < j, k < l < j))$
chain response($[a \wedge b]$, c)	$(\forall ev_i \mid ev_i.type = a, (\exists ev_j \mid ev_j.type = b, i < j) \rightarrow (ev_{j+1}.type = c)) \wedge$
	$(\forall ev_i \mid ev_i.type = b, (\exists ev_j \mid ev_j.type = a, i < j) \rightarrow (ev_{j+1}.type = c))$
chain precedence(c, $[a \wedge b]$)	$(\forall ev_i \mid ev_i.type = a, (\exists ev_j \mid ev_j.type = b, j < i) \rightarrow (ev_{j-1}.type = c)) \wedge$
	$(\forall ev_i \mid ev_i.type = b, (\exists ev_j \mid ev_j.type = a, j < i) \rightarrow (ev_{j-1}.type = c))$

refer to the source and target set of c, respectively. The relation $sat_\mathcal{L} : \mathcal{CB} \times C$ captures the satisfaction of a constraint c by a set of traces in an event log $\mathcal{L}$ identified by the trace ids $Z \subseteq C$, e.g., $(c_1, \{i_1, i_5\})$ signifies that the constraint c_1 is satisfied by the traces with trace ids i_1 and i_5. In the remainder, for brevity and when not ambiguous, we adopt a flattened representation of a constraint $c = \langle t, S, T, I \rangle$ where we omit the branching type b and policy p and introduce I as the set of trace ids satisfying c.

Tables 1, 2 show the semantic of the target-branched and source-branched constraint types, respectively. For simplicity, these definitions are provided for the $bf = 2$. The definitions can be easily generalized for $bf > 2$. The proposed semantics is not the only possible one and alternative choices could have been made, as we briefly discuss below. When creating the semantics of the branched constraints, we considered the following three design principles: (i) the semantics of a CBDECLARE constraint should be a *reasonable* extension of the one of the corresponding DECLARE constraint with $|T| = |S| = 1$, (ii) similar branching policies, no matter whether in the source or the target set, should be defined consistently across the different branched constraints, and (iii) the semantics of the OR policy target-branched constraints should be consistent with the one defined in TBDECLARE [5].

The impact of the second design principle on the constraint semantics can be understood considering, for example, the AND policy constraint types in Table 1. The semantics of the *chain response*$(a, [b \land c])$ constraint type is such that it is satisfied by traces in which the event type a is immediately followed by b (or c) and in which the event type c (b) occurs eventually before the end of the trace, that is, using regular expressions on event types, the constraint type would be satisfied by traces matching [^a]*ab*c* or [^a]*ac*b*. This semantics for the AND branching policy is interpreted in exactly the same way for all constraint types. For example, the *response*$(a, [b \land c])$ constraint type is satisfied by traces matching [^a]*a*b*c* or [^a]*a*c*b* so that the expression $b \land c$ always represents the patterns b*c* or c*b*.

4 Discovering Branched Constraints

For discovering the CBDECLARE constraints, we leverage the discovery algorithms of the non-branched (standard) constraints implemented on top of the SIESTA's framework, as explained in [16]. We introduce three core algorithms. EXTRACTTARGETBRANCHEDCONSTRAINTS (ETBC, Alg. 1) orchestrates the standard constraints transformation to branched ones by leveraging two auxiliary algorithms: FINDDISJUNCTIVEEVENTS (FDE, Alg. 2) and FINDCONJUNCTIVE-OREXCLUSIVEEVENTS (FCOEE, Alg. 3). We keep utilizing SIESTA's features in terms of indexing and persistent state of mined constraints in an S3-compatible storage; however, we modify the state type, as explained below. We start by presenting the discovery algorithms of the target-branched (TB) constraints, and then we generalize them to the source-branched (SB) ones.

4.1 Algorithmic Rationale for Target Set Branching

Algorithm 1 outlines the fundamental methodology for extending the discovery of standard constraints to branched ones, leveraging the mining output of [16]. We first modify the output format of [16], and, instead of storing non-branched constraints in the format $\langle t, S, T, support \rangle$[3], we store individual entries in the form $\langle t, S, T, I \rangle$. This improves the efficiency of the detailed procedures discussed next. By retrieving constraints stored in S3, the algorithm proceeds with the

---

[3] Support refers to the number of traces in which the constraint is satisfied.

Algorithm 1. EXTRACTTARGETBRANCHEDCONSTRAINTS (**ETBC**)

Require: Standard Constraint set $SC = \{(t_1, source_1, target_1, i_1), \ldots, (t_m, source_m, target_m, i_m)\} \neq \emptyset$, support threshold $sup \geq 0$, branching factor $bf \in \mathbb{N}^+ \cup \{\text{null}\}$, branching policy $p \in \{\wedge, \vee, \oplus\}$, drop factor $d \in \mathbb{N} \cup \{\text{null}\}$.
Ensure: Target Branched Constraints set TBC.
1: $groups \leftarrow SC.\text{groupBy}(t_i, source_i)$ ▷ group by constraint type and source
2: **for all** $(t_i, source_i) \in groups$ **do**
3:    $T \leftarrow \emptyset$
4:    **for all** $(target_k, i_j) \in groups[(t_i, source_i)]$ **do**
5:       $I_k \leftarrow \bigcup i_j$ ▷ group traces of the same constraint type
6:       $T \leftarrow T \cup \{target_k : I_k\}$ ▷ convert target tuples to maps
7:    $groups[(t_i, source_i)] \leftarrow T$ ▷ $(t_i, source_i) : \{target_1 : I_1, \ldots, target_n : I_n\}$
8: **for all** $(t_i, source_i) \in groups$ **do**
9:    **if** $p = \vee$ **then**
10:      $(targets, traces) \leftarrow \textbf{FDE}(groups, (t_i, source_i), sup, d, b)$
11:    **else**
12:      $(targets, traces) \leftarrow \textbf{FCOEE}(groups, (t_i, source_i), sup, d, b, p)$
13:    $TBC \leftarrow TBC \cup \{(t_i, source_i, targets, |traces|) \mid |targets| = bf\}$
14: **return** BC

discovery by combining target events according to the selected branching policy. Note that we only retrieve the constraint types defined in Sect. 3. Specifically, given a set of discovered standard constraints, along with a branching policy, the algorithm discovers TB constraints with support higher than an threshold sup.

The process starts by grouping the discovered standard constraints by constraint type and source event. Then, each constraint type t_i and source event $source_i$ is associated with all possible targets $target_k$ and the corresponding set I_k of traces that satisfy the constraint with constraint type t_i, source event $source_i$ and target $target_k$. According to the specified branching policy, the algorithm traverses each identified group and employs a greedy strategy to identify a set of combined, correlated targets that produce branched constraints with support higher than the input minimum support threshold sup.

If the branching policy is OR ($p = \vee$) on target events, we aim to identify disjunctive sets of targets. The FDE algorithm (Alg. 2) identifies a disjunctive, non-trivial[4] set of target events for each $(t_i, source_i)$. Starting with the complete set of target events, we iteratively identify the target event with the minimal contribution in support and remove it. In this process, the support provided by the remaining events in the target set gradually decreases, as removing a target $target_k$ either leaves the support unchanged (if the corresponding set I_k fully overlaps with the set of satisfying traces of another target), or reduces it. For example, consider the constraints $response(a, b)$ holding in traces $\{1, 2, 3\}$, $response(a, c)$ holding in $\{1, 4\}$ and $response(a, d)$ holding in $\{5\}$. This process terminates when there are no further events to remove from the evolving set or when it reaches a predefined cardinality. This behavior is determined by the user-selected $mode$—either unbounded or bounded—based on the parameter bf, as described in Sect. 4.2. An example of a TB constraint that may emerge under the OR target-branching policy is $response(a, [b \vee c \vee d])$, which is interpreted as

---

[4] A non-trivial target set for a pair $(t_i, source_i)$ refers to a minimal subset of targets related to that pair with the largest possible support.

Algorithm 2. FINDDISJUNCTIVEEVENTS (**FDE**)

Require: Grouped standard constraints $groups = \{(t_i, source_i) : \{target_1 : I_1, \ldots, target_n : I_n\}\} \neq \emptyset$, constraint type key $(t, source)$, support threshold $sup \geq 0$, drop factor $d > 0$, branching factor $bf \in \mathbb{N}^+ \cup \{\text{null}\}$.
Ensure: Tuple of target-set and covered trace-set (T, C).
1: $events \leftarrow groups[(t, source)]$
2: $support \leftarrow \bigcup_{(target, I) \in events} I$
3: $T \leftarrow keys(events)$
4: $C \leftarrow support$
5: $sumRed \leftarrow 0, sumRedSq \leftarrow 0, countRed \leftarrow 0, stdDev \leftarrow 0$
6: **while** $|events| > bf \lor bf = \text{null} \land |events| > 1$ **do**
7: $\quad contributions \leftarrow \{(target, |support| - |\bigcup_{(target', I') \in events \setminus \{(target, I)\}} I'|) \mid (target, I) \in events\}$
8: $\quad (minEvent, minCont) \leftarrow \arg\min_{(target, c) \in contributions} c$
9: $\quad$ **if** $bf = \text{null}$ **then**
10: $\quad\quad sumRed \leftarrow sumRed + minCont, sumRedSq \leftarrow sumRedSq + minCont^2$
11: $\quad\quad countRed \leftarrow countRed + 1$
12: $\quad\quad meanRed \leftarrow sumRed/countRed$
13: $\quad\quad stdDev \leftarrow \sqrt{(sumRedSq/countRed) - meanRed^2}$
14: $\quad\quad$ **if** $|minCont - meanRed| > dropFactor \cdot stdDev$ and $stdDev \neq 0$ **then**
15: $\quad\quad\quad$ **return** (T, C)
16: $\quad$ **if** $|support| \leq sup$ and $stdDev \neq 0$ **then**
17: $\quad\quad$ **return** (T, C)
18: $\quad events \leftarrow events \setminus \{(minEvent, \cdot)\}, support \leftarrow \bigcup_{(target, I) \in events} I$
19: $\quad T \leftarrow keys(events), C \leftarrow support$
20: **return** (T, C)

event type a being followed by any or all activities b, c, and d, holding in traces 1, 2, 3, 4 and 5.

In contrast, the FCOEE algorithm (Alg. 3) takes an expansion-driven approach in which each $(t_i, source_i)$ is associated with all possible targets $target_k$ and the corresponding set I_k of traces that satisfy the constraint with constraint type t_i, source event $source_i$ and target $target_k$. Pairs $(target_k, I_k)$ are iteratively combined into larger sets. Similarly to FCE, the evolution of the target set is terminated according to the bounding *mode* determined by the bf value (see Sect. 4.2). The way these candidates are formed and extended depends on both the examined constraint type and the chosen branching policy—AND ($p = \land$) or XOR ($p = \oplus$), based on the semantics of Table 2. This distinction is handled by the method findCandidates, which is referenced in the algorithm but not explicitly presented, as its implementation follows naturally from the described logic below.

For the AND policy ($p = \land$), when the constraint type does not involve chain response or chain precedence, the algorithm initializes its candidate set with all combinations of two target events included in *events*. Each candidate consists of a pair of targets along with the intersection of their respective sets of satisfying traces. The iterative process then unfolds by selecting the most supported target pair and expanding it with an additional event chosen from the remaining targets. In the expansion, the current set of satisfying traces is intersected with the set of satisfying traces of the new target. As this target set grows in size, the support of the corresponding branched constraint can only remain stable—if the new set of satisfying traces is fully contained in the existing

Algorithm 3. FINDCONJUNCTIVEORExclusiveEVENTS (**FCOEE**)

Require: Grouped standard constraints $groups = \{(t_i, source_i) : \{target_1 : I_1, \ldots, target_n : I_n\}\} \neq \emptyset$, constraint type key $(t, source)$, support threshold $sup \geq 0$, drop factor $d > 0$, branching factor $bf \in \mathbb{N}^+ \cup \{null\}$, branching policy $p \in \{\wedge, \oplus\}$, frequent traces set $TS_f = \{tf_1, \ldots, tf_m\} \supseteq \emptyset$.
Ensure: Tuple of target-set and covered trace-set (T, C).
1: $events \leftarrow groups[(t, source)]$
2: $events \leftarrow \{(target_i, I_i) \in events \mid I_i \cap TS_f \neq \emptyset \vee TS_f = \emptyset\}$
3: $candidates \leftarrow$ **findCandidates**$(events, p, t, groups)$
4: $(T, C) \leftarrow \arg\max_{(S,I) \in candidates} |I|$
5: **if** $|C| \leq sup$ **then return**(T, C)
6: $sumRed \leftarrow 0, sumRedSq \leftarrow 0, countRed \leftarrow 0, stdDev \leftarrow 0$
7: **while** $|T| < bf \vee bf =$ null $\wedge |candidates| > 0$ **do**
8: $\quad candidates \leftarrow \{(S \cup \{target\}, (I, I_{target})_p \mid (S, I) \in candidates, (target, I_{target}) \in events, target \notin S, (I, I_{target})_p \neq \emptyset\}$
9: $\quad (S_{\text{top}}, I_{\text{top}}) \leftarrow \arg\max_{(S,I) \in candidates} |I|$
10: $\quad$ **if** $bf =$ null **then**
11: $\quad\quad sumRed \leftarrow sumRed + |I_{\text{top}}|, sumRedSq \leftarrow sumRedSq + |I_{\text{top}}|^2$
12: $\quad\quad countRed \leftarrow countRed + 1$
13: $\quad\quad meanRed \leftarrow sumRed/countRed$
14: $\quad\quad stdDev \leftarrow \sqrt{(sumRedSq/countRed) - meanRed^2}$
15: $\quad\quad$ **if** $||I_{\text{top}}| - meanRed| > dropFactor \cdot stdDev$ and $stdDev \neq 0$ **then**
16: $\quad\quad\quad$ **return** (T, C)
17: $\quad$ **if** $|I_{\text{top}}| \leq sup$ **then return**(T, C)
18: $\quad (T, C) \leftarrow (S_{\text{top}}, I_{\text{top}})$
19: **return** (T, C)

---

sets—or decrease. For example, consider the constraints $response(a, b)$ holding in traces $\{1, 2, 3\}$, $response(a, c)$ in $\{1, 2, 4\}$ and $response(a, d)$ in $\{5\}$. An example of a TB constraint that may emerge under the AND target-branching policy is $response(a, [b \wedge c])$, which is interpreted as event type a being followed by both activities b and c, holding in traces 1 and 2.

When dealing with chain response (or precedence, regarding t) constraints, a different initialization strategy is followed. Instead of forming candidate sets by combining target events, the algorithm ensures that each target $target_c$ of the chain response (or precedence) constraint is grouped with the targets of the standard response (or precedence) constraints where $target_c$ serves as the source, retrieving them from $groups$. As the iterations progress, the target set grows by incorporating targets of response (or precedence) constraints that involve the latest event added as source. This method guarantees that the resulting branched constraints maintain their sequential dependencies.

For the XOR policy ($p = \oplus$), when the constraint type does not involve chain response or chain precedence, the algorithm follows an iterative approach similar to the one adopted for the AND policy, but operates on the basis of exclusion rather than intersection. The candidate set is initialized with combinations of two target events, where each candidate consists of a pair of targets along with the set of all traces included in the set of satisfying traces of one of the two targets but not both. During the expansion phase, only the most supported target set is extended at each step, incorporating additional events by combining the current set of satisfying traces with the set of satisfying traces of the new event while removing the common traces. For example, consider the constraints $response(a, b)$ holding in traces $\{1, 2, 3\}$, $response(a, c)$ in $\{1, 2, 4\}$, and $response(a,d)$ in $\{2,5\}$. An example of a target-branched (TB) constraint that may emerge under the XOR target-branching policy is $response(a, [b \oplus d])$, which is interpreted as event type a being followed by b or d but not both, holding in traces 1, 3, and 5.

This exclusion-based approach naturally minimizes overlap in trace coverage among targets, thereby making XOR-branching particularly effective for identifying alternative process paths. In the case of chain response or precedence constraints, the algorithm further adapts the initialization phase to preserve the positional ordering of events. Specifically, candidate pairs are constructed by exploring mutually exclusive-on-second-step execution paths, where each target may serve as source for additional chain response (or precedence) constraints. This recursive expansion ensures that sequential conflicts are avoided while maximizing the coverage of disjoint trace sets.

4.2 Algorithm Variants and Walk-Through Examples

The previously introduced algorithms discover branched constraints by iteratively expanding or contracting target sets. The termination criterion of this process depends on whether a *bounded* or an *unbounded* strategy is applied. These modes define the conditions under which target restructuration ceases, ensuring that the resulting constraints maintain the minimum support.

In the *bounded* mode, a user-defined branching factor $bf \in \mathbb{N}^+$ limits the size of the target set, ensuring that only constraints with an exact number of targets (or sources) are considered. Specifically, bf enforces a strict cardinality for the branched constraints under discovery. If no candidate set satisfies the minimum support threshold while conforming to this branching factor, the algorithm returns the largest feasible set that meets the support requirement[5]. In the FDE algorithm, where the initial event set serves as a superset of candidate targets, the reduction of cardinality is controlled by the bf threshold. Similarly, in the FCOEE algorithm, which starts with the most supported event pair as the initial candidate set, bf restricts the growth of the target set to the specified upper bound.

In contrast, the *unbounded* mode ($bf = $ null) dynamically determines when to halt expansion based on observed trends in support reduction. As the candidate target set evolves, the algorithm continuously evaluates the effect of adding or removing targets. A significant drop in support indicates that further expansion would significantly compromise the validity of the constraint in the log. To formalize this stopping criterion, the algorithm computes the mean and standard deviation of support changes throughout the iterative process. If a prospective modification results in a support reduction exceeding a drop factor d of the standard deviation (std) from the mean ($mean$), the expansion process is terminated. This strategy prioritizes the greedy identification of the most informative

---

[5] This is a design decision rather than an inherent limitation; alternatively, such candidates could be discarded.

branched constraints in the event log, rather than exhaustively enumerating all possible target sets.

Example scenario for the bounded mode. We illustrate the OR-branching process by considering an event log $\mathcal{L}$ consisting of the following three traces[6]: σ_1: $\langle a, b, c, d \rangle$, σ_2: $\langle a, d, e \rangle$, and σ_3: $\langle f, b, c, d \rangle$, and focus on standard constraints of type response(a, b). The constraints and their satisfying traces are as follows: $c_1 = $ response(a, b), $sat_{\mathcal{L}}(c_1) = \{1\}$, $c_2 = $ response(a, d), $sat_{\mathcal{L}}(c_2) = \{1, 2\}$, and so on. The first step in the ETBC is to group constraints by their constraint type and source event. For the source event, e.g., a, we obtain the grouped constraints: $(response, a) : \{b : \{1, 2, 3\}, c : \{1, 2\}, d : \{1, 3\}, e : \{2\}, f : \{3\}\}$. We consider the inputs $bf = 2$ and $sup = 0.5$ and indicate $abs\_sup = sup \cdot |\mathcal{L}| = 1.5$. Initially, the full set of targets is considered, i.e. $\{b, c, d, e, f\}$ with support $|\{1, 2, 3\}| = 3 > abs\_sup$. The algorithm iteratively removes the least contributing target, where the contribution is determined by the number of traces covered. Since e and f each appear in only one trace, one of them (e.g., f) is removed first. Now, the target set is $\{b, c, d, e\}$ with the same support (3). The process continues removing e, and then c (or d), until only the top-two contributing-to-support targets $\{b, c\}$ remain and still meet the support threshold, since $|\{1, 2, 3\}| > abs\_sup$. This means that in at least half of the traces (since the support threshold is set to 0.5), when a appears, it is followed by any or all of b and c, satisfying the response constraint.

Example scenario for unbounded mode. Consider the source event a and the following target events with their trace sets: $b : \{1, 2, 3, 4\}$, $c : \{3, 4, 5, 6\}$, $d : \{5, 6, 7, 8\}$, $e : \{7, 8, 9, 10\}$, and $f : \{1, 2, 9, 10\}$. FCOEE initializes with $bf = $ null, $d = 1$, and XOR branching. It starts with $b \oplus d$, forming a trace set $\{1, 2, 3, 4, 5, 6, 7, 8\}$ of cardinality 8. Adding c reduces the support to 6, and adding e further reduces it to 4. The mean reductions (4, 3.5) and standard deviations (1.5, 1.2) remain within the drop factor limit, so the process continues. However, attempting to add f drops the trace set to $\emptyset$. The mean reduction is now 2.5, with a standard deviation of 1.3. Since $|0 - 2.5| > 1.3$, f is rejected, and the process terminates. Finally, the algorithm returns the TB constraint: $response(a, [b \oplus d \oplus c \oplus e])$, which holds for traces $\{1, 2, 9, 10\}$.

Mitigating increased complexity issues. In Alg. 3, identifying all potential target sets of length two or more, along with computing their corresponding trace intersections, imposes substantial computational overhead (although our experimental results in the next section provide evidence regarding the algorithm practicality). To address the challenge of search space explosion, an optimized variant of the algorithm has been developed as well. This version incorporates an initial filtering step designed to preselect candidate target events, ensuring that the search space remains manageable and does not increases excessively with the number of mined constraints for each activation event. Specifically, for each $(t, source)$, the most frequently occurring traces (TS_f) are first identified. These high-frequency traces serve as a heuristic filter, retaining only target events that

---

[6] For simplicity, events ev_i of a trace σ are directly represented by their type, i.e. $ev_i \equiv ev_i.type$.

Table 3. Characteristics of evaluated datasets, TL = Trace Length.

Log	#Trace	#Event	#Event Type	Duplicate events	TL_{min}	TL_{avg}	TL_{max}
SEPSIS	1,050	15,214	16	< 0.01%	3	14	185
BPIC 2012	13,087	262,200	36	< 0.01%	3	20	175
BPIC 2020	6,449	72,151	34	0.04%	3	11	27

frequently co-occur within them. By prioritizing traces that are common among multiple targets, this approach increases the likelihood of identifying meaningful co-occurrences while significantly reducing unnecessary computations. For simplicity, the computation of TS_f is not explicitly included in Algorithm 1; however, it is referenced as an optional parameter in Alg. 3.

4.3 Generalization to Source-Branched Constraints

The ETBC algorithm, as described in Alg. 1, is designed to handle target branching by grouping constraints based on the constraint type and source event. However, the same algorithmic framework can be adapted to handle source branching with minimal modifications. The key insight is that the FDE and FCOEE algorithms operate on a set of events and their associated trace ids, which satisfy a given constraint type. These algorithms are agnostic to whether the events represent target or source events, as their input includes a constraint type, an event key (a source event for TB), and a set of events with associated trace ids. In order to adapt ETBC for source branching, the initial grouping step in the algorithm needs to be modified. Instead of grouping constraints by constraint type and source event, we group them by constraint type and target event. The output of the adapted ETBC algorithm will then consist of a set of source events for each constraint type, with a single target event (the one used in the grouping key). This adaptation allows the algorithm to seamlessly handle source branching while maintaining the same underlying logic and structure as the original target branching implementation.

5 Experimental Evaluation

In this section, we evaluate the performance of different methods and branching policies used for mining CBDECLARE constraints.

Setting and Competitors. We employ publicly available event logs as shown in Table 3 that exhibit diversity in trace volume, event types, and trace length. We compare the performance of our CBDECLARE constraint discovery approach against two existing methods. The first competitor is TBDECLARE as implemented within MINERful [6], which exclusively supports the OR branching of targets. The second competitor is an SQL-based framework introduced in [17], originally designed for discovering standard DECLARE constraints using SQL queries, which we extend to support 2-branched target discovery for this evaluation and engineered within a PostgreSQL DBMS. Since both competitors are limited to OR branching, our comparative analysis focuses on the 2-branched OR-policy setting, to evaluating the three systems under identical conditions. The experiments presented below are conducted on a single machine running Ubuntu 20.04, equipped with 64GB of RAM and an 8-core (16-thread) CPU clocked at 2.10GHz. Each approach is allocated 50GB of RAM and granted full access to the CPU resources.

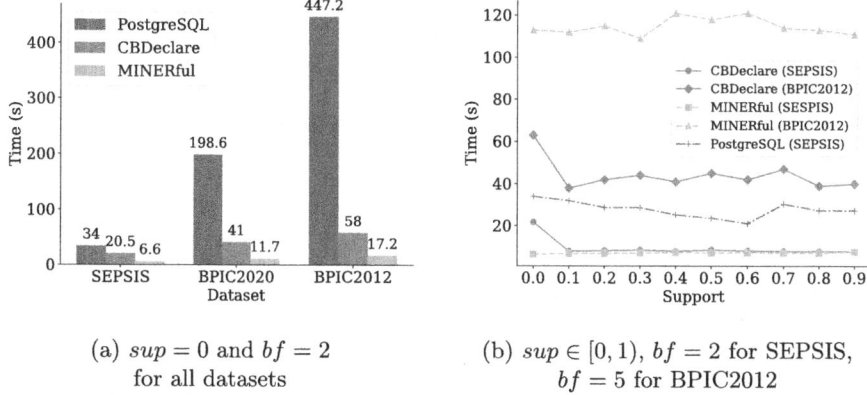

Fig. 1. Execution time comparison among PostgreSQL, CBDECLARE, and MINERful. (a) Aggregated indexing and TB discovery time across different datasets. (b) Total TB discovery time for ten support thresholds.

Experiment 1: CBDECLARE vs. Competitors. We first compare the efficiency of MINERful, CBDECLARE, and SQL-based methods for discovering OR-TBDECLARE constraints with a zero-support threshold. To ensure fairness, we set the branching factor parameter of MINERful (b) and CBDECLARE (bf) to 2, matching the extended SQL queries we provided. Additionally, we disabled any optional pruning or filtering in both systems for consistency. Figure 1 shows the total execution times, including indexing phases, as MINERful operates fully in memory, making it difficult to isolate the mining task duration. As shown in Fig. 1a, the SQL-based method has the longest execution times across all datasets, mainly due to the slow indexing process —specifically, the overhead of writing to the database— and the absence of optimizations for handling combinations in SQL queries, leading to significant computational overhead during the discovery process.

In contrast, CBDECLARE benefits from heuristic-based processing during indexing and discovery, enhancing efficiency, though it still lags behind MINERful on smaller-scale tasks. This difference stems from MINERful's in-memory operation [8], which allows fast constraint discovery for manageable data sizes. CBDECLARE builds on top of SIESTA and, as such, is designed with scalability in mind ([15,16]), using batch-processing techniques and incremental indexing to handle larger event logs – an essential feature when in-memory approaches become impractical. CBDECLARE inherits this scalability property of SIESTA. Interestingly, even for the datasets examined, which easily fit into the main memory, as shown in Fig. 1b, CBDECLARE performance suffers only in the initial query. More specifically, both CBDECLARE and PostgreSQL experience significant overhead during the initial mining query due to writing-to-disk operations, resulting in a noticeable performance gap relative to MINERful. However, in subsequent queries, MINERful and CBDECLARE exhibit equal performance for smaller tasks, i.e., small datasets and $bf = 2$. Increasing bf on larger datasets yields running times for CBDECLARE up to 2.5 lower than MINERful. In addition, CBDECLARE supports more options (source-branched along with AND and XOR target-branched policies; unbounded mode) than [6].

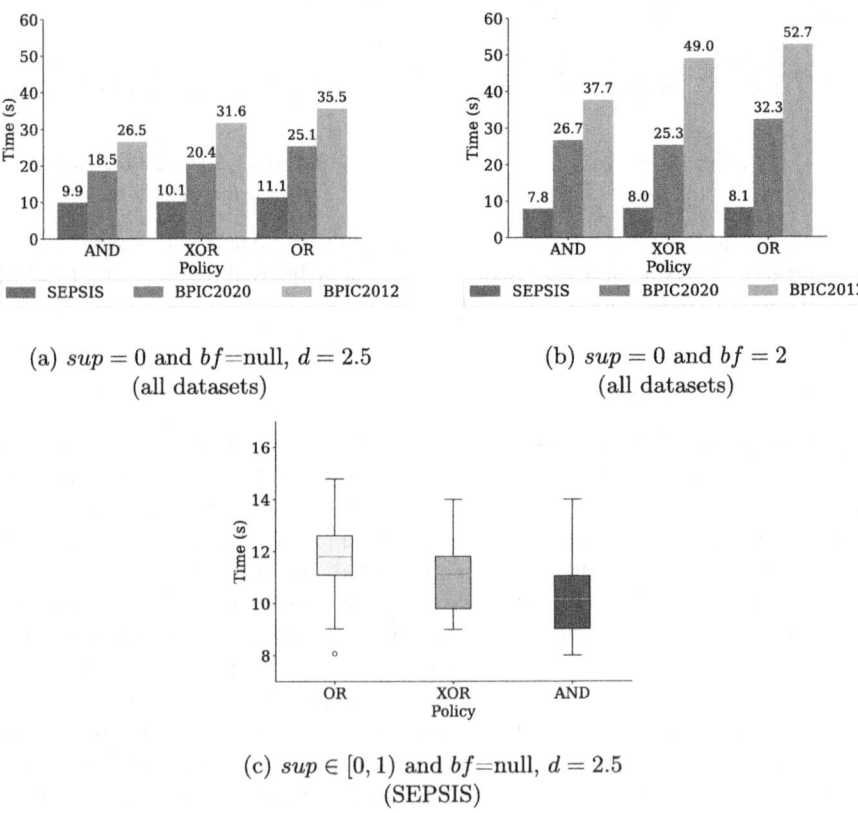

Fig. 2. Discovery time comparison of TB policies across different scenarios.

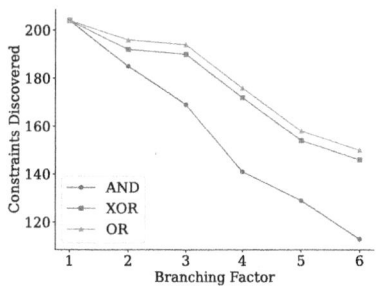

Fig. 3. TB Constraints mined for different branching factors and $sup = 0$ (BPIC2020).

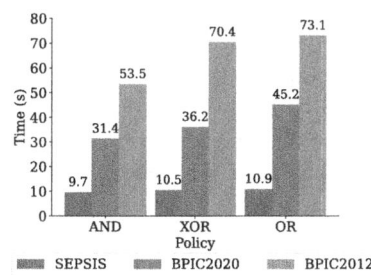

Fig. 4. Discovery time for different source branching policies with $sup = 0$ and $bf = 2$.

Experiment 2: Branching Policies. In this experiment, we evaluate the performance of different branching policies—AND, OR, and XOR— when discovering CBDECLARE constraints. The experiment shows how discovery time increases with the log size, reflecting the varying efficiencies of each policy in handling the search space. As shown in Fig. 2, the time required to discover constraints (excluding indexing) increases with the input log size, demonstrating consistent behavior of the policies.

Figures 2a and 2b show that OR-branching consistently requires more time for discovery, as expected, due to a higher number of single-target constraints and larger target supersets for each activation event (see Alg. 2). In both discovery modes, OR-branching has the slowest reduction in support. This is because the number of traces covered by the target set sees minimal reduction when removing individual targets, as many traces are still covered. Additionally, OR requires more computations during the target set reduction phase. These results highlight two key performance factors for the SIESTA-based approach: the number of single-target constraints discovered initially and the extent of event subsequence replication across logs, leading to target overlaps.

The AND policy, relying on trace-set intersection operations, is efficient regardless of log size, and becomes more efficient with larger logs. As seen in both datasets, it is the most efficient approach in terms of execution time, especially as event numbers increase. This suggests that the target-set expansion strategy of AND is well-suited to larger logs, benefiting from higher support. As shown in Alg. 3, intersection operations on small target sets are computationally efficient, as their length remains small. In unbounded branching (Fig. 2a), the threshold is quickly reached, and adding events to the target set rapidly reduces support. A similar trend is seen in bounded discovery (Fig. 2b), where the branching limit or support threshold is met quickly.

The XOR policy shows execution times in between those of the AND and OR policies in all experimental scenarios. As shown in the figure, XOR's dual behavior contributes to its performance: it shares the intersection-driven nature of AND but converges faster toward the branching limit due to the exclusion principle (see Sect. 4.1). While XOR explores combinations like AND, its convergence is faster, reducing computational cost relative to OR. XOR branching is less impacted by chain-type constraints, contributing to faster convergence compared to AND in some datasets. Figure 2c shows that, although XOR has a higher average execution time than AND, it runs faster than OR. The wider 25%-quartile for AND suggests that variations in event log structure and support similarity among candidate target subsets affect its performance.

Experiment 3: Number of Constraints. In this experiment, we assess the impact of the branching factor on the number of discovered constraints (see Fig. 3). Setting $bf = 1$ corresponds to the standard constraints initially discovered by the SIESTA framework, meaning that all policies return the original set of non-branched constraints. As bf increases, constraints are progressively merged following each policy's rationale. Given that $sup = 0$ (accepting all constraints with support greater than 0), only combinations that inherently reduce their support to zero are eliminated. The results illustrate a clear declining trend in the number of discovered constraints across all policies as the branching factor grows. This reduction occurs due to the merging process invalidating some combinations that no longer satisfy the minimum support criterion after branching. The AND policy demonstrates the steepest decline, as its strict requirement for full trace set overlap among targets causes many combined constraints to result in zero support. XOR shows a moderate reduction, whereas OR maintains a relatively higher number of constraints due to its permissive trace union condition. This trend is also reflected in the branched mining execution time, where the computational cost correlates directly with constraint reduction.

Experiment 4: Source-Branching. The goal of this experiment is to determine if there is a notable difference in execution times between the source and the target branching discussed in previous experiments. The results in Fig. 4 show that source branching consistently requires slightly more time across all policies. This is mostly due to the nature of process traces. In many real-world scenarios, processes begin in a relatively uniform way (e.g., standardized initial steps), leading to broader branching opportunities at the source level, i.e., there are more candidate combinations. In contrast, target branching is more constrained, as reaching a particular point in a trace suggests the process has already followed a path in a manner that limits possible continuations thus reducing branching complexity. Despite the higher overhead of source branching, the difference remains moderate, and the relative ranking of policies remains the same: XOR has the fastest execution times, followed by AND, with OR incurring the highest cost. However, this behavior warrants further study, which is outside the scope of this work.

6 Conclusion and Future Work

In this paper, we introduce CBDECLARE, a generalization of branched DECLARE covering three policies (OR, XOR, AND) and provide a comprehensive solution for the discovery of these types of constraints. We observe that the discovery of branched DECLARE constraints and, in particular, the discovery of source-branched DECLARE provides also an instrument for the discovery of *disjunctive* DECLARE models, a well-known problem in declarative process mining [9]. For example, the discovery of the two source-branched constraints $response([a \vee b], c)$ and $response([d \vee e], f)$ is equivalent to the discovery of the disjunction of the four DECLARE models: $response(a, c) \wedge response(d, f)$, $response(a, c) \wedge response(e, f)$, $response(b, c) \wedge response(d, f)$, $response(b, c) \wedge response(e, f)$. In the future, we aim to (i) extend the discovery approach to the full set of DECLARE constraints, (ii) investigate the implications of the new semantics on the compliance monitoring [14] of this type of constraints, and (iii) extend the approach to constraints including additional perspectives like data and time.

References

1. Back, C.O., Slaats, T., Hildebrandt, T.T., Marquard, M.: Discover: accurate and efficient discovery of declarative process models. Int. J. Softw. Tools Technol. Transfer **24**(4), 563–587 (2022)
2. van Beest, N., Groefsema, H., García-Bañuelos, L., Aiello, M.: Variability in business processes: automatically obtaining a generic specification. Inf. Syst. **80**, 36–55 (2019)
3. Burattin, A., Cimitile, M., Maggi, F.M., Sperduti, A.: Online discovery of declarative process models from event streams. IEEE Trans. Serv. Comput. **8**(6), 833–846 (2015)
4. Cecconi, A., De Giacomo, G., Di Ciccio, C., Maggi, F.M., Mendling, J.: Measuring the interestingness of temporal logic behavioral specifications in process mining. Inf. Syst. **107**, 101920 (2022)
5. Di Ciccio, C., Maggi, F.M., Mendling, J.: Discovering target-branched declare constraints. In: BPM, pp. 34–50. Springer (2014)
6. Di Ciccio, C., Maggi, F.M., Mendling, J.: Efficient discovery of target-branched declare constraints. Inf. Syst. **56**, 258–283 (2016)
7. Di Ciccio, C., Maggi, F.M., Montali, M., Mendling, J.: On the relevance of a business constraint to an event log. Inf. Syst. **78**, 144–161 (2018)
8. Di Ciccio, C., Mecella, M.: On the discovery of declarative control flows for artful processes. ACM Trans. Manage. Inf. Syst. **5**(4) (2015)
9. Di Francescomarino, C., Donadello, I., Ghidini, C., Maggi, F.M., Rizzi, W., Tessaris, S.: Making sense of temporal event data:a framework for comparing techniques for the discovery of discriminative temporal patterns. In: CAiSE. vol. 14663, pp. 423–439. Springer (2024)
10. Maggi, F.M., Bose, R.J.C., van der Aalst, W.M.: Efficient discovery of understandable declarative process models from event logs. In: CAiSE, pp. 270–285 (2012)
11. Maggi, F.M.: Declarative Process Mining, pp. 625–632. Springer International Publishing (2019)

12. Maggi, F.M., Dumas, M., García-Bañuelos, L., Montali, M.: Discovering data-aware declarative process models from event logs. In: BPM, pp. 81–96 (2013)
13. Maggi, F.M., Montali, M., Di Ciccio, C., Mendling, J.: Semantical vacuity detection in declarative process mining. In: BPM, pp. 158–175 (2016)
14. Maggi, F.M., Westergaard, M., Montali, M., van der Aalst, W.M.P.: Runtime verification of LTL-based declarative process models. In: RV 2011, vol. 7186, pp. 131–146. Springer (2011)
15. Mavroudopoulos, I., Gounaris, A.: SIESTA: A Scalable InfrastructurE of Sequential paTtern Analysis. IEEE Trans. Big Data 1–16 (2022)
16. Mavroudopoulos, I., Varvoutas, K., Kougka, G., Gounaris, A., Comuzzi, M.: Exploiting general purpose big-data frameworks in process mining: The case of declarative process discovery. In: BPM, pp. 185–202 (2024)
17. Riva, F., Benvenuti, D., Maggi, F., Marrella, A., Montali, M.: An SQL-based declarative process mining framework for analyzing process data stored in relational databases, pp. 214–231 (09 2023)

SimBank: From Simulation to Solution in Prescriptive Process Monitoring

Jakob De Moor[1(✉)], Hans Weytjens[1,2], Johannes De Smedt[1], and Jochen De Weerdt[1]

[1] Research Centre for Information Systems Engineering, KU Leuven, Leuven, Belgium
{jakob.demoor,johannes.desmedt, jochen.deweerdt}@kuleuven.be
[2] School of Computation, Information and Technology, TUM, Munich, Germany
hans.weytjens@tum.de.be

Abstract. Prescriptive Process Monitoring (PresPM) is an emerging area within Process Mining, focused on optimizing processes through real-time interventions for effective decision-making. PresPM holds significant promise for organizations seeking enhanced operational performance. However, the current literature faces two key limitations: a lack of extensive comparisons between techniques and insufficient evaluation approaches. To address these gaps, we introduce SimBank: a simulator designed for accurate benchmarking of PresPM methods. Modeled after a bank's loan application process, SimBank enables extensive comparisons of both online and offline PresPM methods. It incorporates a variety of intervention optimization problems with differing levels of complexity and supports experiments on key causal machine learning challenges, such as assessing a method's robustness to confounding in data. SimBank additionally offers a comprehensive evaluation capability: for each test case, it can generate the true outcome under each intervention action, which is not possible using recorded datasets. The simulator incorporates parallel activities and loops, drawing from common logs to generate cases that closely resemble real-life process instances. Our proof of concept demonstrates SimBank's benchmarking capabilities through experiments with various PresPM methods across different interventions, highlighting its value as a publicly available simulator for advancing research and practice in PresPM.

Keywords: Prescriptive Process Monitoring · Simulation · Optimization

1 Introduction

Prescriptive Process Monitoring (PresPM) offers a valuable opportunity for businesses to enhance decision-making. Its methods leverage event log data from business processes and use machine learning to automatically prescribe interventions, such as machine maintenance, customer calls to maximize turnover, or loan application cancellations [16]. The goal is to optimize the specified target variable(s), such as delivery time (of an order), acceptance (of a loan offer), recovery (of a patient), availability (of a product), or a default rate (in manufacturing). The benefits of these prescriptive systems include product and service quality improvements, cost savings, and increased staff satisfaction.

Despite their potential, significant limitations are present in current PresPM methodologies, particularly in benchmarking practices. Firstly, there is a notable lack of comprehensive comparisons among PresPM approaches. To the best of our knowledge, only one study has thoroughly compared multiple methods [36], though it only examines two. Additionally, existing research predominantly focuses on a limited range of intervention types, neglecting the development and testing of effective approaches for more complex intervention optimization problems, such as optimizing multiple sequential interventions. Secondly, accurate method evaluation remains a persistent challenge. Most evaluations rely on historical datasets [3,4,9,18,35], which lack counterfactual outcomes, i.e. the results of actions not taken or recorded. Method performance has to be estimated, since not every prescribed action for a test case has been observed. Moreover, these estimates may be misleading as the (test) datasets reflect the policies under which they were collected (e.g., a bank's loan approval criteria), yielding reliable performance estimates only for frequently observed parts of the optimization space.

To address these challenges in benchmarking and evaluation, we propose SimBank, a first-of-its-kind synthetic data generator designed specifically for PresPM research. SimBank simulates cases inspired by a bank's loan application process and addresses the two main limitations in existing methodologies. First, SimBank ensures accurate method evaluation by providing a controlled environment where the underlying process-generating mechanisms are explicitly known. This allows for precise calculation of outcomes based on prescribed actions, eliminating the need to rely on historical data. Second, SimBank facilitates comprehensive comparisons between PresPM methods by enabling controlled experimentation. Researchers can adjust process parameters to simulate diverse scenarios, supporting both offline training (with pre-generated event logs) and online training (in dynamic, real-time environments). Moreover, SimBank incorporates three main prescriptive intervention optimization problems, making it possible to compare methods across varying complexity levels. By combining these intervention types, SimBank enables testing of sequential intervention strategies—an area that remains underexplored in PresPM research. The simulator, based on commonly used logs, generates cases that closely reflect real-world process flows by including activities that occur in loops and in parallel. To validate SimBank's benchmarking capabilities, we conduct experiments with four different PresPM methods across the defined intervention scenarios. These experiments assess whether SimBank provides meaningful comparisons, enabling researchers and practitioners to accurately identify scenarios where specific PresPM methods excel or underperform. This approach bridges the gap between simulated and real-world environments, supporting the practical deployment of PresPM solutions.

The paper is structured as follows. Section 2 covers the background. Section 3 details the simulator development. Section 4 presents a proof of concept to validate SimBank's effectiveness in comparing PresPM methods. We conclude this paper and suggest avenues for future work in Sect. 5.

2 Background

2.1 Preliminaries

PresPM is an emerging extension of Process Mining, which is dedicated to discovering, monitoring, and improving real-life process models [1]. PresPM advances this goal by recommending case-specific interventions, aiming to directly optimize process outcomes. Recommendations are often derived from observational event logs, making them particularly useful in situations where traditional methods like randomized controlled trials (RCTs) or A/B testing are too costly or risky, such as a bank testing a loan assignment strategy. While still in its infancy, the field is steadily gaining momentum [4,9,16,32].

Building on the framework in [36], interventions in PresPM can be characterized along two dimensions: action width and action depth. Action width refers to the number of possible actions during an intervention, while action depth indicates the number of possible timing points for the intervention. When multiple interventions are performed in sequence, they form an intervention sequence. Table 1 illustrates these dimensions with examples from marketing interventions aimed at increasing turnover through customer contact. These dimensions are instrumental in defining the complexity of an intervention.

Table 1. Intervention dimensions and examples

Example: contact a customer	Action width	Action depth
Intervene with a *call* or *email* at a given point in the process.	2	1
Intervene with a *call*, *email* or *visit* and decide at which of 4 given points in the process to intervene.	3	4
Intervene with a *call* or *email* and decide at which of 2 given points in the process to intervene. Next, intervene with a *visit* or *post card* and decide at which of 4 given points in the process to intervene (intervention sequence).	$2 \to 2$	$2 \to 4$

In this study, we focus on optimizing interventions for individual process instances in isolation, an assumption that simplifies the problem by disregarding interdependencies between cases. In real-world scenarios, such interdependencies—often due to shared resource constraints—can significantly impact outcomes. Recent research has begun to address these complexities [26,32], but they remain outside the scope of this work—a simplification also made in the majority of existing studies.

2.2 Prescriptive Process Monitoring Approaches

The first major approach adopted in PresPM is Causal Inference (CI), which focuses on estimating the effect of an intervention action on a target variable for each individual case using offline data, whether synthetic or real [14]. A core challenge in CI is that only the outcome of the chosen intervention action is recorded for each case, while the counterfactual outcome remains unknown. One way to estimate this counterfactual outcome is to analyze similar cases in the data that received a different intervention action. However, real-world data-gathering policies often introduce confounding, where other variables influence both treatment/intervention and outcome. This can lead to treated and untreated cases differing strongly in their distributions, unlike the balanced and confounding-free setup of RCTs. Traditional CI methods attempt to correct for such distortions [19]. An example of CI in PresPM is the method described in [9], which incorporates CI by augmenting logs with estimated counterfactuals to identify optimal causal effect estimators. RealCause, a generative AI framework originally used as a CI evaluation tool, facilitates this process. Other studies include [3,30,31], and [32], although CI is not necessarily the main focus in the latter two.

Reinforcement Learning (RL) is the other most commonly adopted approach in PresPM. In RL, an agent learns to make (sequential) decisions by interacting with an environment, observing states, taking actions, and receiving rewards to develop a policy that maximizes cumulative rewards over time. A popular variant is Q-learning, where state-action values are collected in a Q-table, defining the policy [34]. Deep Learning (DL) is frequently integrated with RL in applications with large state spaces, exemplified by Deep Q-learning (DQN), where a neural network approximates the Q-table [24]. Online RL trains in real time, posing risks with its trial-and-error approach, but, like RCTs, avoids confounding. In contrast, offline RL uses potentially confounded pre-collected data, but is gaining popularity as a safer alternative. Examples of online RL in PresPM include studies [22,32], though real-life event logs are used to create the online environment, making their methods akin to offline RL despite online agent activity. Offline RL examples include [4], where similarity-based approaches are used to create training environments, and [3], where RealCause is applied to estimate counterfactual results.

2.3 Limitations in Prescriptive Process Monitoring

Current approaches in PresPM mainly rely on historical offline datasets, with BPIC12 and BPIC17 [11,12] being by far the most commonly used in works such as [3,4,9,32]. Although insightful, historical datasets lack counterfactual outcomes. While we can estimate the impact of interventions, their true effects remain unknown, resulting in only approximate evaluations. RealCause is the most widely used evaluation method to obtain these approximate evaluations [3,9,31–33]. For example, in [3], RealCause not only generates alternative outcomes to train an RL agent, but also defines the evaluation metrics, assuming that these generated outcomes are the true outcomes. Furthermore, comparison studies of PresPM methods are scarce. To the best of our knowledge, [36] is the only work to compare two implementations comprehensively by using a synthetic setup for precise evaluation. This scarcity is partly due to the lack of accurate evaluation mechanisms and unclear intervention dimensions, making comparisons difficult. These dimensions, i.e., action width and depth, are crucial to know, as the effectiveness of PresPM methods depends on them. For example, [36] shows that their CI implementation, though safer, performs worse than RL in scenarios with action depths larger than 1 (timing choice). It is also worth noting that most research focuses on interventions with action width 2 and action depth larger than 1 [9,31,32]. For example, in BPIC12 and BPIC17, the act of sending multiple offers is considered an intervention and follows these dimensions. Optimization of intervention sequences is underexplored. Only the methods in [18,35] address this by deploying multiple predictive models. However, their approaches may accumulate errors, leaving room for improvement. Lastly, commonly used logs often lack empirically validated causal relationships. Researchers typically select interventions without external domain expert confirmation, especially regarding their heterogeneous effects across different cases. For instance, while the intervention in BPIC12 and BPIC17 is correlated with higher loan acceptance rates, its causal impact has not been extensively studied or externally validated.

Manually defined simulators offer a solution to these limitations. In such a simulator, the process-generative function is fully known by design, allowing for the calculation of outcomes for all possible interventions, thus supporting accurate evaluation by generating counterfactuals. This setup also enables controlled experimentation, e.g., parameters such as confounding level can be adjusted systematically. Fully specifying the simulator from the ground up provides the flexibility to clearly define a range of interventions with varying complexity within the same process, and explicitly model their causal effects on outcomes. Combining these interventions also supports research for intervention sequences.

2.4 Current Simulations in Process Mining

Given the benefits of a manually defined simulator for PresPM, it is important to understand why such a new simulator may be preferred over existing simulators. While CI and RL studies outside of Process Mining often use simulations [2,34], these do not generate business process data. Simulators have also gained traction within Process Mining. For example, Data-Driven Process Simulation (DDPS) approaches include [7,15,27]. Others train and employ DL models to generate simulated event logs, such as in [6]. Examples of approaches that combine both, or hybrid simulations, are [5,21]. Extending these simulations to PresPM research would be useful, as they can be applied to any event log and could enable realistic experiments. However, they have a major limitation in this context: these simulations are based on historical datasets, and thus still suffer from the same counterfactual limitations. They are typically developed to reflect the policy that generated the original event log (e.g., bank policy), making them unlikely to accurately generate outcomes when a PresPM method recommends actions that differ from past practice. Additionally, they do not support multiple intervention types or allow controlled experiments relevant to PresPM research, such as manipulating confounding levels. Even if certain components are manually defined to extend these simulations, the evaluation accuracy remains uncertain due to the limitations in the parts that are not manually defined but approximated.

In contrast, manually defined simulators provide the opposite trade-off: they offer accuracy in evaluation and enable fully customizable experiments, including support for multiple interventions and confounding scenarios. However, they are not automatically generalizable to any event log and may lack realism. Currently, only one such simulator exists for PresPM [36], but it is overly simplistic, lacking loops, parallelism, and a variety of interventions. This reveals a clear gap and opportunity for developing a more sophisticated, realistic specified simulator tailored to PresPM needs that provides customizable and fully accurate experiments.

3 Simulator Development

In this section, we introduce the development of SimBank. We begin by outlining the high-level requirements derived from our review of the literature, followed by a description of the process and its interventions. We then delve into the implementation details, highlighting SimBank's capabilities.

3.1 Requirements

As discussed in Sects. 1 and 2, current research in PresPM faces two major limitations: a) challenges for accurate model evaluations, and b) an absence of comparisons of PresPM methods, both in evaluating them against each other and in assessing their effectiveness for different intervention types. To address these issues, we identified four requirements for SimBank:

- **Complete evaluation.** The simulator should be able to calculate the target variable of a case for every possible action of an intervention. This overcomes the limitations of historical data in evaluating PresPM models, which lack counterfactuals. Moreover, it would allow for a direct comparison between true method performance and performance as approximated by commonly used evaluation techniques in PresPM and CI research, such as RealCause [25].

- **Inter-intervention method comparison.** The simulator should allow for a comparison of methods across various interventions with differing dimensions. There is significant potential for further research on intervention sequences, and our simulator should enable exploration of these more advanced optimization challenges.
- **Intra-intervention method comparison.** The simulator should support experiments within each intervention type to assess method performance across varying dataset characteristics (e.g., level of confounding). It should also accommodate methods requiring offline or online training, as both are commonly used (see Sect. 2).
- **Balancing realism and simplicity.** A simulator for generating realistic datasets must balance complexity and interpretability. Business processes often include parallel activities, loops, and a high degree of structural variety, reflected in the diversity of trace variants [28]. Additionally, a simulator should create realistic variables relevant to its business context, e.g., a loan application process. At the same time, the generated datasets should remain simple enough to allow clear evaluation of PresPM methods. Too many dimensions or variables can obscure conclusions.

3.2 Description of the Process and Interventions

The design of the simulator's process model centers on a fictional loan application process in a bank. The model is illustrated in Fig. 1, which includes activity costs, durations, the bank's loan processing policy, and external environmental factors. The process begins with an application submission, followed by the bank selecting a procedure, potentially contacting HQ and the customer, calculating an interest rate, or canceling the application. If an interest rate is offered, the client's response follows. Table 2 lists the data attributes guiding the bank policy and the client's (actual but unobservable) quality, an external factor beyond the bank's control. The target variable to optimize is the loan profit, considering variables like the amount requested, interest rate, elapsed time, costs, (actual) client quality, and cancellation reasons. The client's decision also affects profit, but is in itself influenced by elapsed time, interest rate, and amount. For simplicity, potential client default is excluded. We have integrated three main interventions into the loan application process: *Choose procedure*, *Set interest rate*, and *Time contact HQ*. The intervention dimensions and details are outlined in Table 3.

3.3 Implementation and Capabilities

Tooling. The simulator is built in Python, primarily using the *simpy* library [23]. While the initial code is based on the PNSIM framework [27], it is extensively modified to suit the requirements of our research, incorporating the specific features discussed below. We use random number generators to generate variables. The code and comprehensive guidance on using SimBank can be found at Zenodo [10].

Requirements Realization. The first requirement aims to ensure a <u>complete evaluation</u>. In the simulator, the generative functions are predefined, allowing us to precisely assess the outcomes of every possible action at an intervention point. To maintain consistency across methods, we use a fixed seed for the random number generators. This guarantees that the same recommended actions for a specific test case always leads to the same outcome for that action, both in control-flow and data-flow, enabling a fair comparison of PresPM methods.

The second requirement aims to enable <u>inter-intervention comparisons</u>. This is achieved by incorporating the three distinct interventions, each varying in action width and depth. Additionally, the ability to combine these interventions facilitates further research into intervention sequences. SimBank allows users to modify interventions, such as adjusting the logic of an intervention function (e.g., setting the interest rate), as well as introduce new ones—either by adding func-

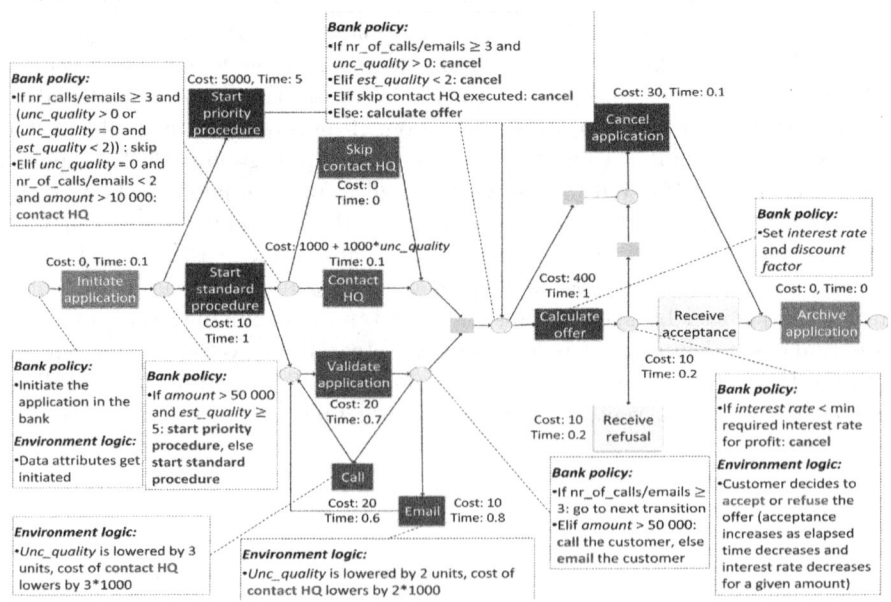

Fig. 1. The simulator process model along with the bank policy and environment logic.

Table 2. Data attributes generated by the simulator. The unobserved attribute 'quality' cannot be used to improve the policy.

Attribute	Description
Activity	Executed activity
Timestamp	Activity timestamp
Case_nr	Case number
Cost	Cumulative cost of the case
Amount	Requested loan amount
Est_quality	Quality of the client, estimated by the bank
Unc_quality	Uncertainty of the bank regarding the estimated quality
Interest rate	Interest rate offered to the client
Discount factor	Discount factor, taking into account the estimated quality
Quality	Actual quality of the client, unobserved by the bank

Table 3. The main interventions included in the simulator.

Intervention	Action width	Action depth	Description
Choose procedure	2	1	Choose between the standard procedure or the priority procedure. The priority option is faster and skips activities but costs more and ignores client quality uncertainties.
Set interest rate	3	1	Set an appropriate interest rate (0.07, 0.08, or 0.09), balancing profitability with the risk of the client declining the loan offer.
Time contact HQ	2	4	Time the contact with HQ, which runs concurrently with customer contact. Contacting HQ is costly and time-consuming, so it should be initiated early to minimize additional delays alongside customer contact. However, the higher the client's quality uncertainty, the greater the cost of contacting HQ. Since uncertainty lowers after each customer contact, contacting HQ early is not always optimal. Additionally, HQ contact should only ever occur when the bank is confident the application will not be canceled, avoiding excess costs. If HQ contact is not made, it will be skipped, leading to automatic cancellation. Interventions after 'validate application' are excluded, reducing the action depth from 8 to 4, as customer variables remain unchanged during validation, rendering the intervention points directly after validation irrelevant.

tions that influence specific variables during an activity or by creating entirely new activities.

The third requirement focuses on enabling <u>intra-intervention method comparisons</u>. The simulator allows for generating datasets based on existing bank policies (to test offline methods using fixed datasets) or by randomly selecting actions during interventions. The latter approach mimics an RCT, ensuring unconfounded data collection. By adjusting the proportion of RCT data versus (bank) policy-based data in the dataset, we can control the level of confounding, ranging from 0 to 1, which we denote as δ. This enables us to assess the robustness of offline methods against confounding introduced by existing policies across a specified range. The bank policy can also be adjusted to examine how the performance of the data-gathering policy impacts method performance. Additionally, we support comparisons between offline and online methods. The simulator operates in two main modes: offline simulation, which returns a complete event log, and online mode. In online mode, also used for method evaluation, a case is generated up to a specified intervention point where an external policy dictates the action to be taken. Afterwards, the case simulation resumes until the next potential intervention point, repeating this cycle. If no further intervention points exist, the simulation continues until the outcome can be calculated.

The final requirement aims to balance realism with simplicity, which we achieve in several implementation choices. SimBank includes true activity concurrency, previously not present in the PNSIM simulator. This advancement enables activities in different parallel branches to be executed before, after, or simultaneously with one another, making the simulator more realistic. We accomplish this by designating which activities belong to the various parallel branches and by explicitly setting the timestamps to indicate whether an activity occurs before, during, or after an activity in another branch. Some activities are also arranged into loops, as shown in Fig. 1. This greatly enhances the structural diversity of the process, particularly when compared to the simpler configurations outlined in [36], which remains the only other study employing a fully synthetic setup. The underlying logic of SimBank draws inspiration from real-world banking practices. For example, the risk-free rate, a key part of the discount rate, is modeled after the 10-year Belgian government bond yield at the time of the simulator's development. Profit, the target variable, is calculated using the net present value, reflecting the time-related value of money [8]. The activities in SimBank are also based on the most commonly used event logs from PresPM: BPIC12 and BPIC17. Examples of activities that appear in both SimBank and these logs include *call, validate application, calculate offer, initiate application,* and *receive acceptance/refusal.* At the same time, we intentionally maintain simplicity to ensure the datasets remain interpretable. This includes aggregating complex factors, such as client quality, into single variables and limiting the overall number of activities. While we incorporate more activities than in [36], we avoid excessive loops or overly complex parallel structures, maintaining a balance that supports clear conclusions. If users feel the process is still too simplified, they have the option to add more activities in the same way as the existing ones.

4 Proof of Concept

This section focuses on validating SimBank for PresPM benchmarking, demonstrating its ability to highlight method differences, support both online and offline approaches, and enable diverse experimental scenarios. we emphasize the importance of SimBank's accurate evaluation, which stands out compared to existing evaluation methods. The focus is on showcasing SimBank's capabilities rather than optimizing a PresPM method's performance. We begin with the experimental setup and then present the results.

4.1 Experimental Setup

We experiment with 4 PresPM methods—2 based on CI and two based on RL—to assess their effectiveness in optimizing interventions. We also use the widely adopted RealCause evaluation method to compare its approximated evaluation results with the accurate SimBank evaluation. First, we explain the RealCause implementation, followed by the 4 PresPM methods.

- **Evaluation method: RealCause.** RealCause was developed to benchmark CI approaches by generating counterfactuals that reflect a real dataset, helping to assess whether a CI method recommends the optimal actions. RealCause has been used in PresPM research, such as in [3,9,31–33]. Given the intervention action recommended by the PresPM approach for a test case, RealCause generates the corresponding outcome. The method models two key distributions. The first is the selection distribution, which determines which intervention action is assigned to a case and is modeled here by a Bernoulli distribution for interventions *Choose procedure* and *Time contact HQ*, and a categorical distribution for *Set interest rate*. The second is the outcome distribution, modeled by a Sigmoid Flow distribution with discrete value concentrations [13]. The distributions are approximated using a TARNet-based structure, which includes a neural network to learn treatment-agnostic representations and two sub-networks tailored to specific treatment groups [29]. This design could be effective as it utilizes all data for the initial representation network while emphasizing treatment through the separate networks. The models were tested using the statistical tests of the original RealCause paper [25].
- **PresPM method 1: Standard S-learner (CI).** As a standard CI implementation, we adopt the established S-learner framework from Causal Machine Learning [17], where a single model is trained with the intervention action as an input variable. We employ an LSTM network, which takes as input the case variables, actions, and event variables of a prefix up to an intervention point. The output is the predicted outcome (profit) for the given prefix and current action. For an intervention sequence, we can extend this approach by training separate models for each intervention. Preprocessing includes one-hot encoding for categorical features, standardization of numeric variables, and sequence padding. We use early-stopping during training to prevent overfitting. For interventions with action depth 1 (fixed), the optimal action is chosen by predicting and comparing outcomes across all possible actions. For interventions with an action depth larger than 1 (such as *Time contact HQ*), the intervention is triggered when the predicted profit difference between intervening and not intervening exceeds a threshold determined on a validation set. This approach is referred to as the standard S-learner.
- **PresPM method 2: RealCause-based S-learner (CI).** The second CI approach is inspired by [3,9], and combines RealCause with the S-learner framework. RealCause is first used to generate counterfactuals by estimating profits for actions other than those observed in the real training dataset (from

SimBank). We then train an XGBoost model instead of an LSTM, as Real-Cause requires non-sequential encoding. Preprocessing follows the approach in [9], which includes aggregation encoding. A threshold for timed interventions is again determined on a validation set. In summary, this approach uses RealCause as a data augmentation tool, as in [3], while integrating it with a CI framework, as in [9]. We refer to this CI approach as RealCause-based S-learner.

- **PresPM method 3: Deep Q-learning (RL).** We adopt the DQN approach as standard RL implementation, employing two neural networks—an online network and a target network—for training stability. Both networks share the same LSTM architecture as the standard S-learner for consistency. During inference, the online network processes the current state, including all prior event variables, case variables, and actions, to output Q-values for each possible action. The action with the highest Q-value is chosen. Rewards are based on the case's final profit, with penalties for interventions at non-intervention points. We refer to this RL method as Deep Q-learning.
- **PresPM method 4: K-means-based Q-learning (RL).** Our second RL approach is inspired by [4]. Their method involves preprocessing prefixes by analyzing activity counts and positions, then using K-means clustering to group them. An event log is replayed to build an RL training environment, where states are defined by the prefix cluster and the last activity. In our implementation, we simplify this by using only K-means clustering and the state representation constructed from an offline dataset. Traditional Q-learning is applied to generate a Q-table online using cases from SimBank, with the same reward structure as the Deep Q-learning approach. Unlike [4], we skip replaying an event log to construct the environment. We refer to this approach as K-means-based Q-learning.

To provide a general overview of SimBank's capabilities, we begin by applying the standard PresPM methods (S-learner and DQN) across all interventions. Additionally, we apply the standard S-learner to an intervention sequence combining *Choose procedure* and *Set interest rate*, but exclude this for DQN due to its high computational demands. Next, we extend our analysis for the *Time contact HQ* intervention, as its dimensions are the most studied in PresPM research (action width 2 and action depth > 1, see Sect. 2.3). For this intervention, we incorporate the methods inspired by existing approaches: the RealCause-based S-learner and K-means-based Q-learning. Finally, to assess the relevance of the confounding parameter in SimBank, we adjust δ for offline methods (CI-based and RealCause), testing two values for *Time contact HQ* ($\delta = 0$ and $\delta = 0.999$), and six levels for the standard S-learner. All other experiments use $\delta = 0$.

All methods are trained on 100,000 cases, with 1,000 for validation and 10,000 for testing. RealCause and the K-means model are trained on an extra unrelated dataset of 10,000 cases. The evaluation metric, denoted as *Gain*, is the relative change in total profit compared to the bank's baseline policy. A policy therefore outperforms the bank policy if its *Gain* > 0. Each experiment is repeated 5 times.

4.2 Results

Table 4 presents the results for the *Choose procedure* and *Set interest rate* interventions. Note that the performance of the bank policy is not in the table, as its *Gain* is naturally 0. In both cases, the standard S-learner and Deep Q-learning outperform the random policy and bank policy. For the *Choose procedure* intervention, the standard S-learner demonstrates the best performance and greater stability, with a lower standard deviation. For the *Set interest rate* intervention, the standard S-learner also slightly outperforms Deep Q-learning and remains the more stable method. While Q-learning theoretically offers optimal solutions [34], its practical implementation underperforms. Potential improvements include exploring alternative state representations, architectures, or efficient sampling techniques [20].

Table 4. Comparison of approaches for interventions *Choose procedure* and *Set interest rate*, using a training set for CI generated with no confounding ($\delta = 0$).

Approach	Set-up	Choose procedure		Set interest rate	
		Gain	StDev	Gain	StDev
Random	-	-0.209	0.008	-0.297	0.009
CI	Standard S-learner	**0.445**	0.011	**0.144**	0.001
RL	Deep Q-Learning	0.323	0.015	0.131	0.007

Table 5 shows results for the *Time contact HQ* intervention on confounding-free datasets ($\delta = 0$), matching an RCT. All methods outperform the random and bank policies, with Deep Q-learning achieving the highest gain. The standard S-learner performs less effectively and is less stable, aligning with the findings in [36]. Methods inspired by prior works perform worse than standard implementations. The RealCause-based S-learner likely loses temporal information due to aggregation encoding and accumulates errors due to dual-model use (RealCause and XGBoost). Similarly, K-means-based Q-learning likely sacrifices temporal detail by solely relying on cluster labels and the last activity for state representation. Interestingly, method rankings are consistent between the RealCause and true evaluations, as the RCT training data ensures accurate counterfactuals. However, RealCause underestimates absolute performance.

Table 6 presents results for the *Time contact HQ* intervention with $\delta = 0.999$, creating a heavily confounded training dataset. Note that the performance of the random policy remains unchanged from the unconfounded setting, since it does not rely on the training data. All methods outperform the random and bank policies, though the standard S-learner only marginally exceeds the bank policy. True evaluation shows the RealCause-based S-learner performs best, while the standard S-learner is more stable. However, performance suffers across the board, as the dataset is confounded and lacks diversity. The RealCause-based S-learner performs better, likely due to the TARNet-based structure in RealCause which

Table 5. Comparison of approaches for intervention *Time contact HQ*, using a training set for CI and RealCause with no confounding ($\delta = 0$).

Approach	Set-up	True		RealCause	
		Gain	StDev	Gain	StDev
Random	–	0.235	0.016	−0.230	0.033
CI	Standard S-learner	0.691	0.072	0.619	0.075
	RealCause-based S-learner	0.565	0.001	0.457	0.101
RL	Deep Q-Learning	**0.766**	0.013	**0.675**	0.064
	K-means-based Q-Learning	0.661	0.037	0.584	0.055

emphasizes treatment selection. Moreover, the RealCause evaluation model leads to inaccurate evaluations when trained on a confounded dataset. It overestimates the RealCause-based S-learner, due to the alignment between this learner and the RealCause evaluation model. This underscores the risk of relying on Real-Cause for both data augmentation during training and for defining evaluation metrics, as seen similarly in [3], albeit with an RL agent instead. More critically, RealCause ranks the random policy higher than the standard S-learner, despite the true evaluation indicating otherwise. These findings emphasize the importance of accurate and unbiased evaluation methods, such as those provided by SimBank. Since most real-world event logs are collected under existing, non-randomized policies (unlike RCTs), they are likely to be confounded. In such cases, relying solely on RealCause can lead to misleading conclusions.

Table 6. Comparison of approaches for intervention *Time contact HQ*, using a training set for CI and RealCause with strong confounding ($\delta = 0.999$).

Approach	Set-up	True		RealCause	
		Gain	StDev	Gain	StDev
Random	–	-0.235	0.016	0.091	0.640
CI	Standard S-learner	0.053	0.375	-0.008	0.858
	RealCause-based S-learner	**0.198**	0.454	**2.249**	4.551

Figure 2 contains the results of the experiments on the standard S-learner where we vary δ at 6 different values. The minimum value of 0.95 is chosen since performance stays more or less consistent between 0 and 0.95 for a dataset of 100,000 cases. The error bars indicate a confidence interval of the performance at each level. Except for the *Set interest rate* intervention, a general trend is that both performance and stability decrease for higher values of δ. Especially the bank's original policy for the *Time contact HQ* intervention seems undermining, given the high sensitivity of the standard S-learner to δ. For the sequence of *Choose procedure* and *Set interest rate*, sequential optimization proves beneficial, as our setup outperforms optimizing only one of the interventions. However, since our implementation consists of one model per intervention, each model optimizes its intervention independently, assuming the bank policy for the other. This likely limits overall performance. More research is needed to improve the optimization of intervention sequences, which is especially relevant in business processes.

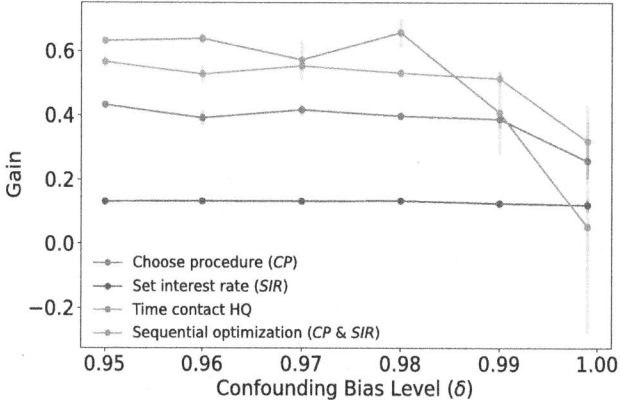

Fig. 2. Performance of the standard S-learner across 6 levels of δ for the 3 main interventions and 1 intervention sequence. A clear trend emerges, showing decreasing performance as δ increases.

These results show that using SimBank for method comparisons offers valuable insights. By leveraging SimBank's synthetic data, we can pinpoint the strengths and weaknesses of both online and offline methods. This analysis can span various intervention dimensions and scenarios where challenges might emerge from data-gathering policies. The findings also highlight the critical need for accurate evaluation in PresPM. We have shown that evaluation methods like RealCause can be validated using SimBank.

5 Conclusion and Future Work

We introduce SimBank, a synthetic data simulator designed to benchmark methods in PresPM research. Current research in this field often lacks comprehensive comparisons and accurate evaluations, and SimBank aims to address these gaps. SimBank simulates a bank's loan application process and supports both online and offline method comparisons. It includes three key intervention problems relevant to business processes, with possible combinations of interventions, and allows researchers to adjust data-generating parameters like confounding levels to create experiments tailored to PresPM research. In our proof of concept, we demonstrate that SimBank can provide valuable insights by enabling comparisons between interventions of varying complexity and between different setups for the same intervention. We also emphasize the importance of precise evaluation, which SimBank facilitates by allowing for fully accurate assessments.

A limitation of our simulator is that SimBank is only partially realistic. While a fully synthetic simulator can never completely replicate real-world scenarios, we have tried to minimize this limitation by incorporating features like true concurrency, loops, and basing it on real-life logs and variables. Another limitation is the assumption that there are no interdependencies between process instances. However, this can be addressed in future work, as we used PNSIM as our starting point, which can potentially account for, e.g., delays in one process instance caused by others.

Future research will focus on optimizing intervention sequences, particularly for offline methods like CI. The approach outlined in [2] shows promise in this regard, but requires further investigation to determine its effectiveness in highly variable and complex business process environments. Additionally, we plan to extend the simulator with case interdependencies by including the potential delays and resource limitations. Lastly, Causal Discovery (CD) for PresPM can be explored to identify interventions [37]. CD methods can be tested with SimBank. Interventions and their dimensions are often assumed predefined in PresPM, which may not always apply in real-world business processes.

Acknowledgments. This work was supported the Research Foundation Flanders (FWO) under grant number G039923N and 11A6J25N, and Internal Funds KU Leuven under grant number C14/23/031.

References

1. van der Aalst, W.M.P.: Process Mining, pp. 2171–2173. Springer (2009). https://doi.org/10.1007/978-0-387-39940-9_1477
2. Bica, I., Alaa, A.M., Jordon, J., van der Schaar, M.: Estimating counterfactual treatment outcomes over time through adversarially balanced representations (2020). https://arxiv.org/abs/2002.04083
3. Bozorgi, Z.D., Dumas, M., Rosa, M.L., Polyvyanyy, A., Shoush, M., Teinemaa, I.: Learning when to treat business processes: Prescriptive process monitoring with causal inference and reinforcement learning (2023). https://arxiv.org/abs/2303.03572
4. Branchi, S., Buliga, A., Francescomarino, C.D., Ghidini, C., Meneghello, F., Ronzani, M.: Recommending the optimal policy by learning to act from temporal data (2023). https://arxiv.org/abs/2303.09209
5. Camargo, M., Báron, D., Dumas, M., González-Rojas, O.: Learning business process simulation models: a hybrid process mining and deep learning approach. Inform. Syst. **117** (2023).https://doi.org/10.1016/j.is.2023.10224
6. Camargo, M., Dumas, M., González-Rojas, O.: Learning accurate LSTM Models of business processes. In: Hildebrandt, T., van Dongen, B.F., Röglinger, M., Mendling, J. (eds.) BPM 2019. LNCS, vol. 11675, pp. 286–302. Springer, Cham (2019). https://doi.org/10.1007/978-3-030-26619-6_19

7. Camargo, M., Dumas, M., González-Rojas, O.: Automated discovery of business process simulation models from event logs. Decision Support Syst. **134** (2020) https://doi.org/10.1016/j.dss.2020.113284
8. Dahlquist, J., Knight, R.: Principles of Finance. OpenStax, Houston, Texas (2022)
9. Dasht Bozorgi, Z., Teinemaa, I., Dumas, M., La Rosa, M., Polyvyanyy, A.: Prescriptive process monitoring based on causal effect estimation. Inform. Syst. **116** (2023). https://doi.org/10.1016/j.is.2023.102198
10. De Moor, J.: Simbank (2025). https://doi.org/10.5281/zenodo.15574272
11. van Dongen, B.: Bpi challenge (2012). https://doi.org/10.4121/uuid:3926db30-f712-4394-aebc-75976070e91f
12. van Dongen, B.: Bpi challenge (2017). https://doi.org/10.4121/uuid:5f3067df-f10b-45da-b98b-86ae4c7a310b
13. Huang, C.W., Krueger, D., Lacoste, A., Courville, A.: Neural autoregressive flows. In: International conference on machine learning. PMLR (2018)
14. Imbens, G.W., Rubin, D.B.: Causal Inference for Statistics, Social, and Biomedical Sciences: An Introduction. Cambridge University Press (2015)
15. Kirchdorfer, L., Blümel, R., Kampik, T., van der Aa, H., Stuckenschmidt, H.: Agentsimulator: An agent-based approach for data-driven business process simulation. In: 6th International Conference on Process Mining. IEEE (2024)
16. Kubrak, K., Milani, F., Nolte, A., Dumas, M.: Prescriptive process monitoring: Quo vadis? (2021). https://arxiv.org/abs/2112.01769
17. Künzel, S., Sekhon, J., Bickel, P., Yu, B.: Meta-learners for estimating heterogeneous treatment effects using machine learning. Proc. National Acad. Sci. **116** (2017).https://doi.org/10.1073/pnas.1804597116
18. Leoni, M.d., Dees, M., Reulink, L.: Design and evaluation of a process-aware recommender system based on prescriptive analytics. In: 2020 2nd International Conference on Process Mining (2020). https://doi.org/10.1109/ICPM49681.2020.00013
19. Lousdal, M.L.: An introduction to instrumental variable assumptions, validation and estimation. Emerging Themes Epidemiol. **15** (2018).https://doi.org/10.1186/s12982-018-0069-7
20. Mai, V., Mani, K., Paull, L.: Sample efficient deep reinforcement learning via uncertainty estimation (2022). https://arxiv.org/abs/2201.01666
21. Meneghello, F., Francescomarino, C.D., Ghidini, C., Ronzani, M.: Runtime integration of machine learning and simulation for business processes: Time and decision mining predictions. Inform. Syst. **128** (2025). https://doi.org/10.1016/j.is.2024.102472
22. Metzger, A., Kley, T., Rothweiler, A., Pohl, K.: Automatically reconciling the trade-off between prediction accuracy and earliness in prescriptive business process monitoring. Inform. Syst. **118** (2023). https://doi.org/10.1016/j.is.2023.102254
23. Meurer, A., et al.: Sympy: symbolic computing in python. PeerJ Comput. Sci. **3** (2017). https://doi.org/10.7717/peerj-cs.103
24. Mnih, V., et al.: Playing atari with deep reinforcement learning (2013). https://arxiv.org/abs/1312.5602
25. Neal, B., Huang, C., Raghupathi, S.: Realcause: Realistic causal inference benchmarking. CoRR **abs/2011.15007** (2020). https://arxiv.org/abs/2011.15007
26. Padella, A., Mannhardt, F., Vinci, F., de Leoni, M., Vanderfeesten, I.: Experience-based resource allocation for remaining time optimization. In: Business Process Management. Springer (2024)
27. Pourbafrani, M., Vasudevan, S., Zafar, F., Xingran, Y., Singh, R., van der Aalst, W.M.P.: A python extension to simulate petri nets in process mining (2021), https://arxiv.org/abs/2102.08774

28. Schreiber, C., Abbad-Andaloussi, A.: Structural process variety and standardization. In: 6th International Conference on Process Mining (2024). https://doi.org/10.1109/ICPM63005.2024.10680658
29. Shalit, U., Johansson, F.D., Sontag, D.: Estimating individual treatment effect: generalization bounds and algorithms (2017). https://arxiv.org/abs/1606.03976
30. Shoush, M., Dumas, M.: Prescriptive process monitoring under resource constraints: A causal inference approach. In: Process Mining Workshops. Springer (2022)
31. Shoush, M., Dumas, M.: When to intervene? prescriptive process monitoring under uncertainty and resource constraints. In: Business Process Management Forum. Springer (2022)
32. Shoush, M., Dumas, M.: Prescriptive process monitoring under resource constraints: a reinforcement learning approach (2024). https://arxiv.org/abs/2307.06564
33. Shoush, M., Dumas, M.: White box specification of intervention policies for prescriptive process monitoring. Data & Knowledge Engineering (2025)
34. Sutton, R.S., Barto, A.G.: Reinforcement Learning, second edition: An Introduction. MIT Press (2018)
35. Weinzierl, S., Dunzer, S., Zilker, S., Matzner, M.: Prescriptive business process monitoring for recommending next best actions. In: Business Process Management Forum. Springer (2020)
36. Weytjens, H., Verbeke, W., De Weerdt, J.: Timed process interventions: Causal inference vs. reinforcement learning. In: Business Process Management Workshops. Springer (2024)
37. Zanga, A., Stella, F.: A survey on causal discovery: Theory and practice (2023). https://arxiv.org/abs/2305.10032

Detecting Undesired Process Behavior by Means of Retrieval Augmented Generation

Michael Grohs[1,2](✉), Adrian Rebmann[2], and Jana-Rebecca Rehse[1]

[1] University of Mannheim, Mannheim, Germany
{michael.grohs,rehse}@uni-mannheim.de
[2] SAP Signavio, Berlin, Germany
adrian.rebmann@sap.com

Abstract. Conformance checking techniques detect undesired process behavior by comparing process executions that are recorded in event logs to desired behavior that is captured in a dedicated process model. If such models are not available, conformance checking techniques are not applicable, but organizations might still be interested in detecting undesired behavior in their processes. To enable this, existing approaches use Large Language Models (LLMs), assuming that they can learn to distinguish desired from undesired behavior through fine-tuning. However, fine-tuning is highly resource-intensive and the fine-tuned LLMs often do not generalize well. To address these limitations, we propose an approach that requires neither a dedicated process model nor resource-intensive fine-tuning to detect undesired process behavior. Instead, we use Retrieval Augmented Generation (RAG) to provide an LLM with direct access to a knowledge base that contains both desired and undesired process behavior from other processes, assuming that the LLM can transfer this knowledge to the process at hand. Our evaluation shows that our approach outperforms fine-tuned LLMs in detecting undesired behavior, demonstrating that RAG is a viable alternative to resource-intensive fine-tuning, particularly when enriched with relevant context from the event log, such as frequent traces and activities.

Keywords: Process Mining · Conformance Checking · Retrieval Augmented Generation

1 Introduction

Conformance checking is a sub-discipline of process mining which compares process executions that are captured as traces recorded in an event log to process behavior captured in a process model [9]. As input, conformance checking techniques require an event log and a dedicated process model that explicitly specifies what behavior is desired in the process [7]. Based on this, they can identify which traces diverge from the desired behavior and, subsequently, locate the undesired behavior within the trace [7]. This undesired behavior has negative impact on any objective of the process, including but not limited to compliance, performance, and outcome. In this paper, we focus on undesired control-flow of processes, i.e., activity executions that are not envisioned.

The applicability of conformance checking hence depends on the availability of high-quality models for the process in question [21, p.11]. However, creating and maintaining such models requires substantial effort and stakeholder involvement [11], which means that they are often unavailable, incorrect, or out of date [9]. Further, the process at hand might be too complex to be effectively captured in a model [2] or the organization might decide not to model its processes to spare itself the efforts [11]. In all of these cases, conformance checking is not applicable. Nevertheless, organizations might still want to know whether their processes contain undesired behavior [8].

To alleviate this issue, several approaches have been proposed that detect undesired behavior in event logs without a dedicated process model. Unsupervised anomaly detection techniques identify statistically infrequent traces, assuming these are undesired [17,22]. However, since low frequency does not always indicate undesirability, some approaches instead leverage the natural language semantics of activities to detect undesired process behavior [1]. Recent approaches leverage the natural language understanding capabilities of Large Language Models (LLMs) for this task. The key idea is that fine-tuning an LLM on process behavior enables it to learn how processes function, allowing it to differentiate between desired and undesired behavior [6,8,15]. However, fine-tuning is highly resource-intensive and the resulting models often do not generalize well, especially to processes that differ considerably from those used for fine-tuning [6].

To overcome these limitations, we propose an approach for detecting undesired process behavior that requires neither a dedicated process model nor resource-intensive fine-tuning. Instead, we use Retrieval Augmented Generation (RAG), which builds on the idea that LLMs answer more accurately if they are supplied with explicit task-related knowledge that has not been used in training [19]. Concretely, we provide an LLM with direct access to a knowledge base that captures both desired and undesired behavior of a broad range of processes. In addition, we provide it with explicit knowledge about the event log at hand in form of frequent traces and activities. Then, the LLM can use this task-related knowledge to find undesired behavior in previously unseen traces. In particular, our approach can detect instances of five established patterns of undesired behavior [16]: *inserted, skipped, repeated, replaced,* and *swapped* activities (or sequences thereof). Thus, our approach can identify more complex instances of undesired behavior compared to existing approaches, which are limited to deviations involving at most two activities [6,8], even though undesired behavior can span any number of activities. Our evaluation experiments show that our approach outperforms both general-purpose LLMs and LLMs that have been fine-tuned for detecting undesired process behavior.

The remainder of this paper is structured as follows. We first discuss related work in Sect. 2. Then, we present our approach in Sect. 3 and evaluate it in Sect. 4. We discuss our findings and conclude in Sect. 5.

2 Related Works

Our work relates to conformance checking, unsupervised anomaly detection, and semantic anomaly detection. Also, it relates to RAG-usage in other process mining tasks.

Conformance Checking. Traditional conformance checking techniques compare event log traces to process models that define desired process behavior [7]. They all require a process model as input [9]. State-of-the-art techniques are trace alignments [9], which identify inserted and missing activities. Based in these alignments, it is also possible to identify high-level patterns of undesired behavior such as repetitions or swaps [12,13]. Furthermore, declarative conformance checking techniques check the satisfaction of constraints, identifying violations of temporal patterns like precedence or response [18]. Unlike our approach, all corresponding techniques require a process model as input.

Unsupervised Anomaly Detection. Unsupervised anomaly detection tries to find undesired behavior without relying on a process model, by identifying statistically rare behavior in a given event log. Corresponding techniques are, e.g., based on association rule mining [5] or likelihood graphs of traces [4]. Deep learning methods like GRU autoencoder neural networks combined with an anomaly threshold [22] or LSTM networks [17] have also been proposed. All these techniques determine whether a process instance is statistically unlikely. However, likeliness does not necessarily correspond to desirability, so infrequency alone does not suffice to detect deviations from desired process behavior. This constitutes a conceptual difference to the goal of our approach.

Semantic Anomaly Detection. Rather than relying on infrequency, several approaches employ language models, which allows them to consider the natural language semantics of activities to identify undesirable behavior. This has been done on trace-level and activity-level. On trace-level, semantically anomalous traces can be detected [24], but this does not allow for locating undesired activities. On activity-level, some approaches identify semantically anomalous ordering relationships between pairs of activities [8,24], but can therefore not detect more complex undesired behavior encompassing more than two activities. Another approach on activity-level (called *xSemAD*) detects violated declarative constraints by fine-tuning a sequence-to-sequence model [6]. While constraints explain why certain activity pairs are undesired in traces, they are limited to just two activities. When violations span more than two activities, xSemAD detects multiple individual constraint violations, making it difficult to identify more complex instances of undesired behavior. Furthermore, despite requiring substantial resources for fine-tuning, xSemAD struggles with precision, due to a many false positive violations. Finally, an activity-level approach, called *DABL*, fine-tunes an LLM to detect insertions, skips, repetitions, and swaps of activities in traces [15]. Since the authors have not evaluated their approach regarding its ability to detect which activities deviate, the efficacy and efficiency of such a fine-tuning strategy remains uncertain. In contrast to xSemAD and DABL, our approach does not require any fine-tuning.

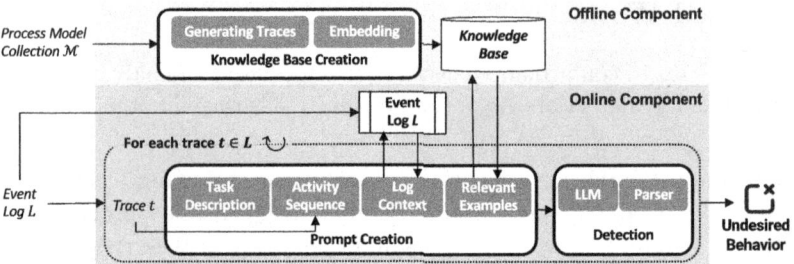

Fig. 1. Overview of our approach for detecting undesired behavior

Retrieval Augmented Generation. The idea behind RAG is to provide an LLM with relevant information that it has not seen during training. This is particularly helpful for knowledge-intensive tasks [19]. The relevant information is **R**etrieved from a knowledge base that contains documents, textual reports, or other related data objects, then **A**ugments an LLM prompt, which eventually **G**enerates the desired output [19]. Compared to fine-tuning the LLM with the relevant information directly, this is less computationally expensive and can lead to less overfitting on the training data [19]. In process mining, RAG has been rarely used so far. One application uses it in a conversational agent based on process model retrieval and fine-tuning to support human decisions [3]. Another uses it to support the deployment of processes and their operations [20]. Our approach is the first to use RAG to detect deviations from desired behavior.

3 Approach

This section presents our approach for detecting undesired behavior in an event log by means of RAG. Section 3.1 gives an overview of the approach, before Sect. 3.2 and Sect. 3.3 detail its components and phases.

3.1 Overview

In this section, we give an overview of our approach's input, design, and output.

Input. As input, our approach takes a process model collection $\mathcal{M}$ and an event log L. To define them, let $\mathcal{A}$ be the universe of possible activities that can be performed organizational processes. A process model M is a finite set of desired execution sequences over a set of activities $A \subseteq \mathcal{A}$, where each sequence $s \in M$, with $s = \langle a_1, \ldots, a_n \rangle$ and $a_i \in \mathcal{A}$, should lead the process from an initial to a final state. A process model collection $\mathcal{M}$ is a finite set of process models. Further, focusing on the control-flow of a process, we define a trace $t = \langle a_1, \ldots, a_n \rangle$, with $a_i \in \mathcal{A}$, as a sequence of activities that have been executed during the execution of a single instance of an organizational process. An event log L is a finite multi-set of such traces.

Design. The general idea of our approach is to enable an LLM to detect undesired behavior in an event log by providing it with desired and undesired behavior from similar processes as well as explicit information about that log. To achieve that, our approach consists of an offline and an online component (cf. Figure 1).

(1) *Offline Component* (Sect. 3.2): Given a process model collection $\mathcal{M}$, the offline component populates a knowledge base of desired and undesired process behavior. This behavior can be accessed by an LLM by means of RAG and used to detect undesired behavior in a process that is not contained in the knowledge base.

(2) *Online Component* (Sect. 3.3): The online component detects undesired behavior in event log L, analyzing each trace $t \in L$ individually. For that, it retrieves similar behavior from the knowledge base and extracts relevant context from L in form of frequent behavior. Then, it combines retrieved behavior, extracted log context, and a task description to prompt an LLM which identifies any undesired parts of t.

Output. Our approach outputs undesired behavior per trace t from event log L. To characterize this undesired behavior, we employ a set of five established patterns, so-called deviation patterns [16]: *inserted, skipped, repeated, replaced,* and *swapped* activities.[1] These patterns account for undesired behavior that spans any number of activities, thus providing a more comprehensive view of deviations. In particular, each pattern relates to a specific type of undesired fragment, i.e., one or multiple consecutive activities:

- The *inserted* pattern indicates that a trace contains a fragment which was not supposed to occur at that point in the trace.
- The *skipped* pattern indicates that a trace misses a fragment which was supposed to occur at some point in the trace.
- The *repeated* pattern indicates that a fragment is re-executed in a trace although only the first execution was desired.
- The *replaced* pattern indicates that one fragment in a trace was executed although a different fragment was supposed to occur.
- The *swapped* pattern indicates that two fragments of a trace were performed in the wrong order.

If a trace t is found to be undesired, our approach outputs a set of instantiations of these deviation patterns. If t is found to be desired, the output contains no behavior for t.

3.2 Offline Component

The offline component populates a knowledge base of desired and undesired behavior, which can be accessed by the online component, allowing an LLM to

---

[1] Note that the original source [16] also proposes a *loop* pattern, but in the context of undesired behavior, this is simply a specific version of the *repeated* pattern.

use this knowledge to detect undesired behavior in previously unseen processes. The idea is that the LLM can better detect the five types of deviation patterns when having access to traces from similar processes with known patterns of these types. Complementarily, the LLM might better identify desired behavior when exposed to traces from similar processes that are known to be desired. To populate the knowledge base, our approach performs two steps. First, it generates traces for each model $M \in \mathcal{M}$, some of which conform to the model (desired behavior) whereas others are enriched with known deviations from the model (undesired behavior). Second, it transforms the traces into vector representations to allow the online component to retrieve them. We next elaborate on these steps.

Generating Traces. To generate the traces, we use the desired executions defined in $\mathcal{M}$ and additionally create undesired executions by adding noise to a share of them. First, we obtain the desired behavior by creating traces that correspond to the activity sequences captured in all $M \in \mathcal{M}$, resulting in one event log L^M per model.[2] Then, we add instantiations of the five deviation pattern types, referred to as *deviations*, to a subset of the traces in L^M, aiming for a balance between conforming and deviating traces. These deviations correspond to known undesired behavior. The result is a collection of conforming and deviating traces from all $M \in \mathcal{M}$, including information about which (if any) deviations occur where in these traces.

Embedding. The next step in the offline component involves creating numerical representations (*embeddings*) of the generated traces. These embeddings are essential for quantifying the similarity between individual traces during retrieval from the knowledge base. To ensure that the similarities factor in both the natural language semantics of the individual activities as well as their ordering relations in the trace, we represent each trace as a sentence (a single string that concatenates the activities). We then employ a pre-trained language model that was specifically designed to capture the meaning of entire sentences to obtain one embedding per trace [23]. We then store these embeddings, along with any known deviations in the respective trace, in the knowledge base. This knowledge base is then used by the online component to search for the most similar embeddings and retrieve them along with its known deviations, as we explain next.

3.3 Online Component

The online component takes an event log L as input and then performs two steps for each trace $t \in L$. In the *prompt creation* step, it populates a prompt template with process knowledge. The resulting prompt is then used in the *detection* step so that a general-purpose LLM can use it to detect undesired behavior in the trace.

---

[2] Note that we limit the number of loop executions to two.

Prompt Creation. The first step of the online component populates a prompt template for the task to detect undesired behavior in a trace $t \in L$. In addition to a textual task description, the template includes the activity sequence of t, explicit log context from L, and relevant example traces retrieved from the knowledge base. Except for the task description, all other parts are added to the template dynamically for each trace. In the following, we elaborate on the parts of the prompt template, illustrated in Fig. 2.[3]

Task Description. This first part introduces and explains the five deviation pattern types to be detected. For example, it states that the *inserted* pattern refers to a fragment which was not supposed to occur at a certain point in the trace. We specify that any detected undesired behavior in t must be an instance of one of these pattern types, affecting specific activity fragments—either a single activity or multiple consecutive ones. Finally, it asks to output "No Deviation." if no undesired behavior is found. In the following, we describe the dynamic parts of the prompt in detail.

Activity Sequence. This part is dynamic and corresponds to the sequence of activities in t, separated by commas. The prompt states that the previously defined patterns should be applied to this particular activity sequence.

Log Context. This part contains information from L that can be used to reason about undesired behavior in t. Concretely, we include activities that were executed in many traces in L, giving the LLM an idea of what other important activities could have happened in t. Further, we include the most frequent traces from L, giving it an idea of what order of activities was often executed and indicating that these traces can be considered a reference for desired behavior.

Relevant Examples. This final part contains traces from the knowledge base that are similar to t plus known deviations in them. To retrieve these traces, we create an embedding of t with the same procedure as in the offline component. Then, we calculate the cosine similarity of all traces in the knowledge base and retrieve a configurable number of traces with the highest similarity to t as well as all known deviations (if any) in them. These can be used to generalize and transfer (un)desired behavior to t.

---

[3] The complete prompt template can be found in our repository: https://doi.org/10.6084/m9.figshare.29125898.

Detection. The second step of the online component detects undesired behavior in t by feeding the prompt into an LLM and parsing the output into a structured format.

LLM. We use a general-purpose LLM, designed to generate accurate responses to a wide range of questions. This LLM is not fine-tuned for the specific task at hand, but it has the potential to transfer its general natural language understanding capabilities for this purpose. We want to enable this transfer by providing the LLM with the prompt from the previous step, which contains explicit information from the log as well as examples from other processes that illustrate desired and undesired behavior (retrieved from our populated knowledge base). The output of the LLM is parsed in the next step.

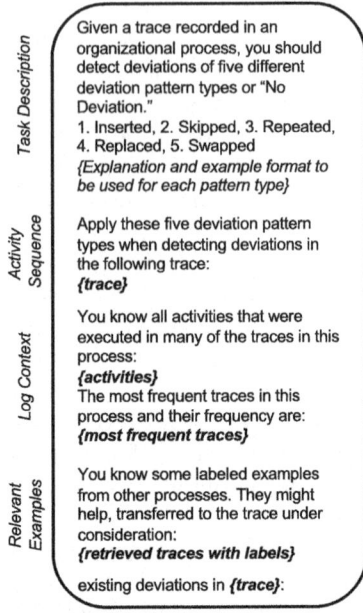

Fig. 2. Excerpt from prompt template. Boldness indicates dynamic parts that are adjusted per trace.

Parser. Finally, we parse the output string that the LLM generated into in a standardized output format. This is necessary because the output of the LLM is not guaranteed to be structured, leading to problems when the output should be used in downstream analyses. An example of this standardized format is shown in Fig. 3. To this end, we define a data model for the output of our approach in which we represent each deviation pattern type as a separate entity type. The entity types can be instantiated with trace fragments that are affected by a respective deviation.[4] For example, skipped activities should be parsed into the Skip() type that contains a list of the activities in the affected fragment. Swapped and replaced patterns affect two fragments, so they contain two distinct lists of activities (e.g., Swap(['Approve'], ['Ship'])). Data model and the LLM output string from the previous step are handed back to LLM, which is tasked to parse the string into an instantiation of the data model, enforcing that the defined format is returned.

Our approach outputs one instantiation of the data model per trace $t \in L$. In case no deviations were found in t, the output does not contain any instantiated deviation.

---

[4] The complete data model can be found in our repository.

LLM Output

Skipped: 'Create Order', 'Assign Order'

Repeat: 'Pay', Swapped: 'Approve' with 'Ship'

Parsed Output

Skip(['Create Order', 'Assign Order'])

Repeat(['Pay']), Swap(['Approve'],['Ship'])

Fig. 3. Exemplary parsed output

4 Evaluation

In this section, we evaluate the capability of our approach to detect undesired behavior, comparing it to four baseline approaches. In the following, we first describe the data collection (Sect. 4.1) and the experimental setup (Sect. 4.2). Then, we present the results of our experiments in terms of accuracy, robustness, and efficiency, demonstrating that our approach outperforms four baseline approaches in these aspects (Sect. 4.3). Finally, we showcase the practical applicability of our approach, illustrating that it can identify undesired behavior in a real-life event log (Sect. 4.4). Following the guidelines for LLM-based process analysis from [10], our Python implementation, sources, model configurations, process stages, and outcomes are available in our repository[5].

4.1 Data Collection

To evaluate our approach, we require (i) a process model collection $\mathcal{M}$ to populate the knowledge base in the offline component and (ii) event logs with known deviations of the five deviation pattern types to assess performance in a controlled setting.

Process Model Collection. As process model collection, we use part of the SAP Signavio Academic Models (SAP-SAM) [25] dataset which is the largest publicly available collection to date. This model collection includes very diverse processes, ranging from business processes to education processes. From SAP-SAM, we select English models that meet specific requirements to ensure that the corpus includes a wide variety of valid process behavior. In particular, we require that a model can be transformed into a sound workflow net to allow for a playout of the model and that no two models have the same activity set to ensure the exclusion of duplicates. Last, we require a model to contain at least five different activities to ensure a minimum level of process complexity. The result is a collection of 7,791 process models.

We split these models into two sets, *SAP-SAM-train*, encompassing 80% (6,232) of the models, and *SAP-SAM-test*, including the remaining 20% (1,559). The former becomes the process model collection $\mathcal{M}$ that we use to populate the knowledge base, whereas the latter is used to generate event logs for testing, as explained below.

---

[5] https://doi.org/10.6084/m9.figshare.29125898.

Event Logs for Test Data. We use three sources of test data: (1) the process models in SAP-SAM-test, (2) a smaller, proprietary collection of 396 process models (PROP), and (3) a manually created event log with known deviations [16]. (2) contains diverse process models from an internal collection, which therefore passed more thorough quality checks than those in SAP-SAM. (3) is a concrete log which contains manifold deviation patterns. By definition, this log cannot be found in the SAP-SAM data, which only consists of process models.

For (1) and (2), we synthetically create event logs with known deviations as ground truth. Concretely, we obtain an event log L^M per model M by creating one trace in the log for each sequence $s \in M$. If M includes less than 100 sequences, we ensure a minimum number of desired traces per log by randomly duplicating traces in L^M until a size of 100 is reached.[6] Then, we add undesired behavior to 55% of the traces to create balanced data. For each deviating trace, we add up to three deviations of a randomly chosen pattern type and affected fragments of one to three random activities. Then, we verify that the deviation with all its activities is an actual deviation as defined by M, and otherwise disregard the deviation and select another until we find a valid one.[7] Through that, we obtain an event log with known deviations per process model that serves as ground truth.

Last, we also use an event log (3) from [16] as test data in addition to the synthetically generated event logs. This log contains manually crafted and hence more realistic deviations. Concretely, we draw on a purchase-to-pay process (P2P) created for the (manual) identification of deviation patterns in traces [16]. For this process, the authors created a set of 18 deviating traces and a corresponding ground truth that specifies which deviation patterns of the five types exist in the traces. In addition, the authors defined which traces are desired according to a process model. Combining both desired and deviating traces, we obtain the event log for P2P and use this log as test data.

Table 1 shows descriptive statistics of the generated event logs for SAP-SAM-test and PROP as well as statistics for P2P. Note that the deviation patterns are slightly imbalanced, since randomly generated replacements and swaps may not constitute undesired behavior according to the process model, meaning that they are disregarded and another deviation type is chosen.

Table 1. Descriptive statistics for used event logs.

	Event Logs	Number of Traces			Avg. Devs. per Trace	Number of Deviation Patterns				
		total	conforming	deviating		insert	skip	repeat	replace	swap
SAP-SAM-test	1,559	180,077	82,990	97,087	0.76	34,201	28965	31,536	21,233	21,534
PROP	396	39,600	20,402	19,198	0.66	6,702	4,073	5,837	4,749	4,666
P2P [16]	1	58	40	18	0.51	6	5	12	3	4

---

[6] We note that most models in (1) and all models in (2) do not contain more than 100 sequences.

[7] We limit the number of retries to 10 per trace to guarantee termination and randomly select the deviation type again for each retry.

4.2 Experimental Setup

In this section, we describe details on the implementation of our approach, the optimization of hyperparameters, evaluation metrics, and used baselines.

Implementation. We describe the configuration of the knowledge base population, the selection of language models and output parser, and the execution environment.

Knowledge Base Population. To populate the knowledge base in the offline component, we use the process model collection SAP-SAM-train. For the generation of traces, we follow the same procedure as for the generation of test data from process models, described in the previous section. The resulting traces are embedded and stored in the knowledge base along with their known deviations, as described in Sect. 3.2. As a result, we obtain a knowledge base with embeddings of 354,127 desired traces and 349,487 undesired traces, which contain 120,827 insertions, 99,133 skips, 117,627 repetitions, 71,738 replacements, and 77,508 swaps.

Language Models and Output Parser. We select a pre-trained embedding LLM in the offline component, as well as a general-purpose LLM and a parser in the online component of our approach, configuring seeds for reproducibility. As pre-trained embedding LLM, our approach uses Nomic Embed [23]. This model has been shown to capture long context dependencies which can be of high relevance for processes with long term dependencies and also promises computational efficiency [23]. As general-purpose LLM, we select Gemma 2 in its 9B version [26]. This specific LLM is chosen due to its practical size in combination with an open-sourced API and promising capabilities as shown in multiple benchmarks [26]. For facilitating the parsing of the LLM's output, we employ the BAML API,[8] an open-source library that allows to define an output schema for the LLM to adhere to. This schema is provided as described in Sect. 3.3.

Environment. We conducted experiments on a machine with 256GB of RAM and an NVIDIA A10G Tensor Core GPU with 24 GB of GPU memory.

Hyperparameter Optimization. We conducted a grid search to identify the most promising hyperparameters for our approach. To this end, we used 100 process models that are not part of the test set. In particular, we optimized the following three hyperparameters, with the best performing one highlighted in bold:

(1) Number of most frequent traces to add to prompt: [0,1,2,**3**,4,5]
(2) Number of retrieved relevant trace examples to add to prompt: [0,3,**5**,8,10,15]
(3) Relative frequency of activities (in traces) included in the prompt: [0%, **10%**, 20%]

---

[8] https://docs.boundaryml.com/ref/overview.

Metrics. We measure the capability to detect undesired behavior using precision, recall, and F1-score (Eq. 1 a.–c.). If a trace contains a certain deviation, it is classified as a true positive (TP) if the complete fragment of the deviation is identified correctly and as a false negative (FN) otherwise. If a deviation is detected but does not exist in the ground truth, it is a false positive (FP). In addition, detected deviations might be only partially correct when only parts of the affected fragment are identified. We account for that by assigning partial TPs, FPs, and FNs on activity level. For example, if one deviation consists of three skipped activities but only two of them are detected, TP is increased by $\frac{2}{3}$ and FN is increased by $\frac{1}{3}$. Further, pattern types can be interpreted differently since all five used types can be described as (a combination of) insertions and skips. In particular, repetitions are a special case of insertions, replacements consist of a skip (the replaced activities) and an insertion (the replacing activities), and swaps indicate that activities are skipped at one point of the trace and inserted at another point. We account for this by also granting (partial) TPs when an alternative but still (partially) correct interpretation of the deviation patterns is detected. For example, if the approach returns a skip of the replaced activities along with an insertion of the replacing activities instead of a direct replacement, we still consider it correct.

In addition to the capability to detect the five pattern types, we also measure the ability to detect *conform* traces. For that, a trace is a TP if it does not contain any deviations in the ground truth and no deviations are detected. It is a FN if there are deviations in the ground truth but none are detected and it is a FP in the opposite case.

$$a.\ Prec. = \frac{TP}{TP+FP} \quad b.\ Rec. = \frac{TP}{TP+FN} \quad c.\ F1 = \frac{2 \times Prec. \times Rec.}{Prec. + Rec.} \quad (1)$$

Baselines. To put the performance of our approach into context, we compare it to approaches that employ expensive task-specific fine-tuning instead of RAG as well as to approaches that prompt general purpose (*vanilla*) LLMs without RAG.

Fine-Tuned Approaches. We compare against two approaches that use fine-tuned LLMs.

(a) *xSemAD* [6]: Our first baseline is the state-of-the-art approach for semantic anomaly detection. xSemAD detects undesired behavior in event logs by learning constraints from a process model collection and checking whether these constraints hold in the logs. This learning involves fine-tuning of a sequence-to-sequence model. In contrast to our work, xSemAD does not detect deviation patterns but constraint violations. Although this is a different representation, we can compare it to our approach by interpreting both patterns and constraint violations on activity level. All constraint violations indicate either a skipped or inserted activity. Similarly, all deviation patterns can be seen as inserted and skipped activities, as outlined above. Thus, we transform all deviations in the test data to (combinations of) insertions and skips and only evaluate those. We report the performance of xSemAD with the required threshold parameter θ that produced the best results for the set of process models we used for hyperparameter optimization.

(b) *DABL* [15]: To relate our approach to a fine-tuning strategy that detects the same deviation patterns as our approach, we compare it to DABL [15]. The idea behind DABL is to fine-tune Llama 2 (13B version) on the same task as our approach with the difference that replacements are not included. Thus, we treat them as a combination of insertions and skips, as done for xSemAD.

Both xSemAD and DABL require fine-tuning of an LLM. We use the fine-tuned LLMs provided by the authors of both approaches since they are publicly available. This ensures no diversions from the intended approaches. Both relied on the SAP-SAM data set for fine-tuning, using a significantly higher number of process models than is contained in SAP-SAM-train. This minimizes the chance that our approach has access to information which is unknown to xSemAD or DABL. Rather, it is likely that both fine-tuned LLMs have seen parts of our test set SAP-SAM-test during training, which is why the performance of xSemAD and DABL for SAP-SAM can be seen as an upper bound.

Vanilla LLM. We compare against two approaches solely based on Vanilla LLMs.

(c) *No Log Context*: We compare our approach to an LLM without any access to explicit knowledge of the event log at hand. This baseline follows a so-called zero-shot strategy, which means that it only provides the task description to the LLM, without log context and retrieved examples in the prompt.

(d) *Log Context*: To show the value of RAG for our approach, we compare it to a Vanilla LLM with only access to the log context contained in our prompt. Thus, the LLM does not receive examples retrieved from the knowledge base. This baseline received the same hyperparameter optimization as our approach.

4.3 Experimental Results

This section presents the results of our evaluation experiments. We report on the accuracy, robustness, and computational efficiency in detecting undesired behavior.

Accuracy. Table 2 shows precision, recall, and F1 scores for the different test data sets and compares the performance of our approach to the four baselines. The numbers for SAP-SAM-test and PROP are averages across all logs. The results for each individual log can be found in our repository. We first discuss overall results before shifting the focus to the comparison against the individual baselines.

Table 2. Precision, recall, and F1 for used datasets. Boldness indicates best score.

Data	Pattern	Fine-Tuned Approaches						Vanilla LLM						Our Approach		
		xSemAD [6]			DABL [15]			No Log Context			Log Context					
		Prec.	Rec.	F1	Prec.	Rec.	F1	Prec.	Rec.	F1	Prec.	Rec.	F1	Prec.	Rec.	F1
SAP-SAM-test	inserted	0.22	0.44	0.23	**0.78**	0.45	0.57	0.69	0.16	0.24	**0.78**	0.81	**0.79**	0.77	**0.82**	0.79
	skipped	0.14	0.69	0.20	0.15	0.02	0.03	0.02	0.00	0.00	**0.68**	0.30	**0.39**	0.58	0.32	0.38
	repeated	-	-	-	0.92	0.48	0.61	0.83	0.78	0.78	0.84	0.78	0.80	0.88	**0.81**	**0.83**
	replaced	-	-	-	-	-	-	0.28	0.06	0.09	**0.68**	0.42	0.51	0.64	**0.44**	**0.52**
	swapped	-	-	-	0.28	**0.33**	**0.28**	0.48	0.06	0.09	0.58	0.13	0.19	**0.60**	0.16	0.23
	conforming	0.11	0.11	0.11	0.63	0.72	0.66	0.54	0.82	0.64	**0.89**	0.94	0.90	0.88	**0.95**	**0.91**
PROP	inserted	0.18	0.41	0.17	**0.85**	0.35	0.49	0.70	0.06	0.11	0.74	0.61	0.64	0.80	**0.79**	**0.79**
	skipped	0.12	**0.66**	0.17	0.11	0.01	0.02	0.00	0.00	0.00	**0.60**	0.33	**0.39**	0.55	0.31	0.36
	repeated	-	-	-	0.95	0.48	0.61	0.82	0.76	0.76	0.84	0.77	0.78	0.87	**0.79**	**0.80**
	replaced	-	-	-	-	-	-	0.32	0.02	0.04	0.67	0.37	0.47	**0.70**	**0.46**	**0.55**
	swapped	-	-	-	0.20	**0.26**	**0.19**	0.41	0.04	0.07	0.61	0.11	0.17	**0.63**	0.13	0.19
	conforming	0.05	0.07	0.06	0.50	0.67	0.57	0.56	0.86	0.66	**0.81**	0.91	0.85	**0.81**	**0.92**	**0.86**
P2P	inserted	0.14	**0.69**	0.24	0.83	0.35	0.50	**1.00**	0.30	0.46	**1.00**	0.30	0.46	**1.00**	0.60	**0.75**
	skipped	0.05	**0.70**	0.10	**1.00**	0.10	0.18	0.00	0.00	0.00	0.40	0.43	0.41	**1.00**	0.57	**0.73**
	repeated	-	-	-	0.92	0.50	0.65	0.95	0.54	0.69	**1.00**	0.79	0.88	0.98	**1.00**	**0.99**
	replaced	-	-	-	-	-	-	**1.00**	0.40	**0.57**	0.38	0.40	0.39	**1.00**	**0.40**	**0.57**
	swapped	-	-	-	0.57	**0.50**	**0.53**	**1.00**	0.04	0.07	0.07	0.11	0.09	**1.00**	0.11	0.19
	conforming	0.00	0.00	0.00	0.87	**1.00**	0.93	0.85	**1.00**	0.92	0.97	0.83	0.89	**0.98**	**1.00**	**0.99**

Overall Results. Our approach achieves a relatively high F1-score for all deviation pattern types and conforming traces across all used event logs. Thereby, it accurately detects repeated activities, likely because these are recognizable with only a general idea of what repetitions are. Insertions are also detected reasonably well, whereas skips and replacements are detected less consistently. We suspect that these are harder to detect within many possibly missing activities, whereas inserted activities must be observed in the trace, leading to fewer possibly inserted activities. Further, our approach struggles to detect swaps. The reason could be that capturing an incorrect order is more challenging as this can be specific to the process. Our approach detects conforming traces accurately, reaching an F1-score above 0.85. In most cases, it achieves higher precision than recall, meaning that not all deviations are detected but there are few false alarms.

xSemAD [6]. xSemAD detects a large amount of deviations that constitute false positives, as indicated by the low precision. This also leads to low recall and precision of conforming traces. That is since it outputs declarative constraint violations rather than the five deviation patterns directly. Consequently, not traces but pairs of activities are analyzed. Thus, one wrong constraint can lead to many false positives. This shows that the goal of xSemAD differs significantly from the goal of our approach. Also, this indicates that traces should be analyzed individually for accurate feedback on trace-level.

DABL [15]. This fine-tuning strategy is often effective in detecting insertions and repetitions. However, skips are rarely recognized. We suspect that this is since the fine-tuned model overfits skipped activities in the training data and is not able to generalize towards unseen processes. Notably, the F1 of swaps is similar to our approach and higher for P2P and SAP-SAM, indicating that fine-tuning can help to detect re-ordered activities.

No Log Context. In contrast to our approach, this baseline achieves lower F1 scores across all deviation pattern types. Especially skipped and replaced activities are not detected accurately. This is because it is nearly impossible to detect skipped activities with no idea which activities could have occurred. The only pattern type that is recognized often is *repeated*, likely since the LLM has some encoded idea of what repetitions are.

Log Context. Only supplying the general-purpose LLM with log context achieves relatively high F1-scores across all deviation pattern types. For skips in SAP-SAM and PROP, the F1 is even higher than our approach. Together with better accuracy than the other baselines, this indicates that log context already helps to detect undesired behavior. However, for P2P, the performance is significantly worse than our approach. We believe that only log context is not sufficient in more complex processes that contain meaningful rather than randomized deviations, such as P2P.

Table 3. Standard deviation of precision, recall, and F1 for P2P over three random seeds.

Pattern	Fine-Tuned Approaches						Vanilla LLM						Our Approach		
	xSemAD [6]			DABL [15]			No Log Context			Log Context					
	Prec.	Rec.	F1	Prec.	Rec.	F1	Prec.	Rec.	F1	Prec.	Rec.	F1	Prec.	Rec.	F1
inserted	±.04	±.01	±.05	±.00	±.00	±.00	±.00	±.00	±.00	±.24	±.00	±.04	±.00	±.05	±.04
skipped	±.02	±.07	±.03	±.00	±.00	±.00	±.47	±.07	±.12	±.04	±.02	±.03	±.00	±.00	±.00
repeat	-	-	-	±.00	±.00	±.00	±.02	±.02	±.01	±.01	±.06	±.03	±.00	±.00	±.00
replace	-	-	-	-	-	-	±.14	±.00	±.03	±.14	±.00	±.05	±.00	±.00	±.00
swap	-	-	-	±.00	±.00	±.00	±.00	±.00	±.00	±.01	±.08	±.03	±.00	±.00	±.00
conforming	±.00	±.00	±.00	±.00	±.00	±.00	±.00	±.00	±.00	±.00	±.03	±.01	±.00	±.00	±.00

Robustness. To evaluate how robust the approaches perform across multiple runs, we apply them three times with different random seeds to the P2P log. Table 3 shows the standard deviation of all metrics over these three runs. We see that DABL performs most consistently; our approach shows some variety in recall of insertions (± 0.05). Both Vanilla LLMs differ considerably across the three runs, especially in precision. Across all three runs, our approach outperforms all baselines. This indicates that RAG both improves performance and reduces variability in the results.

Computational Efficiency. We assess computational efficiency based on the time for training (if required) and inference. xSemAD and DABL require training in form of LLM fine-tuning, whereas our approach only requires the population of the knowledge base in the offline component. The Vanilla LLMs are used without any training. Since we did not train xSemAD and DABL ourselves, we report the training times from the original papers, which used considerably stronger GPUs than we had available.

Table 4 shows the training time and average inference time per event log. Training of xSemAD took 227 h (more than 9 d), which is even surpassed by DABL with 267 h. In contrast, our approach only required 1.30 h for the knowledge base population. This difference is significant, especially considering that we used less computational resources. xSemAD and DABL used more process models during training than our approach, leading to an increase in training time that is not justified by a better performance. For the average inference time, xSemAD is the quickest by far. This is because not every single trace of the event log is checked by the LLM. Rather, the LLM uses all activities to generate a set of constraints once, after which constraint fulfillment is checked per trace. Although this decreases inference time, the accuracy is impacted heavily. DABL requires the most inference time due to its LLM with a size of around 30 GB. Our approach takes less time than DABL but more than the Vanilla LLM baselines. This inference time of our approach can be considered long, but our experiments showed that each trace has to be analyzed individually for accurate performance.

Table 4. Training and average inference time for used datasets.

	xSemAD [6]	DABL [15]	No Log Context	Log Context	Our Approach
Training (h)	227.00	267.51	-	-	1.30
Avg. inference (min)	0.40	22.50	4.01	4.11	9.30

4.4 Real-Life Application

In this section, we demonstrate the practical applicability of our approach by showing which undesired behavior it detects in a well-known real-life event log. To this end, we apply it to the event log from the BPI Challenge 2019, focusing on the consignment sub-process[9]. This purchase order handling process only contains activities related to the ordering and receiving of goods, whereas payment is handled separately. The three most frequent traces that the approach provides to the LLM as part of our prompt, are:

---

[9] https://doi.org/10.4121/uuid:d06aff4b-79f0-45e6-8ec8-e19730c248f1.

(1) ⟨Create Purchase Order (PO) Item, Record Goods Receipt⟩
(2) ⟨Create Purchase Requisition (PReq) Item, Create PO Item, Record Goods Receipt⟩
(3) ⟨Create PO Item, Record Goods Receipt, Record Goods Receipt⟩

Since true deviations are unknown, we conducted a qualitative assessment, similar to other works [1,6,8]. In particular, we illustrate our approach based on three traces t_1, t_2, and t_3, which we show in Table 5. For each of these traces, the table displays the trace itself, one exemplary retrieved trace from the knowledge base including the corresponding retrieved deviation, and the output of our approach.

Table 5. Exemplary illustration of our approach in BPI Challenge 2019. Shown are three analyzed traces, one retrieved trace-deviation pair, and the output of our approach.

Trace	Retrieved Example from Knowledge Base		Approach Output
	Retrieved Trace	Retrieved Deviation	
t_1 ⟨Create PO Item, Change Quantity, Delete PO Item⟩	⟨Receive Order, Check Order, ..., Receive Goods⟩	No Deviation.	Replace(['Record GR'], ['Delete PO Item'])
t_2 ⟨Create PO Item, Change Storage Location, Record GR⟩	⟨Create PO Item', 'Record Invoice Receipt⟩	Skipped: 'Receive Goods'	Insert(['Change Storage Location'])
t_3 ⟨Create PO Item, Change Quantity, Change Quantity, Record GR⟩	⟨Create PO, Print and Send PO, Change Price, Change Price, Receive Goods, Scan Invoice, Book Invoice⟩	Repeated: 'Change Price'	Repeat(['Change Quantity'])

PO = Purchase Order; GR = Goods Receipt

In the first trace t_1, our approach detects a replacement of *Record GR*—which occurs in the three most frequent traces—by *Delete PO Item* at the end of the trace. This matches the assumption that deleting an order item has negative impact on the process goal to successfully complete the order and is consequently undesired. The retrieved trace ends with *Receive Goods*, which is similar to the replaced activity. For trace t_2, our approach recognizes a problem with *Change Storage Location*, which is inserted according to the output. This aligns with the expectation that changing the storage location is probably inefficient and hence undesired. The insertion might have been detected with the help of the three most frequent traces, in which *Change Storage Location* does not occur. In the last trace t_3, our approach detects a repetition of *Change Quantity* as it occurs twice, indicating undesired inefficiency. The retrieved trace has a similar deviation as *Change Price* is repeated. For all traces, the retrieved examples contain similar behavior. In summary, this analysis reveals insights into the process flow of purchasing goods, highlighting undesired behavior with potential impact on the organization.

5 Conclusion

In this paper, we proposed an approach for detecting undesired behavior in event logs by means of RAG. The approach does not rely on any fine-tuning but rather utilizes the capabilities of a general-purpose LLM, supplied by means of RAG with relevant examples from other processes as well as dynamically selected context from the event log. Our experiments demonstrate that this approach not only significantly reduces computational resource requirements but also enhances performance compared to plain fine-tuning. In particular, the additional log context our approach provides to the LLM was essential for detecting missing activities. In addition, RAG improved the results, especially in more complex processes, indicating that it is possible to generalize behavior from other processes. Finally, the output of our approach is able to capture deviations that involve any number of activities rather than constraint violations which refer to at most two activities such as xSemAD.

Our approach is subject to limitations. First, our approach returns slightly different answers across multiple random seeds, leading to variety in its performance. Although this is undesired, our approach has been shown to outperform the baselines, meaning that the variety is not negatively impacting the performance to an extend where our approach performs worse. Second, our approach extracts explicit information from the event log, assuming that frequent activities and frequent traces help to identify undesired behavior. Although this assumption turned out to be true in the used experiments, it might not hold for all cases. We acknowledge that the findings for our generated logs as well as the manually designed process from [16] might not be generally applicable. Third, swaps are not accurately detected by our approach, potentially because re-ordered activities are hard to detect for LLMs. Moreover, our experiments showed that swaps are the only pattern type where DABL performs as good or better than our approach, indicating that fine-tuning on particular process behaviors could be helpful. Fourth, our approach does require process models to be available for the offline component. Thus, users of the approach must have access to such process models, which can be done based on, e.g., a suitable subset of SAP-SAM. To achieve best performance, these underlying processes should be similar to the investigated ones. However, this is not a pre-requisite as our approach has been shown to work without ensuring process similarity. Future works might investigate the effects of different model collections in the offline component. Fifth, real-world processes might be customized in detail. Thus, knowledge from other processes could not be applicable out of the box. In these cases, organization-specific adaption could help to incorporate the process customization. Last, the inference time of our approach per event log is relatively long, possibly decreasing practical applicability. However, only this strategy has proven to achieve accurate performance, especially in contrast xSemAD which has a lower inference time but also many false positive detections.

Our approach represents a novel direction in the application of LLMs in process mining research, being the first to employ RAG as a means to provide LLMs access to process knowledge to solve a process mining task. In the future, we want to investigate whether a more sophisticated retrieval of information from other processes might improve the performance. For example, we could experiment both with more sophisticated trace embeddings as well as embeddings and retrieval of not just traces but additional information from the process model collection. This might help to extend our approach towards undesired behavior from perspectives other than the control-flow, to which our current approach is limited to. Also, the inference time of our approach might be reduced with the batching of traces, utilizing the full allowed input tokens of the LLM. Further, given the variety of process models in SAP-SAM, a more detailed investigation of these models such as a cross-fold validation could help to show the robustness of our approach. In addition, we want to investigate how imbalanced processes with significantly more desired than undesired traces influence our approach, a setting more likely in reality [14]. In such cases, it could be even more helpful to include log context into the prompt as the most frequent traces represent good references for desired behavior. For the current experiments, we created for a balanced dataset to investigate many different and diverse undesired executions. Further, we want to assess whether a combination of fine-tuning, RAG, and the inclusion of explicit context from the event log might result in more accurate detections. Our findings suggest that this might be a valuable direction which could benefit from the advantages of these three ideas in conjunction. Using the insights gained through our work, we aim to also employ LLMs in combination with RAG in other process mining tasks. For instance, process improvement opportunities could be derived based on access to knowledge of similar processes.

References

1. van der Aa, H., Rebmann, A., Leopold, H.: Natural language-based detection of semantic execution anomalies in event logs. Inf. Syst. **102**, 101824 (2021)
2. van der Aalst, W.M.: What makes a good process model? Lessons learned from process mining. SoSyM **11**(4), 557–569 (2012)
3. Bernardi, M.L., Casciani, A., Cimitile, M., Marrella, A.: Conversing with business process-aware large language models: the BPLLM framework. J. Intell. Inf. Syst., 1–23 (2024)
4. Böhmer, K., Rinderle-Ma, S.: Multi instance anomaly detection in business process executions. In: Carmona, J., Engels, G., Kumar, A. (eds.) Business Process Management, pp. 77–93. Springer International Publishing, Cham (2017). https://doi.org/10.1007/978-3-319-65000-5_5
5. Böhmer, K., Rinderle-Ma, S.: Mining association rules for anomaly detection in dynamic process runtime behavior and explaining the root cause to users. Inf. Syst. **90**, 101438 (2020)
6. Busch, K., Kampik, T., Leopold, H.: xSemAD: explainable semantic anomaly detection in event logs using sequence-to-sequence models. In: BPM, pp. 309–327. Springer, Cham (2024). https://doi.org/10.1007/978-3-031-70396-6_18
7. Carmona, J., van Dongen, B., Solti, A., Weidlich, M.: Conformance Checking. Springer, Cham (2018). https://doi.org/10.1007/978-3-319-99414-7
8. Caspary, J., Rebmann, A., van der Aa, H.: Does this make sense? Machine learning-based detection of semantic anomalies in business processes. In: BPM, pp. 163–179. Springer, Cham (2023). https://doi.org/10.1007/978-3-031-41620-0_10

9. Dunzer, S., Stierle, M., Matzner, M., Baier, S.: Conformance checking: a state-of-the-art literature review. In: S-BPM ONE, p. 1–10. ACM (2019)
10. Estrada-Torres, B., del Río-Ortega, A., Resinas, M.: Mapping the landscape: exploring large language model applications in business process management. In: BPMDS, pp. 22–31. Springer, Cham (2024). https://doi.org/10.1007/978-3-031-61007-3_3
11. Friedrich, F., Mendling, J., Puhlmann, F.: Process model generation from natural language text. In: King, R. (ed.) Active Flow and Combustion Control 2018: Papers Contributed to the Conference Active Flow and Combustion Control 2018, September 19–21, 2018, Berlin, Germany, pp. 482–496. Springer International Publishing, Cham (2019). https://doi.org/10.1007/978-3-642-21640-4_36
12. García-Bañuelos, L., Van Beest, N., Dumas, M., La Rosa, M., Mertens, W.: Complete and interpretable conformance checking of business processes. Trans. Softw. Eng. **44**(3), 262–290 (2017)
13. Grohs, M., van der Aa, H., Rehse, J.-R.: Beyond log and model moves in conformance checking: discovering process-level deviation patterns. In: Marrella, A., Resinas, M., Jans, M., Rosemann, M. (eds.) Business Process Management: 22nd International Conference, BPM 2024, Krakow, Poland, September 1–6, 2024, Proceedings, pp. 381–399. Springer Nature Switzerland, Cham (2024). https://doi.org/10.1007/978-3-031-70396-6_22
14. Grohs, M., Pfeiffer, P., Rehse, J.R.: Proactive conformance checking: an approach for predicting deviations in business processes. Inf. Sys. **127**, 102461 (2025). https://www.sciencedirect.com/science/article/pii/S0306437924001194
15. Guan, W., Cao, J., Gao, J., Zhao, H., Qian, S.: DABL: detecting semantic anomalies in business processes using large language models (2024). https://arxiv.org/abs/2406.15781
16. Hosseinpour, M., Jans, M.: Auditors' categorization of process deviations. J. Inf. Sys. **38**(1), 67–89 (2024)
17. Lahann, J., Pfeiffer, P., Fettke, P.: LSTM-based anomaly detection of process instances: benchmark and tweaks. In: Montali, M., Senderovich, A., Weidlich, M. (eds.) Process Mining Workshops: ICPM 2022 International Workshops, Bozen-Bolzano, Italy, October 23–28, 2022, Revised Selected Papers, pp. 229–241. Springer Nature Switzerland, Cham (2023). https://doi.org/10.1007/978-3-031-27815-0_17
18. de Leoni, M., Maggi, F.M., van der Aalst, W.M.P.: Aligning event logs and declarative process models for conformance checking. In: BPM, pp. 82–97. Springer, Cham (2012). https://doi.org/10.1007/978-3-642-32885-5_6
19. Lewis, P., et al.: Retrieval-augmented generation for knowledge-intensive NLP tasks. Adv. Neural. Inf. Process. Sys. **33**, 9459–9474 (2020)
20. Monti, F., Leotta, F., Mangler, J., Mecella, M., Rinderle-Ma, S.: NL2ProcessOps: towards LLM-guided code generation for process execution. In: Marrella, A., Resinas, M., Jans, M., Rosemann, M. (eds.) Business Process Management Forum: BPM 2024 Forum, Krakow, Poland, September 1–6, 2024, Proceedings, pp. 127–143. Springer Nature Switzerland, Cham (2024). https://doi.org/10.1007/978-3-031-70418-5_8
21. Munoz-Gama, J.: Conformance checking and diagnosis in process mining: comparing observed and modeled processes. Springer, Cham (2016). https://doi.org/10.1007/978-3-319-49451-7
22. Nolle, T., Luettgen, S., Seeliger, A., Mühlhäuser, M.: BINet: multi-perspective business process anomaly classification. Inf. Syst. **103**, 101458 (2019)

23. Nussbaum, Z., Morris, J.X., Duderstadt, B., Mulyar, A.: Nomic embed: training a reproducible long context text embedder (2024). https://arxiv.org/abs/2402.01613
24. Rebmann, A., Schmidt, F.D., Glavaš, G., van Der Aa, H.: Evaluating the ability of LLMs to solve semantics-aware process mining tasks. In: ICPM, pp. 9–16 (2024)
25. Sola, D., Warmuth, C., Schäfer, B., Badakhshan, P., Rehse, J.-R., Kampik, T.: SAP Signavio academic models: a large process model dataset. In: Montali, M., Senderovich, A., Weidlich, M. (eds.) Process Mining Workshops: ICPM 2022 International Workshops, Bozen-Bolzano, Italy, October 23–28, 2022, Revised Selected Papers, pp. 453–465. Springer Nature Switzerland, Cham (2023). https://doi.org/10.1007/978-3-031-27815-0_33
26. Team, G.: Gemma2 (2024). https://doi.org/10.34740/KAGGLE/M/3301, https://www.kaggle.com/m/3301

Layouting Object-Centric Directly Follows Graphs

Deoksang Lee[1], Minseok Song[1(✉)], and Wil M. P. van der Aalst[2]

[1] Department of Industrial and Management Engineering, Pohang University of Science and Technology (POSTECH), Pohang, South Korea
{duksang4834,mssong}@postech.ac.kr
[2] Process and Data Science, RWTH Aachen University, Aachen, Germany
wvdaalst@pads.rwth-aachen.de

Abstract. Process mining extracts insights from event logs to analyze and optimize business processes. Object-Centric Process Mining (OCPM) extends traditional process mining by utilizing Object-Centric Event Logs (OCEL), explicitly capturing interactions among multiple objects within business processes. Despite its advantages, visualizing object-centric process models remains challenging due to increased complexity, leading to denser graphs with more nodes and edges compared to traditional process mining. To address this challenge, this study proposes an advanced process layout generation method designed to improve the visualization clarity of Object-Centric Directly Follows Graph (OC-DFG), a widely used representation of object-centric process models. Our approach systematically assigns distinct axes to each object type, enhancing readability by clearly separating interactions. Additionally, the method incorporates edge cross-minimization and spatial optimization techniques to further improve layout efficiency. We validate the proposed method quantitatively using both traditional layout evaluation metrics and newly developed metrics specifically designed for OCPM. Experimental results indicate notable improvements in layout metrics associated with readability for OC-DFG. This work contributes to process mining, modeling, and analytics by offering a structured visualization approach for complex object-centric process models, which may support stakeholders in analyzing processes and making informed decisions.

Keywords: OC-DFG · Object-Centric Directly Follows Graph · Layout · Visualization · OCPM · Object-Centric Process Mining

1 Introduction

In the era of data-driven process automation, business process management is critical in optimizing workflows, ensuring compliance, and supporting decision-making [1, 2]. Process mining, a key discipline within business process management, enables organizations to extract insights from event logs recorded in information systems [3]. Traditionally, process mining has relied on case-centric event logs, where a single case type represents an entire process. However, real-world business processes involve multiple interacting objects such as orders, customers, suppliers, and machines. The case-centric paradigm fails to capture these interactions, leading to information loss and limiting its applicability to complex business environments.

To overcome this limitation, Object-Centric Process Mining (OCPM) emerged as a new paradigm that models business processes from a multi-object perspective [4, 5]. Unlike traditional process mining, OCPM explicitly represents interactions between multiple objects using Object-Centric Event Logs (OCEL), providing a more realistic and granular view of processes.

Following the rise of OCPM, various object-centric process model discovery techniques have been developed [6–8]. However, a significant challenge remains: the absence of effective layout generation methods for visualizing object-centric process models. Prior research established that effective visualizations enhance stakeholders' decision-making in business contexts [9–12]. Despite this, existing studies have relied on general graph layouts that prioritize aesthetic placement of nodes and edges rather than representing processes well. This approach hinders users' ability to understand business processes. Furthermore, they do not differentiate object types, preventing users from distinguishing object types and understanding interactions between object types. As depicted in Fig. 1(a), these limitations impede process analysts from using OCPM in real-world business environments. Consequently, a more advanced layout generation method is needed to differentiate object types and visualize interactions clearly, as shown in Fig. 1(b).

To address this challenge, we propose an advanced process layout generation method to enhance the visualization of object-centric process models. Specifically, we focus on Object-Centric Directly-Follows Graphs (OC-DFG), one of the most widely used object-centric process model representations [5]. Our approach assigns a distinct object type axis to each object type, positioning frequently interacting object types closer together while placing less related types farther apart. Additionally, we introduce cross-minimization and node positioning methods tailored to OC-DFG layouts. To evaluate the proposed method, we conduct a quantitative evaluation, introducing new metrics to measure the quality of OC-DFG layouts.

The proposed OC-DFG layout generation method may help users better interpret OC-DFG and extract process-related insights more efficiently, thereby accelerating the adoption of OCPM in practice. While this study focuses on OC-DFG, the method can also be applied to other object-centric process models such as Object-Centric Petri Nets (OC-PN) and Object-Centric Business Process Modeling Notations (OC-BPMN), broadening its applicability in process mining and business process management.

2 Background

This section introduces fundamental concepts for understanding the proposed method.

2.1 Universe

Let $\mathbb{U}_{ev}$ be the universe of events, $\mathbb{U}_{act}$ be the universe of activities, $\mathbb{U}_o$ be the universe of objects, $\mathbb{U}_{ot}$ be the universe of object types, $\mathbb{G}$ be the universe of the graph, $\mathbb{Z}$ be the universe of integers, and $\mathbb{F}$ be the universe of real numbers.

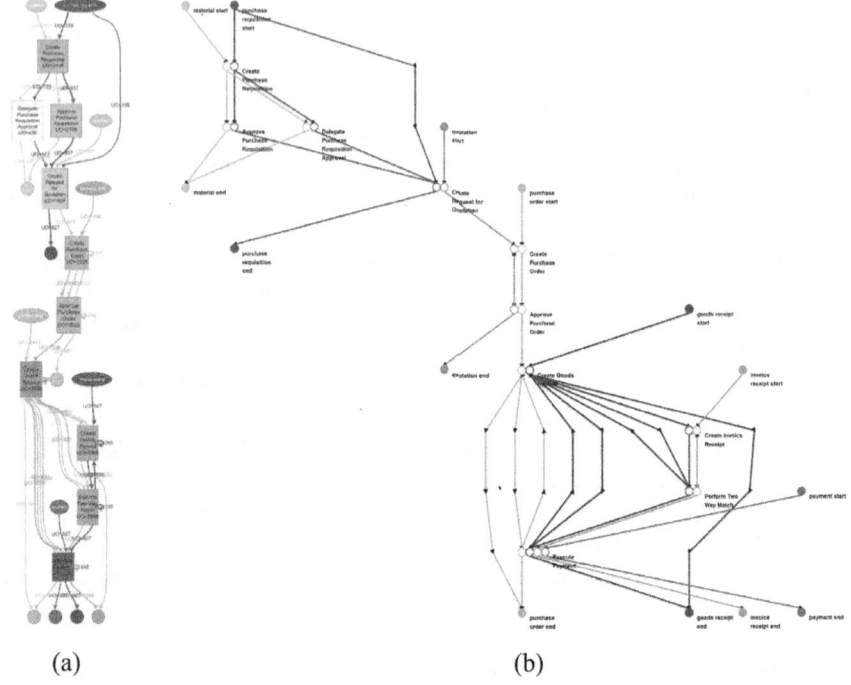

Fig. 1. (a) A visual representation of an object-centric process model using a general graph layout. (b) The same object-centric process model visualized using the proposed method.

2.2 Backbone-Based DFG Layout, Backbone, Rank, and Order

Backbone-based DFG layout is a visualization method designed to enhance DFG readability and clarity [13], as illustrated in Fig. 2.

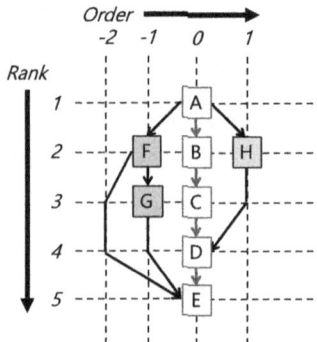

Fig. 2. An example of the backbone-based DFG layout. Red nodes and edges represent the backbone structure.

It identifies a backbone that represents the most frequent or significant process flow and serves as the central structure. For example, in Fig. 2, the sequence $A \rightarrow B \rightarrow C \rightarrow D \rightarrow E$, colored red, forms the backbone of the layout.

In the backbone-based DFG layout, node positions are determined by rank and order. Rank refers to vertical placement, where nodes with lower ranks appear higher in the layout. Conversely, order determines horizontal positioning, with nodes of lower order placed to the left of those with higher order.

2.3 Object-Centric Event Log (OCEL)

An OCEL extends traditional event logs to capture multi-object interactions within processes [4, 5, 14], as shown in Table 1. According to prior studies [4, 5], the OCEL is defined as a tuple $OCEL = (EV, ET, O, OT, \pi_{act}, \pi_{time}, \pi_{omap}, \pi_{otype}, \pi_{atype} \leq)$, where $EV \subseteq \mathbb{U}_{ev}$ is a set of events, $ET \subseteq \mathbb{U}_{act}$ is a set of activities, $O \subseteq \mathbb{U}_{ot}$ is a set of objects, and $OT \subseteq \mathbb{U}_{ot}$ is a set of object types. $\pi_{act} : EV \rightarrow ET$ maps events to activities, $\pi_{time} : EV \rightarrow \mathbb{U}_{time}$ assigns timestamps to events, $\pi_{omap} : EV \rightarrow O$ relates events to objects, $\pi_{otype} : O \rightarrow OT$ assigns objects to object types, and $\pi_{atype} : ET \rightarrow 2^{OT}$ maps each activity to a set of object types, and $\leq$ defines a total order on events based on their timestamps.

In Table 1, $EV = \{ev_1, ev_2, ev_3, ev_4, ev_5\}$, $ET = \{po, ca, pi\}$, $O = \{o_1, o_2, i_1, i_2\}$, $OT = \{Order, Item\}$, $\pi_{act}(ev_1) = po$, $\pi_{act}(ev_2) = ca$, $\pi_{time}(ev_1) = $ 09-03-2025 15:34, $\pi_{omap}(ev_2) = \{o_1, i_1\}$, $\pi_{otype}(o_1) = Order$, $\pi_{otype}(i_1) = Item$, $\pi_{atype}(po) = \{Order\}$, and $\pi_{atype}(ca) = \{Order, Item\}$.

Table 1. Example of an OCEL

Event ID	Activity name	Timestamp	Objects involved	
			Order	Item
ev_1	place_order (po)	09-03-2025 15:34	$\{o_1\}$	–
ev_2	check_availability (ca)	09-03-2025 15:40	$\{o_1\}$	$\{i_1\}$
ev_3	place_order (po)	09-03-2025 16:00	$\{o_2\}$	–
ev_4	check_availability (ca)	09-03-2025 16:15	$\{o_2\}$	$\{i_2\}$
ev_5	pick_item (pi)	09-03-2025 16:45	$\{o_1\}$	$\{i_1\}$

2.4 Object-Centric Directly Follows Graph (OC-DFG)

An OC-DFG is a process model that intuitively represents interactions among multi-objects using nodes and edges [5]. To discover the OC-DFG, the OCEL is flattened for each object type, generating an individual DFG per object type. Then, these DFGs are merged to form the final OC-DFG, integrating interactions across object types.

3 OC-DFG Layout Generation Method

In this section, we propose an OC-DFG layout generation method consisting of data preparation, backbone determination, rank assignment, DFG layout generation for each object type, OC-DFG layout generation, cross-minimization, and node positioning.

3.1 Data Preparation

In this step, we construct the sets of nodes N, edges E, and variants V from the given $OCEL = (EV, ET, O, OT, \pi_{act}, \pi_{time}, \pi_{omap}, \pi_{otype}, \pi_{atype}, \leq)$.

To ensure that each object type has clearly defined start and end points, we introduce synthetic start and end nodes, denoted as $N_{start} = \{start_{ot}|ot \in OT\}$ and $N_{end} = \{end_{ot}|ot \in OT\}$. We update π_{atype} as $\pi_{atype} = \pi_{atype} \cup \{(start_{ot}, ot)|ot \in OT\} \cup \{(end_{ot}, ot)|ot \in OT\}$. In Table 1, $N_{start} = \{start_{Order}, start_{Item}\}$, $N_{end} = \{end_{Order}, end_{Item}\}$, $\pi_{atype}(start_{Order}) = Order, \pi_{atype}(start_{Item}) = Item$, $\pi_{atype}(end_{Order}) = Order$, and $\pi_{atype}(end_{Item}) = Item$.

Each object $o \in O$ is associated with a trace τ_o, which is the ordered sequence of activities linked to that object. Let $ev_1, \ldots, ev_n \in EV$ be events such that $o \in \pi_{omap}(ev_i)$ for all $1 \leq i \leq n$, and the sequence $ev_1 \leq ev_2, \cdots \leq ev_n$ respects the total order $\leq$ defined over the OCEL. Then, the trace τ_o is constructed as $\tau_o = \langle start_{ot}, \pi_{act}(ev_1), \pi_{act}(ev_2), \ldots, \pi_{act}(ev_n), end_{ot}\rangle$ where $ot = \pi_{otype}(o)$. For instance, in Table 1, the trace of o_1 is $\tau_{o_1} = \langle start_{Order}, po, ca, pi, end_{Order}\rangle$.

Based on the collection of traces, we define three sets. First, the node set includes all activities recorded in the event log, as well as the synthetic start and end nodes, i.e., $N = ET \cup N_{start} \cup N_{end}$. Second, the edge set E is constructed from all transitions between consecutive activities in each trace. Each edge is labeled with the corresponding object type. Formally, the edge set is defined as $E = \{(a_i, a_{i+1}, ot)|o \in O \wedge ot = \pi_{otype}(o) \wedge \tau_o = \langle start_{ot}, a_1, \ldots, a_n, end_{ot}\rangle \wedge 0 \leq i \leq n\}$. In Table 1, the edges are $E = \{(po, ca, Order), (ca, pi, Order), (ca, pi, Item)\}$. Third, the variant set V is built by collecting all unique traces found in the event log, i.e., $V = \{\tau_o|o \in O\}$.

After constructing these sets, we organize the nodes, edges, and variants by object type. For each object type $ot \in OT$, we define N_{ot} as the set of nodes associated with ot, i.e., $N_{ot} = \{n \in N|ot \in \pi_{atype}(n)\}$, E_{ot} as the subset of edges labeled with ot, i.e., $E_{ot} = \{(a_1, a_2, ot') \in E|ot' = ot\}$, and V_{ot} as the set of traces for object type ot, i.e., $V_{ot} = \{\tau_o|o \in O, \pi_{otype}(o) = ot\}$.

3.2 Backbone Determination

Given a set of nodes N, a set of edges E, and a set of variants V over an $OCEL = (EV, ET, O, OT, \pi_{act}, \pi_{time}, \pi_{omap}, \pi_{otype}, \pi_{atype}, \leq)$, we determine the backbone for each object type $ot \in OT$. The backbone is intended to represent the core sequence of activities within the process. In this study, we derived the backbone from the most frequent variant.

Let $mv_{ot} = \langle a_1, a_2, \ldots, a_n\rangle$ be the most frequent variant of ot, where each $a_i \in N$. The sequence of backbone nodes is $BN_{ot} = \langle b_1, b_2, \ldots, b_k\rangle$, where each $b_i \in N$, all elements b_i are distinct, and their relative order in mv_{ot} is preserved. In other words,

BN_{ot} consists of the unique activities from mv_{ot}, preserving the order in which they first appear. The set of backbone edges is defined as the set of directed transitions between consecutive backbone nodes that are also observed in the original variant. Formally, $BE_{ot} = \{(b_i, b_{i+1}, ot) | 1 \leq i < k \land j \in \{1, \ldots, n-1\} \land a_j = b_i \land a_{j+1} = b_{i+1}\}$.

For example, consider the most frequent variant for an object type ot as $mv_{ot} = \langle A, B, A, C, D, E, D \rangle$. The backbone node sequence is $BN_{ot} = \langle A, B, C, D, E \rangle$ and the set of backbone edges is $BE_{ot} = \{(A, B, ot), (C, D, ot), (D, E, ot)\}$. The edge (B, C, ot) is excluded as it does not appear in the original variant.

For convenience, we denote the i-th element of a sequence S as $S[i]$. Accordingly, elements of the backbone node sequence can be accessed using the notation $BN_{ot}[i]$.

3.3 Rank Assignment

In this step, node ranks are assigned using Integer Programming (IP) and node rank mapping function is obtained. The prior study [13] suggested a manual rank assignment method using traditional event logs. However, it has two limitations when applied to OCPM: it may introduce unnecessary back edges and result in longer edge lengths.

To address these issues, we introduce an IP-based rank assignment method. The model employs four decision variables: r_n, y_e, z_e, and α_e. Here, r_n is an integer variable representing the rank of node n ($r_n \geq 1$), and y_e is an integer variable capturing the length of edge e ($y_e \geq 0$). The binary variable z_e equals 1 if the start node's rank exceeds the end node's rank for edge e, and 0 otherwise. Precedence violation α_e is an integer variable ($\alpha \geq 0$) that accounts for cases where the start node's rank is higher than the end node's rank. Functions $snode: E \rightarrow N$ and $enode: E - N$ return the start and end nodes of an edge $e, freq: E \rightarrow \mathbb{Z}$ returns the frequency of the edge.

$$\text{Min} \quad \sum_{e \in E} freq(e) \times \alpha_e + y_e \quad (1)$$

$$s.t. \quad r_{snode(e)} + 1 \leq r_{enode(e)} + \alpha_e \forall e \in E \quad (2)$$

$$r_{start_{ot}} + 1 \leq r_n \forall n \in N_{ot}, n \neq start_{ot} \quad (3)$$

$$r_{end_{ot}} \geq r_n + 1 \forall n \in N_{ot}, n \neq end_{ot} \quad (4)$$

$$y_e \geq r_{snode(e)} - r_{enode(e)} \forall e \in E \quad (5)$$

$$y_e \geq -(r_{snode(e_i)} - r_{enode(e_i)}) \forall e \in E \quad (6)$$

$$r_{BN_{ot}[i]} + 1 \leq r_{BN_{ot}[i+1]} \forall i = 1, 2, \ldots, |BN_{ot}| - 1, \forall ot \in OT \quad (7)$$

$$x_{snode(e)} - x_{enode(e)} \leq -\varepsilon + M z_e \forall e \in E \quad (8)$$

$$x_{snode(e_i)} - x_{enode(e_i)} \geq \varepsilon - (1 - z_e) M \forall e \in E \quad (9)$$

The objective function (1) minimizes the precedence violations and edge length. Constraint (2) ensures that the end node is placed below its start node. However, in some cases, this may not be feasible. To handle such cases, we introduce α_e in constraint (2). Constraints (3) and (4) enforce that the start and end nodes of object types are positioned above and below all other nodes of the same type, respectively. Constraints (5) and (6) calculate the edge length. Constraint (7) ensures that the ranks of nodes in the backbone increase, placing backbone nodes linearly. Constraints (8) and (9) enforce that start and end nodes of an edge occupy different ranks to avoid horizontal edges (ε: a very small constant; M: a very large constant). After determining the node ranks, we define the set of node ranks $R \subseteq \mathbb{Z}^+$ as $R = \{r_n | n \in N\}$ and mapping function $\pi_{rank} : N \to \mathbb{Z}^+$ as $\pi_{rank} = \{(n, r_n) | n \in N\}$.

3.4 DFG Layout Generation for Each Object Type

Definition 1 (DFG layout). *A DFG layout is a tuple* $L = (N, E, \pi_{rank}, \pi_{order})$ *where* N *is a set of nodes,* $E \subseteq N \times N$ *is a set of edges,* $\pi_{rank} : N \to \mathbb{Z}^+$ *is a node rank mapping function, and* $\pi_{order} : N \to \mathbb{Z}$ *is a node order mapping function.*

In this stage, we construct a DFG layout for each object type by determining the node orders. To achieve this, we adopt the method proposed in [13], which determines node orders based on nodes, edges, backbone, and node ranks.

Given an $OCEL = (EV, ET, O, OT, \pi_{act}, \pi_{time}, \pi_{omap}, \pi_{otype}, \pi_{atype}, \leq)$, we have already extracted the required elements for each object type ot, as described in Sect. 3.1 and 3.2: a set of nodes N_{ot}, a set of edges E_{ot}, a sequence of backbone nodes BN_{ot}, and a set of backbone edges BE_{ot} for each object type. Furthermore, in Sect. 3.3, we determined the rank mapping function $\pi_{rank} : N \to \mathbb{Z}^+$. From this, we define the node mapping function for each object type as $\pi_{rank}^{ot} = \{(n, \pi_{rank}(n)) | n \in N_{ot}\}$. Using these elements, we compute the node orders $O_{ot} \subseteq \mathbb{Z}$ and obtain mapping function $\pi_{order}^{ot} : N_{ot} \to \mathbb{Z}$ following the method from [13]. As a result, the finalized DFG layout for each object type is constructed as $L_{ot} = (N_{ot}, E_{ot}, \pi_{rank}^{ot}, \pi_{order}^{ot})$.

3.5 OC-DFG Layout Generation

After generating DFG layouts for each object type, these individual layouts are incrementally merged to construct an OC-DFG layout. This merging process consists of four steps: (1) initializing the OC-DFG layout with the main object type, (2) selecting the target object type, (3) generating OC-DFG layout candidates, and (4) selecting the optimal layout. Steps (2) through (4) are repeated until no object types remain to be merged.

Recall that we have already obtained the set of nodes N, the set of nodes for each object type N_{ot}, the set of edges E, the set of edges for each object type E_{ot}, the node rank mapping function π_{rank}, the node rank mapping function for each object type π_{rank}^{ot}, the set of node order mapping function for each object type π_{order}^{ot}, and the DFG layout for each object type L_{ot}. Before detailing the merging procedure, we introduce several concepts.

Definition 2 (Object type axis). *Let* OT *be a set of object types. An object type axis of an object type* $ot \in OT$ *is defined as a non-negative integer, i.e.,* $ota \in \mathbb{Z}_{\geq 0}$.

The object type axis indicates the horizontal index of the object type within the OC-DFG layout. For example, suppose an OC-DFG layout assigns axis values of 1, 0, and 2 to ot_A, ot_B, and ot_C. This implies that ot_B appears on the leftmost side of the layout, ot_A is placed between ot_B and ot_C, and ot_C is positioned on the rightmost side.

Definition 3 (Node object type axis function). *Let N be a set of nodes, OT be a set of object types, $\pi_{atype} : N \to OT$ be a node object type mapping function, OTA be a set of object type axes, $\pi_{ota} : OT \to OTA$ be an object type axis mapping function. Node object type axis function $\pi_{nota} : N \to OTA$ is $\pi_{nota} = \{(n, \pi_{ota}(\pi_{atype}(n)))\}$.*

For example, suppose an OC-DFG layout includes nodes of object types ot_A, ot_B, and ot_C, assigned to axes 1, 0, and 2. If a node n is positioned along the axis of ot_B, then $\pi_{atype}(n) = ot_B$, $\pi_{ota}(ot_B) = 0$, and thus $\pi_{nota}(n) = 0$.

Definition 4 (OC-DFG layout). *An OC-DFG layout is defined as a tuple $OL = (N, E, OT, OTA, \pi_{rank}, \pi_{order}, \pi_{ota}, \pi_{nota})$ where N is a set of nodes, $E \subseteq N \times N \times OT$ is a set of edges, OT is a set of object types, OTA is a set of object type axes, $\pi_{rank} : N \to \mathbb{Z}^+$ is a node rank mapping function, $\pi_{order} : N \to \mathbb{Z}$ is a node order mapping function, $\pi_{ota} : OT \to OTA$ is an object type axis mapping function, and $\pi_{nota} : N \to OTA$ is a node object type axis function.*

1) **Initializing the OC-DFG layout with the main object type.** In this step, an OC-DFG layout OL is initialized using a main object type. The main object type is identified as the highest number of shared nodes among object types. Let $OCEL = (EV, ET, O, OT, \pi_{act}, \pi_{time}, \pi_{omap}, \pi_{otype}, \pi_{atype}, \leq)$ be an OCEL and N_{ot} be a set of nodes of the object type $ot \in OT$. The main object type is determined as $ot = argmax_{ot' \in OT} |\{n | n \in N_{ot'} \wedge |\pi_{atype}(n)| > 1\}|$.

    Then, the OC-DFG layout $OL = (N, E, OT, OTA, \pi_{rank}, \pi_{order}, \pi_{ota}, \pi_{nota})$ is initialized. Let ot be a main object type and $L_{ot} = (N_{ot}, E_{ot}, \pi_{rank}^{ot}, \pi_{order}^{ot})$ be the DFG layout of the main object type. Then, the OC-DFG OL is initialized as $N = N_{ot}$, $E = \{(a_1, a_2, ot) | (a_1, a_2) \in E_{ot}\}$, $OT = \{ot\}$, $OTA = \{0\}$, $\pi_{rank} = \pi_{rank}^{ot}$, $\pi_{order} = \{(n, -\min_{n \in N}(\pi_{order}^{ot}(n)) + \pi_{order}^{ot}(n)) | n \in N_{ot}\}$, $\pi_{ota} = \{(ot, 0)\}$, and $\pi_{nota} = \{(n, 0) | n \in N_{ot}\}$.

2) **Selecting the target object type.** Next, a target object type ot is selected. Before explaining how to select the target object type, we introduce the distance between two set of nodes.

Definition 5 (Distance between two set of nodes). *Let N_A and N_B be sets of nodes. The distance between two sets is defined as $dist(N_A, N_B) = 1 - (2 \times |N_A \cap N_B|)/(|N_A| + |N_B|)$.*

This measure reflects the dissimilarity between N_A and N_B with a value of 0 indicating identical sets and 1 indicating no overlap.

The target object type is the one whose DFG layout is the most like the current layout OL. Let $OCEL$ be an OCEL, OL be a current OC-DFG layout, and $L_{ot'}$ be a DFG layout of object type ot'. Then, the target object type ot is defined as $ot = argmin_{ot' \in OT(OCEL), ot' \notin OT(OL)} dist(N(OL), N(L_{ot'}))$.

3) **Generating OC-DFG layout candidates.** After determining the target object type ot, the target DFG layout $L_{ot} = (N_{ot}, E_{ot}, \pi_{rank}^{ot}, \pi_{order}^{ot})$ is merged with the current OC-DFG layout $OL = (N, E, OT, OTA, \pi_{rank}, \pi_{order}, \pi_{ota}, \pi_{nota})$. The target axis $ta \in \mathbb{Z}_{\geq 0}$ indicates the horizontal position where L_{ot} is to be inserted. There are $|OT(OL)| + 1$ possible values for ta. For example, if OL contains object types ot_A, ot_B, and ot_C assigned to axes 0, 1, and 2, then ta can take on values 0 (inserting to the left of ot_A), 1 (between ot_A and ot_B), 2 (between ot_B and ot_C), or 3 (to the right of ot_C).

For each possible target axis ta, we construct temporary layout. Then, we choose the layout with the lowest cost in the next step. The temporary OC-DFG layout $OL' = (N', E', OT', OTA', \pi'_{rank}, \pi'_{order}, \pi'_{ota}, \pi'_{nota})$ is constructed as $N' = N \cup N_{ot}$, $E' = E \cup \{(a_1, a_2, ot_{target}) | (a_1, a_2) \in E_{ot}\}$, and $\pi'_{rank} = \pi_{rank} \cup \pi_{rank}^{ot}$. The function π'_{order} is updated to maintain consistency in the layout. First, three parameters are computed: the left margin of OL at ta as $lm = \max(\{\pi_{order}(n) | n \in N \wedge \pi_{nota}(n) < ta\}) + 1$; the order of the leftmost node in L_{ot} as $lmo = \min(\{\pi_{order}^{ot}(n) | n \in N_{ot}\})$; and the width of L_{ot} as $w = \max(\{\pi_{order}^{ot}(n) | n \in N_{ot}\}) - \min(\{\pi_{order}^{ot}(n) | n \in N_{ot}\})$. Nodes with an object type axis less than ta retain their order. If a node's object type axis equals ta, its order is updated using lm, lmo, and its local order. Nodes with an axis greater than ta adjust their order based on its global order in OL and the width of $L_{ot_{target}}$: $\pi'_{order} = \{(n, \pi_{order}(n)) | n \in N \wedge \pi_{nota}(n) < ta\} \cup \{(n, \pi_{order}(n) + w - 1) | n \in N \wedge \pi_{nota}(n) \geq ta\} \cup \{(n, lm + \pi_{order}^{ot}(n) - lmo + 1) | n \in N_{ot} \wedge n \notin N\}$. The π'_{ota} is adjusted based on the position of the target object type. If an object type is the target, it is assigned the target axis ta. Object types positioned to the left of ta retain their positions, whereas those at or beyond ta shift to the right as $\pi'_{ota} = \{(ot', ta) | ot' \in OT \wedge ot' = ot\} \cup \{(ot', \pi_{ota}(ot')) | ot' \in OT \wedge \pi_{ota}(ot') < ta\} \cup \{(ot', \pi_{ota}(ot') + 1) | ot' \in OT \wedge \pi_{ota}(ot') > ta\}$. The function π'_{nota} is updated to reflect the inclusion of the new object type. If a node exists only in L_{ot}, it is assigned to ta. If a node was positioned to the left of ta, it retains its position. Otherwise, it is shifted to the right as $\pi'_{nota} = \{(n, ta) | n \in N_{ot} \wedge n \notin N\} \cup \{(n, \pi_{nota}(n)) | n \in N \wedge \pi_{nota}(n) < ta\} \cup \{(n, \pi_{nota}(n) + 1) | n \in N \wedge \pi_{nota}(n) \geq ta\}$.

As a result, the set of OC-DFG layouts $C = \left\{OL'_0, OL'_1, \ldots, OL'_{|OT|}\right\}$ is obtained. The example of merging OL with L_{ot} at the target axis zero is shown in Fig. 3.

4) **Selecting the layout with the lowest cost.** After generating OC-DFG layout candidates $C = \left\{OL'_0, OL'_1, \ldots, OL'_{|OT|}\right\}$, the cost for each candidate is calculated. Then, the layout with the lowest cost is selected and assigned to new OL.

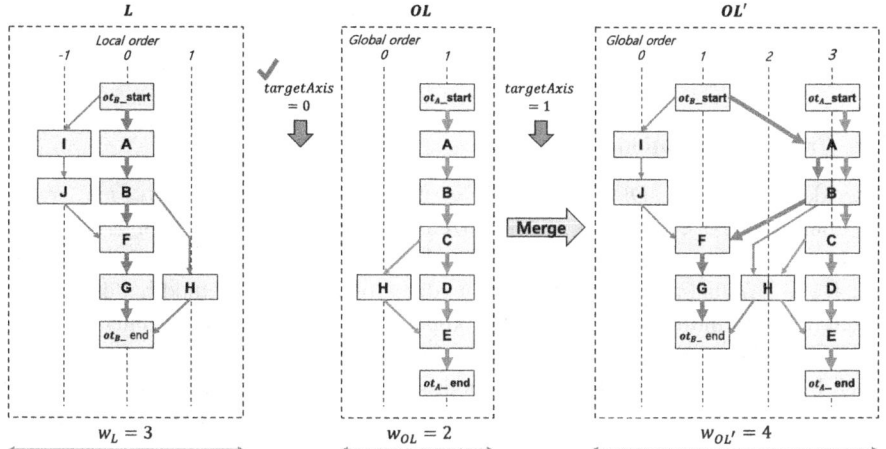

Fig. 3. A graphical explanation of merging the current OC-DFG layout OL with the target object type layout L at the target axis zero.

Definition 6 (Distance between two set of nodes in OC-DFG layout). Let $OL = (N, E, OT, OTA, \pi_{rank}, \pi_{order}, \pi_{ota}, \pi_{nota})$ be an OC-DFG layout. Let $N_A, N_B \subseteq N$ be two sets of nodes. The layout-based distance between two sets of nodes is defined as $dist_{lay}(N_A, N_B) = |avg(\{\pi_{order}(n) | n \in N_A\}) - avg(\{\pi_{order}(n) | n \in N_B\})|$.

Definition 7 (Distance rank function). Let OT be a set of object types, N be a set of nodes, N_{ot} be a set of nodes of type $ot \in OT$, $\pi_{atype} : N \rightarrow OT$ be an object type mapping function, and $d : N \times N \rightarrow \mathbb{F}$ be a distance function. The distance rank function $dr_d : OT \times OT \rightarrow \mathbb{Z}^+$ is defined as $dr_d(ot_A, ot_B) = |\{(ot_i, ot_j) | ot_i, ot_j \in OT \wedge ot_i \neq ot_j \wedge d(N_{ot_i}, N_{ot_j}) < d(N_{ot_A}, N_{ot_B})\}|$.

Definition 8 (Cost of an OC-DFG layout). Let $OL = (N, E, OT, OTA, \pi_{rank}, \pi_{order}, \pi_{ota}, \pi_{nota})$ be an OC-DFG layout, $N_{ot} \in N$ be a set of nodes of type $ot \in OT$, $d_{ocel} : N \times N \rightarrow \mathbb{F}^+$ be the distance function defined in Definition 5, $d_{lay} : N \times N \rightarrow \mathbb{F}^+$ be the distance function defined in Definition 6, $dr_d : OT \times OT \rightarrow \mathbb{Z}^+$ be the distance rank function over the distance function d, $ec : \mathbb{G} \rightarrow \mathbb{Z}$ a function that returns the number of edge crossings in graph, and $\lambda_1, \lambda_2 \in [0, 1]$ are parameters. Cost of the given OC-DFG layout OL is defined as:

$$cost(OL) = \lambda_1 \sqrt{\frac{\sum_{i=1}^{|OT|} \sum_{j=i+1}^{|OT|} \left(dr_{d_{ocel}}(ot_i, ot_j) - dr_{d_{lay}}(ot_i, ot_j)\right)^2}{\sum_{i=1}^{|OT|} \sum_{j=i+1}^{|OT|} dr_{d_{ocel}}(ot_i, ot_j)^2}} + \lambda_2 \frac{ec(OL)}{|E|^2}$$

The first term measures the deviation between the ranked distances of object types in the OCEL and their corresponding ranked distances in the layout. This ensures that object types closely related in the OCEL are also positioned near each other in the layout, while those with larger ranked distances remain farther apart. The second term penalizes layouts with a high number of edge crossings.

Finally, the current layout OL is updated as $OL = argmin_{OL' \in C} cost(OL')$ where $C = \left\{ OL'_0, OL'_1, \ldots, OL'_{|OT|} \right\}$ is the set of OC-DFG layout candidates.

3.6 Cross-Minimization

After deriving the OC-DFG layout $OL = (N, E, OT, OTA, \pi_{rank}, \pi_{order}, \pi_{ota}, \pi_{nota})$, the number of edge crossings is minimized as explained in Algorithm 1. The algorithm takes as input OL, a maximum number of iterations $maxIter$, and a function $ec : \mathbb{G} \to \mathbb{Z}_{\geq 0}$ that returns the number of edge crossings. Our method builds upon the edge-crossing minimization techniques proposed in [13, 15], which evaluate whether swapping adjacent nodes within the same rank reduces crossings. We refine this approach by restricting swaps to adjacent nodes of the same object type. This constraint preserves the alignment of nodes along their respective object type axes, enhancing interpretability while optimizing layout quality.

Algorithm 1 *crossMinimization*
Input: $OL, maxIter$
Output: OL_{best}

---

1    $OL_{best} \leftarrow OL$
2    $maxRank \leftarrow \max_{n \in N} \pi_{order}(n)$
3    $\textbf{for } i \leftarrow \{1, \ldots, maxIter\} \textbf{ do}$
4      $\textbf{for } r \leftarrow \{1, \ldots, maxRank - 1\} \textbf{ do}$
5        $currRankN \leftarrow sort_{order}(\{n \mid n \in N \land \pi_{rank}(n) = r\})$
6        $postRankN \leftarrow sort_{order}(\{n \mid n \in N \land \pi_{rank}(n) = r + 1\})$
7        $\textbf{for } j \leftarrow \{1, \ldots, |postRankN| - 1\} \textbf{ do}$
8          $n_j, n_{j+1} \leftarrow postRankN_j, postRankN_{j+1}$
9          $\textbf{if } \pi_{ota}(n_j) = \pi_{ota}(n_{j+1}) \textbf{ do}$
10          $OL_{swap} \leftarrow swap(OL, n_1, n_2)$
11          $\textbf{if } ec(OL_{swap}) < ec(OL_{best}) \textbf{ do}$
12            $OL_{best} \leftarrow OL_{swap}$
13        $\textbf{end}$
14      $\textbf{end}$
15    $\textbf{end}$
16    $\textbf{return } OL_{best}$

3.7 Node Positioning

In this step, the node coordinates are computed as described in Algorithm 2. The algorithm uses $OL = (N, E, OT, OTA, \pi_{rank}, \pi_{order}, \pi_{ota}, \pi_{nota})$, $maxIter$, the set of nodes N_{ot} and edges E_{ot} for each object type. First, the x-coordinates of the nodes are initialized using their orders. Then, the x-coordinates of the non-backbone nodes are updated by calculating the median of their adjacent nodes' x-coordinates. Next, the non-backbone nodes are left-packed within each object type to optimize spacing. Subsequently, backbone nodes undergo the same left-packing.

After computing the x-coordinates, the algorithm evaluates whether the total edge length along the x-axis is reduced. If an improvement is found, the updated x-coordinates are recorded as the optimal values. This process is iteratively repeated for a predefined number of iterations. Finally, the set of x-coordinates of nodes $X = \{x_n | n \in N\}$ is determined. From now, we define the function $x : N \rightarrow \mathbb{Z}_{\geq 0}$ to obtain x-coordinates of nodes, i.e., $x = \{(n, x_n) | n \in N\}$. The y-coordinates of nodes are computed using rank of nodes as $y_n = \pi_{rank}(n) \times k$ where k is a given gap between two adjacent ranks. We define the function $y : N \rightarrow \mathbb{Z}_{\geq 0}$ to obtain y-coordinates of nodes, i.e., $y = \{(n, y_n) | n \in N\}$.

Algorithm 2 *nodePositioning*
Input: $OL, maxIter, N_{ot}, BN_{ot}$
Output: *xbest*

1    $xcoord \leftarrow initxcoord(OL)$
2    $xbest \leftarrow xcoord$
3   **for** $i = \{1, ..., maxIter\}$ **do**
4   **for** $ot \in OT$ **do**
5           $nonBN_{ot} \leftarrow N_{ot} - BN_{ot}$
6           $xcoord \leftarrow medianpos(xcoord, nonBN_{ot})$
7           $xcoord \leftarrow packcutNonBN(xcoord, nonBN_{ot})$
8           $xcoord \leftarrow packcutBN(xcoord, BN_{ot})$
9   **if** $xlength(xcoord) < xlength(xbest)$ **do**
10           $xbest \leftarrow xcoord$
11   **return** *xbest*

---

4 Evaluation and Result

4.1 Event Logs

In the evaluation, we used seven OCELs, as detailed in Table 2.

Table 2. OCEL description

Event log	#event types	#object types	#events	#objects
Ethereum blockchain [16]	26	5	96	93
Hinge manufacturing [17]	11	12	38,528	23,771
Hiring [18]	24	6	18,119	3,768
Hospital [18]	9	8	14,642	3,688
Logistics [19]	14	7	35,372	13,882
Order-to-cash (O2C) [18]	22	9	28,278	8,819
Procure-to-pay (P2P) [18]	10	7	14,671	9,054

4.2 Benchmark

We selected the OC-DFG layout from the web-based tool developed by [6] as a benchmark for its well-organized structure, including node ranking, color-coded object types, and clearly visualized start and end nodes, which closely resembles our proposed layout. For evaluation, we parsed the benchmark layout from its SVG elements, calculating node positions and orders based on x/y coordinates.

4.3 Metrics

For evaluation, we employ seven metrics categorized into two aspects: general layout evaluation and object-centric process layout evaluation. For object-centric process layouts, we propose two new metrics object type compactness and object type interaction preservation. Let $OL = (N, E, OT, OTA, \pi_{rank}, \pi_{order}, \pi_{ota}, \pi_{nota})$ be a finalized OC-DFG layout, N_{ot} be the set of nodes of type $ot \in OT$, $x : N \to \mathbb{Z}_{\geq 0}$ be a node x-coordinate mapping function, and $y : N \to \mathbb{Z}_{\geq 0}$ be a node y-coordinate mapping function.

- Number of back edges (QM_{be}): Measures the flow consistency of the layout, defined as $|\{e | e \in E, \pi_{rank}(snode(e)) > \pi_{rank}(enode(e))\}|$
- Number of edge crossings (QM_{ec}): Measures the number of edge intersections.
- Total edge length (QM_{el}): Computes the cumulative length of all edges in OL as:
- $\sum_{e \in E} \sqrt{(x(snode(e)) - x(enode(e)))^2 + (y(snode(e)) - y(enode(e)))^2}$
  where $snode : E \to N$ and $enode : E \to N$ return the start and end nodes of the edge.
- Edge orthogonality (QM_{eo}): Measures the alignment of edges to an imaginary Cartesian grid as $1 - \frac{1}{|N|} \sum_{i=1}^{|E|} \min(\theta_i, 90° - \theta_i, 180° - \theta_i)/45°$ where θ_i denotes the angular deviation of the edge e_i from the horizontal or vertical grid lines ($0 \leq QM_{eo} \leq 1$).
- Node orthogonality (QM_{no}): Evaluates the efficiency of rank and order space utilization as $|\{n | n \in N\}|/(w \times h)$ where $w = \max(\{\pi_x(n) | n \in N\})$ and $h = \max(\{\pi_y(n) | n \in N\})$ represent the layout width and height, respectively. ($0 \leq QM_{no} \leq 1$).
- Object type compactness (QM_{OTC}): Measures how well nodes of the same object type are clustered, extending the silhouette method proposed in [20]. The silhouette score is $S_i = (b_i - a_i)/\max(a_i, b_i)$ where a_i is the mean distance to nodes of the same object type, and b_i is the minimum mean distance to nodes of other types. $a_i = \frac{1}{|N_{ot_i}| - 1} \sum_{n_a, n_b \in N_{ot_i}, a \neq b} |x(n_a) - x(n_b)|$ and $b_i = \min_{ot_j \neq ot_i} \frac{1}{|N_{ot_j}|}$ $\sum_{n_a, n_b \in N_{ot_j}} |x(n_a) - x(n_b)|$. The final score is the mean of S_i, ranging from -1 to 1, where higher values indicate better clustering.
- Object type interaction preservation (QM_{OTIP}): Assesses whether the object types that frequently interact in the event log are positioned close to each other in the layout. For this, we extend Non-Metric Multidimensional Scaling (NMDS) method [21] and quantified as:

$$\sqrt{\frac{\sum_{i=1}^{|OT|} \sum_{j=i+1}^{|OT|} \left(dr_{d_{ocel}}(ot_i, ot_j) - dr_{d_{lay}}(ot_i, ot_j)\right)^2}{\sum_{i=1}^{|OT|} \sum_{j=i+1}^{|OT|} dr_{d_{ocel}}(ot_i, ot_j)^2}}.$$

4.4 Result

The result highlights a trade-off between the general layout and object-centric process layout as summarized in Table 3. While our method has weakness in QM_{ec}, QM_{el}, and QM_{no}, it excels in QM_{be} and QM_{eo}. For OCPM-specific metrics, our method outperforms the benchmark, demonstrating better object type separation and representation of interactions. In other words, the benchmark provides a more compact and aesthetic layout, it does not account for object-centric processes.

Table 3. Overview of quantitative evaluation results

Method	Metric (Count of dominant metrics)						
	Metrics for general layout					Metrics for OCPM	
	QM_{be}	QM_{ec}	QM_{el}	QM_{eo}	QM_{no}	QM_{OTC}	QM_{OTIP}
Our	7 (−66%)	0	2	6 (+0.08)	0	7 (+0.29)	5 (−0.10)
Benchmark	0	7 (−57%)	5 (−10%)	1	7 (+0.05)	0	2

For general layout metrics, our method significantly reduces back edges (QM_{be}) by 66% due to its rank assignment, which explicitly minimizes back edges. In contrast, the benchmark prioritizes aesthetic node placement without considering flow consistency. Additionally, our approach improves edge orthogonality (QM_{eo}) by arranging backbones linearly. However, reducing back edges increases edge length (QM_{el}) and node orthogonality (QM_{no}) since additional ranks are needed to reduce back edges. Edge crossings (QM_{ec}) are higher in our method due to object type axis constraints, which require nodes of the same object type to stay together. Additionally, our cross-minimization restricts node swaps within the same object axis, preserving layout consistency but increasing crossings. The benchmark, which does not impose these constraints, achieves fewer crossings but sacrifices object type separation, which makes understanding object interactions harder.

For OCPM-specific metrics, our method outperforms the benchmark. It achieves 29% higher object type compactness (QM_{OTC}) by placing nodes of the same object type closer together. Additionally, it performs better in object type interaction preservation (QM_{OTIP}) by positioning frequently interacting object types closer together while keeping less related types farther apart, preserving the relationships observed in the event log.

4.5 OC-DFG Layouts Visualized

We present the OC-DFG layouts generated by both the benchmark and our method for P2P and Logistics OCELs. Figure 1 shows the layouts for the P2P OCEL, while Fig. 4 illustrates the layouts for the Logistics OCEL.

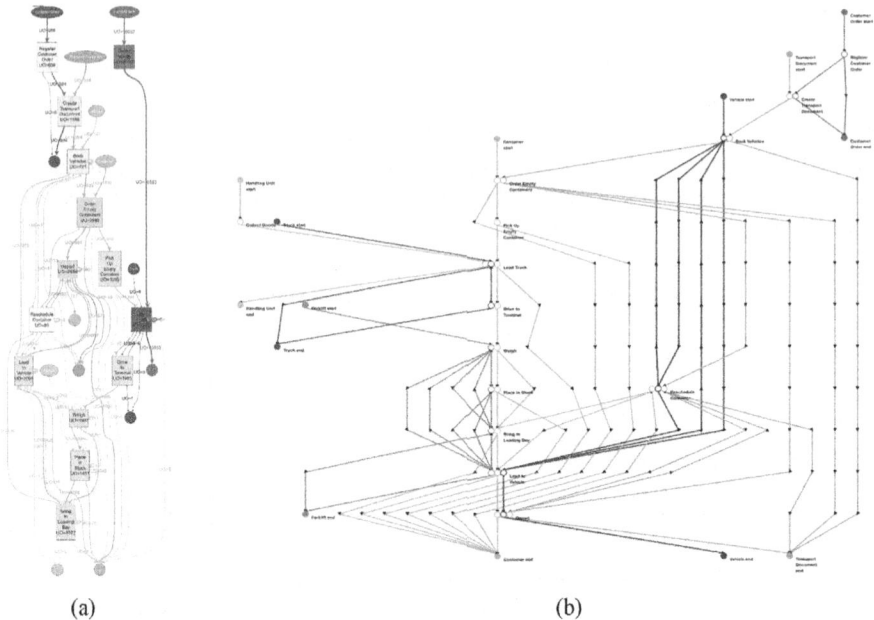

Fig. 4. (a) An OC-DFG layout generated by the benchmark and (b) an OC-DFG layout generated by our method using P2P OCEL.

5 Conclusion

In this study, we proposed an advanced OC-DFG layout generation method, assigning a distinct object type axis to each object type to better distinguish object types and reveal their interactions in OC-DFG. While the method has some limitations in general layout metrics such as edge crossings and node orthogonality, it outperforms the benchmark in other general and OCPM-specific metrics such as back edges, edge orthogonality, object type compactness, and object type interaction preservation.

Beyond academic relevance, the proposed approach may have practical value in domains such as manufacturing, logistics, healthcare, and finance, where understanding multi-object interactions is essential for process optimization and automation. While further validation is required, this structured visualization approach may assist practitioners in analyzing object interactions and interpreting complex processes more effectively in practical settings.

Nevertheless, some limitations remain. Our evaluation was conducted on relatively small OCEL datasets and limited to a comparison with a single benchmark method. Broader validation with large-scale logs and additional tools such as Celonis and PM4Py is necessary. Furthermore, while we introduced several quantitative evaluation metrics— including newly designed ones—we did not provide clear empirical evidence on how these metrics relate to users' actual understanding of the process model. The extent to which improvements in these metrics translate into better interpretability remains to be investigated through user studies. The method also currently focuses solely on

OC-DFGs, without support for OC-Petri Nets or OC-BPMN. Lastly, the influence of hyperparameters used in the layout cost function (λ_1 and λ_2) has not yet been analyzed.

For future work, we plan to conduct qualitative user studies to evaluate how the proposed layout supports process comprehension, addressing the current lack of empirical validation. We also aim to extend our benchmark comparisons to include tools like Celonis and PM4Py, and to test our method on larger OCEL datasets to ensure scalability. Additionally, we will analyze the sensitivity of layout quality to cost function parameters (λ_1 and λ_2) and explore extending our approach to support OC-Petri Nets and OC-BPMN to increase applicability in diverse modeling contexts.

Acknowledgments. This work was supported by the National Research Foundation of Korea (NRF) grant funded by the Korean government (MSIT) (RS-2024-00357330).

References

1. Nofal, M.I., Yusof, Z.M.: Integration of business intelligence and enterprise resource planning within organizations. Procedia Technol. **11**, 658–665 (2013)
2. Vogel-Heuser, B., Wimmer, M.: Digital Transformation: Core Technologies and Emerging Topics from a Computer Science Perspective, 1st edn. Springer, Heidelberg (2023)
3. van der Aalst, W.M.P., et al.: Business process mining: An industrial application. Inf. Syst. **32**, 713–732 (2007)
4. van der Aalst, W.M.P.: Object-centric process mining: dealing with divergence and convergence in event data. In: Ölveczky, P., Salaün, G. (eds.) SEFM 2019. LNCS, vol. 11724, pp. 3–25. Springer, Cham (2019). https://doi.org/10.1007/978-3-030-30446-1_1
5. van der Aalst, W.M.P.: Object-centric process mining: an introduction. In: Cerone, A. (ed.) ICTAC 2021. LNCS, vol. 13490, pp. 73–105. Springer, Cham (2023). https://doi.org/10.1007/978-3-031-43678-9_3
6. Berti, A., van der Aalst, W.: OC-PM: analyzing object-centric event logs and process models. Int. J. Softw. Tools Technol. Transfer **25**(1), 1–17 (2023)
7. van der Aalst, W.: Object-centric process mining: unraveling the fabric of real processes. Mathematics **11**(12), 2691 (2023)
8. van der Aalst, W., Berti, A.: Discovering object-centric petri nets. Fund. Inform. **175**(1–4), 1–40 (2020)
9. Moore, J.: Data visualization in support of executive decision making. Interdiscip. J. Inf. Knowl. Manag. **12**, 125 (2017)
10. Park, S., Bekemeier, B., Flaxman, A., Schultz, M.: Impact of data visualization on decision-making and its implications for public health practice: a systematic literature review. Inform. Health Soc. Care **47**(2), 175–193 (2022)
11. Eberhard, K.: The effects of visualization on judgment and decision-making: a systematic literature review. Manag. Rev. Q. **73**(1), 167–214 (2023)
12. Kim, S.H.: Understanding the role of visualizations on decision making: a study on working memory. Informatics **7**(4), 53(2020)
13. Mennens, R.J.P., Scheepens, R., Westenberg, M.A.: A stable graph layout algorithm for processes. Comput. Graph. Forum **38**(3), 725–737 (2019)
14. Object-Centric Event Log 2.0. https://ocel-standard.org. Accessed 09 Mar 2025
15. Gansner, E.R., Koutsofios, E., North, S.C., Vo, K.P.: A technique for drawing directed graphs. IEEE Trans. Softw. Eng. **19**(3), 214–230 (1993)

16. Comprehensive Ethereum Execution Data for Object-Centric Process Mining of Decentralized Applications (DApps). https://zenodo.org/records/11065366. Accessed 09 Mar 2025
17. sOCEL 2.0: A Sustainability-Enriched OCEL of a Hinge Production Process. https://zenodo.org/records/13638681. Accessed 09 Mar 2025
18. Simulated Object-Centric Event Logs (OCEL 2.0) for Order-to-Cash, Procure-to-Pay, Hiring, and Hospital Patient Lifecycle Processes. https://zenodo.org/records/13879980. Accessed 09 Mar 2025
19. Container Logistics Object-centric Event Log. https://zenodo.org/records/8428084. Accessed 09 Mar 2025
20. Rousseeuw, P.J.: Silhouettes: a graphical aid to the interpretation and validation of cluster analysis. J. Comput. Appl. Math. **20**, 53–65 (1987)
21. Kruskal, J.B.: Multidimensional scaling by optimizing goodness of fit to a nonmetric hypothesis. Psychometrika **29**(1), 1–27 (1964)

Leveraging the Diamond Pattern for Scalable and Upgradeable Blockchain-Based Business Process Management Applications

Victor Lemaire[1], Tiphaine Henry[1(✉)], Álvaro García-Pérez[1], Walid Gaaloul[2], and Sara Tucci-Piergiovanni[1]

[1] Université Paris-Saclay, CEA, List,
F-91120 Palaiseau, France
tiphaine.henry@cea.fr
[2] Telecom SudParis, UMR 5157 Samovar,
Institut Polytechnique de Paris, Palaiseau, France
walid.gaaloul@telecom-sudparis.eu

Abstract. The integration of blockchain smart contracts in Business Process Management (BPM) enhances trust and transparency in cross-organizational process execution. However, traditional smart contract-based solutions face scalability challenges when handling large or complex business process models in Solidity-compatible blockchains. Additionally, existing architectures lack flexibility, requiring full redeployment and state migration for updates to model specifications or execution engines. To address these issues, we propose a novel smart contract architecture based on the diamond software pattern. This architecture enables the execution of large-scale BPM models, as it maintains a constant contract size regardless of the model's complexity. The modularized execution logic also improves upgradeability by allowing updates to specific components without redeploying the entire system. We evaluate our approach using a variety of process models. Our experiments demonstrate that the diamond-based architecture can successfully handle models with up to 100 elements, while state-of-the-art solutions struggle with models exceeding 50 elements. Furthermore, the gas costs associated with deploying the diamond infrastructure are mitigated after running three instances, making the architecture particularly suitable for multi-tenant collaborations, where participants can share the same execution engine across models and instances.

Keywords: Blockchain · Smart Contract · Diamond Pattern · BPM

1 Introduction

Cross-organizational business process management (BPM) increasingly relies on decentralized execution engines that follow predefined choreography protocols. This distributed setting introduces challenges of trust, accountability, and coordination [18], particularly when multiple organizations must agree on the execution state of a shared process. To address these concerns, recent work has proposed using blockchain technology and smart contracts to implement verifiable execution engines. Smart contracts—programs deployed on a blockchain that enforce rules in an immutable and transparent way [21]—enable tamper-resistant and auditable process execution.

Existing blockchain-based BPM engines typically encode the process model, execution logic, and instance state in smart contracts [7,9,13,14,22,23]. These approaches fall into two broad categories based on their architecture: monolithic contracts and factory-based contracts.

In the monolithic design, each process instance is represented by a single smart contract that embeds the full model, logic, and state. While straightforward, this architecture does not scale: Ethereum and compatible blockchains impose bytecode size limits (24,576 bytes) that restrict the complexity of deployable contracts. Models with a moderate number of elements, such as 14 tasks, 7 gateways, and 18 events, can exceed this limit. Furthermore, spawning multiple instances duplicates both model and logic, resulting in redundant deployments and high gas costs. Updates to the model or logic require full contract redeployment, complicating maintenance and limiting flexibility.

The factory-based architecture addresses some of these issues by encoding the model and execution logic once in a factory contract. Process instances are then created as lightweight contracts holding only state. Each executable element (e.g., task, gateway) is handled through dedicated start and callback methods in the factory. While this approach improves instance scalability, it still suffers from model size limitations: the number of methods grows linearly with the number of model elements. For instance, a model with 50 tasks and 2 events requires encoding 104 methods [13], which can again exceed contract size limits. Additionally, updating the model or logic requires factory redeployment and instance migration—an expensive and error-prone process [2,8].

To overcome these limitations, we introduce a new execution engine architecture based on the diamond standard (EIP-2535) [15,16]. Our approach decouples model specification from execution logic. The model is encoded as a structured set of variables, while the logic is implemented once across all instances using a fixed set of reusable methods (facets). Both the model specification and instance states are stored in a central smart contract called the diamond contract. Each facet provides a subset of the execution logic, following the principle of separation of concerns. This modular architecture allows new process instances to be created without deploying new contracts: a simple method call to the diamond contract suffices.

This design improves scalability by avoiding method duplication and reducing bytecode growth. Unlike the factory approach, the number of methods does not depend on the number of model elements, enabling the encoding of larger and more complex models. The architecture also supports upgradeability at two levels: (i) models can be updated directly in storage (provided no active instances are running), and (ii) execution logic can evolve over time by replacing facets, without disrupting existing processes. This provides long-term maintainability and adaptability to evolving requirements, standards, or security practices.

We validate our architecture with a prototype implementation on Ethereum, using BPMN models from the SAP-SAM dataset [20] and domain-specific models inspired by our lab's projects, ranging from 20 to 100 executable elements. Our results confirm that the diamond-based engine overcomes the model size limits of previous designs. The additional deployment cost is amortized after just three instances, due to shared logic reuse. We also demonstrate successful upgrades of the model and logic, without redeploying the entire architecture or interrupting running instances.

The remainder of this paper is structured as follows: Sect. 2 provides background on blockchain, smart contracts, and the diamond pattern. Section 3 proposes our architecture. Section 4 presents our experimental setup and results. Section 5 reviews related work. Finally, Sect. 6 presents our conclusions and outlines directions for future research.

2 Key Concepts on Smart Contracts and the Diamond Pattern

2.1 Blockchain, Smart Contracts and Gas Costs

A blockchain is an immutable, tamper-resistant distributed ledger maintained by a peer-to-peer network of computing nodes [21]. While blockchains were initially designed to record digital currency transactions, they can store arbitrary data, such as digital assets, access rights, or business process events. Each piece of data is broadcast to the network as a transaction, digitally signed by the user's private key. Users are identified by addresses derived from their public keys, ensuring pseudonymous but verifiable interactions. Transactions are grouped into blocks and validated by network participants through a consensus mechanism, such as Proof of Work or Proof of Stake. Each block includes a header, a list of verified transactions, and the hash of the previous block. This chaining of blocks ensures that any modification to past data would invalidate subsequent blocks, making the ledger effectively immutable. Once confirmed, blocks are replicated across all nodes, forming a consistent and tamper-proof record of events.

A smart contract is an immutable program deployed on a blockchain and executed by all nodes in response to incoming transactions [21,24]. It encapsulates both data (state variables) and executable logic (methods). When a user invokes a method via a transaction, the code is run by each node, and any state changes are validated and committed through the blockchain's consensus protocol. This guarantees consistent behavior across the network, without relying on a central authority. Smart contracts must first be deployed via a special transaction that uploads the contract code to the blockchain which incurs a gas cost proportional to the size of the code. Once deployed, the contract receives a unique address derived from the deployer's address and transaction nonce, which serves as its permanent identifier on the network. The deployer becomes the contract's initial owner. Interaction with a smart contract involves sending a transaction to its address, specifying the method to invoke and any input parameters. Each method call also consumes gas, covering the computational resources required

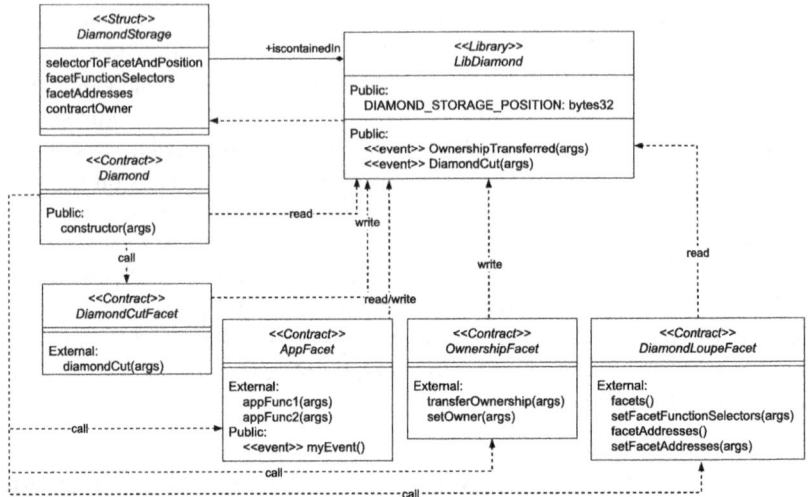

Fig. 1. Class Diagram of the Diamond Standard (EIP-2535).

for execution. Smart contracts can interact with one another by calling external contract methods using message calls, enabling modular and composable logic. These inter-contract calls incur an additional fixed gas fee plus the execution cost of the called method, which must be accounted for during design and deployment.

2.2 Diamond Pattern

The diamond pattern (EIP-2535)[1], introduced by Nick Mudge [16], extends the proxy pattern [19] to overcome the size and immutability limitations of smart contracts. It enables modular and upgradeable architectures by distributing contract logic across components called facets, while maintaining a single entry point—the diamond contract—for all interactions. As illustrated in Fig. 1, the diamond contract acts as a proxy that routes incoming function calls to the appropriate facet, based on the function's four-byte selector (derived from its signature in Solidity). A special facet, DiamondCut, manages the contract's upgradeability by allowing the addition, replacement, or removal of facets. Utility facets such as Ownership (for access control) and Loupe (for method inspection) are typically included to support administration and introspection [17]. To ensure consistent access to application state, the architecture relies on a centralized storage layout called DiamondStorage. This shared state is accessible across all facets using a fixed storage slot and helper methods from the LibDiamond library, which guarantees data consistency throughout the system. Upgrades are performed via the diamondCut function, which takes as input a

---

[1] https://eips.ethereum.org/EIPS/eip-2535.

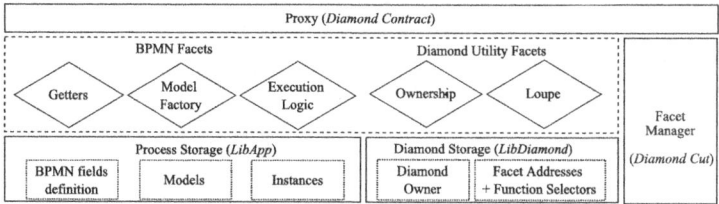

Fig. 2. Diamond-based Architecture.

list of function selectors, facet addresses, and actions—Add, Replace, or Remove. This allows fine-grained, low-cost updates to the contract logic without affecting the underlying state or requiring redeployment. Although the initial deployment of a diamond-based contract is more expensive than traditional or proxy-based patterns, the cost is amortized over time. Its modularity, reuse of logic, and ability to support safe upgrades make it well suited for large-scale and evolving smart contract applications [3].

3 Proposed Architecture

This section outlines the architecture of the proposed diamond-based architecture, detailing its core components (Sect. 3.1), the deployment of the diamond (Sect. 3.2), the BPMN lifecycle management (Sect. 3.3) and BPMN-related facet updates (Sect. 3.4).

3.1 Overall Architecture

The architecture, depicted in Fig. 2, follows the diamond pattern introduced in Sect. 2.2. We specialize the diamond pattern with three facets comprising methods related to BPMN process management. The *Getters* facet provides data access points for off-chain interactions (i.e., read requests of the state of the workflow). The *Execution Logic* facet manages instance execution, with only one off-chain callable method, *Invoke*, which initiates execution requests within the model instance. Finally, the *Model Factory* facet handles model management and instance creation.

Data storage is divided between two smart contracts. Diamond storage, defined in *LibDiamond*, holds the address of the diamond owner taking of the role of the process manager responsible for the execution engine, and a mapping that links methods to the corresponding facet addresses. This storage is shared across diamond utility facets and the *Facet Manager*. Process storage, maintained by *LibApp*, stores models and instance implementation specifications. A model is represented by a structure containing the model name, owner, its set of executable elements and dependencies, and instance counts, and mappings that associate elements and instances with identifiers. Executable elements are represented by structures defining attributes like identifier, type, XML identifier,

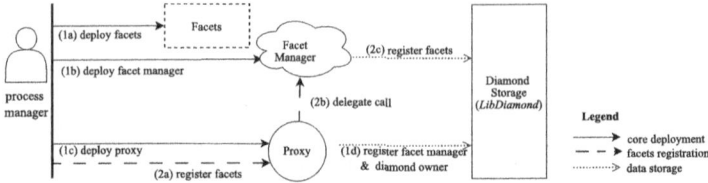

Fig. 3. Key steps for deploying the diamond-based architecture.

name, lane, input data, and output data. Instances are defined by structures containing the model name, instance identifier, role-address mappings. Process storage also keeps track of running instances. Their state is managed using a set of state-tracking arrays, i.e., three binary arrays keeping track of activated, pending, or executed elements.

3.2 Diamond Deployment

The deployment of the diamond-based architecture, as shown in Fig. 3, follows a two-stage process designed to ensure modularity and maintainability of the methods. In the first stage, the core diamond structure is established through the deployment of smart contracts. The process manager sequentially deploys each facet (Step 1a) and the *Facet Manager* (Step 1b), followed by the *Proxy*, which is initialized with the *Facet Manager*'s address (Step 1c). The *Proxy* then registers the *Facet Manager* in diamond storage and assigns the process manager as the diamond owner, ensuring controlled management of facet modifications. The second stage links the deployed facets to the diamond, finalizing the architecture's modular composition. The process manager collects all facet addresses and submits them to the *Proxy* (Step 2a), which forwards the request to the *Facet Manager* (Step 2b). After verifying the process manager's ownership, the *Facet Manager* registers the facet addresses in diamond storage. This structured approach maintains the integrity of the architecture while enabling seamless upgrades and modifications.

3.3 BPMN Smart Contract Lifecycle

The BPMN execution engine encoded in the facets can be used to register and execute instances of models such as the BPMN collaboration model of a multi-party workflow for constructing a 3D plan from laser scans depicted in Fig. 5. The model involves several participants with distinct roles: two participants gather laser scan data, a central entity orchestrates the processing and generation of the 3D model, another participant runs simulations to validate its accuracy, and a certification authority verifies the final model before issuing an approval certificate. The model is inherently collaborative, relying on frequent message exchanges between participants while also containing internal elements that each

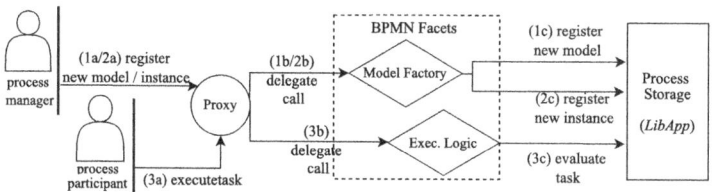

Fig. 4. Illustration of the BPMN model registration, instantiation, and execution.

participant selectively disclose for accountability—for example, to prove compliance with industry standards. In this scenario, blockchain is adopted not only for accountability but also to ensure tamper-proof logging, verifiable data provenance, and immutable certification—all without relying on a trusted central authority. We describe below each step of the BPMN smart contract lifecycle, namely model registration, instance deployment, and instance execution (c.f., Fig. 4).

Step i: Registration of a new BPMN model. Registering a BPMN model involves parsing the XML representation of the BPMN model and translating it into a directed graph, where nodes represent executable process elements (e.g., tasks, events), and edges represent control-flow or message-flow dependencies. In this graph-based view, a parent-child relationship corresponds to a directed edge from a predecessor (parent) element to its successor (child). This structure captures the dynamic behavior of the process model. The process manager initiates the registration by calling the *addModel* method via a proxy (Step 1a), which forwards the request to the *Model Factory* facet (Step 1b). This method requires the following parameters: (1) the model name (e.g., "LaserScan") which is a unique identifier for the BPMN model; (2) an array T of executable elements, where each element is defined by its ID, type, and name; (3) a 2D array C encoding child relationships, i.e., for each node, the list of its successors in the control-flow graph (e.g., C maps "3D-Compute" to "Landscape-Reconstruction"); (4) a 2D array P encoding parent relationships, i.e., for each node, the list of its predecessors (e.g., P maps "3D-Compute" to "Landscape-Reconstruction"); (5) a 2D array Mi representing incoming message flows, i.e., message dependencies where the given element is a recipient; (6) a 2D array Mo representing outgoing message flows, i.e., messages emitted by the element. The method first verifies the uniqueness of the model name. If no conflict is found, it persists the model's control and message flow graph into the application's storage. Once registered, the BPMN model acts as a template from which runtime instances can be instantiated.

Step ii: Deployment of a BPMN Instance of the Model. With the model registered, the next step is to deploy an instance, creating a runtime representation of the BPMN model. This instance maintains three state-tracking arrays: executed, included, and pending. These arrays monitor the execution status of each executable element within the instance, and enable real-time updates, unlike tradi-

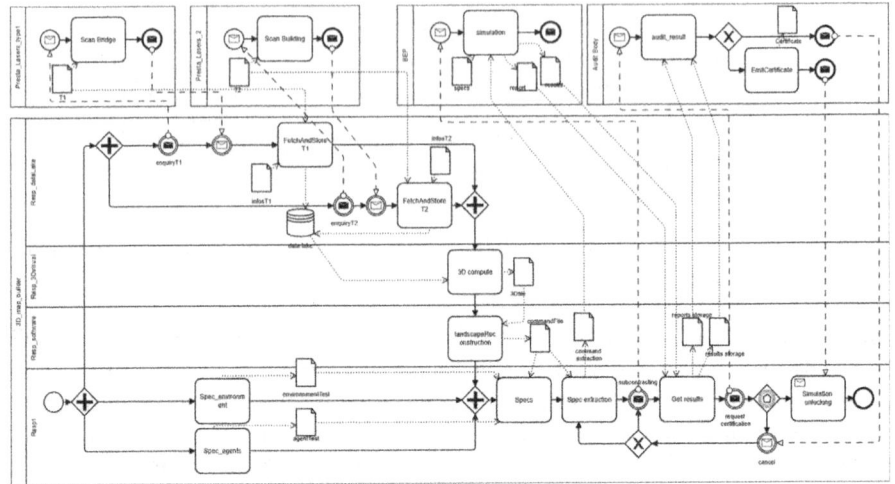

Fig. 5. BPMN of the LaserScan motivating example.

tional BPMN where states are managed externally. The deployment begins when the process manager triggers the proxy, which calls the *newInstance* method in the *Model Factory* facet (Fig. 3, Step 2). The *newInstance* method requires the model name as input. It verifies that the specified model exists and initializes the three instance's state-tracking arrays. The included array marks which elements are active or ready for execution. Tasks corresponding to start events are initialized to 1, indicating they are included and ready to begin. The pending array and executed arrays mark elements that are being processed or completed. Both are initialized with 0 for all elements, representing no elements are pending or completed at the start. An event is emitted to notify listeners of the creation of the new instance, providing its unique identifier. In our motivating example, invoking *newInstance("LaserScan")* for the LaserScan model creates an instance with id 0. The method initializes the state-tracking arrays such that the included array activates the "startOrder" start event, while the pending and executed arrays remain at 0. The system then increments the instance counter, guaranteeing unique identifiers for subsequent instances.

Step iii: Execution of the instance. Once deployed, the instance progresses to execution, where process participants invoke executable elements according to BPMN rules. The *invoke* method, located within the *Execution Logic* facet, allows process participants to execute a specified element. This method takes the model name, element and instance identifiers as parameters (Fig. 3, Step 3). If the element is executable (i.e., it has a value of 1 in either the pending or included array), the method sets the element to executed and sets child elements to included by updating the corresponding instance state-tracking arrays. In out motivating example, invoking "3D-Compute" within the LaserScan model sets the included value for "3D-Compute" to 0 and that for "Landscape-

Reconstruction" to 1, while updating the executed array to mark "3D-Compute" as completed. Different element types affect execution dynamics following traditional BPMN orchestration logic already described in [22,23]: user tasks depend on explicit user input and remain pending until the user completes them, gateway tasks (e.g., exclusive or parallel) control the flow by evaluating conditions or activating multiple paths concurrently, event-based tasks wait for triggers such as messages, transitioning to executed only when the specified event occurs. As execution progresses, tasks continue transitioning through their lifecycle. Once all tasks in an instance are completed, the process reaches its final state, marking the successful execution of the BPMN model instance.

3.4 Diamond-Based Architecture Updates

This section categorizes diamond-based architecture updates into two types: (i) model specification updates, e.g., adding, removing, or replacing elements in the model, and (ii) execution engine encoding updates, e.g., changing the smart contract implementation of the model or execution logic.

3.5 Model Specification Updates

Changes to a BPMN model specification encompass modifications to the control flow, such as adding or removing elements, adjusting element dependencies, or modifying element properties (e.g., type, name, lane, input/output data). In our LaserScan motivating example, runtime updates to the model specification are required to adapt to regulatory changes introducing new validation steps, to reassign tasks following participant unavailability or data anomalies, or to support different certification variants depending on clients or jurisdictions.

Such model specification udpates can be triggered by the model owner, or set of owners in case of a change negotiation. To maintain consistency and prevent execution anomalies, model specification updates are permitted only for models without running instances as modifying the model specification of a running instance could introduce state inconsistencies, potentially leading to deadlocks or livelocks [5].

A model specification update can be one of the three operations of *replace*, *add*, or *remove*. These operations apply to elements within the model, including elements, events, gateways, and dependencies. Modifications are executed via the *updateModel* method on a model stored in *LibApp*.

The *replace* operation substitutes an element $e \in E$ with an updated element e' while preserving e's dependencies. It is defined as $replace(M, e, e') = (E', D', S')$ where $E' = (E \setminus \{e\}) \cup \{e'\}$, $D' = (D \setminus D_e) \cup D_{e'}$, with D_e and $D_{e'}$ representing the dependencies of e and e', respectively, and where S' updates the element state $S(e)$ to $S(e')$. Given and element e, the element state $S(e)$ is the state-tracking array for e, which consists of three components: the first tracks whether e is executable, the second indicates whether e is pending execution, and the third tracks whether e has been executed. An example of a *replace* operation is adding a data object to the "Scan-Bridge" activity of Fig. 5. The "Scan-Bridge" specification stored in LibApp is updated with the new activity.

The *add* operation inserts a new element $e' \notin E$ and updates dependencies accordingly. It is defined as $add(M, e', P, C) = (E', D', S')$ where $E' = E \cup \{e'\}$ and $D' = D \cup \{(p, e') \mid p \in P\} \cup \{(e', c) \mid c \in C\}$, and where P and C are the sets of parent and child elements of e'. Additionally, the three elements in the state-tracking array $S'(e')$ are extended to keep track of the inclusion, execution, and pending state of the new element e'. An example of the *add* operation is adding a new activity, "Retrieve3DSimulations", to the RespDataLake lane and linking it to the two gateways in Fig. 5. The new activity is added to the model's initial set of elements, and the dependencies of the two gateways are updated to reflect the change.

The *remove* operation deletes an element $e \in E$ and updates dependencies. It is defined as $remove(M, e) = (E', D', S')$ where $E' = E \setminus \{e\}$ and $D' = D \setminus \{(p, e) \mid p \in E\} \setminus \{(e, c) \mid c \in E\}$. Additionally, the state-tracking array is updated by setting the index of e to zero, i.e., $S'(e) = (0, 0, 0)$. An example of the *remove* operation is removing the *Resp1* activity, "SpecAgents". This involves removing the two dependencies of "SpecAgents" on the gateways, and vice versa, removing "SpecAgents" from the dependencies of both gateways. Finally, the inclusion state of "SpecAgents" is set to 0.

3.6 Execution Engine Encoding Updates

During the redesign of the BPMN execution system, implementation choices at the smart contract level may change. Execution engine encoding updates can accommodate these changes by (i) modifying the general encoding of models into smart contracts, or (ii) modifying the execution logic contained in the diamond facets. An execution engine encoding update begins by optionally modifying the model encoding, by adjusting how BPMN elements, dependencies, and states are represented in *LibApp*. This modifications can only be applied before instantiation. The update proceeds by the process manager modifying the model encoding of *LibApp*, by deploying a new version of this library. Subsequently, the update proceeds by modifying the execution logic, either changing the BPMN facet methods affected by the model encoding change, or enriching the execution logic itself to reflect revised logic for model registration, execution, and instance access. For instance, the execution engine can be extended with additional mechanisms such as a timeout strategy that uses block numbers instead of timestamps. Finally, the update concludes by the process manager applying the *diamondCut* function in order to reference the new methods, while existing stored data is preserved.

One example of a model encoding update is replacing state-tracking arrays with mappings to improve instance state access. To implement this change, the LibApp storage contract is redeployed with the revised encoding, and the affected facets are updated accordingly. Figure 6 illustrates this process through a concrete update to the addModel method in the *Model Factory* facet, providing

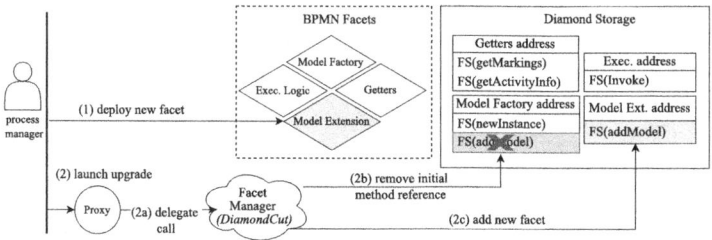

Fig. 6. Illustration of a model encoding update.

an end-to-end view of a single method upgrade. The update begins by deploying a new version of the *Model Factory* facet that includes the revised logic (Step 1). This introduces a new function selector, derived from the hash of the updated method signature. To avoid disrupting existing facets, the new method is encapsulated within a new facet. The process manager then prepares the `diamondCut` payload to replace the outdated method (Step 2), specifying the new facet address, the action type (Replace), and the updated function selector. The diamond proxy forwards this request to the *Facet Manager* (Step 2a), which handles the update in three substeps: removing the old function selector (Step 2b) and linking the new selector to the updated facet (Step 2c). This procedure is applied to all methods affected by `LibApp` modification.

In conclusion, the diamond-based architecture provides a modular framework for BPMN process updates, enabling both pre-instantiation modifications and dynamic execution logic updates. By leveraging *diamondCut*, the system ensures adaptability while preserving workflow continuity.

4 Evaluation

4.1 Implementation and Experimental Setup

The evaluation of the diamond-based architecture's performance, scalability, and upgradeability is conducted through three primary experiments: (1) measuring the gas costs associated with the instantiation and execution of BPMN models of varying complexities to quantify the overhead introduced by the diamond structure; (2) assessing the scalability of the architecture in handling multiple process instances and models; and (3) examining the upgradeability of the architecture through an access control update scenario. The prototype is based on the optimized reference implementation of the Diamond Standard[2], which includes a mapping linking each method selector to its corresponding facet address. This enables dynamic addition, removal, and modification of methods. A monolithic

---

[2] https://github.com/mudgen/diamond-2-hardhat.

Table 1. BPMN Models used for the experiments. *(Ex = External, In = Internal)*

Model ID	Name	Task	Gate	Events Start	End	Throw	Catch	Flows Ex.	In.	Pool	Data Obj.
A	OrderToCash	1	6	1	4	0	1	0	21	1	0
B	BulkBuyer	4	0	5	5	6	6	10	21	5	0
C	LCA	7	2	2	1	5	3	7	18	2	12
D	SecuRoutiere	14	6	4	4	0	0	4	28	4	0
E	Certification	10	5	5	8	1	1	8	25	5	16
F	LaserScan	14	7	5	6	4	3	8	40	5	15
G	Seadmete Garantii	21	5	3	2	0	0	0	34	6	0
H	Candidature Proc.	21	7	4	8	0	2	13	38	4	1
I	Zad2Kolab	8	8	4	9	3	8	11	38	4	0
J	SPARK Logistics	24	8	1	12	2	9	12	48	5	0
K	ScanConsortium	43	11	18	20	3	3	20	80	18	58

approach is also implemented for comparison. This monolithic approach consists in generating a custom smart contract implementation per model that encapsulates the model specification, instance states, and the execution logic within a single contract, providing a baseline for performance comparison. We use the Gas-Reporter plugin within Hardhat for recording gas consumption during deployments and method calls. To assess the architecture's capacity to manage models of varying complexity, we utilize a set of BPMN models outlined in Table 1. This includes models from both academic sources (e.g., OrderToCash and BulkBuyer models from Weber et al. [23]) and industry (e.g., SAP-SAM dataset and the LCA, Certification, LaserScan, and ScanConsortium models from our lab), with model sizes ranging from 20 to 100 elements. These models test both the scalability and efficiency of the diamond-based approach in managing complex BPMN processes. The diamond smart contracts, BPMN model sources, and evaluation scripts are available at https://doi.org/10.5281/zenodo.15490989 [10].

4.2 Experiments

Gas Cost Analysis for BPMN Instantiation and Execution. The objective of this experiment is to compare the gas consumption of BPMN instantiation and execution between the diamond-based and monolithic architectures. Specifically, the focus is on the gas costs associated with both deploying BPMN models (instantiation) and invoking elements within these models (execution). Gas costs for both the deployment and invocation phases are recorded. Results are presented in Fig. 7. In terms of deployment, the monolithic architecture typically incurs lower initial costs than the diamond-based approach. However, the diamond architecture's cost structure remains stable as the number of instances or models increases. Execution costs are similar for both architectures, ranging between

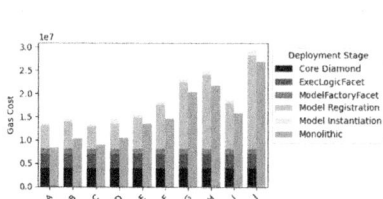
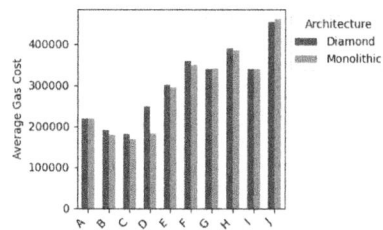

(a) Deployment Gas Costs per Scenario

(b) Average Invoke Gas Costs per Scenario

Fig. 7. Diamond vs Monolithic BPMN Instantiation and Execution Gas Costs.

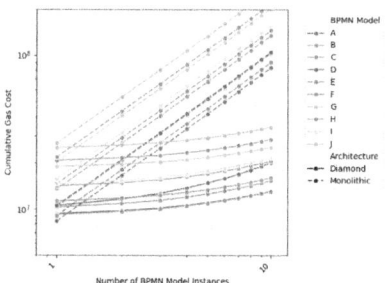
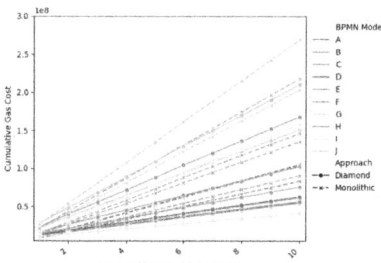

(a) Gas cost per number of deployed instances

(b) Gas cost per number of model registrations

Fig. 8. Single and Multi-Model Scalability Experiments.

180,000 and 450,000 gas per element invocation. Despite this, the diamond-based architecture shows advantages for larger models (the monolithic approach fails for contracts with complexities greater than process J), where the cost per invocation decreases across multiple calls as model and execution logic is shared between instances. Thus, although the monolithic approach may be more economical for smaller, single-use models, the diamond-based architecture appears better suited for larger, multi-instance BPMN processes. The diamond structure optimizes gas costs over multiple executions, making it a more scalable and cost-efficient choice for complex BPMN elements.

Scalability Evaluation. To assess the benefits of shared BPMN logic, we compare the gas costs of deploying multiple process instances under both architectures. In the monolithic setup, each new instance requires redeploying the full BPMN smart contract. In contrast, the diamond-based architecture deploys the diamond structure once; subsequent instances are created by calling a lightweight `newInstance` method. As shown in Fig. 8a, the diamond-based approach initially incurs higher gas costs, but becomes more efficient after two instances. At around two instances (12M gas), both approaches converge; beyond that, the

Table 2. Gas costs for model specification and execution engine updates.

ID	Replace	Add	Remove
B	281,091	398,564	38,928
F	284,655	398,528	38,892
K	268,336	398,660	39,024

(a) Model Specification

Execution Engine Update	3,274,972
LibApp-v2 Deployment	72,227
Facet Deployment	2,085,646
Diamond Cut	1,116,099

(b) Execution Engine Update

diamond-based architecture scales more efficiently by avoiding redundant redeployments. This highlights its cost advantage in managing multiple instances. We also evaluate multi-model deployments using the BPMN models from Table 1. Figure 8b shows that while the diamond approach requires an upfront cost to set up its infrastructure, each new model can be added via a simple addModel call. By contrast, the monolithic approach redeploys the full contract for each model, leading to a steeper rise in costs. The crossover point occurs after four deployments, after which the diamond-based approach proves more economical. Overall, these results confirm the scalability benefits of the diamond-based architecture. By reusing shared execution logic and separating model registration, it amortizes the initial deployment cost and offers a more sustainable solution for blockchain-based BPMN management.

Upgradeability Evaluation. This experiment evaluates the feasibility and cost of upgrading a diamond-based architecture, focusing on both model specification and execution engine updates. To assess model updates, we measure the gas cost of three operations—add, replace, and remove—across three BPMN models of increasing size: BulkBuyer (small), LaserScan (medium), and ScanConsortium (large). Updates are applied using the updateModel function, which takes the operation type and relevant parameters, such as the index of the target element or the new activity specification. No facet upgrades are involved in this step. As shown in Table 2, remove is the most efficient operation, consuming approximately ten times less gas than replace and twelve times less than add. Replace is more efficient than add, as it modifies existing entries without increasing storage size. Gas costs remain stable across models, suggesting that the operation type, rather than model size, drives cost. These findings suggest that replace should be preferred over add when possible, and that deactivation (a soft delete) may be more gas-efficient than full removal. Further investigation is needed to evaluate the long-term storage impact of such strategies. To evaluate execution logic upgrades, we simulate structural changes to the engine: introducing a new metadata field into the model specification and replacing three state-tracking arrays with a mapping structure. The upgrade involves deploying a new application storage contract (LibApp-v2) and three updated BPMN facets reflecting the modified logic for model registration, execution, and instance access. The upgrade is finalized through a diamond cut operation. As reported in Table 2, the total gas cost for this upgrade is 3,274,972 gas. This includes 72,227 gas for

Table 3. State of the art comparison. (B.S. = Byte Size)

Ref	Approach	Language	Execution	B.S. Limit	Upgrad.	Flex.	Implem.
[9,23]	Factory	BPMN	On-chain	X	X	X	✓
[14,22]	Monolithic	BPMN, DCR	On-chain	X	X	X	✓
[4]	Sidechains	BPMN	Hybrid	✓	X	X	✓
[6]	Sidechains	BPMN	Hybrid	X	X	X	✓
[1]	Federated	BPMN	Off-chain	✓	✓	✓	✓
[12]	Fragments	BPMN	On-chain	X	X	✓	✓
[11]	Proxy-based	BPMN	On-chain	✓	✓	X	✓
Our work	Diamond	BPMN	On-chain	✓	✓	✓	✓

deploying LibApp-v2, 2,085,646 gas for the three facets, and 1,116,099 gas for the diamond cut itself. These results highlight that while the diamond architecture supports cost-effective minor updates—such as modifying a method in isolation—larger structural changes, especially those involving shared storage libraries and multiple facets, entail significantly higher costs. The architecture thus provides strong upgradeability for small and medium-scale modifications, but substantial updates remain computationally expensive.

5 Related Work

Table 3 compares leading blockchain-based BPM approaches across six dimensions: contract language, execution model, byte-size handling, upgradeability, runtime flexibility, and implementation status. Factory-based solutions [9,23] address Solidity's byte-size limit by modularizing models into subprocesses, each in a separate contract. This supports larger models but leads to high gas costs and lacks upgrade or runtime adaptation capabilities. Monolithic designs [14,22], which embed the entire model and logic into one contract, are simpler but constrained by byte-size caps and offer no upgrade path or flexibility. Sidechain-based approaches [4,6] offload execution or coordination to external chains like Quorum or Polygon, reducing gas costs and alleviating size constraints. However, they introduce synchronization challenges and lack native upgradeability. Proxy-based strategies [11] support versioning through contract indirection (e.g., Unstructured Proxy, Eternal Storage), allowing runtime updates but only at contract-level granularity, with per-participant deployment. Control-flow update methods [12] add operations like link-process and choose-path for late binding and decentralized agreement but lack systematic upgrade support. Federated models [1] separate execution from blockchain, using smart contracts as event logs, which improves adaptability but introduces trust and synchronization issues. In contrast, our approach builds on the Diamond Standard (EIP-2535) to unify scalability, upgradeability, and runtime flexibility in a single on-chain architecture. By separating concerns across facets and centralizing state in a diamond contract, it enables logic reuse, fine-grained upgrades, and dynamic model evolution without compromising verifiability or coordination overhead.

6 Discussion and Conclusion

This study investigates the applicability of the diamond pattern in enhancing scalability and upgradeability for blockchain-based BPMN execution, addressing limitations in current approaches. By isolating and representing the execution logic in a fixed number of methods shared among model instances, our architecture overcomes byte-size constraints and enables selective updates to execution logic. Gas cost analyses comparing diamond-based, monolithic, and factory-based BPMN smart contracts demonstrate that our approach significantly improves model and instance scalability. While deploying the diamond core incurs higher initial costs, shared model registration and instantiation methods make it cost-efficient for managing multiple processes and instances. Additionally, the modular design facilitates operational logic updates, such as modifying or refining methods of the execution engine.

Despite its advantages, some issues remain open. First, ensuring semantic correctness in process evolution remains an open issue, requiring formal verification techniques to validate updates dynamically. Second, while the architecture scales better than monolithic contracts, complex BPMN semantics (e.g., event-driven subprocesses, data objects) may introduce performance bottlenecks. Moreover, the proposed implementation is yet to be optimized in terms of gas costs. Nonetheless, balancing modularity and gas efficiency is crucial, as frequent inter-facet interactions increase overhead. Future research should optimize this trade-off while exploring hybrid on/off-chain storage solutions and Layer-2 scaling mechanisms to further reduce computational costs. Third, dynamic updates require strict access control to prevent unauthorized modifications. Governance models that ensure trustworthy process evolution in decentralized environments must be developed, balancing upgradeability and security.

References

1. Adams, M., Suriadi, S., Kumar, A., ter Hofstede, A.H.: Flexible integration of blockchain with business process automation: a federated architecture. In: CAiSE Forum, pp. 1–13. Springer (2020)
2. Bandara, H.D., Xu, X., Weber, I.: Patterns for blockchain data migration. In: EuroPLoP, pp. 1–19 (2020)
3. Benedetti, A., Henry, T., Tucci-Piergiovanni, S.: A comparative gas cost analysis of proxy and diamond patterns in EVM blockchains for trusted smart contract engineering. In: FC. Springer (2025)
4. Bodorik, P., Liu, C.G., Jutla, D.: Tabs: transforming automatically BPMN models into blockchain smart contracts. BCRA 4(1), 100115 (2023)
5. Brahem, A., et al.: A trustworthy decentralized change propagation mechanism for declarative choreographies. In: BPM, pp. 418–435. Springer (2022)

6. Corradini, F., et al.: Blockchain-based execution of BPMN choreographies with multiple instances. DLT (2023)
7. Corradini, F.: A flexible approach to multi-party business process execution on blockchain. Futur. Gener. Comput. Syst. **147**, 219–234 (2023)
8. Di Sorbo, A., Laudanna, S., Vacca, A., Visaggio, C.A., Canfora, G.: Profiling gas consumption in solidity smart contracts. J. Syst. Softw. (2022)
9. García-Bañuelos, L., Ponomarev, A., Dumas, M., Weber, I.: Optimized execution of business processes on blockchain. In: BPM, pp. 130–146. Springer (2017)
10. Henry, T., Gaaloul, W., Álvaro Álvarez García, Tucci-Piergiovanni, S.: Leveraging the diamond standard for scalable and upgradeable blockchain-based business process management applications—code, and experimental scenarios (2025). https://doi.org/10.5281/zenodo.15490989
11. Klinger, P., Nguyen, L., Bodendorf, F.: Upgradeability concept for collaborative blockchain-based business process execution framework. In: ICBC, Springer (2020)
12. López-Pintado, O., Dumas, M., García-Bañuelos, L., Weber, I.: Controlled flexibility in blockchain-based collaborative business processes. IS (2022)
13. López-Pintado, O., Dumas, M., García-Bañuelos, L., Weber, I.: Interpreted execution of business process models on blockchain. In: EDOC, pp. 206–215 (2019)
14. Madsen, M.F., et al.: Collaboration among adversaries: distributed workflow execution on a blockchain. In: FAB, vol. 20 (2018)
15. Malik, S., Bandara, H.D., van Beest, N.R., Xu, X.: Smart contracts' upgradability for flexible business processes. In: BPM, pp. 55–70. Springer (2024)
16. Mudge (@mudgen), N.: ERC-2535: Diamonds, Multi-Facet Proxy, Ethereum Improvement Proposals, no. 2535 (2020). https://eips.ethereum.org/EIPS/eip-2535
17. Mudge (@mudgen), N.: How diamond upgrades work (2022). https://dev.to/mudgen/how-diamond-upgrades-work-417j
18. Müller, M.: Trust mining: analyzing trust in collaborative business processes. IEEE Access **9**, 65044–65065 (2021)
19. Qasse, I., Hamdaqa, M., Þór Jónsson, B.: Immutable in principle, upgradeable by design: exploratory study of smart contract upgradeability (2024). https://arxiv.org/abs/2407.01493
20. Sola, D., Warmuth, C., Schäfer, B., Badakhshan, P., Rehse, J.R., Kampik, T.: Sap Signavio academic models: a large process model dataset (2022). https://doi.org/10.48550/ARXIV.2208.12223
21. Szabo, N.: Formalizing and securing relationships on public networks. First Monday **2**(9) (1997). https://doi.org/10.5210/fm.v2i9.548, https://firstmonday.org/ojs/index.php/fm/article/view/548
22. Tran, A.B., Lu, Q., Weber, I.: Lorikeet: a model-driven engineering tool for blockchain-based business process execution and asset management. In: BPM Forum, pp. 56–60 (2018)
23. Weber, I., et al.: Untrusted business process monitoring and execution using blockchain. In: BPM, pp. 329–347. Springer (2016)
24. Wood, G., et al.: Ethereum: a secure decentralised generalised transaction ledger. Ethereum Proj. Yellow Pap. **151**(2014), 1–32 (2014)

Balancing Confidentiality and Transparency for Blockchain-Based Process-Aware Information Systems

Alessandro Marcelletti[1](✉), Edoardo Marangone[2], Michele Kryston[3], and Claudio Di Ciccio[3]

[1] University of Camerino, Camerino, Italy
alessand.marcelletti@unicam.it
[2] Sapienza University of Rome, Rome, Italy
edoardo.marangone@uniroma1.it
[3] University of Utrecht, Utrecht, The Netherlands
{m.kryston,c.diciccio}@uu.nl

Abstract. Blockchain enables novel, trustworthy Process-Aware Information Systems (PAISs) by enforcing the security, robustness, and traceability of operations. In particular, transparency ensures that all information exchanges are openly accessible, fostering trust within the system. Although this is a desirable property to enable notarization and auditing activities, it also represents a limitation for such cases where confidentiality is a requirement since interactions involve sensitive data. Current solutions rely on obfuscation techniques or private infrastructures, hindering the enforcement capabilities of smart contracts and the public verifiability of transactions. Against this background, we propose CONFETTY, an architecture for blockchain-based PAISs to preserve confidentiality and transparency. Smart contracts enact, enforce and store public interactions, while attribute-based encryption techniques are adopted to specify access grants to confidential information. We assess the security of our solution through a systematic threat model analysis and evaluate its practical feasibility by gauging the performance of our implemented prototype in different scenarios from the literature.

Keywords: Business process management · Distributed ledger technologies · Blockchain · Attribute-based encryption · Security · Privacy

1 Introduction

Blockchain enables new forms of trustable Process-Aware Information Systems (PAISs) and inter-organizational collaborations [27]. This is possible thanks to the characteristics of public permissionless blockchains, which provide strong guarantees removing the need for third-party authorities [22]. Starting from the foundation, the distributed ledger and consensus mechanisms permit distributed and immutable information storage. These features, combined with the cryptographic primitives, ensure security properties and accountability of recorded operations. Smart contracts are programs immutably stored inside the blockchain that can encode and enforce business logic. On top of this, transparency is a key aspect, as information stored in the ledger and performed operations are accessible to everyone within the system.

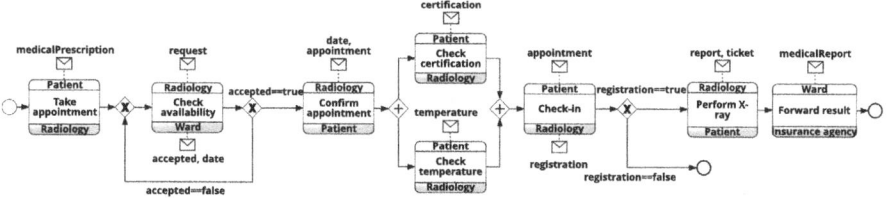

Fig. 1. BPMN choreography diagram of an X-ray diagnostic analysis [7]

Business logic enforcement and transparency are highly desirable properties for collaborative process execution and monitoring [10]. However, public observability of the whole set of exchanged data hinders the adoption of blockchain-based solutions in contexts where confidentiality is a critical requirement. Striking a balance between these two aspects is crucial in highly regulated sectors wherein sensitive data are treated (consider, for instance, the domains of the pharmaceutical supply-chain and healthcare [11]). Existing solutions to solve this conundrum employ private permissioned blockchains [6,30]. Nevertheless, these solutions may hide data from auditors and assume high robustness and security guarantees of the consortium's system. Other approaches propose encryption techniques to restrict visibility of data stored on-chain only to authorized parties [15,16,20,21]. However, their integration with process management systems is limited or non-existent.

Against this background, we propose an approach and an architecture for **blockchain-based process enactment that preserves the confidentiality of exchanged information while keeping public enforcement and transparency of process execution.** We name our approach CONFidentiality EnforcemenT TransparencY (CONFETTY). In particular, we rely on smart contracts to encode and execute business process logic while logging the interactions between parties, and we resort to Multi-Authority Attribute-Based Encryption (MA-ABE) [4] to control the access of different parties to the activities' data payloads and information artifacts, thus safeguarding confidentiality. We demonstrate the security of our approach against a threat model for the proposed architecture. We implemented our artifact and evaluated the feasibility of our solution by showing its application in the context of a healthcare scenario taken from the literature. Also, we analyzed the execution performance with collaborative processes presented in the related scientific literature.

The remainder of the paper is organized as follows. Section 2 illustrates the example and highlights the motivation for this work. Section 3 provides the analysis of the state of the art and introduces the fundamental concepts on which CONFETTY relies. Section 4 describes the proposed architecture. Section 5 evaluates our solution and provides an assessment of its security and feasibility. Finally, Sect. 6 concludes the work and draws future research directions.

2 Example, Problem Illustration and Requirements

To illustrate the problem we aim to tackle, we introduce a running example in the healthcare domain inspired by [7]. Figure 1 depicts the management process for X-ray diagnostics in the form of a BPMN choreography diagram. The patient first presents the

Table 1. Requirements and corresponding actions in the approach

Requirement	Approach
R1 Public enforcement and transparency of process execution should be guaranteed	The control-flow of the process is managed by smart contracts deployed on public blockchain
R2 The process should be independently auditable with low overhead while guaranteeing authenticity	On-chain information retains publicly available process execution tracking information and hashes with locators for associated encrypted data
R3 Information artifacts should be written in a permanent, tamper-proof and non-repudiable way	Messages are stored in a tamper-proof distributed off-chain file storage, and the resource locators are stored on-chain to keep track of information
R4 Access to information artifacts should be controlled while ensuring their overall integrity and availability	MA-ABE encrypts information artifacts stored in a distributed environment. Decryption is possible only if the requester holds the necessary attributes.
R5 User-defined policies should control access levels for authenticated users with a fine-granular scheme	MA-ABE policies are associated with messages within individual activities

medical prescription, asking for an appointment. The radiology department checks the availability of the ward, and if a date is available, the appointment is confirmed. Otherwise, there is a new tentative. After obtaining the registration details, the temperature and vaccine certification of the patient are verified. A radiology clerk checks whether the appointment can be confirmed. If so, the X-ray exam is done, and the results are provided to the ward with a final report. In the end, an insurance agency receives the patient's results. The insurance agency can access the patient's medical report to follow up with compensatory actions in the subscribed contract (e.g., the reimbursement of medical bills). Beyond the boundaries of the choreography diagram in Fig. 1, we assume the presence of an inspector of the local Ministry of Health. They are not among the process participants, yet they should have full access to the information exchanged even after the conclusion of the instance for auditing purposes.

Table 1 lists the requirements derived from the above use case. To begin with, ensuring public execution and tracking of healthcare procedures is critical to improving process efficiency and trust in the public towards the healthcare system. We concretize this aspect in (Req. R1). Enabling auditing of those procedures is crucial since it permits investigations on misconduct or lack of compliance with regulations (Req. R2). This is allowed by the data-based evidence, which should be ensured by information storage in permanent, tamper-proof and non-repudiable information artifacts (Req. R3). The guarantees offered by a public blockchain-based PAIS are fit for purpose from this standpoint.However, sensitive patient data must be shielded from unauthorized inspections (Req. R4). For instance, although the insurance agency is involved in the process, it should not have full access to all exchanged data (e.g., the vaccine certification). Publicly disclosing the result of the diagnostic analysis or information about vaccine certification would severely violate the privacy sphere of the patient. To prevent this, access should be restricted only to designated and authorized entities through a fine-grained access control mechanism (Req. R5). Balancing a secure, publicly verifiable and decentralized infrastructure with the need for data confidentiality remains an open challenge, which motivates our research.

3 Background and State of the Art

In this section, we first describe the core concepts underlying our approach, namely blockchain technology and Attribute-Based Encryption (ABE), and then illustrate how these pillars serve as the foundation for existing approaches in the literature relevant to our investigation.

3.1 Background

Blockchain technology. A ledger is an append-only singly-linked list of transactions, representing asset transfers among accounts. A blockchain collates ledger segments into blocks, with headers storing block numbers and hashes linking to previous blocks, ensuring transaction order and data integrity. Selected nodes append new blocks and receive cryptocurrency as rewards. The blockchain is replicated across all nodes, ensuring transparency. Consensus protocols and cryptographic primitives enable public verification and decentralization without third-party oversight. Several platforms implement the blockchain protocol (e.g., Ethereum [31]). Some blockchains allow for the deployment of *smart contracts* [13] which are programs deployed onto the blockchain and executed within a dedicated virtual machine (e.g., the Ethereum Virtual Machine, EVM). Smart contracts functions are invoked through transactions and their core features are code immutability and invocation storage on the blockchain, making such information available for auditing purposes [10]. Smart contracts execution incurs a cost on the invoker. The cost is measured in gas units and depends on the complexity of the called function and the size of the exchanged data. To save on the latter, distributed tamper-proof storage systems are often employed to store the actual data in a non-modifiable file, while a permalink to that file is saved in the smart contract state. The InterPlanetary File System (IPFS) [1] is a typical example of such a system. Based on a peer-to-peer network wherein each node stores chunks of the data, IPFS binds every file with a string used both as a unique identifier and resource locator. The string is based on the hash of the file content itself so that if the file is altered, the permalink does not match. As we will see next, these features fostered the development of blockchain-based PAISs using smart contracts as the backbone for business (process) rule enforcing (in our example, the BPMN choreography is translated into smart contract code, and its steps correspond to invocations to its functions) [7,9].

Attribute-based Encryption. ABE is a public key encryption where the ciphertext, an encrypted plaintext, and the corresponding decryption key are linked through custom boolean attributes [25]. In Ciphertext-Policy Attribute-Based Encryption (CP-ABE) [2], a variant of ABE, every user bears a set of attributes encoded in their key. For example, the owner of the blockchain account 0xB0...1AA1 can be endowed with attributes `Patient` and `PID476948` specifying role and process instance 476948, respectively. Data owners write policies to be attached to the CP-ABE-encrypted document (the *ciphertext*). Those policies express conditions for granting in-clear access to the data. Policies are propositional logic formulae employing ABE attributes as literals. The policy is evaluated on attributes associated with a potential reader: Only if the formula is satisfied can the user's key decrypt the data. For instance, the key of the user 0xB0...1AA1 can decrypt data locked with a policy like `PID476948 and PATIENT`, but cannot unlock the contents of a document associated with a policy like `PID476949 and RADIOLOGY`. Attributes are typically assigned by one authority: Multi-Authority Attribute-Based Encryption (MA-ABE) [4] is a variant of ABE that removes this single point of failure using a multi-authority method. In MA-ABE, every authority creates a part of the decryption key. Once collected the user merges all parts to obtain the final key. In this paper, we employ the Ciphertext-Policy variant of MA-ABE [17]. It integrates MA-ABE with the aforementioned policy-based approach.

3.2 State of the Art

Over the last few years, several solutions that automate collaborative processes using blockchain technology have been designed [9,18,27,29]. Such approaches demonstrated the effectiveness of blockchain-based solutions in improving trust among participants in multi-party collaborations and allowing for monitoring and auditing [8,10]. These studies enhance the integration of blockchain technology with process management, unlocking security and traceability benefits. However, they primarily focus on the control-flow perspective and lack mechanisms for secure access control to the stored data on public platforms.

A relevant area of research for our work pertains to data privacy and integrity within blockchain systems. Several studies examine the application of encryption techniques to achieve this goal. Hawk [14] employs user-defined private smart contracts to automate the implementation of cryptographic protocols. Pham et al. [24] propose a decentralized storage solution that integrates InterPlanetary File System (IPFS), MA-ABE, and blockchain technology. Hong et al. [12] introduce a data-sharing solution based on decentralized ABE (MA-ABE), blockchain, and IPFS. Bramm et al. [3] present BDABE, an access control mechanism that applies a distributed ABE scheme within a consensus-driven infrastructure for real-world applications. Yan et al. [33] introduce a scheme for fine-grained access control by implementing proxy encryption and decryption while supporting policy hiding and attribute revocation. The encrypted data is stored on IPFS, and the metadata is stored on the blockchain. All these studies, along with the aforementioned [19–21], cater for secure access control on blockchains but lack the integration of this technology with process management systems.

To the best of our knowledge, this is the first work that focuses on combining blockchain-based process execution engines with encryption schemes that guarantee data confidentiality and access control over data shared on public blockchain platforms. In the next section, we provide a description of our solution.

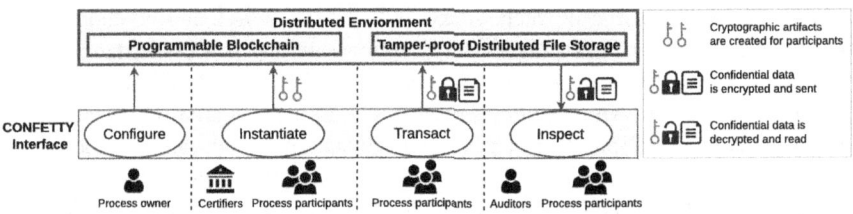

Fig. 2. A graphical sketch of the CONFETTY functionalities, users, artifacts, and interfaces

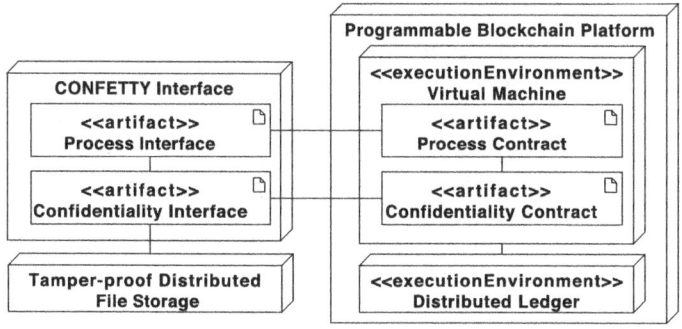

Fig. 3. The deployment diagram of CONFETTY

4 The CONFETTY Architecture

In this section, we describe the proposed CONFETTY architecture. We first introduce its core components and functionalities (Sect. 4.1). Then, we describe how components interact and manage information to support confidential-preserving blockchain-based process execution while meeting Reqs. R1 to R5 (Sect. 4.2). Due to space restrictions, we do not enter the technical details of our approach but provide an overarching presentation of our solution's rationale and key concepts, motivating them with the requirements they aim to meet.

4.1 Functional Viewpoint

Figure 2 provides an overview of the CONFETTY approach, which caters for the following main functionalities. *(i)* The **configure** functionality helps the process owners specify the process to run on-chain. It requires the registration of process information, including activities to execute, data to exchange, control-flow routing, the expected participants' roles (like the information enclosed in the choreography diagram in Fig. 1) and how they can operate with confidential data (via (MA-ABE) policies) to control information access. *(ii)* The **instantiate** functionality lets certifiers authenticate the participants, define their role in the process (e.g., that user 0xB0...1AA1 is a patient), and kickstart a new process instance on-chain by setting its public state. *(iii)* The **transact** functionality allows process participants to interact, following the process, to exchange data at run time. This functionality manages both public and confidential data, updating and enforcing the process state according to the logic in the blockchain. *(iv)* The **inspect** functionality allows participants and auditors to access public and confidential data.

The functionalities above are realized by the CONFETTY main components depicted in Fig. 3. CONFETTY relies on two external components used as architectural buttresses: *(i)* A **Programmable Blockchain Platform** (like Ethereum), *(ii)* and a **Tamper-proof Distributed File Storage** (like IPFS). The former maintains the public state, allowing for the creation of logic and constraints while notarizing the executed operations. It comprises two core elements: The Distributed Ledger, storing the transactions, and the Virtual Machine, hosting and running smart contracts. The latter (which we will henceforth name as Distributed FS for brevity) stores business data and thus improves the performance and costs of a blockchain as in common practice. The data is saved on the external storage (upon encryption, if it must remain secret), while a resource locator with the hash of that data is kept on-chain for notarization and future retrieval. Notice that we take inspiration from known patterns adopting Distributed FSs for large data storage [32]. Based on this known pattern, we devise new techniques to preserve secrecy and control access to that process data at runtime. The primary components of CONFETTY are the following: *(i)* The **Process Interface** handles all the operations involving the participants and related to the process. This is done by acting as an interface for the blockchain, abstracting from the technical implementation and exposing functionalities to the users. In particular, the Process Interface instantiates the process on the blockchain on given process specification data. To support the *instantiate* functionality, it provides capabilities for role assignment to participants. *(ii)* The **Process Contract** is a smart contract that manages the process instances and their execution. It elaborates and manages all process specifications and their instances' public states, acting as a software realization both of factory and proxy patterns [32]. It supports the *instantiate* functionality by enforcing the control flow, the data flow, the assignment of tasks to participants, and the routing of decision points.In this way, the Process Contract updates the public state of the process and generates transparent records that are accessible and verifiable by interested auditors. *(iii)* The **Confidentiality Interface** is in charge of handling the participants' authorizations and the operations over confidential data. It operates as a façade towards the blockchain to store access grants, distributes decryption keys to the users solely based on their roles and involvement in the process, and handles the storage and retrieval of encrypted data. *(iv)* The **Confidentiality Contract** is a smart contract handling the notarization of authorizations for confidential data, as well as for their writing and access. We conclude this subsection with a few considerations about the users and trust model. We assume that the users we mentioned while discussing the functionalities (i.e., process owners, certifiers, process participants, and auditors) possess a blockchain account with a key pair to be able to publicly identify themselves (as with the aforementioned user 0xB0...1AA1) and sign blockchain transactions built through the CONFETTY modules. We also remark that we consider the CONFETTY platform as a trusted party, while users are assumed to be *honest but curious*. Finally, notice that the involvement of a pre-appointed third-party certifier to authenticate process participants is necessary as a user alone can confirm their identity (e.g., via the aforementioned public-private key scheme) but cannot attest alone to the truthfulness of assertions regarding themselves. The selection and appointment of certifiers transcends the goals of this paper, but in our motivating scenario, e.g., certifiers can be trusted third parties like the Public Ministry of Health (for hospital personnel), the Public Registry (for citizens), and the Insurance Registrar (for agencies), or subsidiaries thereof. In our architecture, certifiers operate like blockchain oracles [23].

4.2 Information Viewpoint

After providing an overview of the CONFETTY architecture, we describe how information is handled in realizing the aforementioned functionalities by the CONFETTY components.

Table 2. Access policies. `$PID` represents a placeholder for the identifier of the process instance before its assignment

Message	Access policy
Medical prescription	`MINISTRY-INSPECTOR or ($PID and (PATIENT or RADIOLOGY))`
Registration	`MINISTRY-INSPECTOR or ($PID and (RADIOLOGY or PATIENT))`
Report	`MINISTRY-INSPECTOR or ($PID and (RADIOLOGY or WARD))`

The first functionality is *configure*. Here the process owner sends the process specification files (like the diagram in Fig. 1) to the Process Interface. The Process Interface, in turn, processes the input data, storing the behavioral and business constraints it reports. Also, it builds a set of parametric access policies, associating roles of the process with the data they can access via encryption directives (meeting Req. R5). Once created, the policies are returned to the process owner. The one associated with the medical prescription (sent by the patient to the radiology clerk to take an appointment in Fig. 1), e.g., reads "`$PID and (PATIENT or RADIOLOGY)`". Here, `PATIENT` and `RADIOLOGY` are the attributes representing the actors' roles in the process. Not all patients and radiology clerks should access any medical prescription. Hence `$PID`, an attribute placeholder for the process instance identifier (a piece of information that can be known only at runtime), restricting access to the sole actors involved in that process instance. We give the process owner the option to revise the policies by adding custom roles that are not expected active parties in the process but need access to data (as per Req. R2). This is the case, for instance, of external auditors who need to certify data and its compliance with norms, laws and regulations. Considering the medical prescription, the revised policy specifies that a user has to be attested to have the role of a `MINISTRY-INSPECTOR`, or again to be a `PATIENT` or `RADIOLOGY` within the running process instance: "`MINISTRY-INSPECTOR or ($PID and (PATIENT or RADIOLOGY))`". Table 2 shows an extract from the final access policies of our running example. Once policies are confirmed, they are sent to the Confidentiality Interface, which stores them in the Distributed FS. Their permalink is passed back to the Confidentiality Interface to save it.

Figure 4 depicts the *instantiate* functionality. Differently from the *configure*, needed once for each process, this one is designed for every new instantiation thereof. To initialize a new process instance (e.g., number 476948), participants send their data to the Process Interface (1) to authenticate themselves as participants in the process with their role (e.g., that user `0xc5...Fd43` operates as a `WARD` clerk in a new process instance `PID476948`). (2) The authentication is attested to by certifiers confirming the user's self-declared roles. When all process participants are identified and associated

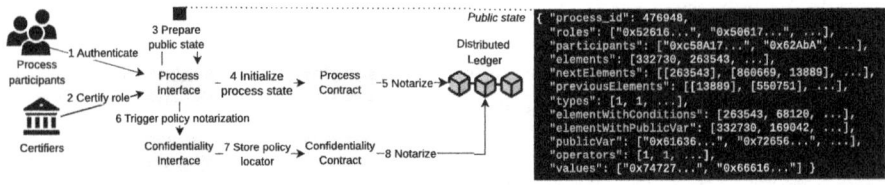

Fig. 4. Communication diagram of the *instantiate* functionality, with a snapshot of public state

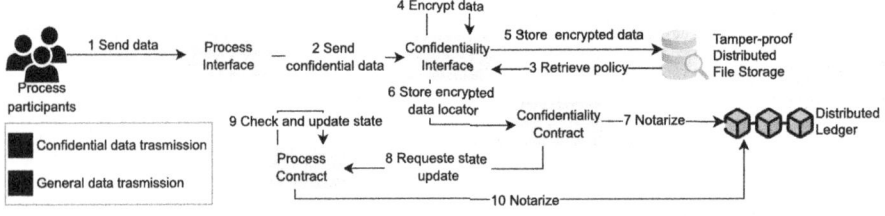

Fig. 5. Communication diagram of the *transact* functionality

with a role, (3) the Process Interface merges the received information with the process data acquired in the configuration. Thereupon, (4) a selected participant kickstarts a new process instance through the Process Interface associated with the aforementioned information on the Process Contract, thereby initializing the instance's public state. This step is done by sending a transaction to the Process Contract with the public state of the process, including conditions driving the control flow, such as exclusive gateway choices (as per Req. R1). Notice that this crucial step is notarized on chain (5) (meeting Req. R2). The selection of the participant, or participants, on behalf of whom the transaction is signed, transcends the scope of this paper and depends on the decision process and guidelines of the consortium. After initializing the process state, (6) the Process Interface triggers the Confidentiality Interface, which (7) stores the policy locator previously saved on the Confidentiality Contract. Finally, (8) the Confidentiality Interface notarizes the operation.

Figure 5 displays the *transact* functionality, which represents the exchange of information during the execution of an activity in the process instance. It encompasses a public and a confidential case, depending on the process participants and their willingness to classify information as disclosable or not. In both cases, this functionality updates the public state in the blockchain according to previous enforcement as per Req. R1 (e.g., activity to execute, expected participant) and manages accessibility (meeting Req. R4).

Let us begin with the exchange of public data (and the update of the public state). The main steps are highlighted in blue in Fig. 5 and salient on-chain operations are reported in Alg. 1. Initially, a user sends the execution data to the Process Interface in the context of an activity (Req. R1). For example, the WARD clerk identified by 0xc5...Fd43 replies to the "Check availability" activity request by the RADIOLOGY user within process instance 476948. To do so, the clerk sends a "Check availability" response message wherein the accepted variable is assigned with true

Algorithm 1: On-chain logic for public state update and next element activation

```
1  Function updatePublicState(instanceID, messageID, variables):
2      authUser ← instances[instanceID].elements[messageID].role
3      element ← instances[instanceID].elements[messageID]
4      if sender is authUser & element.type is MESSAGE & element.state is ENABLED then
5          foreach variable in variables do  instances[instanceID].publicVars[variable.name] ← variable.value
6          activateNext(instanceID, messageID)

7  Function activateNext(instanceID, messageID):
8      currentElement ← instances[instanceID].elements[messageID]
9      instances[instanceID].elements[messageID].state ← COMPLETED
10     foreach nextElement in currentElement.nextElements do
11         if nextElement is GATEWAY then  executeGateway(instanceID, nextElement.ID)
12         else  nextElement.state ← ENABLED
```

since the appointment is confirmed alongside an available date. The boolean answer (accepted) is public data that is used later to make control-flow decisions and advance the state of the process, defining the next activity to perform. In our scenario, the exclusive choice gateway is based on the value of the accepted variable. Then, the Process Interface invokes the Process Contract to request a state update, sending the execution data (Line 1). In turn, the Process Contract enforces the business and control-flow logic through the run of its routine (based on previously agreed upon and immutable code, see *configure* functionality), which checks (Line 4) if the executed activity is enabled and if the message sender (in this case, user 0xc5...Fd43) is the expected one (a WARD), finally updating the state (Line 5) if the verification outcome is positive. This step advances also the state of the process (Line 6), updating the expected next element (here, "Confirm appointment") and participant (RADIOLOGY). In case the element is a gateway, its logic is automatically executed to advance the control flow of the process (Line 11). In the end, this piece of information is notarized into the Distributed Ledger (Req. R2).

For the confidential case, we refer again to Fig. 5 (see the marks in red). Overall, (1) once the Process Interface has received the data, (2) it forwards it to the Confidentiality Interface. The latter (3) retrieves the policy from the Distributed FS, and (4) uses it to encrypt the data via MA-ABE (as per Req. R4). For example, considering Fig. 1 and Table 2, the medical prescription of the patient is encrypted with the following policy: {MINISTRY-INSPECTOR or (PID476948 and (PATIENT or RADIOLOGY))}. In particular, the Confidentiality Interface retrieves the specific authorization for the data to be stored and embeds it in the encryption mechanism, ensuring that only authorized participants can access it. This step is needed only in confidential cases, as in public ones, information is directly stored on the blockchain and intentionally kept publicly readable. Once encrypted, (5) the Confidentiality Interface stores data in the Distributed FS, (6) its locator in the Confidentiality Contract, and (7) notarizes the operation in the Distributed Ledger (meeting Req. R2 and Req. R3). In this case, the resource locator that contains the encrypted medicalPrescription (e.g., LXBw...2DAx) is Qmo9...0weI.). In this case, the resource locator that contains the encrypted medicalPrescription (e.g., LXBw...2DAx) is Qmo9...0weI.

The execution of this routine by the Confidentiality Contract includes a call to the Process Contract (8) to request a state update enclosing the execution data. Following the example, this data is the activity being executed ("Check availability"), the user that performs it (e.g., PATIENT, along with their corresponding address 0x2e...6dd9), and the process instance (e.g.,476948). Then, (9) the Process Contract checks the current public state to update it. The operations involving the smart contracts (10) are notarized in the Distributed Ledger.

The *inspect* functionality pertains to the access to message data. The public data can be freely retrieved, while the confidential data is restricted to the sole authorized users included in the access policy provided during the encryption phase (see the *configure* functionality above) Req. R5. To read public data, a process participant or an auditor may send a request to the Process Interface, which in turn retrieves the necessary information from the Process Contract and sends it back to the requester. This is the case, e.g., for the accepted field of the "Check availability" task. Reading confidential data requires more passages and a preliminary step. All users should request at least once (and at every attribute update) a decryption key, to be used for every message they are going to read. For the key request, the process participant (e.g., a WARD clerk) or an auditor (e.g., a HEALTHCARE-INSPECTOR) ask for a decryption key to the Confidentiality Interface. The latter forwards the request to the Confidentiality Contract, which is thus notarized on the Distributed Ledger. Then, the Confidentiality Interface generates the attribute-based key (or *a-b key* for short) based on the attributes of the user requesting it (e.g., WARD and PID476948, or HEALTHCARE-INSPECTOR) via MA-ABE, and sends it back. The aforementioned step enables the decryption of confidential data. Once the process participants have received the a-b key, they can use it to decrypt the confidential data. To do so, a user makes a request to the Confidentiality Interface. Then, the Confidentiality Interface retrieves the encrypted data from the Distributed FS and forwards it to the user. At this point, the latter uses the a-b key to decrypt the confidential data via MA-ABE. Only if the user's attributes satisfy the policy previously used to encrypt the data can the user's a-b key decrypt the confidential data (and the user inspects it in clear).

A note on decryption keys. We remark that an a-b key stays with the user and lasts for as long as their attributes are not updated (if a user is involved in a new process instance, say 476949, a new attribute gets associated with them: PID476949). The same key will be used to access all messages. For example, consider user 0xB0...1AA1 (who holds the attributes PID476948 and PATIENT). Per the policies in Table 2, their a-b key can decrypt the medical prescription and the registration, but not the report. The a-b key of the user 0xc5...Fd43 (PID476948 and WARD) can decrypt the report instead, but not the other two documents. Notice that a HEALTH-INSPECTOR, instead, can use their unique a-b key to access all the aforementioned documents regardless of the process instance they pertain to. The key, in other words, is associated with the user, and it is the sole instrument to access a document. Therefore, the Confidentiality Interface does not need to generate a different decryption key for every document and give the same copy to multiple users, nor create numerous copies of the same document for different decryption keys (one per user), thus saving unnecessary data replications and key distributions, and ensuring integrity (Req. R4). Also, as the key stays with the users, and the documents are stored and notarized on tamper-proof external infrastructures (the Distributed FS and Blockchain Platform, respectively), the information remains available even should the CONFETTY system stop functioning (Req. R3). These aspects have an impact on security considerations, as we discuss next.

5 Evaluation

To assess the feasibility of the CONFETTY architecture, we perform a two-step evaluation. First, we analyze the CONFETTY threat model, evaluating it against the proposed architecture. Then, we describe the implemented prototype and analyze its performance. The source code of CONFETTY, alongside the experimental data we use and performance analysis results, are available at https://doi.org/10.5281/zenodo.15482587.

5.1 Threat Model

In this section, we identify potential security threats that could compromise data availability (Req. R2) and confidentiality (Req. R5) within our solution, and outline how the CONFETTY architecture addresses them. Our threat analysis is based on the STRIDE framework [26], a theoretical model that categorizes threats into six distinct groups: spoofing (i.e., impersonation of a legitimate entity), tampering (i.e., unautho-

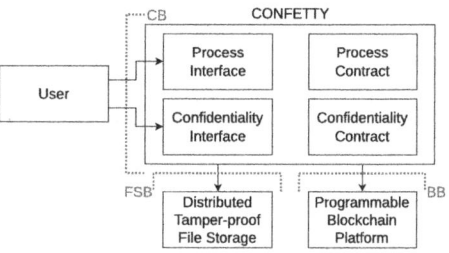

Fig. 6. The CONFETTY threat model, based on the components in Fig. 3

rized modification of data to compromise its integrity), repudiation (i.e., the denial of having performed a particular action), information disclosure (i.e., the unauthorized exposure of sensitive data), denial of service (DoS, i.e., the disruption or degradation of system availability), and elevation of privileges (i.e., the unauthorized acquisition of higher-level access rights).We assume users and components to be honest but curious and trusted, respectively. The possible threats thus reside in the communication between them, subject to potential attacks (see Sect. 4.1). Figure 6 displays a Data Flow Diagram (DFD), in which the orange dotted lines indicate the trust boundaries crossing data exchange points between different perimeters, represented by black arrows connecting different elements. The opening side of each boundary indicates the trusted component in the information exchange.

The **CONFETTY Boundary (CB)** indicates that CONFETTY is the trusted component in the communication with a user (let it be a certifier, a process participant or owner, or an auditor). In this case, a spoofing attack could result in a user impersonating another. Still, this attack is prevented since we assume that each user is linked to the Blockchain Platform credentials. As for tampering, trust is based on the fairness of the user who sends the data. It cannot be the case for a user to repudiate an action because all of them are recorded on the blockchain with a transaction signed with the user's private key. An attacker could sniff the data a user sends to CONFETTY. However, we resort to TLS channels for those communications, specifically designed to prevent information leakage [28]. DoS attacks are, in principle, feasible unless we pre-filter the range of potential clients invoking the CONFETTY platform. Indeed, all functionalities foresee an initial data input from users (see, e.g., step 1 in Fig. 5) which malicious nodes could exploit with multiple calls and bulky payloads. However, DoS attacks would lead to a disruption or degradation of the system's availability. Yet, all data, state, and other process instances' information can be retrieved from the Blockchain Platform and the Distributed Tamper-proof File Storage to reactivate or replicate the system without any data loss. Information disclosure is prevented because all sensitive information is encrypted via MA-ABE. Elevating a user's privileges is impossible since attributes and permissions are linked to a blockchain account.

The Blockchain Platform is the trusted component in the **Blockchain Boundary (BB)**. In this case, a spoofing attack could result in a malicious version of CONFETTY impersonating the original software object. This attack is prevented since CONFETTY operates as a gateway for transactions ultimately signed by users via their private keys. Information disclosure can be achieved only on public data or resource locators of sensitive data. However, the latter is stored upon MA-ABE-encryption, making its content unreadable by malicious users. Other possible attacks (DoS, tampering, repudiation, elevation of privileges) are counteracted by the design of blockchain protocols. As for the Distributed Tamper-proof **File Storage Boundary (FSB)**, tampering or DoS attack and elevation of privileges are counteracted by design, too. Spoofing and repudiation attacks are ineffective since the authorship of the provided information is proven by signatures of transactions stored on-chain. Since we resort to MA-ABE encryption, information disclosure is negated since the sensitive data stored is encrypted.

5.2 Implementation

CONFETTY is deployed on three main tiers. An application server hosts the Confidentiality Interface and the Process Interface, developed in Python 3.6.9 and Java SDK-13, respectively. Users communicate with it via TLS communication channels [28]. We utilize IPFS[1] as the Distributed Tamper-proof File Storage. We employ Sepolia and Ganache as Programmable Blockchain Platforms with their Ethereum Virtual Machine (EVM) as a test environment and local RPC, respectively, connected via Web3.js[1].

5.3 Execution Analysis

Motivated by Req. R2, we evaluated our prototype in terms of time and cost according to the four key functionalities of CONFETTY (see Sect. 4.1). We set the number of (MA-ABE) authorities to four and performed the experiments on a machine equipped with a processor Intel(R) Core(TM) i7-13700H CPU @ 2.40 GHz with 14 cores and 32GB of RAM. To isolate the architecture functionalities, we excluded human interactions and the time needed for blockchain-specific operations. Such operations strictly depend on the blockchain protocol and the platform adopted, representing technical choices that are out of the scope of this work.

---

[1] IPFS: https://ipfs.tech/; Sepolia: https://sepolia.dev/; Ganache: https://archive.trufflesuite.com/ganache/; Web3.js: https://web3js.readthedocs.io/en/v1.10.0/. Accessed: August 24, 2025.

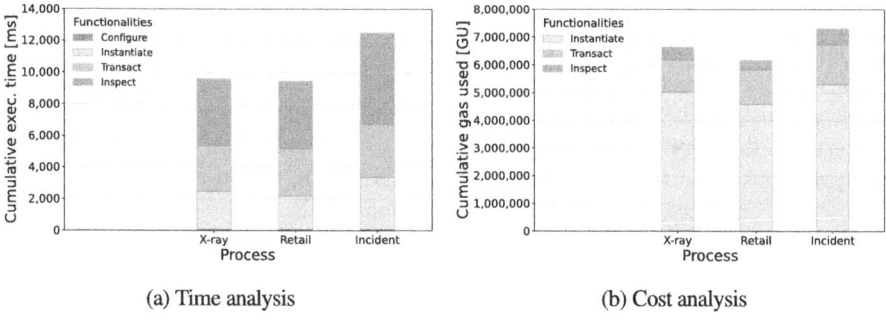

(a) Time analysis (b) Cost analysis

Fig. 7. Performance comparison

Execution Performance and Expenditure. We gauged the time and cost of our prototypical implementation against data simulating the execution and message exchange of processes from the literature. Figure 7 illustrates the results, averaging over runs that took the longest path. We gathered data using a local RPC as a stub for the Programmable Blockchain Platform.

Firstly, let us focus on simulations of our motivating scenario (Sect. 2), which we will henceforth name X-ray process. For the time evaluation, we ran an instance for the longest path five times. We employed the Sepolia testnet as the Blockchain Platform and passed input data based on real-world files in the healthcare domain upon anonymization (available in our codebase). The *configure* functionality turns out to be the fastest, requiring an average of 71 ms. For the *instantiate* functionality, we let four process participants interact with the tool as specified in the choreography and it averaged 2689 ms. This higher amount depends on the need to authenticate process participants and prepare the data to send on-chain.

For the *transact* functionality, the runtime required an average of 2476 ms in total, with the encryption of single messages averaging 247 ms. Finally, the *inspect* functionality required an average of 4390 ms including the key request management. However, notice that this step was performed only once and required 2695 ms. Once the key is received, the single reading and decryption operations for a message averaged 169 ms. Overall, if compared to the expected time for Ethereum block creation (around 12 s), the overhead introduced by CONFETTY and related operations (e.g., encrypt and decrypt of message payloads) is orders of magnitude smaller (milliseconds) and can therefore be considered acceptable. To analyze the cost, we considered the sole operations that involve the blockchain, thus, *instantiate*, *transact*, and *inspect*. The highest cost is associated with the *instantiate* functionality, consuming around 5 million Gas Units (henceforth, GU), which corresponds to the total amount for the instantiation of the public state (2931698 GU) and other related operations (each averaging 90577 GU). *Transact*ing consumed 1689190 GU in total for all the information exchanges, with a cost for a single transaction averaging at 115096 GU. Another relevant cost is the deployment of smart contracts, consuming around 5 million GU. However, this operation is performed only once during the bootstrapping of CONFETTY and does not pertain to single process instantiations. Finally, *inspect* required 89535 GU for the key request management.

The reading operations do not produce transactions, thus incurring no fees. In total, the execution of the X-ray process, without considering the deployment, consumed a total of 6884376 GU which translates to 0.105035017 ETH in fees. However, to evaluate the actual cost in fiat currency, the fee value has to be related to a specific blockchain platform, as this aspect significantly influences the multiplicative factor. Considering Polygon blockchain, the above fee of 0.105035017 POL would result in 0.02 US$[2], representing an affordable cost considering the obtained benefits. Taking as a reference another blockchain-backed entity like a standard Ethereum Non-Fungible Token (NFT), it requires around 2 million GU for deployment and 50000 to 200000 GU for transfer or minting operations [5]. In this context, CONFETTY exhibits a higher deployment cost (performed only once for kickstarting the entire CONFETTY system) given by the more complex smart contracts logic, while its execution cost is in line with an NFT, resulting in a cost-efficient performance.

Let us now compare the above results with two other processes from the literature, namely retail [6] and incident management [8]. The X-ray, incident and retail processes have similar structures in the number of messages (10, 13 and 12, respectively), gateways (5, 3 and 2) and involved participants (4, 5 and 3) and exhibit similar trends from both runtime and cost standpoints. The incident management process is the slowest and most expensive one with an 8cost compared to the X-ray process, respectively. The experiments show that CONFETTY performs comparably to literature tools and scenarios. Indeed, ChorChain [8] required approximately 5 million GU for the incident management process, while MultiChain [6] required around 6 million GU for the retail process. Notice that such approaches only focus on public execution and enforcement without incorporating any confidential mechanisms and that they deploy a contract for each new instance.

Scalability Assessment. Lastly, we analyzed the scalability of CONFETTY. To this end, we considered the dependence of time and cost taking as independent variables the dimensions of the process affecting the input size for our tool. In this section, we focus on the number of process participants and the choreography model size due to space restrictions as they evidence the most salient results. Further experiments including other variables like the number of gateways (parallel and exclusive) and the message payload size can be found in the online repository. The default values for the variables adhere to the example provided in Fig. 1. For every test, we let one variable change the value and keep the others fixed, assigned with the respective defaults. Figures 8 and 9 dissect the registered timings and costs to serve the different functionalities concurring to the total by stacking the respective subtotals.

Figure 8 depicts the performance trend varying the number of process actors actively sending messages from two (the minimum for a multi-party process), to ten (the maximum admissible within the X-ray choreography, with one different sender per message and activity). Figure 8(a) depicts the trend for cumulative time performance, showing a total variation of 3.9 s, mostly related to the *instantiate* functionality as it includes the authentication of all the participants. The costs shown in Fig. 8(b) follow a comparable linear trend.

---

[2] The conversion rate is of 0.21 US$ at the time of writing (March 14, 2025).

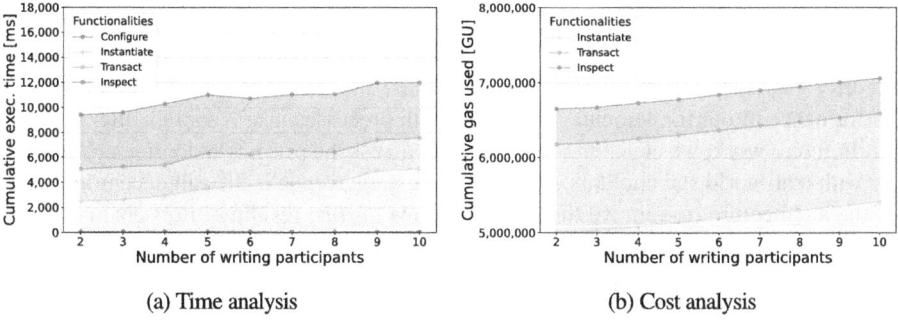

Fig. 8. Performance with increasing process participants

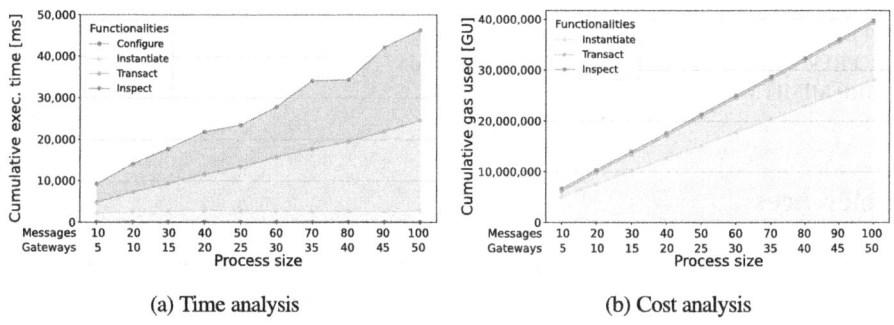

Fig. 9. Performance with increasing choreography size

Figure 9 displays the results observed varying the process model size. We progressively increased it by sequentially concatenating the whole choreography diagram multiple times to derive a new one, thus scaling the size up to ten times the original one. Figure 9(a) shows a linear growth in the runtime of the *transact* and *inspect* functionalities, without a significant impact on the remaining one, *instantiate*. The execution time increases with the number of messages due to additional read and write operations. Figure 9(b) shows the cost increase, with a linear growth registered for the *instantiate* and *transact* functionalities. More elements increase on-chain storage needs and hence gas cost. Read operations remain constant as they only handle one-time key management on-chain (hence the flat cost associated with *inspect*).

6 Conclusions

In this work, we presented CONFETTY, a blockchain-based architecture for PAISs. CONFETTY enables transparency and public enforcement for process execution while preserving the confidentiality of sensitive exchanged data. A public state saved on-chain maintains process data for enforcement and auditability purposes. The evolution of the public state is mediated by smart contracts implementing the business process control- and data-flow logic. We use MA-ABE encryption to hide sensitive information and grant access only to authorized parties based on their roles, encoded as attributes in access policies and embedded directly in their decryption keys. We confirmed its security guarantees with a threat model analysis, and demonstrated the time and cost performance of our implemented prototype with processes taken from the literature.

In future work, we envision a field study to assess the practical adoption and usability with real-world stakeholders. Another interesting avenue is the full decentralization of the architecture (to remove the existing points of trust residing in off-chain components) and governance (to cater for multiple process owners during the instantiation phase). Conducting a security analysis of our approach against stronger adversarial models is an intriguing future endeavor, too. Finally, we envision a study and design of a solution to enable automated decision support via computation held on encrypted data.

Acknowledgments. work was partly funded by projects PINPOINT (B87G22000450001) under the PRIN MUR program, and Health-e-Data, funded by the EU-NGEU under the Cyber 4.0 NRRP MIMIT programme.

References

1. Benet, J.: IPFS - content addressed, versioned, P2P file system. CoRR **abs/1407.3561** (2014)
2. Bethencourt, J., Sahai, A., Waters, B.: Ciphertext-policy attribute-based encryption. In: SP, pp. 321–334 (2007)
3. Bramm, G., Gall, M., Schütte, J.: BDABE - blockchain-based distributed attribute based encryption. In: SECRYPT, pp. 99–110. INSTICC, SciTePress (2018)
4. Chase, M.: Multi-authority attribute based encryption. In: TCC, pp. 515–534. Springer (2007)
5. Choi, W., Woo, J., Hong, J.W.K.: Gas cost analysis of fractional NFT on the ethereum blockchain. In: International Conference on Blockchain and Cryptocurrency, pp. 1–6. IEEE (2023)
6. Corradini, F., et al.: Model-driven engineering for multi-party business processes on multiple blockchains. Blockchain: Res. Appl. **2**(3), 100018 (2021)
7. Corradini, F., Marcelletti, A., Morichetta, A., Polini, A., Re, B., Tiezzi, F.: A flexible approach to multi-party business process execution on blockchain. Fut. Gen. Comp. Syst. **147**, 219–234 (2023)
8. Corradini, F., et al.: Engineering trustable and auditable choreography-based systems using blockchain. ACM Trans. Manage. Inf. Syst. **13**(3) (2022)
9. Di Ciccio, C., et al.: Blockchain support for collaborative business processes. Informatik Spektrum **42**, 182–190 (2019)
10. Di Ciccio, C., Meroni, G., Plebani, P.: On the adoption of blockchain for business process monitoring. Softw. Syst. Model. , 1–23 (2022). https://doi.org/10.1007/s10270-021-00959-x
11. Ghadge, A., Bourlakis, M., Kamble, S., Seuring, S.: Blockchain implementation in pharmaceutical supply chains: A review and conceptual framework. IJPR **61**(19), 6633–6651 (2023)
12. Hong, L., Zhang, K., Gong, J., Qian, H.: A practical and efficient blockchain-assisted attribute-based encryption scheme for access control and data sharing. Sec.Commun.Netw. **2022**(1), 4978802 (2022)

13. Khan, S.N., Loukil, F., Ghedira-Guegan, C., Benkhelifa, E., Bani-Hani, A.: Blockchain smart contracts: applications, challenges, and future trends. Peer-to-Peer Netw. Appl. **14**(5), 2901–2925 (2021). https://doi.org/10.1007/s12083-021-01127-0
14. Kosba, A., Miller, A., Shi, E., Wen, Z., Papamanthou, C.: Hawk: The blockchain model of cryptography and privacy-preserving smart contracts. In: SP, pp. 839–858. IEEE (2016)
15. Köpke, J., Meroni, G., Salnitri, M.: Designing secure business processes for blockchains with SecBPMN2BC. Future Gener. Comput. Syst. **141**, 382–398 (2023)
16. Lin, C., He, D., Zeadally, S., Huang, X., Liu, Z.: Blockchain-based data sharing system for sensing-as-a-service in smart cities. ACM Trans. Internet Technol. **21**(2) (2021)
17. Liu, Z., et al.: Multi-authority ciphertext policy attribute-based encryption scheme on ideal lattices. ISPA/IUCC/BDCloud/SocialCom/SustainCom,pp. 1003–1008 (2018)
18. López-Pintado, O., García-Bañuelos, L., Dumas, M., Weber, I., Ponomarev, A.: Caterpillar: a business process execution engine on the ethereum blockchain. Softw., Pract. Exper. **49**(7),pp. 1162–1193 (2019)
19. Marangone, E., Di Ciccio, C., Friolo, D., Nemmi, E.N., Venturi, D., Weber, I.: Enabling data confidentiality with public blockchains (2023)
20. Marangone, E., Di Ciccio, C., Friolo, D., Nemmi, E.N., Venturi, D., Weber, I.: MARTSIA: Enabling data confidentiality for blockchain-based process execution. In: International Conference on Enterprise Design, Operations, and Computing, pp. 58–76 (2023)
21. Marangone, E., Di Ciccio, C., Weber, I.: Fine-grained data access control for collaborative process execution on blockchain. In: BPM Blockchain and RPA Forum, pp. 51–67. Springer (2022)
22. Mendling, J., et al.: Blockchains for business process management - challenges and opportunities. ACM Trans. Manag. Inf. Syst. **9**(1), pp.4:1–4:16 (2018)
23. Mühlberger, R., et al.: Foundational oracle patterns: connecting blockchain to the off-chain world. In: BPM: Blockchain and RPA Forum, pp. 35–51 (2020)
24. Pham, V.D., et al.: B-box - a decentralized storage system using IPFS, attributed-based encryption, and blockchain. In: International Conference On Computing And Communication Technologies (RIVF), pp. 1–6 (2020)
25. Sahai, A., Waters, B.: Fuzzy identity-based encryption. In: EUROCRYPT, pp. 457–473. Springer-Verlag (2005)
26. Scandariato, R., Wuyts, K., Joosen, W.: A descriptive study of microsoft's threat modeling technique. Requir. Eng. **20**(2), 163–180 (2015)
27. Stiehle, F., Weber, I.: Blockchain for business process enactment: a taxonomy and systematic literature review. In: BPM Blockchain and RPA Forum. LNBIP, vol. 459, pp. 5–20. Springer (2022)
28. Thomas, S.: SSL and TLS essentials: securing the web (2000)
29. Tran, A.B., Lu, Q., Weber, I.: Lorikeet: a model-driven engineering tool for blockchain-based business process execution and asset management. In: BPM Demos, pp. 56–60. Springer (2018)
30. Uddin, M.: Blockchain medledger: hyperledger fabric enabled drug traceability system for counterfeit drugs in pharmaceutical industry. Int. J. Pharm. **597**, 120235 (2021)
31. Wood, G.: Ethereum: a secure decentralised generalised transaction ledger (2014)
32. Xu, X., Weber, I., Staples, M.: Blockchain Patterns, pp. 113–148. Springer Int. Publishing (2019)
33. Yan, L., Ge, L., Wang, Z., Zhang, G., Xu, J., Hu, Z.: Access control scheme based on blockchain and attribute-based searchable encryption in cloud environment. JCC **12**(1), 61 (2023)

A Rollout-Based Algorithm and Reward Function for Resource Allocation in Business Processes

Jeroen Middelhuis[✉], Zaharah Bukhsh, Ivo Adan, and Remco Dijkman

Eindhoven University of Technology, Department of Industrial Engineering and Innovation Sciences, Eindhoven, Netherlands
{j.middelhuis,z.bukhsh,i.adan,r.m.dijkman}@tue.nl

Abstract. Resource allocation plays a critical role in minimizing cycle time and improving the efficiency of business processes. Recently, Deep Reinforcement Learning (DRL) has emerged as a powerful technique to optimize resource allocation policies in business processes. In the DRL framework, an agent learns a policy through interaction with the environment, guided solely by reward signals that indicate the quality of its decisions. However, existing algorithms are not suitable for dynamic environments such as business processes. Furthermore, existing DRL-based methods rely on engineered reward functions that approximate the desired objective, but a misalignment between reward and objective can lead to undesired decisions or suboptimal policies. To address these issues, we propose a rollout-based DRL algorithm and a reward function to optimize the objective directly. Our algorithm iteratively improves the policy by evaluating execution trajectories following different actions. Our reward function directly decomposes the objective function of minimizing the cycle time, such that trial-and-error reward engineering becomes unnecessary. We evaluated our method in six scenarios, for which the optimal policy can be computed, and on a set of increasingly complex, realistically sized process models. The results show that our algorithm can learn the optimal policy for the scenarios and outperform or match the best heuristics on the realistically sized business processes.

Keywords: Business process optimization · Resource allocation · Deep Reinforcement Learning · Reward function · Rollouts

1 Introduction

Efficient resource allocation plays a crucial role in minimizing the mean cycle time within business processes. In data-driven business process optimization (BPO), methods aim to optimize the process with respect to some objective. Many BPO techniques rely on heuristics or rule-based strategies (e.g., [7,26]), which are non-adaptive to the dynamicity of business processes. More recently, learning-based methods, such as reinforcement learning (RL), have emerged as a powerful tool for process optimization, leveraging a sequential decision-making framework to allocate resources effectively [11,12]. However, existing RL-based BPO methods rely on engineered reward signals to guide the learning process.

Engineered rewards are designed to guide decision-making agents toward achieving a specific objective. These engineered rewards are approximations of the objective, and maximizing this reward should also optimize the objective. However, engineered rewards, when misspecified, can lead to unintended agent behavior [1]. One risk is reward hacking, where an agent exploits loopholes in the reward function, resulting in unintended or undesirable behaviors without completing the intended objective [16]. For example, a cleaning robot may be given points for picking up trash. Instead of cleaning up trash, it can also knock over trash, thus maximizing the reward it earns, but failing to achieve the objective of cleaning an area. Similar unintended behaviors can arise in business processes where an agent optimizes a different objective than intended.

In the domain of BPO, three different reward functions for cycle time minimization have been proposed, but each has inherent limitations that can lead to suboptimal policies. The first reward function rewards an agent +1 for each case completion but does not account for the realized cycle time of the case [9,25]. The second reward function penalizes the agent equivalent to the cycle time of a completed case [5]. However, this approach may inadvertently incentivize the agent to delay or avoid completing cases to minimize penalties. The third reward function returns the inverse of the cycle time when a case completes, ensuring that fast case completions yield a higher positive reward [11,12]. However, this reward function can be gamed by the agent, resulting in suboptimal policies. While these methods can lead to efficient policies for certain processes, they also require problem-specific engineering of reward functions and algorithms, leading to suboptimal policies or failure to learn a policy. Furthermore, the typical DRL algorithms, such as proximal policy optimization (PPO) [19], used in the aforementioned studies are not specifically designed to handle dynamic environments, such as a business process, making it challenging to learn effective policies [23].

This paper introduces a rollout-based DRL algorithm and a reward function to optimize the mean cycle time. Our algorithm directly determines the sum of rewards following different actions, removing the need to approximate state-action values. Our reward function eliminates the need for extensive reward engineering, ensuring that the learned policies directly optimize the desired objective. We provide a comprehensive evaluation of our algorithm and benchmark it with existing reward functions using the state-of-the-art PPO algorithm [19]. The results show that our algorithm successfully learns the optimal policy across for typical business process pattenrs and outperforms the PPO algorithm.

The remainder of the paper is structured as follows. Section 2 provides an overview of the related work. Section 3 introduces background concepts. Section 4 explains our algorithm and reward function. Section 5 evaluates our method and Sect. 6 concludes our work.

2 Related Work

This section reviews existing algorithms and reward functions for resource allocation in business processes, highlighting their limitations and motivating the gap in the literature. Section 2.1 discusses existing algorithms and identifies their shortcomings for business processes. Section 2.2 details existing reward functions and highlights the need for a more effective reward function.

2.1 Algorithms for Business Process Optimization

BPO algorithms are designed to improve overall process efficiency by optimizing key performance indicators [6]. Several BPO methods rely on static resource rankings based on event logs [2,7,26] or process models [20,21]. While these methods offer a structured approach, they lack adaptability in dynamic environments.

Dynamic approaches assign resources at runtime using process state data. Park and Song [17] employ predictive modeling to minimize costs. Recently, RL has emerged as a powerful tool for BPO due to its ability to adapt to observed process dynamics. Early work applied Q-learning to minimize cycle time [5]. However, Q-learning struggles with scalability as it stores each state-action pair separately, making it inefficient for large-scale, continuous business environments. DRL deals with this issue with approximators [22] and has been used for optimizing business processes [9,10,25], demonstrating effectiveness in large-scale processes [11,12,15].

Popular DRL algorithms have been successfully applied in game environments [14,19]. However, these environments differ from BPO settings, as they often have well-defined, deterministic rules. For instance, in chess, a move has a predictable and deterministic outcome. In contrast, business processes are driven by random events, such as unpredictable case arrivals and variable activity durations. Standard DRL algorithms are not explicitly designed to handle such stochastic environments [23]. Within the BPO domain, applications of these algorithms often do not converge to optimal policies [5,11,12,25]. To address this issue, we propose a rollout-based algorithm to learn optimal policies for business processes. Our approach is inspired by other rollout-based algorithms [8,23] and we extended this idea to fit the requirements of resource allocation for business processes. For a more extensive review of the resource allocation algorithms, we refer to Pufahl et al. [18].

2.2 Reward Functions for Business Process Optimization

While DRL offers a promising framework for resource allocation, its success depends on the design of the reward function, which remains a key challenge in BPO. An effective reward function should provide a feedback signal that guides the agent toward optimizing a specific objective function.

In BPO, three reward functions have been proposed for cycle time minimization: (1) a reward of +1 per completed case [9,25]; (2) a penalty equal to cycle time per completed case [5,15]; and (3) a reward of $\frac{1}{1+CT}$ per completed case, where CT is cycle time [11,12]. While these reward functions encourage cycle time minimization, they are prone to reward hacking [1,16], in which the agent optimizes a different objective than intended.

The first reward function encourages case completions but ignores the temporal aspect, which may lead to inefficiency. The second reward function more directly targets the cycle time but only returns penalties. Consequently, avoiding case completions by, for example, postponing, results in zero rewards, which are better than penalties. The third reward function rewards short cases disproportionately, which can lead to unintended behavior. For example, completing two cases with cycle times $\{1, 10\}$ yields a higher cumulative reward than two cases with cycle times $\{5, 5\}$ (a cumulative reward of $\frac{13}{22}$ compared to $\frac{1}{3}$, respectively), despite the latter having a shorter mean cycle time. While a postpone penalty was introduced [12] to mitigate this issue, it required additional engineering and did not fully eliminate reward hacking.

A broader limitation shared by all three reward functions is the sparse and delayed nature of the reward function. The agent only receives a reward after a case is completed, and intermediate actions receive no rewards, making learning effective policies challenging [3]. These limitations of the existing engineered rewards highlight the need for a reward function that guides the agent toward cycle time minimization and does not require problem-specific reward engineering.

In this paper, we introduce a reward function that decomposes the contribution to the sum of cycle times and integrates it with a rollout-based DRL algorithm to learn optimal resource allocation policies.

3 Background

This section introduces the resource allocation problem and how it can be formulated and solved as an MDP. Section 3.1 defines resource allocation concepts. Section 3.2 presents the Markov Decision Process (MDP) framework and how DRL can be applied to learn a resource allocation policy on an MDP.

3.1 Resource Allocation in Business Processes

In a business process, cases from a set $\mathcal{C}$ arrive according to some arrival process. A case consists of a series of activities from a set $\mathcal{A}$ that must be executed by resources from a set $\mathcal{R}$. When a specific case $c \in \mathcal{C}$ enters the process, it is added to the set of ongoing cases $C \subseteq \mathcal{C}$ and initiates an activity instance $k \in \mathcal{K}$. An activity instance represents an execution of an activity for a specific case, and a case may involve multiple such instances throughout its lifecycle. Subsequent activity instances are generated during the execution of the case. In this paper, we assume a Poisson arrival process and exponential processing times. Figure 1 shows an example of a business process.

Activity instances that are awaiting execution form the set of unassigned activity instances, denoted as $K \subseteq \mathcal{K}$. For a specific activity, $act \in \mathcal{A}$, the subset of activity instances and unassigned activity instances are denoted by $\mathcal{K}_{act} \subseteq \mathcal{K}$ and $K_{act} \subseteq \mathcal{K}_{act}$, respectively. To execute an unassigned activity instance $k \in K$, a resource $r \in \mathcal{R}$ must be allocated. The resource set $\mathcal{R}$ is divided into available resources $\mathcal{R}^+ \subseteq \mathcal{R}$ and unavailable resources $\mathcal{R}^- \subseteq \mathcal{R}$, where $\mathcal{R}^+ \cup \mathcal{R}^- = \mathcal{R}$ and

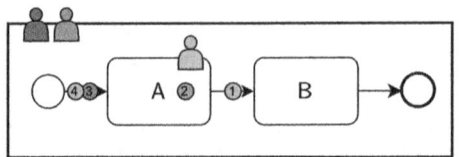

Fig. 1. Example of a process model with two activities, three resources, and four active cases. Activities A and B have unassigned activity instances ($K_A = \{3, 4\}, K_B = \{1\}$). The yellow resource is processing activity instance 2 ($\mathcal{R}^- = \{r_{\text{yellow}}\}, B = \{(2, r_{\text{yellow}})\}$). The red and green resources are available ($\mathcal{R}^+ = \{r_{\text{red}}, r_{\text{green}}\}$).

$\mathcal{R}^+ \cap \mathcal{R}^- = \emptyset$. Not all resources are qualified to perform every activity, often due to varying expertise or authorization levels. Therefore, for each activity, $act \in \mathcal{A}$, the set of eligible resources, as defined by the process model, is $\mathcal{R}_{act} \subseteq \mathcal{R}$.

Using these components, we can form a resource-to-activity assignment (r, act). Considering resource eligibility, we define the set of allowed assignments as $\mathcal{D} = \{(r, act) \mid act \in \mathcal{A}, r \in \mathcal{R}_{act}\}$. We define the set of possible assignments at runtime as $D = \{(r, act) \mid (r, act) \in \mathcal{D}, r \in \mathcal{R}^+, |K_{act}| > 0\}$. Making an assignment (r, act) starts the execution of (r, k), which is the assignment of resource $r \in R_{act}$ to an activity instance $k \in K_{act}$. Then, r and k are taken from $\mathcal{R}^+$ and the set of assigned activity instances K, and instead put into $\mathcal{R}^-$ and set of assigned activity instances Z, and the set of ongoing assignments $B \subseteq Z \times \mathcal{R}^-$.

The current condition of the process can be represented as an execution state $(C, K, \mathcal{R}^+, \mathcal{R}^-, B, t)$, where $t \in T \subseteq [0, \infty)$ is a moment in time. $\mathcal{S}$ is the set of all possible execution states. In each state, the agent can decide either to make an assignment $(r, act) \in D$ or to postpone the decision. After making a decision, the state transitions to the next state $(C', K', \mathcal{R}^{+\prime}, \mathcal{R}^{-\prime}, B', t')$ in two steps, as follows. In the first step, if the agent makes an assignment (r, act), a resource r is assigned to an unassigned activity instance $k \in K_{act}$, i.e., (r, k). Next, changes are made to the resource sets $\mathcal{R}^{+\prime} = \mathcal{R}^+ - \{r\}$ and $\mathcal{R}^{-\prime} = \mathcal{R}^- \cup \{r\}$, the set of unassigned activity instances $K' = K - \{k\}$, and the set of ongoing assignments $B' = B \cup \{(r, k)\}$. If the agent chooses to postpone, nothing happens in this step. In the second step, an event is triggered. This can either be the arrival of a new case or the completion of an assigned activity instance $k \in Z$. We denote the set of events that is possible in a state s after executing action $a \in D \cup \{\text{postpone}\}$ as $E(s, a)$. These events all have a timing probability distribution associated with them, which models how long it will take before they occur (e.g., how long it will take from an action until the next case arrival). Since we assume exponential distributions, the timing distribution of event $e \in E(s, a)$ can be represented by a rate λ_e. When a new case c' arrives, this also generates an initial activity instance k', such that $C' = C \cup \{c'\}$ and $K' = K \cup \{k'\}$. When an assignment (r, k) completes, the state changes such that $B' = B - \{(r, k)\}, \mathcal{R}^{+\prime} = \mathcal{R}^+ \cup \{r\}$ and $\mathcal{R}^{-\prime} = \mathcal{R}^- - \{r\}$. Subsequently, a new activity instance k' may be generated depending on the process model, which changes $K' = K \cup k'$. If the completed activity instance results in the completion of a case, this case is removed from the set of active cases $C' = C - \{c\}$. The cycle time of a case CT_c is equal to the difference between its arrival and departure time.

3.2 Deep Reinforcement Learning

We aim to solve the resource allocation problem using Deep Reinforcement Learning (DRL), which combines deep neural networks with reinforcement learning (RL) to solve complex decision-making tasks. In RL, an agent interacts with an environment, taking actions based on a policy $\pi(a|s)$ to maximize cumulative rewards [22]. The agent observes a state s, selects an action a, receives a reward r, and transitions to a new state s'. We refer to this interaction as a decision step. DRL leverages deep learning to approximate policies, value functions $V^\pi(s)$, or Q-values, such that it can efficiently handle large state spaces. To apply DRL, we must model resource allocation as a Markov Decision Process (MDP).

An MDP provides a mathematical framework for sequential decision-making [22], where outcomes are partly under the control of an agent and partly caused by randomness. Formally, an MDP is a tuple (S, A, P, R), where S is a finite set of states, A a finite set of actions, $P(s'|s,a)$ is the probability that, when an agent takes an action a in state s, the environment transitions into the state s', and $R(s,a)$ is function that models the reward that the agent gets when performing action a in state s. Guided by the reward, The agent learns a policy $\pi(a|s)$ to maximize cumulative rewards over an episode, which is a sequence of interactions between the agent and the environment, during which the agent decides on an action a, and the environment subsequently applies the action to transition into a next state $P(s'|s,a)$ and gives a reward R. The episode starts in an initial state s_0 and ends in a terminal state.

Standard MDPs assume decisions occur at discrete intervals, whereas decision-making in business processes often occurs at random intervals. To address this, a Continuous-Time MDP (CTMDP) extends the MDP framework by allowing decision steps to occur at arbitrary time points.

4 Method

This section presents our proposed algorithm to learn optimal resource allocation policies using DRL. Since DRL uses an MDP model to learn, we first explain in Sect. 4.1 how a resource allocation problem in a business process, as it is described in Sect. 3.1, can be mapped to a (CT)MDP. Subsequently, Sect. 4.2 introduces our proposed rollout-based algorithm to learn the optimal policy. The learning process is guided by our reward function, which is equivalent to the objective, which is described in Sect. 4.3. Finally, we need to compute the optimal policy to ensure that our learned policies are optimal. Section 4.4 describes how we can transform the CTMDP, which is a business process, into a discrete-time MDP such that we can compute the optimal policy.

4.1 The Business Process as a CTMDP

To apply a DRL algorithm to solve the resource allocation problem described in Sect. 3.1, we map it to a CTMDP (S, A, P, R). An overview of the RL based on the (CT)MDP is shown in Fig. 2. In the remainder of this section, we detail how each component S, A, P of the CTMDP is created from the process. Since we propose both an algorithm and reward function, we provide an in-depth overview of the reward function R in Sect. 4.3.

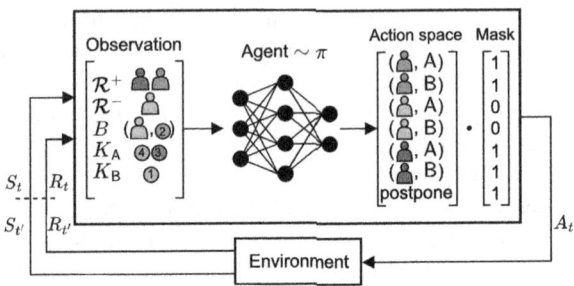

Fig. 2. The agent-environment interaction for the resource allocation problem based on the business process in Fig. 1.

State. The state contains all information related to the condition of the business process, such as the waiting activity instances and assignments that are being processed. Providing the agent with every possible state as an input would cause a state explosion, making the problem computationally infeasible. For that reason, at each decision step, the agent partially observes the state based on features that are relevant to the decision and are derived from the state variables. The information represented by these features is called an observation, which we model as follows:

- A binary mapping $\mathcal{R} \rightarrow \{0, 1\}$, indicating whether the resource is available or not. For resource $r \in \mathcal{R}$, this feature equals 1 if $r \in \mathcal{R}^+$ and 0 if $r \in \mathcal{R}^-$.
- A binary mapping $\mathcal{D} \rightarrow \{0, 1\}$, indicating whether an assignment $(r, act) \in \mathcal{D}$ is being executed or not. For assignment $(r, act) \in \mathcal{D}$, this feature equals 1 if $\exists r \in \mathcal{R}^+, k \in K_{act}$ such that $(r, k) \in B$, and 0 otherwise.
- A continuous mapping $\mathcal{A} \rightarrow [0, 1]$, indicating the number of waiting cases at activity $act \in \mathcal{A}$. Similarly to [12], we scale this feature between 0 and 1 by dividing it by 100 and truncating values above 1. For activity $act \in \mathcal{A}$, this feature equals $\min(\frac{|K_{act}|}{100}, 1)$.

Action. Based on the observation of the state, the agent selects an action from the action space, which includes all possible resource-to-activity assignments $\mathcal{D}$, as well as a postpone action. Postponing to wait for better future decisions can be beneficial when only inefficient assignments are currently available. In some states, no assignments are possible, making postponing the only option. During training, we differentiate between forced and voluntary postponing, allowing the agent to learn when postponing is a strategic choice rather than a necessity.

At each decision step, we check which actions are possible based on the state, and infeasible actions are masked, disallowing the agent to choose these actions. For example, assignments that do not have their respective resource or unassigned activity instances available are masked.

State transition. As explained in Sect. 3, when taking an action $a \in D \cup \{\text{postpone}\}$ in state $s \in S$, the process transitions into a state $s' \in S$, depending on the event $e \in E(s,a)$ from the set of possible events, which happens after executing the action with a rate λ_e. To represent this in the CTMDP, we define $P(s'|s,a)$, which specifies the probability of transitioning from s to s' after taking action a. In the CTMDP, the time until the next decision step is exponential with rate $\sum_{e \in E(s,a)} \lambda_e$ and is independent of the event that triggers it. The probability that it reaches state s' due to event $e' \in E(s,a)$ is:

$$\frac{\lambda_{e'}}{\sum_{e \in E(s,a)} \lambda_e} \quad (1)$$

The CTMDP formulation outlined in this section enables us to train a DRL agent to optimize resource allocation policies.

4.2 Rollout-Based Algorithm

The algorithm that we use to find the best policy for assigning resources to tasks is based on the rollouts of that policy. A rollout refers to the process of simulating an agent's interaction with the environment (i.e., repeatedly performing an action and checking its effect) while following a specific policy for a specific number of decision steps. The benefit of using rollouts is that we can use the sum of rewards, which is also called the return, to represent exactly the objective we aim to optimize, such as the sum of cycle times. We use a rollout to estimate the expected return of a policy, given by:

$$G_t = \sum_{i=0}^{N-1} R_{t+i+1} \quad (2)$$

where $R_j = R(s_j, a_j)$ is the reward of decision step j, and N represents the number of decision steps in each rollout. To improve the reliability of the estimate, we perform M rollouts and compute the average return. We use common random numbers to reduce variance across rollouts actions [23]. Algorithm 1 presents our rollout-based approach.

Our rollout-based algorithm has three steps: *initialization*, *evaluation*, and *improvement*. In the *initialization* step (line 1), we define a bootstrap policy π_0, for which we use a greedy policy. During the *evaluation* step (lines 2, 12–14), we deploy π on the business process and evaluate its performance. In the *improvement* step (lines 3–11), we explore alternative actions and update the policy to π'. The evaluation and improvement steps are repeated for I iterations.

During the improvement step, we use rollouts to estimate the best action to take in a state, i.e., the action that has the highest average return. First, we sample a set of random states $\mathcal{S} \subseteq S$. For each state $s \in \mathcal{S}$, we determine the optimal action a^* from the set of possible actions $a \in A(s)$ in that state using rollouts. The (CT)MDP $\mathcal{M}$ provides the framework for generating the random states and performing the rollouts. Figure 3 illustrates how rollouts assess different actions.

Algorithm 1: Rollout-based algorithm

Input: $\mathcal{M}, \pi_0, I, M, N$
Output: π

1. $\pi = \pi_0$;
2. $V^\pi(s_0) \leftarrow \textbf{evaluate}(\mathcal{M}, \pi)$;
3. **for** $i = 1, 2, \ldots, I$ **do**
4.   Generate states $\mathcal{S}$ from $\mathcal{M}$;
5.   dataset $= \emptyset$;
6.   **foreach** $s \in \mathcal{S}$ **do**
7.   **foreach** $a \in A(s)$ **do**
8. $G^\pi(s, a) \leftarrow \textbf{rollout}(\mathcal{M}, \pi, M, N)$;
9. $a^* = \arg\max_{a \in A(s)} G^\pi(s, a)$;
10.   dataset $\leftarrow$ dataset $\cup \{(s, a^*)\}$;
11. $\pi' \leftarrow \textbf{learn}(\pi, \text{dataset})$;
12. $V^{\pi'}(s_0) \leftarrow \textbf{evaluate}(\mathcal{M}, \pi')$;
13.   **if** $V^{\pi'}(s_0) > V^\pi(s_0)$ **then**
14. $\pi = \pi'$;
15. Return π;

The best state-action pairs identified through rollouts are used to update the policy. To represent the policy, we use a neural network where the input represents the observed state, and the output corresponds to the best action label. After updating the policy π', we evaluate $V^{\pi'}(s_0)$, where s_0 in the initial state of $\mathcal{M}$, which is an empty process. If the new policy performs better, we proceed with $\pi = \pi'$. Otherwise, we revert to the previous policy π.

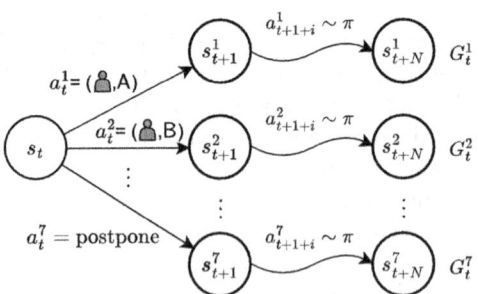

Fig. 3. Rollouts for the example in Fig. 2, representing lines 7–10 of Algorithm 1. For each action $a \in A(s_t)$, a rollout is performed. Consequently, the state transitions to state s_{t+1}. Next, a rollout of $N-1$ decision steps following policy π is performed to estimate the return G_t of each action.

4.3 Reward Function

A fundamental component of the DRL framework is the reward function, which serves as the guiding signal for the agent to learn an optimal policy. If the reward signal does not accurately reflect the objective function, the agent may learn suboptimal policies. We propose a reward function that, when maximized, directly minimizes the total cycle time without requiring reward engineering. The reward function is dense as it provides a reward at each decision step, which reduces the variance in rewards and improves learning efficiency [3]. The reward for an action taken in state s, which occurs at time t is defined as:

$$r_t = -|C|(t' - t) \quad (3)$$

where $|C|$ is the number of active cases in the process at t. In this equation, t and t' and the times of the current and next decision step, respectively (i.e., according to s and s'). We give a negative reward such that the sum of cycle times is minimized when the cumulative reward is maximized. Figure 4 shows an example of how the reward function is used for three cases consisting of one or two activities.

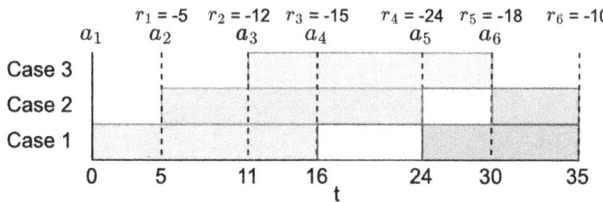

Fig. 4. Example of the proposed reward function for three cases. At each decision step, the reward function returns a reward equal to the contribution to the total cycle time during the state transition. Waiting cases (hatched area) also incur cycle time. The reward is returned after the state transition.

In this example, at $t = 0$, case 1 arrives and is assigned by taking action a_1. The state transitions due to the arrival event of case 2. Since five time units passed and there there is one case in the process, a reward of $r_1 = -5$ is returned. Similarly, for instance, for a_4, eight time units pass before transitioning to the next state while there are three cases in the system, resulting in a reward of $r_4 = -24$. Using this reward function, the agent receives a continuous feedback signal equal to the contribution to the sum of cycle times during that transition.

Since the state transition time cannot be minimized, optimizing our reward function effectively reduces $|C|$, the number of cases in the system. Our reward function explicitly penalizes high values of $|C|$, and with a constant arrival rate λ, minimizing $|C|$ directly leads to a lower cycle time, as $\hat{CT} = \hat{C}/\lambda$, according to Little's law [4], where $\hat{C}$ is the average number of cases in the system. Thus, by minimizing $|C|$, we naturally minimize the mean cycle time $\hat{CT}$.

4.4 Uniformization of the CTMDP and Optimal Policy

To assess if our algorithm can learn the optimal policy, we first have to compute this policy. This can feasibly be done for small models - like the ones we will use in the evaluation - as described in this section.

Computing the optimal policy of a CTMDP using value iteration is not straightforward due to the stochastic transition times on which the rewards depend. Therefore, we first introduce a transformation to uniformize the CTMDP into a discrete-time MDP [24]. We can then apply the value iteration algorithm to the MDP to compute the optimal policy. This policy is also optimal for the CTMDP in terms of average reward [24]. Note that the MDP is only required to compute the optimal policy and is not required for our algorithm presented in Sect. 4.2. First, we choose τ to represent the time between states, such that $t \in T = \{n\tau \mid n \in \mathbb{N}_0\}$:

$$0 < \tau \leq \min_{s,a} \tau_s(a) \text{ where } \min_{s,a} \tau_s(a) = \frac{1}{\max_{s,a} \sum_{e \in E(s,a)} \lambda_e} \quad (4)$$

where $\min_{s,a} \tau_s(a)$ is the minimum expected transition time in the CTMDP. Based on τ, we can transform the CTMDP into a discrete-time MDP with the same state and action space and the following one-step rewards and one-step transition probabilities:

$$\tilde{r}_s(a) = \frac{r_s(a)}{\tau_s(a)} \qquad\qquad\qquad s \in S \text{ and } a \in A(s) \quad (5)$$

$$\tilde{p}_{ss'}(a) = \begin{cases} \frac{\tau}{\tau_s(a)} p_{ss'}(a), & s \neq s', \\ \frac{\tau}{\tau_s(a)} p_{ss'}(a) + (1 - \frac{\tau}{\tau_s(a)}), & s = s', \end{cases} \quad s, s' \in S \text{ and } a \in A(s) \quad (6)$$

where $r_s(a)$ and $\tau_s(a)$ are the expected reward and expected time in state s when taking action a in the CTMDP, respectively. Furthermore, $p_{ss'}(a)$ is the expected transition probability of going from state $s \in S$ to state $s' \in S$ in the CTMDP. Due to the non-uniformity of the transition time in the CTMDP, expected rewards are transformed into reward rates, and transition probabilities include fake transitions to make the times in between decision steps uniform.

Using value iteration, we compute the optimal value function V^* for the MDP. However, due to unbounded queue length features, the state space is infinite. Therefore, we introduce an upper bound of 100 on the queue length features, with substantial penalties for transitions that access states beyond this limit, guiding the agent toward more favorable states. We follow the value iteration algorithm from Tijms [24]. Since we do not discount future rewards, the value function converges when the difference between the minimum and maximum absolute delta between two successive value functions converges to zero.

By applying value iteration, we obtain the *optimal policy* for the bounded MDP. While this formulation introduces an approximation, the upper bound is set sufficiently high to ensure it is never exceeded under a random policy in our evaluations (Sect. 5.1). As a result, the optimal policy derived from the bounded

MDP is unlikely to differ from that of the unbounded MDP. Therefore, we refer to the policy obtained via value iteration on the bounded MDP as the *optimal policy*.

5 Evaluation

This Section evaluates our approach. Section 5.1 introduces our evaluation protocol. Section 5.2 demonstrates the ability of our method to learn optimal policies. Section 5.3 shows the performance of different reward functions and Sect. 5.4 shows an evaluation on more realistically sized models.

5.1 Evaluation Setup

For the evaluation, we use the scenarios from Middelhuis et al. [12] shown in Fig. 5, which represent typical process patterns.

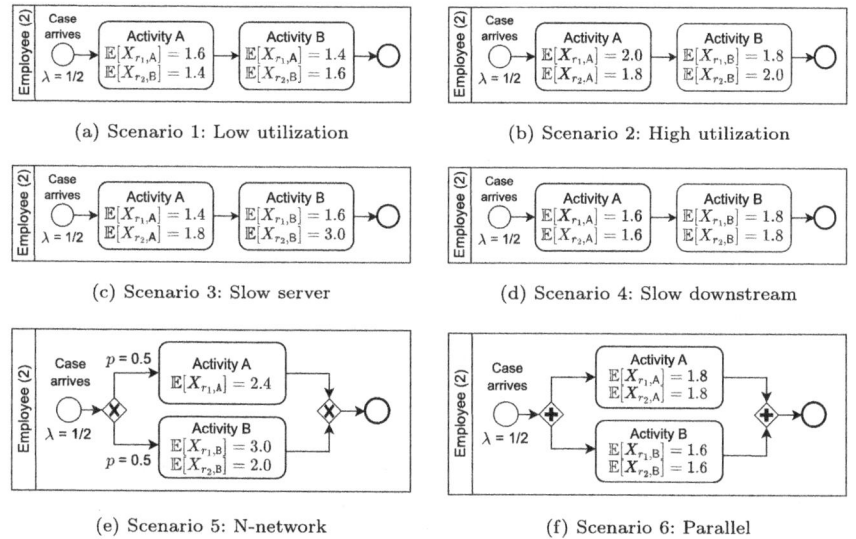

Fig. 5. Six scenarios with two activities and resources. In each scenario, cases arrive with a rate of $\lambda = 1/2$. Activity act is executed by resource r with rate $1/\mathbb{E}[X_{r,act}]$.

Similar to Middelhuis et al. [12], we simulate 2500 cases per episode, which terminates when all cases are completed. For evaluation we use 300 episodes. Before training and evaluating our method, we tune hyperparameters on the MDP through grid search. The selected parameter values are applied when training all policies. We tune the number of rollouts per action, $M \in \{50, 100, 200\}$, and rollout length, $N \in \{25, 50, 100\}$, and the transformation parameter $\kappa \in \{0.1, 0.25, 0.5, 1\}$, which determines the MDP step size $\tau = \kappa \min_{s,a} \tau_s(a)$. To ensure consistency between MDP and CTMDP settings, we scale the rollout length N with τ. Hyperparameter tuning leads to a value of $M = 100$. A longer rollout length of $N = 100$, and a $\kappa = 0.5$, aligning with the recommendation of Tijms [24]. For the neural network architecture, we use two layers of 128 neurons. For each model, we generate 5000 random states in each iteration and perform 10 training iterations, of which we save the best model.

5.2 Evaluation Results of the Proposed Method

We train on both the CTMDP and MDP versions of the six scenarios and evaluate each method on the CTMDP, as it represents the original business process. While the transformation to MDP is not required to use our proposed method, we train an agent on the MDP to investigate if the environment with more predictable rewards impacts the learning process.

Table 1 shows the results, comparing our method, trained on both CTMDP and MDP, to the optimal policy and three heuristics: Shortest Processing Time (SPT), First-In-First-Out (FIFO), and a random policy. We use a t-test ($\alpha = 0.05$) to assess statistical significance. Results not significantly different from the optimal policy shown in bold. For each method, we show the optimality gap: the percentage difference to the optimal policy.

Table 1. Mean cycle time of the optimal policy and optimality gap (%) between the optimal policy and other methods, including our approach and benchmark heuristics.

Scenario	Optimal policy	Our method		SPT	FIFO	Random
		CTMDP	MDP			
Low Utilization	5.7 (0.05)	**-0.41%**	**1.03%**	4.05%	5.05%	17.78%
High Utilization	18.0 (0.55)	**1.31%**	**2.40%**	9.10%	49.23%	100.52%
Slow Server	11.7 (0.23)	**-0.47%**	**1.60%**	151.14%	58.57%	94.90%
Slow Downstream	10.1 (0.20)	**-1.29%**	**-0.34%**	46.79%	**-0.51%**	14.94%
N-system	5.8 (0.06)	**-1.33%**	**-1.15%**	23.88%	**1.31%**	11.71%
Parallel	9.7 (0.19)	**1.56%**	**1.06%**	44.00%	**-2.68%**	16.41%

Table 1 demonstrates that our method successfully learns the optimal policy across all scenarios when trained on either CTMDP or MDP. We trained on both the CTMDP and MDP to investigate if the variable time between decision steps affects performance. However, no significant differences were found. Interestingly, the solution for the FIFO policy in the parallel scenario seems to outperform the optimal policy, though not significantly. This difference occurs because the optimal solution is derived based on the features in our observations, whereas the FIFO policy directly allocates resources according to the state.

In the parallelism scenario, the optimal policy involves executing an activity of a case that already has one of its two activities completed, thereby completing the case with the next activity. Due to the nature of our observation, the agent is unable to learn this specific strategy. Consequently, the optimal policy is optimal for our representation of the (CT)MDP, which is learned by our method.

5.3 Evaluation Results of PPO with Different Reward Functions

PPO is widely used to train DRL agents for resource allocation in business processes [9–12]. However, existing PPO-based methods are trained under varying settings. In this section, we compare PPO's performance using different existing reward functions and our proposed reward function.

Each existing reward function gives a reward when case $c \in \mathcal{C}$ is completed: (a) $+1$, (b) $-CT_c$, and (c) $\frac{1}{1+CT_c}$. In state transition following actions that did not result in a case completion, a reward of zero is returned. A PPO agent combined with reward function (c) is considered the current state-of-the-art method for resource allocation in business processes [11,12].

We use the hyperparameters from Middelhuis et al. [12] for the PPO agents and train each model with the same state and action space. For our proposed reward function (d), we use a discount factor of $\gamma = 1$ since we aim to minimize the cumulative undiscounted return (compared to $\gamma = 0.999$ for the other reward functions). Table 2 presents the optimality gap in the mean cycle time between the optimal policy and a PPO agent with different reward functions.

The code and trained models used to produce our results can be found in our repository: https://github.com/jeroenmiddelhuis/BPO_rollouts.

Table 2. Mean cycle time of the optimal policy and optimality gap (%) between the optimal policy and PPO agent with different reward functions.

	Optimal policy	Reward function				Random
		(a): $+1$	(b): $-CT_c$	(c): $\frac{1}{1+CT_c}$	(d): $-|C|(t-t')$	
Low utilization	5.7 (0.05)	106%	5927%	2%	4%	18%
High utilization	18.0 (0.55)	1384%	4893%	10%	3801%	101%
Slow server	11.7 (0.23)	1940%	7235%	**2%**	6083%	95%
Slow downstream	10.1 (0.20)	866%	5629%	**0%**	2422%	15%
N-system	5.8 (0.06)	26%	4822%	34440%	**1%**	12%
Parallel	9.7 (0.19)	1647%	3419%	15%	41%	16%

Table 2 shows that PPO's performance is highly sensitive to the reward function and scenario, as can be seen by the notable differences in performance. In contrast, our rollout-based algorithm and reward function, as shown in Table 1, can learn the optimal policy for all scenarios without the need for reward engineering, demonstrating their effectiveness and generalization capabilities.

Agents trained with reward functions (a) and (b) fail to outperform a random policy, which means that these reward signals are inadequate for effective learning. Reward function (a) does not account for the timing of case completions, and while the discount factor encourages earlier completions, it is not enough to learn a competitive policy. Reward function (b) considers timing but only provides penalties as a learning signal. Using this reward function, the agent may learn to avoid completing cases such that it receives mostly zero rewards.

On the other hand, an agent with reward function (c) outperforms the random policy and learns the optimal policy in two scenarios. In the N-system scenario, the agent hacks the reward function by completing only one of two activities to maximize rewards but not minimize the mean cycle time (see Sect. 2.2). A postpone penalty in [12] partially mitigates this, but the agent still did not learn a competitive policy. While this behavior is most evident in the N-system, similar undesired strategies can be learned in other processes.

Interestingly, with reward function (d), PPO learns the optimal policy in the N-system scenario but fails to find the optimal policy in the other scenarios. One reason is that our reward function (d) only provides negative rewards, making it difficult for the PPO agent to identify good actions, as all actions are penalized. Additionally, PPO works with episodic returns, so the cumulative reward depends on episode length. As a result, identical state-action pairs that occur later in an episode receive a lower cumulative reward than those that occur earlier, which makes it challenging to learn a good policy.

In conclusion, existing reward functions in combination with a PPO agent are not sufficient to learn competitive policies, while our method learns the optimal policy in all scenarios, outperforming the state-of-the-art in four scenarios and learning a similar policy in the other two scenarios.

5.4 Evaluation Results on Composite Process Models

This section demonstrates our algorithm's capability to learn policies for increasingly complex process models, even as computational complexity grows.

The computational complexity of the algorithm can be measured in terms of the speed at which it learns a good policy, which depends on: the size of the neural network, consequently translating to size of state and action space, and the number of steps between taking an action and seeing the effect of that action (i.e. the moment at which a case completes). These factors are both determined by the number of tasks and the number of resources in a process model. Therefore, we gradually increase the size of a process model and observe the effect on the learning speed.

Specifically, starting with scenario 1, we add scenarios sequentially such that the departures of one scenario are the arrivals of the next, until we have a model that has the same complexity as the composite model from [12]. We train the algorithm with the same hyperparameters only on the CTMDP, as it represents the original business process. We do not train an agent on the MDP as the results show no significant differences with different environment representations (Table 1). We removed the 'postpone' action, because preliminary experiments revealed that it only added noise to the training dataset, and it was rarely selected as the best action. Finally, we report on the learning speed by measuring the number of learning iterations at which our model learns a policy competitive with the heuristics.

Table 3. Mean cycle time with 95% confidence intervals for composite process models comparing our method against heuristic benchmarks, and the number of iterations required for our method to achieve performance equal to or better than the best heuristic.

Scenarios	Our method	Iterations	SPT	FIFO	Random
1	**5.6 (0.05)**	1	5.9 (0.05)	5.9 (0.06)	6.7 (0.08)
1–2	**24.1 (0.67)**	2	25.0 (0.53)	32.7 (1.54)	40.9 (1.73)
1–2–3	**36.7 (0.79)**	1	52.7 (1.52)	47.0 (1.70)	60.9 (2.29)
1–2–3–4	**50.7 (1.00)**	2	71.1 (1.67)	55.8 (1.81)	72.4 (2.46)
1–2–3–4–5	**57.6 (1.38)**	3	81.7 (1.89)	61.4 (1.72)	76.9 (2.19)
1–2–3–4–5–6	**69.6 (1.19)**	6	103.1 (2.56)	**70.8 (2.10)**	87.6 (2.51)

Table 3 shows the results of the comparison. The table shows that for all levels of complexity, our method can learn a policy that performs at least as well as the best heuristic. While the computational complexity seems to increase exponentially, the number of learning iterations for these processes is still small and computationally manageable. In future work, further algorithmic improvements, such as those proposed by Temizöz et al. [23], and code optimization can still be applied to speed up the learning process.

6 Conclusion

In this paper, we introduced a DRL method combining a rollout-based algorithm and a dense reward function to learn optimal policies for business processes. Our algorithm evaluates the execution trajectories of different actions in each state and uses the best-found state-action pairs to update the policy iteratively. Our reward function decomposes the sum of cycle times over the decision steps to provide a continuous feedback signal.

The results show that our proposed method consistently learns the optimal policy for the six evaluated business processes. In contrast, the current state-of-the-art algorithm learns the optimal policy in only two. In four of five composite processes, our method learns a policy that outperforms the best heuristic, and matches it in the fifth. The benefit of using rollouts is that we can directly evaluate the outcome of different actions and do not need to approximate the value function, which is typical for other DRL methods [13,19]. Furthermore, a comparative analysis highlighted that while existing reward functions guide agents toward cycle time minimization, they can often be exploited, leading to suboptimal or unintended behaviors. On the other hand, maximizing our proposed reward function ensures that the objective is minimized, eliminating the need for problem-specific reward engineering.

In this paper, we presented a method to learn optimal policies for business processes. While we can compute the optimal policy on small business processes, we cannot do so for larger processes due to state explosion. In future work, the ability of our method to optimize real-world processes should be explored. Furthermore, the business process with parallelism illustrated that the current state and action representation can be enhanced with case-specific features to learn better policies, which will also be investigated in future research.

Acknowledgments. The research leading to this paper is supported by the Dutch foundation for scientific research (NWO) under the CERTIF-AI project (nr. 17998).

References

1. Amodei, D., Olah, C., Steinhardt, J., Christiano, P., Schulman, J., Mané, D.: Concrete problems in AI safety (2016)
2. Arias, M., Rojas, E., Munoz-Gama, J., Sepúlveda, M.: A framework for recommending resource allocation based on process mining. In: International Conference on Business Process Management Workshops (2016)
3. Arjona-Medina, J.A., Gillhofer, M., Widrich, M., Unterthiner, T., Brandstetter, J., Hochreiter, S.: Rudder: return decomposition for delayed rewards. In: Advances in Neural Information Processing Systems (2019)
4. Hopp, W.J., Spearman, M.L.: Factory physics (2011)
5. Huang, Z., van der Aalst, W.M., Lu, X., Duan, H.: Reinforcement learning based resource allocation in business process management. DKE (2011)
6. Kubrak, K., Milani, F., Nolte, A., Dumas, M.: Prescriptive process monitoring: Quo vadis? PeerJ Computer Science (2022)
7. Kuchař, Š., Vondrák, I.: Automatic allocation of resources in software process simulations using their capability and productivity. J. Simul. (2016)
8. Lagoudakis, M.G., Parr, R.: Reinforcement learning as classification: leveraging modern classifiers. In: ICML (2003)
9. Lo Bianco, R., Dijkman, R., Nuijten, W., van Jaarsveld, W.: Action-evolution petri nets: a framework for modeling and solving dynamic task assignment problems. In: International Conference on Business Process Managemen (2023)
10. Lo Bianco, R., Dijkman, R., Nuijten, W., van Jaarsveld, W.: A universal approach to feature representation in dynamic task assignment problems. In: International Conference on Business Process Management (2024)
11. Meneghello, F., et al.: Optimizing resource allocation policies in real-world processes using hybrid process simulation and deep reinforcement learning. In: International Conference on Business Process Management (2024)
12. Middelhuis, J., Lo Bianco, R., Sherzer, E., Bukhsh, Z., Adan, I., Dijkman, R.: Learning policies for resource allocation in business processes. Inf, Syst (2025)
13. Mnih, V., et al.: Asynchronous methods for deep reinforcement learning. In: International Conference On Machine Learning (2016)
14. Mnih, V., et al.: Playing atari with deep reinforcement learning (2013)
15. Neubauer, T.R., da Silva, V.F., Fantinato, M., Peres, S.M.: Resource allocation optimization in business processes supported by reinforcement learning and process mining. In: Brazilian Conference on Intelligent Systems. Springer (2022)
16. Pan, A., Bhatia, K., Steinhardt, J.: The effects of reward misspecification: mapping and mitigating misaligned models. In: ICLR (2022)
17. Park, G., Song, M.: Prediction-based resource allocation using LSTM and minimum cost and maximum flow algorithm. In: International Conference On Process Mining (ICPM) (2019)

18. Pufahl, L., Stiehle, F., Ihde, S., Weske, M., Weber, I.: Resource allocation in business process executions—a systematic literature study. Inf. Syst. (2025)
19. Schulman, J., Wolski, F., Dhariwal, P., Radford, A., Klimov, O.: Proximal policy optimization algorithms (2017)
20. Schumann, F., Rinderle-Ma, S.: Optimizing resource-driven process configuration through genetic algorithms. In: International Conference on Business Process Management (2024)
21. Si, Y.W., Chan, V.I., Dumas, M., Zhang, D.: A petri nets based generic genetic algorithm framework for resource optimization in business processes. Simul. Model. Pract. Theor. (2018)
22. Sutton, R.S.: Reinforcement learning: an introduction. A Bradford Book (2018)
23. Temizöz, T., Imdahl, C., Dijkman, R., Lamghari-Idrissi, D., van Jaarsveld, W.: Deep controlled learning for inventory control. EJOR (2025)
24. Tijms, H.C.: Stochastic modelling and analysis: a computational approach (1986)
25. Żbikowski, K., Ostapowicz, M., Gawrysiak, P.: Deep reinforcement learning for resource allocation in business processes. In: International Conference on Process Mining Workshops (2023)
26. Zhao, W., Yang, L., Liu, H., Wu, R.: The optimization of resource allocation based on process mining. In: International Conference on Intelligent Computing (2015)

Enhancing Predictive Process Monitoring on Small-Scale Event Logs Using LLMs

Alessandro Padella(✉), Paolo Frazzetto, Nicolò Navarin, and Massimiliano de Leoni

University of Padua, Padua, Italy
{alessandro.padella,paolo.frazzetto}@unipd.it,
{nnavarin,deleoni}@math.unipd.it

Abstract. Predictive Process Monitoring is a process-mining research direction that aims to predict the future of an uncompleted process execution. The vast majority of research work focuses on techniques that are "data greedy" and require a lot of event data to be sufficiently accurate. However, the recent development of Large Language Models presents significant opportunities and potential benefits across various industrial and research domains. They are capable of leveraging their pre-trained knowledge to understand and complete tasks effectively. This paper reports on the design and implementation of a Predictive Process Monitoring framework based on Large Language Models. Experiments on multiple event logs confirm our hypothesis that Large Language Models are capable of providing very accurate predictions, even with as few as 10 training traces.

Keywords: Predictive Process Monitoring · Large Language Models · Few-Shot Prompting · Small-Scale Event Log

1 Introduction

Predictive Process Monitoring (PPM) is a family of techniques that leverages event logs from business processes to generate predictions about the future states or properties of ongoing process instances [12]. PPM methods vary depending on the prediction target, which can include times [35], next activities [18], or process outcomes [31].

Literature has extensively explored Machine and Deep Learning models to enhance prediction quality [4]. However, these models typically require large amounts of data for effective training. When the available event log is limited in size, the applicability of such techniques becomes constrained, reducing the overall potential of PPM. As highlighted in [37], data availability remains one of the most significant challenges faced by researchers and practitioners in this domain.

This paper puts forward a framework for PPM that leverages the potentials of Large Language Models (LLMs) to estimate the duration of process executions. LLMs can leverage their embedded knowledge to generate accurate predictions while demonstrating robustness in handling noisy and unstructured data [23]. Their ability to operate effectively with limited training examples makes them a promising alternative for improving PPM in data-scarce environments [11]. Since LLMs rely on prompts to process and generate responses, designing an effective prompt is crucial for encoding process traces. The prompt must structure the input data in a way that preserves the sequence, dependencies, and relevant attributes of events while remaining concise and interpretable for the LLM. To answer this question, we designed the framework with two possible encodings: *(i)* an encoding in which traces are provided as vectors in which each component refers to one activity and indicates the number of occurrences of that activity in the trace, as extensively employed in the literature [7, 19, 29, 30] *(ii)* a second encoding in which the trace itself is given to the LLM in a novel, purpose-built function, which only maintains global attributes, activities, and times.

The framework was instantiated for Gemini[1]: a freely-available LLM that performs the best, according to the recent standard benchmarks.[2] Experiments were conducted on five processes and event logs, which were temporally divided into training and test logs. Multiple, random sub-logs with 2, 10, and 100 traces were extracted from the training log. These sub-logs have been provided as input to Gemini in order to enable the predictions on the test traces.

Results show that our LLM-based framework is capable of generating accurate predictions with fewer than 100 trace examples. In some cases, accurate predictions were achieved with as few as two traces. By contrast, a Catboost-based benchmark, which is known to be among the best-performing PPM predictors [3, 10, 24], demonstrated low accuracy when trained on the same amount of traces. Notably, even if only employing a few traces, our LLM framework consistently outperformed the benchmark predictor trained on the complete dataset.

The remainder of this paper is organized as follows: Sect. 2 analyzes the relevant literature in LLMs and PPM. Section 3 introduces the preliminary knowledge needed for the development of the approach, outlined in Sect. 4. Experiments on five case studies are reported in Sect. 5, while Sect. 6 resumes the paper and highlights next potential research directions.

2 Related Works

In this Section, we report the literature relevant for PPM and the application of LLMs within business process management. Section 2.1 reports on the relevant literature in PPM, Sect. 2.2 deals with the significant literature of LLMs, while Sect. 2.3 provides an overview of the application of LLMs in the business process management field.

2.1 Predictive Process Monitoring

Over the years, PPM frameworks leveraged Machine learning, Deep Learning, and ensemble methods. Deep learning techniques, such as Long Short-Term Memory networks [1] and Process Graph Transforming models [8], have shown strong performance, with research also focusing on improving training efficiency [26] and robustness [27]. Despite their success, alternative methods, including Local Process Models [34], Support Vector Machines [18], and Random Forest [5], remain relevant, particularly for

---

[1] Gemini Technical Report: https://arxiv.org/abs/2312.11805.
[2] See https://llm-stats.com for an overview of the most popular LLMs.

real-time applications due to their lower computational demands. However, boosting on decision trees, such as XGBoost and CatBoost, has emerged as a best approach, striking a balance between predictive effectiveness and efficiency [3,10,24].

2.2 Large Language Models

LLMs have gained prominence for their performance, versatility, and ability to generate modular responses through chat interfaces. Research shows they can effectively predict time series using text tokenization of numerical data [11], by leveraging contextual information to enhance predictions [32]. These approaches circumvent limitations of traditional forecasting by eliminating models'design and manual encoding of priors. Requeima et al. [23] demonstrated that zero-shot sequence completion can produce accurate forecasts while incorporating textual side information. Additional studies confirm that integrating contextual information in prompts significantly improves LLMs' forecasting capabilities [2].

2.3 Large Language Models in Business Process Management

LLMs are recently attracting growing focus in business process management as well [9]. They have proven to be significantly useful for many process mining tasks, such as process modeling [13], log extraction [6], anomaly detection [33] and they have also been used for assessing the validity of some traces for a given process [22]. Lashkevich et al. in [16] provide a state-of-the-art approach that leverages LLMs for enhancing the optimization of waiting times and, on the other hand, it relies on user-prompted feedback for recommending more effective re-design options. Rebmann et al. [22] present an approach for extracting knowledge from textual data, also providing some textual and synthetically generated datasets for extracting event logs trying to assess if there are missing activities in the extracted processes and eventually generating them. In the meanwhile, the work in [20] leverages LLMs to transform textual data into process representations, followed by training a text-encoding technique BERT-based deep learning model to predict the next-activity in a process. Finally, Kubrak et al. [14] developed an approach based on a chatbot for process analysis in which the LLM is used for explaining the recommendations provided by a model, enhancing their explainability. The method proposed in this study advances the field in two aspects: it introduces a general prompting framework that enables prediction using LLMs and shows that it maintains its predictive efficiency even under conditions of limited log training data.

3 Preliminaries

The starting point for a process mining-based system is an *event log*. An event log is a multiset of *traces*. Each trace is a sequence of events, each describing a particular *process instance* (i.e., a *case*) in terms of the *activities* executed, the associated *timestamps* and other different domain-related *attributes*.

Definition 1 (Events). *Let $\mathcal{A}$ be the set of process activities. Let $\mathcal{T}$ be the set of process timestamps. Let $\mathcal{V} = \mathcal{V}_1 \times \mathcal{V}_2 \times \ldots \times \mathcal{V}_m$ be the Cartesian product of the data attribute sets. An event is a tuple $(a, t_{start}, t_{end}, \vec{v}) \in \mathcal{A} \times \mathcal{T}^2 \times \mathcal{V}$ where a is the event activity, t_{start} and t_{end} the associated timestamps, and $\vec{v}$ the vector of associated attributes.*

A trace is a sequence of events. The same event can occur in different traces. Namely, attributes may be given the same assignment in different traces. This means that the same trace can appear multiple times, although admittedly under extremely rare conditions, and motivates why an event log has to be defined as a multiset of traces:

Definition 2 (Traces & Event Logs). *Let $\mathcal{E} = \mathcal{A} \times \mathcal{T}^2 \times \mathcal{V}$ be the universe of events. A trace σ is a sequence of events, i.e. $\sigma \in \mathcal{E}^*$.[3] An event log $\mathcal{L}$ is a multiset of traces, i.e. $\mathcal{L} \subset \mathbb{B}(\mathcal{E}^*)$.[4]*

Given an event $e = (a, t_{start}, t_{end}, \vec{v})$, the remainder uses the following shortcuts: $activity(e) = a$, $start(e) = t_{start}$, $end(e) = t_{end}$, $duration(e) = t_{start} - t_{end}$, $attr(e) = \vec{v}$. Also, given a single attribute set $\mathcal{V}_i$ it can be classified as *global* or *local*, depending on whether the values in it can vary or not in the same trace. We refer to this attributes as $global(\sigma) = \vec{g}$ and $local(e) = \vec{l}$, and so the equation $global(\sigma) \oplus local(e) = attr(e)$ holds.[5] Furthermore, given a trace $\sigma = \langle e_1, \ldots, e_n \rangle$, $prefix(\sigma)$ denotes the set of all prefixes of σ, including σ, namely $prefix(\sigma) = \{\langle\rangle, \langle e_1 \rangle, \langle e_1, e_2 \rangle, \ldots, \langle e_1, \ldots, e_n \rangle\}$.

The goal of a time prediction framework is to forecast the total execution time of a running process instance that has not completed yet, namely a *running trace*. In this paper, the problem is modeled as the estimation of a **Total Time Function** $\mathcal{T} : \mathcal{X} \to \mathbb{N}_0$ that given a running trace $\sigma' = \langle e_1, \ldots, e_k \rangle$ eventually completing as $\langle e_{k+1}, \ldots, e_n \rangle$, returns the value $end(e_n) - start(e_1)$. The input of the Total Time function is a set $\mathcal{X}$, since not every approach shares the same encoding for event logs. For instance, in [1] the authors encoded traces in an LSTM compatible input, while in [25] the traces are encoded in a *comma-separated values* file suitable for a predictor based on a Decision Tree. This requires defining the **trace-to-instance encoding function** $\rho : \mathcal{E}^* \to \mathcal{X}$ with the goal of accurately translating every trace of the event log into an input suitable for the model. This function has proven to be significantly different based on the chosen predictive approach (cf. [28]).

Figure 1 depicts an example of a trace-to-instance encoding function. In it, the trace is preprocessed by adding the past activities in newly generated columns. For each activity in the trace, the number of previous occurrences of that activity is reported in a dedicated column, encoding the number of past executions of the activity. This encoding allows tracking the frequency of all past activities but does not maintain information about their sequential order, recording only the most recent one.

To formally define this encoding function we have to first define $\rho_{\mathcal{A}}^{hist}(\langle e_1, \ldots, e_n \rangle)$. Here, for each activity $a \in \mathcal{A}$, one dimension exists in $\rho_{\mathcal{A}}^{hist}(\sigma) : \mathcal{E}^* \to (\mathbb{N})^{|A|}$ that

---

[3] The operator * refers to the Kleene star: given a set A, A^* contains all the possible finite sequences of elements belonging to A.
[4] $\mathbb{B}(X)$ indicates the set of all multisets with the elements in set X.
[5] Considering $\oplus$ as the concatenation of vectors e.g.
$[1, 3, 'request\_created'] \oplus [2, True] = [1, 3, 'request\_created', 2, True]$.

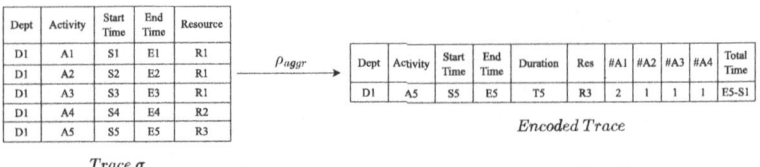

Fig. 1. Output example of a Aggregated History encoding function.

takes on a value equal to the number of events $e_i \in \sigma$ such that $activity(e) = a$ for $i = 1, \ldots, n$. The function ρ_{aggr} is then defined as: $\rho_{aggr}(\langle e_1, \ldots, e_n \rangle) = global(\langle e_1, \ldots, e_n \rangle) \oplus activity(e_n) \oplus start(e_n) \oplus end(e_n) \oplus duration(e_n) \oplus local(e_n) \oplus \rho_{\mathcal{A}}^{hist}(\langle e_1, \ldots, e_n \rangle) \oplus end(e_n) - start(e_1)$. As reported in the Sect. 1, this function has been widely adopted in the literature [3,10,24]

Since this paper aims to leverage a LLM to estimate the total time of a running trace σ', we introduce a general definition of it to establish the necessary conceptual framework:

Definition 3 (Large Language Model). *Let Σ^* be the set of all finite strings over an alphabet Σ. A Large Language Model (LLM) is here modeled as a function $LLM : \Sigma^* \to \Sigma^*$ that, given an input $s \in \Sigma^*$, returns a string $LLM(s)$, which depends on the specific interpretation of the model.*

We are aware that this is a simplification of the actual reality; however, as this paper seeks to propose a method for enhancing prediction quality through the use of an LLM, we provide a broad, simplified mathematical definition of LLMs, solely for the purpose of providing a reference function. This prediction method has been adapted from a more formal framework introduced in [23].

It is worthwhile pointing out that the reminder of this paper uses the term *training* to refer to both LLMs and Machine- and Deep-Learning models, to keep the discussion simple. However, we acknowledge that LLMs are already pre-trained: traces are provided to the LLMs as background, and are not formally used to train its internal parameters.

4 Approach For Small-Scaled Prediction Based on LLMs

This study seeks to leverage the potential of LLMs to develop a framework for PPM, particularly in scenarios where only a small amount of example traces are accessible. Leveraging their embedded knowledge, LLMs can extract and use additional information beyond the event data by incorporating the semantics of events, such as activity names, that traditional models cannot. Figure 2 depicts the proposed approach. Given an event log of completed traces $\mathcal{L}$ and a running trace σ', a trace-to-instance encoding function ρ is applied to transform them into a structured prompt. This prompt, composed of multiple components, is then used to enable the LLM to estimate the Total Time function $\mathcal{T}$.

In the remainder of this Section a new trace-to-instance encoding function ρ_{seq} suitable for LLM is introduced in Sect. 4.1, while Sect. 4.2 defines a context-based prompt suitable for employing an LLM for implementing the Total Time function defined in Sect. 3.

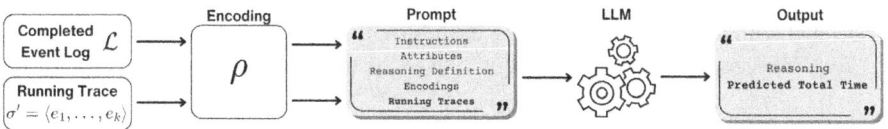

Fig. 2. Pipeline outlining the proposed method using LLMs for PPM.

4.1 An Encoding Function for LLMs

Exploiting an LLM for developing a PPM framework is a topic that has not yet been explored in process mining (cf. Sect. 2). This section introduces a new LLM-suitable trace-to-instance encoding function $\rho_{seq} : \mathcal{E}^* \to \Sigma^*$ that is associated to an encoding that will be referred as **Sequential**.

The input of the ρ_{seq} function is a trace, while the output is a textual prompt that will be later enhanced to become suitable as input for the LLM function defined in Definition 3, that will be used as Total Time function $\mathcal{T}$. Specifically, in this case the generic input set of the Total Time function $\mathcal{X}$ is equal to Σ^*. In ρ_{seq}, each trace $\sigma = \langle e_1, \ldots, e_n \rangle$ is mapped into a string composed of three main elements:

- The values of the global attributes of the trace $global(e_n)$.
- A sequence of tuples $(activity(e_i), duration(e_i))\ for\ i = 1, \ldots, n$.
- The actual value of Total Time $end(e_n) - start(e_1)$.

Formally:

$$\rho_{seq}(\langle e_1, \ldots, e_n \rangle) = global(\langle e_1, \ldots, e_n \rangle) \oplus (activity(e_1), duration(e_1)) \oplus \ldots \oplus$$
$$(activity(e_n), duration(e_n)) \oplus end(e_n) - start(e_1)$$

The sets of local attributes have been intentionally excluded, as it has been demonstrated that LLMs are constrained by two primary factors: **technical** limitations and **methodological** considerations. From a technical perspective, an LLM can only process a certain number of characters; so it becomes necessary to reduce the size of the input to stay within this maximum quantity, namely the Context Length.[6] Note that the Context Length of an LLM is not just a limitation per interaction (e.g., in a chatbot) but an inherent architectural constraint. Additionally, from a methodological standpoint, research demonstrates that the data in an LLM input does not equally impact the model's processing, and the significance of individual data points reduces as the input lengthens, even degrading its performance [15,17]. Therefore, we opted to omit local attributes.

---
[6] See https://llm-stats.com/ for an overview of Context Lengths of the latest models.

Conversely, global attributes were retained since they incorporate domain knowledge and have proved to retain more predictive power than local ones (cf. Galanti et al. [10]).

An example of application of the Sequential trace-to-instance encoding function ρ_{seq} is depicted in Fig. 3, where the example trace reported in Fig. 1 has been processed as a string. The result is a string-form Python object primarily composed of three keys: i) *Dept*, associated with the value of the corresponding global attribute; ii) *ActTimeSeq*, associated with the list of tuples where each activity and its duration are recorded; and iii) *TotalTime*, representing the total duration of the trace.

"{ Dept : D1, ActTimeSeq : { (A1, E1-S1), (A2, E2-S2), (A3, E3-S3), (A4, E4-S4), (A5, E5-S5) }, TotalTime : E5-S1} "

Fig. 3. Output example of a Sequential trace-to-instance encoding function ρ_{seq}.

It is important to note that the Sequential is not the only suitable encoding for LLMs. In fact, the proposed approach also supports the *Aggregated History Encoding*, associated with the ρ_{aggr} trace-to-instance encoding function, as the LLM can predict future values independently of the specific encoding. This flexibility allows the framework to adapt to different encoding strategies.

4.2 Context-Based Prompting Technique for LLMs in Predictive Process Monitoring

In this paper, an LLM is defined as a function that takes as input a string and returns a string containing the predicted value (cf. Definition 3). This Section uses the traces that have been encoded using a trace-to-instance encoding function ρ and incorporates them generating an input suitable for an LLM. In essence we define a prompting technique that allows the model to generate total time predictions along with corresponding reasoning procedure, starting from the encoded traces. The prompting technique is divided into seven key parts, also reported in the Example 1:

- **Initial instruction and Header:** The LLM is introduced to the task with the prompt: *"You are an expert in process mining and machine learning. Your task is to predict the 'Total Time' of process instances based on event logs of activities, where each process instance is a sequence of activities."* (Lines 1–2)
- **Attributes and Encoding description:** Contextual information specific to the process is provided, and the trace-to-instance encoding function is described (ρ_{seq} in the example). (Lines 4–12)
- **Output and Reasoning format specification:** The expected structure for predicted values is defined. (Lines 14–22 and 29–35)
- **Running Trace Format Specification:** The format for describing a running trace is specified to the model. (Lines 24–27)
- **Domain-specific background information:** Additional details about the process from which the data have been extracted. (Lines 37–38)
- **Example Data Provision:** Encoded data are provided as example to the model. (Lines 39–45)

– **Running Trace Provision:** The running trace is provided in the same format as the examples, with a custom last activity referred as "Running", as described at lines 24–27. (Lines 47–50)

Although the proposed encoding is general and applicable to various use cases, certain information within these seven components must be specified by the process analyst and may be optionally removed.

```
1   You are an expert in process mining and machine learning. Your task is to predict the 'total time' of
2   process instances based on event logs, as each process instance is a sequence of activities.
3
4   A event log is a collection of traces, where each trace represents a process instance.
5   Each trace is mapped as a sequence of activities and integers representing the minutes since the start
6   of the process.
7   The log is represented as a python list containing one dictionary for each trace. Included in it are:
8   - the key "ApplicationType", representing the type of application
9   - the key "RequestedAmount", representing the total amount of euros requested in the loan application.
10  - the key "ActTimeSeq", which value is a list of [activity, cumulative elapsed minutes]
11  - The key "total_time", which value is the total execution time in minutes from the start of the activity,
12  that is the value to predict.
13
14  All interactions will be structured in the following way, with the appropriate values filled in.
15
16  [[ ## reasoning ## ]]
17  (your step-by-step reasoning)
18
19  [[ ## answer ## ]]
20  (your predicted total time as an integer)
21
22  [[ ## completed ## ]]
23
24  In adhering to this structure, your objective is to analyze the event log, and apply reasoning to predict
25  the total time for a new case. This case belongs to a not-yet-completed process instance, represented by the
26  label "Running" in "ActTimeSeq", indicating that more activities are expected before reaching the conclusion
27  of the process instance.
28
29  Ensure to articulate each step of your thought process in the reasoning field, detailing how you identify
30  relationships with past cases and leverage your intuition about the meaning of activities to arrive at the
31  solution. The answer should be the final prediction of the total time for the given process instance.
32  Respond with the corresponding output fields, starting with the field [[ ## reasoning ## ]],
33  then [[ ## answer ## ]], and then ending with the marker for [[ ## completed ## ]].
34
35  Your task is to learn from them and predict the 'total time' values for that traces.
36
37  The process deals with a loan application process from a Dutch financial institution. It has been provided
38  in the Business Process Intelligence (BPI) challenge in 2017.
39  The following list shows some completed example cases with their total times:
40
41  {"ApplicationType": "New credit", "RequestedAmount": 5000.0, "ActTimeSeq": [["W_Complete application", 11],
42  ["W_Call after offers", 1464], ["W_Call after offers", 7486]], "total_time": "7486"}
43  {"ApplicationType": "New credit", "RequestedAmount": 15000.0, "ActTimeSeq": [["W_Complete application", 13],
44  ["W_Call after offers", 14], ["W_Validate application", 4328], ["W_Validate application", 8792]],
45  "total_time": "8792"}
46
47  Now predict the total time for this new uncompleted case, considering that the case is still running:
48
49  {"Application_1000386745": {"ApplicationType": "New credit", "RequestedAmount": 18000.0,
50  "ActTimeSeq": [["W_Complete application", 2], ["W_Call after offers", 8571], ["Running"]]}
```

Example 1. Prompting technique example for a loan application process (Bpi17). Lines that have to be provided by the process analyst are marked in bold. In the example, only 2 training traces are provided due to space limitation.

In Example 1 they have been highlighted in bold. They are specifically, (i) the domain-specific background information (Lines 37–38) and (ii) the description of the global attributes (Lines 8–9), as they contain contextual details specific to the process under study. Notably, these informations are optional and can be excluded if necessary. The remaining sections of the prompt are designed to be generic and can be applied to any event log without modification.

This modularity ensures that the framework can be associated with any encoding function and eventually be customized with details about the single process, while minimizing the effort needed for customization when applying the LLM to different process datasets.

5 Evaluation and Results

We assess the effectiveness of our approach in generating accurate predictions using a limited-size training set and comparing its performance against a state-of-the-art benchmark across five distinct case studies. To ensure a robust evaluation, we repeatedly sampled a limited number of traces, trained both a benchmark model and the LLMs on these samples, and conducted multiple experimental runs. The remainder of this section is organized as follows: The case studies are reported in Sect. 5.1, while Sect. 5.2 reports on the experimental setup, dealing with the implementation and the metrics used. Section 5.3 deals with the results and the associated comments. Finally, an example of the LLM's output in terms of values and reasoning is reported in Sect. 5.4.

5.1 Case Studies

The validity of the approaches was assessed using five different processes, for which accordant event logs are available:

Bpi12: This process has been used by the BPI challenge in 2012[7], it contains 8,616 traces, 6 different activities and 1 global attribute: *Requested_Amount*.

Bac: A process referring to a process of a Bank Institution that deals with the closures of bank accounts. It contains 32,429 completed traces, 15 different activities and 2 global attributes: *Closure_Type*, and *Closure_Reason*.[8]

Hospital: This process has been provided by an hospital emergency department. The log is made of 37,945 completed traces, contains 46 different activities and 3 global attributes, that are *Triage_Color*, *Triage_Access* and *Patient_Age*.

Purchasing: A process provided as part of the Fluxicon Disco tool and it is related to a purchase-to-pay (P2P) system, it is synthetic and generated from a model not available to the authors.[9] The extracted event log has 608 traces. It contains 21 different activities and no global attributes.

Bpi17: The subprocess for the workflow-relevant in the 2017 BPI Challenge event data, and it is provided by the same financial institution that provides the log employed in *Bpi12*. It contains 30,276 completed traces, 8 different activities and 2 global attributes that are the *Loan_Goal*, and the *Requested_Amount*.

Note that the Hospital process is not publicly available, due to legal constraints. This also means that a pre-trained LLM cannot have seen them in any form for building a-priori knowledge, although it has likely seen event logs of similar processes, which are supposedly beneficial for generating predictions. This ensures that our model has been tested on both publicly available and non-publicly available event logs, preventing eventual data leakage scenarios.

---

[7] https://doi.org/10.4121/uuid:3926db30-f712-4394-aebc-75976070e91f.
[8] https://github.com/IBM/processmining/tree/main/Datasets_usecases.
[9] https://fluxicon.com/.

Table 1. Summary statistics of the considered case studies.

Case Study	Completed Traces	Activities	Global Attributes	Mean Total Time
Bpi12	8,616	6	Requested_Amount	19,680 min
Bac	32,429	15	Closure_Type, Closure_Reason	23,615 min
Hospital	37,945	46	Triage_Color, Triage_Access, Patient_Age	188 min
Purchasing	608	21	–	115,015 min
Bpi17	30,276	8	Loan_Goal, Requested_Amount	30,240 min

A summary of case studies, associated with the mean total time of each trace, is reported in Table 1.

5.2 Experimental Setup

The whole approach has been implemented in Python and the code is publicly available.[10] For developing the Total Time function, any choice of LLM is valid, and we resorted to **Gemini 2.0 Flash Thinking**: a state-of-the-art LLM developed by Google DeepMind(see footnote 1). The model is built on a multimodal architecture designed for advanced natural language understanding and generation. This specific model has been chosen due to its status as the most powerful and freely available LLM at the time of submission. As the development of LLMs progresses, we anticipate that the performances of upcoming models using our method will likely see improvements. For the purpose of setting a benchmark for prediction, we employed Catboost [21], a state-of-the-art model predictor based on machine learning on decision trees, which has been shown to surpass existing prediction frameworks [25].

Consistently with standard supervised learning practices, we divided the event log $\mathcal{L}$ into training and test, $\mathcal{L}^{comp}$ and $\mathcal{L}^{run}$, respectively. To extract the training log we compute the earliest time t_{split} such that 80% of the identifiers related to traces of $\mathcal{L}$ are completed. This allows us to define $\mathcal{L}^{comp}$ as the set of traces of $\mathcal{L}$ completed at time t_{split}, and consequently, define $\mathcal{L}^{run}$ as $\mathcal{L} \setminus \mathcal{L}^{comp}$. The traces of the test log $\mathcal{L}^{run}$ are truncated to a set $\mathcal{L}^{trunc}$, namely the set of prefixes, that is obtained from $\mathcal{L}^{run}$ by removing every event with a timestamp larger than t_{split}: $\mathcal{L}^{trunc}$ only contains the events that occurred before time t_{split}. This procedure tries to mimic the reality at time t_{split} and it is in line with the principles introduced in [36]. The system is trained on $\mathcal{L}_{comp}$, the predictions are produced for $\mathcal{L}_{trunc}$ and tested using its completed form, $\mathcal{L}_{run}$. Furthermore, to ensure a robust generalization across various process instances and a more balanced comparison with the LLM, we randomly picked 10% of traces and used them as a validation set $\mathcal{L}_{valid}$ to apply a Cross-Validation approach to optimize the following parameters of Catboost: *learning_rate, tree_depth, training_iterations*.

The traces were encoded using two different trace-to-instance encoding functions:

- ρ_{aggr}, the Aggregated History encoding function that proven to be the best option for Catboost (cf. [10]) and introduced in Sect. 3.

---
[10] https://github.com/Pado123/gui_xrecs_presc_analytics.

– ρ_{seq}, the Sequential encoding function introduced in Sect. 4.1.

The results have been reported in terms of **Mean Absolute Error (MAE)**, defined as the mean of the absolute differences between the true values and the predicted values.

This metric is widely used because it allows more interpretability, retaining the original unit of measurement, that for this work is **minutes**. Furthermore, Sect. 5.1 presents each case study along with the mean total time, providing an indication of the typical trace length in relation to the MAE.

As highlighted in Sect. 4.1, the proposed prompting technique includes a component that the process analyst must configure for each case study. This is optional and can be excluded if desired. Therefore, *we report results for both scenarios: when this part that has to be provided by process analyst is included and when it is omitted.*

In addition, due to the fact that the proposed encoding function ρ_{seq} produces a string as output, we were unable to test Catboost using this encoding.

To test the model's ability to maintain accuracy with a reduced training set, we evaluated its performance by progressively shrinking the training data. Specifically, we reduced the number of traces to 100, 10, and 2. This experiment allowed us to assess the model's robustness and accuracy when exposed to increasingly sparse datasets, simulating real-world scenarios where training data may be limited or incomplete. To mitigate the effects of statistical variation due to sampling, experiments have been repeated 20 times, and the reported results represent the mean and standard deviation of these runs, in order to highlight not only the value of MAE but also the possible uncertainty related to the statistical sampling.

5.3 Predictions Results

Table 2 reports the results of the experiments for the different case studies, along with the results from the Catboost banchmark, for which the relevant hyperparameters have been optimized. Results are reported when the training sets are composed of 2, 10, and 100 traces. The column *Model* indicates the predictive model: the benchmark as well as the LLM using the encodings based on aggregated history ρ_{aggr} and the whole trace ρ_{seq} when both the context is and is not used. Recall that, by context, we mean the context-specific background information and the description of global attributes (cf. Sect. 4.2). For completeness, the table also reports on the result of the benchmark when the whole training log has been used, whose size is reported in column *# Train Traces* in terms of the amount of traces. Mean and median values of the each test log are also have been used naïve predictors with the aim to provide a baseline, and results are reported. Recall that, for each process, predictive model, and size of the training set (e.g. 100 traces), the experiments have been repeated 20 times and taken different samples of the same size (e.g. 100 traces), as discussed in Sect. 5.2. The cell numbers represent the mean across the 20 experiments, with the standard deviation indicated after the ± symbol, except in the case of the benchmark on the full log, where the sampling procedure was not performed.

Our LLM-based framework for predictive process monitoring always outperforms the benchmark. Except for the Bpi12 case study, the ρ_{seq} encoding function outperforms the aggregated history ρ_{aggr} encoding function. This shows that, when Gemini

Table 2. Accuracy results in terms of Mean Absolute Error (minutes). For each case study and reduced train set, the lower MAE are highlighted in bold.

Process	Predictive Model	Full log	100 Traces	10 Traces	2 Traces	Mean	Median	# Train Traces
Bpi12	Benchmark	**6,846**	9,394 ± 114	10,811 ± 111	11,594 ± 3,819	9,373	9,374	6,892
	LLM ρ_{aggr} + Context	–	**398 ± 440**	**406 ± 428**	**400 ± 425**			
	LLM ρ_{seq} + Context	–	7,258 ± 878	8,450 ± 1,402	7,549 ± 2,049			
	LLM ρ_{aggr}	–	8,001 ± 2,024	10,205 ± 328	11,067 ± 2,560			
	LLM ρ_{seq}	–	5,328 ± 1,368	5,557 ± 1,696	5,622 ± 1,607			
Bac	Benchmark	**2,647**	6,393 ± 387	8,181 ± 1,633	9,634 ± 1,929	6,172	5,998	25,901
	LLM ρ_{aggr} + Context	–	7,731 ± 1,471	9,481 ± 1,407	12,045 ± 2,302			
	LLM ρ_{seq} + Context	–	2,510 ± 471	6,066 ± 920	6,741 ± 2,692			
	LLM ρ_{aggr}	–	4,894 ± 1,685	8,517 ± 1,218	14,008 ± 1,539			
	LLM ρ_{seq}	–	**2,500 ± 836**	**5,419 ± 1,098**	**5,725 ± 1,273**			
Hospital	Benchmark	**253**	254 ± 1	266 ± 11	277 ± 24	326	254	30,212
	LLM ρ_{aggr} + Context	–	433 ± 418	406 ± 428	400 ± 425			
	LLM ρ_{seq} + Context	–	87 ± 28	98 ± 29	**87 ± 30**			
	LLM ρ_{aggr}	–	401 ± 416	410 ± 431	389 ± 389			
	LLM ρ_{seq}	–	**86 ± 29**	**92 ± 23**	91 ± 32			
Purchasing	Benchmark	**19,639**	26,682 ± 697	56,589 ± 1,733	70,607 ± 4,648	52,377	45,650	486
	LLM ρ_{aggr} + Context	–	46,902 ± 8,299	50,774 ± 12,891	56,277 ± 4,205			
	LLM ρ_{seq} + Context	–	12,081 ± 2,925	14,015 ± 1,651	14,475 ± 4,141			
	LLM ρ_{aggr}	–	37,899 ± 3,657	42,225 ± 3,237	56,127 ± 6,740			
	LLM ρ_{seq}	–	**11,767 ± 2,339**	**12,593 ± 2,722**	**12,071 ± 3,466**			
Bpi17	Benchmark	**9,729**	12,565 ± 43	13,166 ± 783	16,701 ± 4,729	13,189	12,617	24,221
	LLM ρ_{aggr} + Context	–	13,638 ± 3,302	12,732 ± 3,189	14,425 ± 4,011			
	LLM ρ_{seq} + Context	–	8,032 ± 1,338	8,734 ± 1,930	8,787 ± 1,928			
	LLM ρ_{aggr}	–	11,185 ± 3,005	11,043 ± 2,736	14,722 ± 2,958			
	LLM ρ_{seq}	–	**6,931 ± 2,605**	**6,993 ± 2,491**	**7,656 ± 3,107**			

is provided with more information for the predictive task, it is generally capable of discerning the information that is relevant for the specific case study, where an aggregated history imposes – generally wrongly – on the information that is relevant. This is far from being obvious: the feature engineering and the choice of the right aggregation is usually relevant for accuracy when predictive process monitoring is performed using Machine- and Deep-Learning predictors [28]. Remarkably, the context does not seem relevant in all case studies when the ρ_{seq} encoding function is employed, except for the Bpi12 process: the results with and without context are basically equivalent, namely the boundaries set by the standard deviations lie within the same intervals. This is likely due to the fact that the name of the global attributes and/or of the process activities allowed the LLM to derive the context from the pre-trained knowledge. Therefore, the results seem to suggest that the context is particularly relevant in non-conventional settings.

Interestingly, the Gemini predictive accuracy remains remarkably consistent whether it receives two or one hundred traces as input, with the exception of the Bac case study. The reason for this discrepancy in Bac is unclear. However, the general observation is that relying solely on two traces might not consistently yield a reliable predictive model. Also, our Gemini framework notably achieves higher accuracy with less than 100 traces, if compared with the Catboost benchmark trained on the whole event dataset. This superior performance is observed across all case studies, though the results for Bac are less definitive.

5.4 Example of Reasoning to Achieve Predictions

The Gemini's output is not limited to returning the predicted value: Gemini also reports on the reasoning put in place to conclude the prediction. This is in fact a form of explanation of the predictions, although user studies would be necessary to assess whether they are indeed comprehensible by non-LLM and non-AI experts.

```
1   [[ ## reasoning ## ]]
2   The new case is for a 'New credit' application with a 'RequestedAmount' of 18000.
    The process instance
3   has executed activities 'W_Complete application' and 'W_Call after offers'.
4   The process is currently 'Running'.
5   To predict the total time, we will look at historical cases with similar characteristics.
6
7   1. Filter for similar Application Type and Requested Amount:
8      We focus on cases with 'ApplicationType' as 'New credit' and 'RequestedAmount'
9      in a range around 18000.0 (e.g., 10000.0 to 25000.0) to find comparable instances.
10
11  2. Identify relevant historical cases:
12     From the provided log, we select the cases that match 'New credit' application
13     type and have 'RequestedAmount' within the range of 10000.0 to 25000.0.
14     We list their total times:
15     10000.0: 9713, 10000.0: 18803, 11500.0: 4771, 12000.0: 18924, 12000.0: 8102,
16     13000.0: 6645, 13000.0: 21550, 14000.0: 9713, 15000.0: 8049, 15000.0: 17011
17
18  3. Calculate the median total time:
19     To get a robust estimate and reduce the impact of outliers,
20     we calculate the median
21     of these total times.
22     Sorted total times: 4771, 5299, 5975, 5975, 6645, 6645, 7432, 8049, 8102, 9713, 9713,
23     12991, 13056, 17011, 17434, 18803, 18924, 19826, 21550, 71762.
24     Since there are 20 values, the median is the average of the 10th and 11th sorted values,
25     which are 9713 and 9713. Median = (9713 + 9713) / 2 = 9713.
26
27  4. Predict total time:
28     We use the median total time of the similar historical cases as our prediction
29     for the new running case.
30
31  [[ ## answer ## ]]
32  9713
33  [[ ## completed ## ]]
```

Example 2. LLM's output and reasoning returned after the prompting technique example for a loan application process.

Nonetheless, it is a valuable addition to this paper to report and comment on an instance of the prediction's reasoning given as output by Gemini. To do so, Example 2 reports the output returned by the LLM for Example 1.

The output is returned at the end (see line 32 of Example 2). The reasoning of this example is as follows: Gemini has focused on the traces of the provided training log in which for which the 'ApplicationType' is of type *'New credit'* and the request amount is in the range of 10000 to 25000 (see lines 7–9). This leads to 20 traces (see lines 15–16), for which the median value is computed (see lines 24–25), which is used as predicted value (see explanations given at lines 28 and 29).

This reasoning certainly provides a valid explanation of the reasoning behind why this prediction is provided. However, this is only an example of the reasoning procedure performed by Gemini, and we observed that the output prompt can vary on the basis of the context and the number of training traces provided. Effort is necessary to homogenize the output prompt, and to steer towards a solution that is more explainable for process stakeholders. The latter still requires a research investigation that involves users and multiple analysis. We aim to move towards this direction as future work.

The given reasoning explains the prediction, but it is only an example of the reasoning procedure that is performed by Gemini, which changes with context and data. To improve explainability for process stakeholders, a more consistent and standardize output is necessary. This requires in-depth research involving user studies and varied analytical methods, a direction we intend to explore as future work.

6 Conclusions

This research introduces a novel Predictive Process Monitoring (PPM) framework that leverages on the capabilities of Large Language Models (LLMs) to overcome the challenge of limited data. Traditional PPM techniques rely on Machine- and Deep-Learning, which notoriously struggle with small datasets. LLMs have the capability to use the pre-trained knowledge to generalize even when small-scale event logs are provided.

With the premises above, this paper reports on our contribution to design a PPM framework that is based on LLMs. Two alternative prompts have been leveraged to encode the training sets and the contextual information. The framework has been implemented in Python, using Gemini as LLM, which has shown to be the best performing on different benchmarks among those freely available. Experiments have shown that our LLM-based framework enables making accurate predictions with small-scale event logs that are composed by less than 100 traces, with significant accuracy improvements with respect to standard methods from the PPM literature.

Section 5.4 has already reported that our future work will certainly focus on the output prompt generated by LLMs, aiming at its standardization and at its consequent use as explanation method. However, this requires user studies to evaluate alternatives and assess their benefits. While this research focuses on predictive process monitoring, a natural extension is to move toward recommender systems where not only are predictions given but also corrective actions are provided to recover the executions that are predicted to not achieve a satisfactory outcome. In this paper, the outcome is only defined in terms of duration of process executions, but KPIs can generally be of different natures (costs, customer satisfaction, etc.): a natural extension is indeed to extend our predictive framework towards KPIs that are others than execution duration. Finally, an interesting direction of future work is also related to investigating zero-prompting techniques, where the predictive model is provided with no specific examples, in line with the goal of this work to provide a more adaptable and scalable framework.

Acknowledgment. We acknowledge the support of the project "Future AI Research (FAIR) - Spoke 2 Integrative AI - Symbolic conditioning of Graph Generative Models (SymboliG)" funded by the European Union under the National Recovery and Resilience Plan (NRRP), Mission 4

Component 2 Investment 1.3 - Call for tender No. 341 of March 15, 2022 of Italian Ministry of University and Research – NextGenerationEU, Code PE0000013, Concession Decree No. 1555 of October 11, 2022 CUP C63C22000770006.

References

1. Ali, M.A., Dumas, M., Milani, F.: Enhancing the accuracy of predictors of activity sequences of business processes. In: Research Challenges in Information Science. Springer Nature Switzerland (2024)
2. Ashok, A., et al.: Context is key: a benchmark for forecasting with essential textual information. In: NeurIPS Workshop on Time Series in the Age of Large Models (2024)
3. Buliga, A., et al.: Uncovering patterns for local explanations in outcome-based predictive process monitoring. In: Business Process Management - 22nd International Conference, BPM 2024, Krakow, Poland, September 1-6, 2024, Proceedings. LNCS, vol. 14940, pp. 363–380. Springer (2024)
4. Ceravolo, P., Comuzzi, M., De Weerdt, J., et al.: Predictive process monitoring: concepts, challenges, and future research directions. Process Sci. 1(2), 2 (2024)
5. Cunzolo, M.D., et al.: Robust solutions via optimisation and predictive process monitoring for the scheduling of the interventional radiology procedures. Int. Trans. Oper. Res. (2025)
6. Dani, V.S., et al.: Event log extraction for process mining using large language models. In: Cooperative Information Systems - 30th International Conference, CoopIS 2024, Porto, Portugal. LNCS, Springer (2024)
7. Di Francescomarino, C., Ghidini, C.: Predictive Process Monitoring, pp. 320–346. Springer International Publishing, Cham (2022)
8. Elyasi, K.A., van der Aa, H., Stuckenschmidt, H.: PGTNet: a process graph transformer network for remaining time prediction of business process instances. In: Advanced Information Systems Engineering - 36th International Conference, CAiSE 2024, Limassol, Cyprus, 2024, Proceedings. LNCS, vol. 14663, pp. 124–140. Springer (2024)
9. Estrada-Torres, B., del Río-Ortega, A., Resinas, M.: Mapping the landscape: exploring large language model applications in business process management. In: Enterprise, Business-Process and Information Systems Modeling. Springer Nature Switzerland (2024)
10. Galanti, R., et al.: An explainable decision support system for predictive process analytics. Eng. Appl. Artif. Intell. **120**, 105904 (2023)
11. Gruver, N., Finzi, M., Qiu, S., Wilson, A.G.: Large language models are zero-shot time series forecasters. Adv. Neural. Inf. Process. Syst. **36**, 19622–19635 (2023)
12. Kim, J., Comuzzi, M., Dumas, M., Maggi, F.M., Teinemaa, I.: Encoding resource experience for predictive process monitoring. Decision Support Syst. **153** (2022)
13. Kourani, H., Berti, A., Schuster, D., van der Aalst, W.M.P.: Process modeling with large language models. In: Enterprise, Business-Process and Information Systems Modeling, pp. 229–244. Springer Nature Switzerland, Cham (2024)
14. Kubrak, K., Botchorishvili, L., Milani, F., Nolte, A., Dumas, M.: Explanatory capabilities of large language models in prescriptive process monitoring. In: Business Process Management - 22nd International Conference, BPM 2024, Krakow, 2024, Proceedings. LNCS, vol. 14940, pp. 403–420. Springer (2024)
15. Kuratov, Y., et al.: BABiLong: testing the limits of LLMs with long context reasoning-in-a-haystack. Adv. Neural. Inf. Process. Syst. **37**, 106519–106554 (2024)
16. Lashkevich, K., Milani, F., Avramenko, M., Dumas, M.: LLM-assisted optimization of waiting time in business processes: a prompting method. In: Business Process Management - 22nd International Conference, BPM 2024, Krakow, 2024, Proceedings. LNCS, vol. 14940, pp. 474–492. Springer (2024)

17. Li, T., Zhang, G., Do, Q.D., Yue, X., Chen, W.: Long-context LLMs struggle with long in-context learning. Transactions on Machine Learning Research (2024)
18. Oyamada, R., Tavares, G., Barbon Junior, S., Ceravolo, P.: Enhancing Predictive Process Monitoring with Time-Related Feature Engineering. Springer Nature Switzerland (2024)
19. Padella, A., de Leoni, M., Dogan, O., Galanti, R.: Explainable process prescriptive analytics. In: 2022 4th International Conference on Process Mining (ICPM), pp. 16–23 (2022)
20. Pasquadibisceglie, V., Appice, A., Malerba, D.: LUPIN: a LLM approach for activity suffix prediction in business process event logs. In: 2024 6th International Conference on Process Mining (ICPM), pp. 1–8 (2024)
21. Prokhorenkova, L.O., Gusev, G., Vorobev, A., Dorogush, A.V., Gulin, A.: CatBoost: unbiased boosting with categorical features. In: NeurIPS (2018)
22. Rebmann, A., Schmidt, F.D., Glavas, G., van der Aa, H.: Evaluating the ability of LLMs to solve semantics-aware process mining tasks. In: 6th International Conference on Process Mining, ICPM 2024, Kgs. Lyngby, Denmark, October 14-18, 2024, pp. 9–16. IEEE (2024)
23. Requeima, J., Bronskill, J., Choi, D., Turner, R.E., Duvenaud, D.K.: LLM processes: numerical predictive distributions conditioned on natural language. In: Advances in Neural Information Processing Systems (2024)
24. Rizzi, W., Simonetto, L., Di Francescomarino, C., Ghidini, C., Kasekamp, T., Maggi, F.M.: Nirdizati 2.0: new features and redesigned backend. In: Proceedings of the Dissertation Award, Doctoral Consortium, and Demonstration Track at BPM 2019. vol. 4220, pp. 154–158. ceur-ws.org (2019)
25. Shoush, M., Dumas, M.: White box specification of intervention policies for prescriptive process monitoring. Data Knowl. Eng. **155**, 102379 (2025)
26. Snoeck, M., Verbruggen, C., Smedt, J.D., Weerdt, J.D.: Supporting data-aware processes with MERODE. Softw. Syst. Model. **22**(6), 1779–1802 (2023)
27. Stevens, A., De Smedt, J., Peeperkorn, J., De Weerdt, J.: Assessing the robustness in predictive process monitoring through adversarial attacks. In: 2022 4th International Conference on Process Mining (ICPM), pp. 56–63 (2022)
28. Tavares, G.M., Oyamada, R.S., Barbon, S., Ceravolo, P.: Trace encoding in process mining: a survey and benchmarking. Eng. Appl. Artif. Intell. **126**(Part D), 107028 (2023)
29. Tax, N., Verenich, I., La Rosa, M., Dumas, M.: Predictive business process monitoring with LSTM neural networks. In: Dubois, E., Pohl, K. (eds.) Advanced Information Systems Engineering. Springer International Publishing (2017)
30. Taymouri, F., Rosa, M.L., Erfani, S., Bozorgi, Z.D., Verenich, I.: Predictive business process monitoring via generative adversarial nets: the case of next event prediction. In: Fahland, D., Ghidini, C., Becker, J., Dumas, M. (eds.) Business Process Management, pp. 237–256. Springer International Publishing, Cham (2020)
31. Teinemaa, I., Dumas, M., La Rosa, M., Maggi, F.: Outcome-oriented predictive process monitoring: Review and benchmark. ACM Trans. Knowl. Discov. Data **13** (2017)
32. Vacareanu, R., Negru, V.A., Suciu, V., Surdeanu, M.: From words to numbers: your large language model is secretly a capable regressor when given in-context examples. In: First Conference on Language Modeling (2024)
33. van der Aa, H., Rebmann, A., Leopold, H.: Natural language-based detection of semantic execution anomalies in event logs. Inf. Syst. **102**, 101824 (2021)
34. Vazifehdoostirani, M., Genga, L., Dijkman, R.: Encoding high-level control-flow construct information for process outcome prediction. In: 2022 4th International Conference on Process Mining (ICPM), pp. 48–55 (2022)
35. Verenich, I., Dumas, M., La Rosa, M., Maggi, F., Teinemaa, I.: Survey and cross-benchmark comparison of remaining time prediction methods in business process monitoring. ACM Trans. Intell. Syst. Technol. **10** (2019)

36. Weytjens, H., Weerdt, J.D.: Creating unbiased public benchmark datasets with data leakage prevention for predictive process monitoring. In: Business Process Management Workshops - BPM 2021 International Workshops, Rome, Italy, 2021, Revised Selected Papers. Lecture Notes in Business Information Processing, vol. 436, pp. 18–29. Springer (2021)
37. Zimmermann, L., Zerbato, F., Weber, B.: What makes life for process mining analysts difficult? A reflection of challenges. Softw. Syst. Model. 1–29 (2023)

Progression: A Lightweight BPMN Engine Simplifying the Execution and Monitoring of Process Models

Thomas M. Prinz[1][(✉)], Yongsun Choi[2], and Anja Vetterlein[1]

[1] Course Evaluation Service, Friedrich Schiller University Jena, Jena, Germany
{Thomas.Prinz,Anja.Vetterlein}@uni-jena.de
[2] Department of Industrial and Management Engineering, Inje University, Gimhae, Republic of Korea
yschoi@inje.edu

Abstract. Modern workflow management systems are for experts to unleash their power and get process models working. This can be a major obstacle introducing such a system, especially, to public institutions or small and medium-sized enterprises. To improve this situation, this paper introduces a new lightweight system called *Progression*. *Progression* builds up on research outcomes of the last decades (e. g., soundness verification with detailed diagnostic information, liberal OR-join semantics, and loop decomposition). They enable *Progression* only to execute sound acyclic process models interacting with signals (or messages). The system uses a new *Track-and-Trace Semantics*, which allows for local, distributed, and unsupervised executions of process models and increases the security through restricted task accesses. Furthermore, the paths being executed are entirely traceable. Besides the semantics, this paper discusses the architecture and current implementation of *Progression* from process modeling to its execution.

Keywords: Workflow Management System · Business Process Model · Progression · Trace · Security

1 Introduction

Business Process Management (BPM) promises that making implicit business process models explicit and automating them with a *workflow management system* (WMS) allows for prompt reaction on ever-changing business environments and then, as a result, improves the performance of organizations [10]. Of course, unveiling hidden process models and automating them are mostly of benefit [10]. However, the currently available tools are complicated and require a lot of expert knowledge about these tools: In a simulation study of implementing the BPMN process model illustrated in Fig. 1 on modern WMS (such as *Camunda* [7], *Bizagi* [4], *Signavio* [32], *Flowable* [16], and *Bonita Platform* [5]) was ambivalent—the WMS are indeed powerful tools for process model automation, but the sheer power and complexity of these WMS made it difficult and sometimes almost impossible to get a working process instance in a short period [28]. With some training and consulting, we are confident that these tools will unleash their power. However, in our experience-based opinion, time intensive and

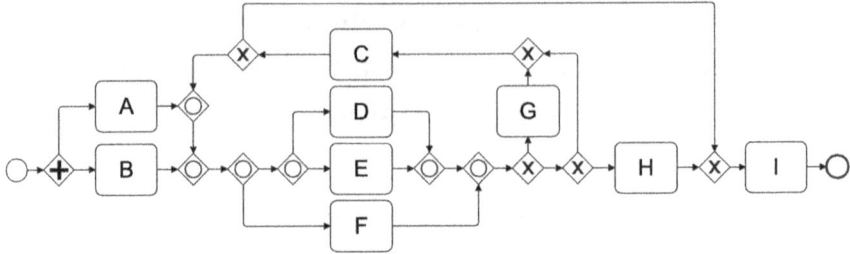

Fig. 1. A (sound) cyclic process model using BPMN (taken from [28]).

costly trainings and consulting are a big hurdle for their adaption, especially for public institutions and small and medium-sized companies. Furthermore, sometimes, structuring general purpose software with process models would be a great benefit, especially regarding the emerging technologies of microservices and cloud environments or just by implementing a UML activity diagram evolved during the design. However, the current heavy-weighted WMS are too powerful to be well usable for such purposes.

In this paper, we will introduce a new lightweight WMS, called *Progression*, with a simple architecture of its execution engine that is based on transformation of the process models rather than executing each element directly. *Progression* evolved out of the *Coast* information system [25], which supports building questionnaires, surveys, and reports for the purpose of teaching quality assurance at the University of Jena, Germany. With diverse growing demands for the *Coast* system, a more flexible and user-centric feature has been requested to design process models, naturally leading to consider existing WMS. However, their "overkill" features related to integration with the *Coast* system led to the design and implementation of a new lightweight WMS. *Progression* is based on BPM research results (as will be shown in Sect. 2) to simplify process execution and, therefore, to reduce the implementation overhead of such a system. This WMS has an extensible architecture and concentrates on the core of a WMS: Process modeling and execution. The mentioned research results (mainly *soundness* [1, 13, 24, 27, 31], *OR-join semantics* [8, 12, 19, 23, 35], and *loop decomposition* [26–28]) enable to only execute *sound* and *acyclic* process models (which do not have cycles). A *verification mechanism* in the process modeler unveils unsound process models during each modification of the model [27, 29]. *Loop decomposition* decomposes sound cyclic process models into acyclic ones with the same behavior and interacting via signals / messages [28]. The focus on *acyclic* process models and the "most liberal semantics" of converging inclusive gateways ("OR-joins") [23] further enable a new entirely *local* [8] execution semantics that we call *Track-and-Trace Semantics* (TTS). The core idea of TTS consists of *six* kinds of tokens for flows, signals, and nodes (cf. Table 1 for their representation): "Unvisited" tokens, "live" and "previously live" tokens, "dead" and "previously dead" tokens, and, finally, "pending" tokens representing nodes under execution. The tokens are, similar to classical *Token-game semantics*, propagated through the process model. However, instead of leaving flows undetermined that do not get the "control", TTS explicitly mark them as *dead*. In addition, *live* and *dead* tokens signalize the flows, nodes, and signals being currently executed. They become *previously live* or

Table 1. Kinds of tokens and their representation.

Kind of token	Representation	Kind of token	Representation
Unvisited / none	⊖	Pending	⊠
Live	•	Previously live	⊤
Dead	†	Previously dead	⊥

previously dead tokens *after* execution (or propagation) such that the user can see in retrospect how a process model was executed. As an important feature, TTS increases the security of process execution as it can detect erroneous or manipulated task executions at runtime.

Besides TTS, this paper explains the software architecture of *Progression* with its core components and how they interact to achieve the execution of process models. The architecture allows for deploying process models as unsupervised choreographies in a network of interacting services. An extendable implementation of *Progression* is currently available in PHP being actively used in *Coast* [25].

This article is structured as follows: Sect. 2 discusses the theoretical background that evolved in the past years to enable a lightweight implementation of a WMS such as *Progression*. This theoretical background is used to define the *Track-and-Trace Semantics* in Sect. 3. Subsequently, Sect. 4 explains how TTS is realized in the *Progression* WMS architecture, whereas Sect. 5 discusses the current state of implementation. Finally, Sect. 6 concludes this paper.

2 Theoretical Background

The architecture of the *Progression* WMS builds up on a strong theoretical foundation being the result of more than 25 years of research in BPM and Petri nets.

Soundness. Since its introduction by Van der Aalst [1], *soundness* is a core property of business process model verification. Soundness was first defined for *workflow nets*—proper Petri nets (a digraph with *transitions, places,* and *flows*) with a single source and a single sink place, in which all places and transitions are on a path from this source to the sink. For a workflow net, soundness ensures the proper termination of each possible execution and that each transition can be fired in at least one execution. Van der Aalst [1] has shown that the soundness of a workflow net correlates with the *liveness* and the *boundedness* of its *short-circuit net*, in which the sink is connected via a transition to the source. *Liveness* ensures that starting from any marking, which is reachable from an initial marking, there is a further reachable marking for each transition, in which this transition is enabled. *Boundedness* ensures that there is a maximal number of tokens for all places in each reachable marking from an initial marking. 1-boundedness (*safeness*) restricts this maximum number of tokens on each place to one.

Sadiq and Orlowska [31] introduced the two errors *deadlock* and *lack of synchronization* for industrial business process models (e. g., BPMN) as *workflow graphs*—graphs with single start and end nodes as well as special nodes for different types of gateways. A *deadlock* is a reachable state of a process model, in which the execution blocks (locally); whereas in a *lack of synchronization*, the same task is executed twice in series unexpectedly [24]. It appeared that the absence of deadlocks and lack of synchronization correlates with the soundness property [11]. This correlation became more obvious after Favre et al. [13] have shown that all workflow graphs (without inclusive gateways) can be translated to *simple free-choice* workflow nets and back. *Simple freechoiceness* of Petri nets ensures that if a place has at least two transitions as direct successors, then these transitions only have that place as the single direct predecessor, i.e., the decision, which transition is fired, is "free". This is comparable to a diverging XOR-gateway in BPMN. Fortunately, sound simple free-choice workflow nets are safe so that the number of tokens in each place is limited by one [34].

Although soundness is considered a minimum quality criterion of business process models [9], the support of ensuring soundness during modeling is weak: None of eight popular investigated WMS (*Camunda* [7], *bpmn.io* [6], *Bizagi* [4], *Signavio* [32], *Activiti* [2], *jBPM* [18], *Flowable* [16], and *Bonita Platform* [5]) either performs a soundness verification or gives diagnostic information. One emerging problem was that most soundness approaches are complex, require expert knowledge, are difficult to implement, and are computationally extensive [26]. In [26], we showed that it is possible to decompose a sound cyclic process model (with *loops* as strongly connected components (SCC)) into a set of process models without loops (called *acyclic*). This decomposition is explained in more detail later in this section. In doing this decomposition, soundness can be efficiently (in cubical computational time complexity) checked by applying five simple rules [27]: (1) Each *loop entry* (where a loop can be entered) must be an XOR- or OR-join, (2) each *loop exit* (where a loop can be left) must be an XOR-split, (3) each *back join* (where the *do-body* of the loop as the region between the loop entries and the loop exits can be entered from inside the loop) must be an XOR- or OR-join, (4) the acyclic transformed process model must be sound, and (5) the acyclic decomposed process model of each loop must be sound. For this reason, existing soundness approaches for acyclic process models can be applied, which provide detailed diagnostic information such as that of Favre and Völzer [14] or that of Prinz and Amme [23]. These foundations in the soundness of industrial process models are used in the architecture to restrict all process models being executed on the engine to be sound.

Semantics of Inclusive Gateways. In general, the previously discussed soundness property is defined on workflow graphs *without* inclusive gateways. Inclusive gateways— also known as *OR-splits* and *OR-joins*—have difficult semantics being discussed intensively in research in the past [8, 12, 19, 23, 35]. Although there is an official semantics described in the BPMN 2.0 specification [21], it has less support in popular WMS since just 5 out of 9 seem to implement it correctly [28]. This coincides with the observations of Corradini et al. [8]. Moreover, the process model in Fig. 1 cannot be executed without running into a deadlock or lack of synchronization on any investigated modern execution engine although it could be executed correctly as shown in [23]. The OR-join semantics of Fahland and Völzer [12] and Prinz and Amme [23] were explicitly designed for sound process models. The semantics of Prinz and Amme was shown to be the "most liberal" one in the sense that there is no other semantics being able to execute more process models in a sound manner (i.e., without deadlock or lack of synchronization) [23]. For this reason, the described *Progression* WMS concentrates on that semantics. Another reason is that it correlates with the OR-join semantics after applying loop decomposition [26]. Actually, as we show later, our architecture does not require a complicated OR-join semantics as it only executes acyclic process models eventually, for which the "waiting" semantics of Völzer [35] is widely accepted. However, this fact is important from a theoretical view as it ensures that the engine does not change the behavior of the original process models being executed [28].

Loop Decomposition. As mentioned before, the architecture of *Progression* only needs to execute *acyclic* process models. Sound acyclic process models have some strong advantages: (1) The semantics of OR-joins is widely accepted and follows a simple approach by waiting until no token may reach it [35], (2) each node can at most be executed once, and (3) each execution of a process model can be represented as a subgraph of the process model (usually called *instance subgraph* or *run*) [30,31]. Furthermore, acyclicity is the foundation of the semantics later used during execution.

Naturally, not every process model occurring in practice is acyclic. Since *Progression* ensures and requires sound process models, the method of *loop decomposition* [26–28] can be applied. *Loop decomposition* identifies *loops* as SCCs in cyclic process models and replaces them with either special *loop nodes* representing the loop (similarly to a sub-process) or with a special structure of throwing and catching intermediate events [28]. Either way, loop decomposition can separate a sound process model with loops into a set of sound process models *without* loops but with the same behavior (even with inclusive gateways). In [28], we describe how to execute any sound cyclic BPMN model with inclusive gateways on most existing process engines (7 out of 9) by using loop decomposition and signal (or message) exchanges. Figure 2 illustrates the decomposition of the process model in Fig. 1. In doing this, the resulting semantics of inclusive gateways coincides with the most liberal semantics in [23]. Loop decomposition ensures that no activity in any resulting acyclic process model can be executed at the same time with one of its copies (as sometimes introduced during decomposition). This allows the global state of the decomposed process models to be represented as usual as token-game-like (if desired) but uses different semantics in the engine.

Since soundness can be checked efficiently with good diagnostic information during each step of modeling and loop decomposition allows for separating a cyclic process model into a set of acyclic process models, *Progression* only focuses on the execution of (sound) acyclic process models. This limitation on executing acyclic process models enables a safer and entirely local execution of the models without the necessity of supervision as we will show later—even for inclusive gateways.

Acyclic Workflow Graphs. Since BPMN is rich in different symbols, artifacts, etc., more foundational considerations are done using *workflow graphs* [31,35]. The following definition extends the usual definition of workflow graphs with *signals* as done in [28] but limits them to be acyclic in the context of this paper:

Definition 1 (Workflow Graph). A *workflow graph* $(N, E, \lambda, \Lambda, \tau, \gamma, M)$ refers to an acyclic directed graph (N, E). N is a set of nodes and $E \subseteq N \times N$ is a set of edges connecting nodes. For an edge $(s,t) = e \in E$, s is called the *source*, and t is called

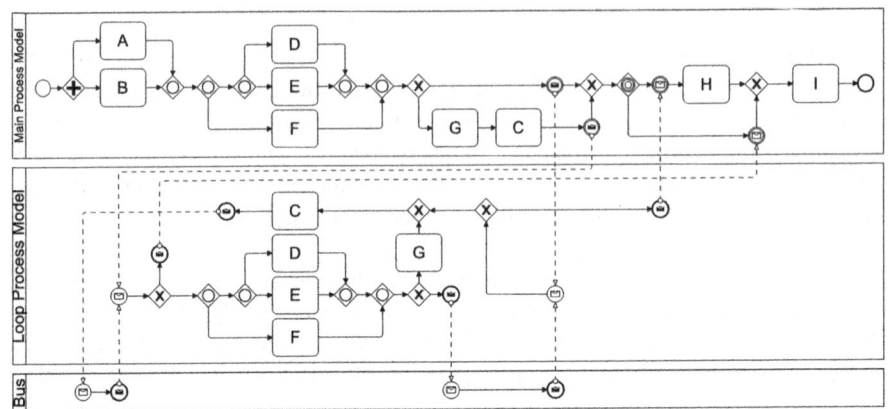

Fig. 2. The process model of Fig. 1 after loop decomposition with signal/message exchanges between the acyclic process models resulting from loop decomposition (taken from [28]).

the *target* of e. A node n has incoming edges, $\triangleright n = \{(s,n) \in E\}$, and outgoing edges, $n\triangleleft = \{(n,t) \in E\}$. Λ is a set of labels $\{Start, Task, AND, OR, XOR, End\}$ and $\lambda: N \mapsto \Lambda$ is a total mapping that assigns a label to each node. An assigned label defines various properties to n:

- There is at least one node with the label *Start* (the *start nodes*) with no incoming but exactly one outgoing edge. There is at least one node with the label *End* (the *end nodes*) with exactly one incoming but no outgoing edge. Each node lies on a path from a start to an end node.
- Nodes with the label *Task* (*tasks*) have exactly one incoming and one outgoing edge. These nodes indicate the specific work that needs to be accomplished.
- All other nodes have labels *AND*, *OR*, or *XOR*. They are separated into *split* and *join* nodes. Split nodes (i.e., AND-split, OR-split, and XOR-split) have exactly one incoming ($|\triangleright n| = 1$) and at least two outgoing edges ($|n\triangleleft| \geq 2$). Join nodes (i.e., AND-join, OR-join, and XOR-join) have at least two incoming edges ($|\triangleright n| \geq 2$) and exactly one outgoing edge ($|n\triangleleft| = 1$).

M is a finite set of *signals* with $\tau: N \mapsto \mathfrak{P}(M)$ is a total mapping of *thrown* and $\gamma: N \mapsto \mathfrak{P}(M)$ is a total mapping of *caught* signals. Signals are specific for tasks, XOR-splits (for representing event-based gateways), start, and end nodes (i.e., $\forall n \in N: \tau(n) \neq \emptyset \implies \lambda(n) \in \{Task, End\}$ and $\forall n \in N: \gamma(n) \neq \emptyset \implies \lambda(n) \in \{Task, Start\} \vee (\lambda(n) = XOR \wedge |n\triangleleft| \geq 2)$).

3 Track-and-Trace Semantics

Resulting from the theoretical background in Sect. 2, *Progression* only needs to execute *sound* and *acyclic* process models. Instead of using classical token-game semantics as described in the BPMN specification [21], we implemented a new semantics that we call

Track-and-Trace Semantics (TTS). TTS reuses ideas of *Dead Path Elimination* of BPEL [20]. *Dead Path Elimination* can be interpreted as using two distinct types of tokens: • (*live*) and † (*dead*). • tokens denote an active control flow, while † tokens represent edges that are not followed. Nodes with only † tokens are *skipped*. This differs from token-game semantics, where edges without tokens are undetermined.

Our proposed TTS adds further *four* kinds of tokens that offer fast validation checks (therefore, increased security) and further, allow for reading token traces *after* a process model has been executed. TTS enables the *local* answering of questions such as "Can an edge still receive a token in a reachable state?" and "Has an edge already carried a token in a previous state?" by focusing on a single node alone. Of course, some questions cannot always be answered until XOR- and OR-splits do not have decided, which of their outgoing edges get a • token.

In total, TTS provides *six* kinds of token (cf. Table 1): ⊖ (none), • (live), † (dead), ⊤ (previously live), ⊥ (previously dead), and ⊠ (pending). †, ⊖, ⊥, and ⊤ all mean that a flow has not the "control" (has 0 tokens in token-games). If a flow carries •, then it has the "control" being similar to having a token in token-games. ⊠ is restricted to nodes to indicate that they are under execution and, therefore, between • and ⊤.

Similarly to token-game semantics or other related execution concepts, TTS describes the current condition of a process model as a *state*. With TTS, however, a state is not a mapping from the set of edges of a workflow graph to a natural number but a mapping from the set of edges, *signals, and nodes* to one of the above six types of tokens. This is possible since the considered acyclic process models are sound (and safe):

Definition 2 (State). A *state* S of a sound workflow graph $WFG = (N, E, \lambda, \Lambda, \tau, \gamma, M)$ is a mapping $S: N \cup E \cup M \mapsto \{\ominus, \bullet, \dagger, \top, \bot, \boxtimes\}, \forall em \in E \cup M: S(em) \neq \boxtimes$. Accessing tokens of a *set* X of nodes, edges, and signals, $X \subseteq N \cup E \cup M$, within a state S, we define $S(X) = \bigcup_{x \in X} \{S(x)\}$.

In an *initial state* S_0, only one start node s has a ⊠ token. If s catches signals (i.e., $\gamma(s) \neq \emptyset$), exactly one has a ⊤ token and the other signals (if any) have ⊥ tokens:

$$\gamma(s) \neq \emptyset \implies \exists m \in \gamma(s): \left(S_0(m) = \top \wedge S_0(\gamma(s) \setminus \{m\}) \subseteq \{\bot\}\right).$$

All other start nodes s' have a ⊥ token with $S_0(\gamma(s')) \subseteq \{\bot\}$ and $S_0(s' \triangleleft) = \{\dagger\}$. The rest of the nodes, edges, and signals of WFG carry ⊖.

In a *termination state* S_t, only one end node e carries a ⊤ token with $S_t(\tau(e)) \subseteq \{\bullet, \top\}$, where all other end nodes e' have a ⊥ token with $S_t(\tau(e)) \subseteq \{\dagger, \bot\}$. The rest of the nodes and edges of WFG have ⊤ or ⊥ tokens whereas signals have •, †, ⊤, or ⊥ tokens.

Of course, at the first moment, six tokens seem confusing to represent states. However, the semantics that will be introduced soon by state transitions only allow for the "finite state automatons" $\ominus \to \bullet \to (\boxtimes \to) \top$ and $\ominus \to \dagger \to \bot$.

Figure 3 shows the acyclic process model of the "Main Process Model" pool of Fig. 2 as a workflow graph in a specific state. This workflow graph differs slightly from the BPMN model. At first, the throwing intermediate events are replaced with tasks

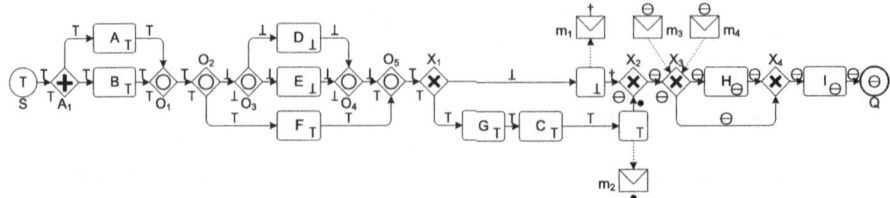

Fig. 3. The acyclic process model within the "Main Process Model" pool in Fig. 2 as a workflow graph in a (TTS) state.

that throw signals (m_1 and m_2). Furthermore, the event-based gateway with the corresponding catching intermediate events is replaced with an XOR-split that catches two signals (m_3 and m_4). The workflow graph is currently in a state marked by the different kinds of tokens within or outside of nodes, above or on the right of edges, and above or below signals (illustrated as envelopes). The "active" execution is at XOR-join X_2: Its left incoming edge carries a † token whereas its below incoming edge has a • token. All nodes and edges with a path to these both edges (all elements "before" those edges) have all ⊤ or ⊥ tokens. All nodes, edges, and signals thereafter have ⊖ tokens.

In contrast to classical token-game semantics, TTS also explicitly considers the tokens of nodes and signals. Signals are included to illustrate their state of processing, e. g., did they already lead to an instantiation of other process models (⊤ instead of •), remained the signal unread (•), or is there no active signal receiving the instance (†)? As a consequence, whether a node is executable or not, does not only depend on the incoming tokens, it also depends on the incoming signals. Since even † tokens are propagated through the process model, executability can be decided *locally* (i. e., without a supervising system). The entire semantics can be characterized as *state transitions*:

Definition 3 (State Transitions). A state S of a sound workflow graph $(N, E, \lambda, \Lambda, \tau, \gamma, M)$ can change into a state S' by handling a node $n \in N$. There are three possible state changes from S to S' considering n:

(1) S changes into S' by *skipping* node n, depicted $S \xrightarrow{n} S'$. Skipping n is only possible if and only if all its incoming edges carry †, i. e., $S(\triangleright n) = \{\dagger\}$. Then, S' is defined as (where signals are ignored and remain "unprocessed" intentionally):

$$\forall x \in (N \cup E \cup M): S'(x) = \begin{cases} \bot, & x \in (\{n\} \cup \triangleright n) \\ \dagger, & x \in (n \triangleleft \cup \tau(n)) \\ S(x), & \text{otherwise} \end{cases}$$

(2) S changes into S' by *executing* node n (and setting it to ⊠), depicted $S \xrightarrow{n}_\boxtimes S'$. This is only possible if n's incoming edges carry • or †, caught signals carry ⊖, •, or †, and at least one incoming edge has • (if any) and at least one caught signal has • (if any):

$$S(\triangleright n) \subseteq \{\bullet, \dagger\} \quad \wedge \quad S(\gamma(n)) \subseteq \{\ominus, \bullet, \dagger\} \quad \wedge$$
$$\triangleright n \neq \emptyset \implies \bullet \in S(\triangleright n) \quad \wedge \quad \gamma(n) \neq \emptyset \implies \bullet \in S(\gamma(n)).$$

In this case, S' follows from S with

$$\forall x \in (N \cup E \cup M): S'(x) = \begin{cases} \top, & x \in (\triangleright n \cup \gamma(n)) \wedge S(x) = \bullet \\ \bot, & x \in (\triangleright n \cup \gamma(n)) \wedge S(x) = \dagger \\ \mathsf{x}, & x = n \\ S(x), & \text{otherwise.} \end{cases}$$

(3) Eventually, S changes into S' by *finishing* the execution of n (and setting n to $\top$), depicted $S \xrightarrow{n}_{\top} S'$. $S \xrightarrow{n}_{\top} S'$ requires $S(n) = \mathsf{x}$. If n is a task, join node, or AND-split, let $O = n\triangleleft$. If n is an XOR-split, let $O = \{o\}, o \in n\triangleleft$. Otherwise, if n is an OR-split, let $O \subseteq n\triangleleft, |O| \geq 1$. Then, S' results from S with

$$\forall x \in (N \cup E \cup M): S'(x) = \begin{cases} \top, & x = n \\ \bullet, & x \in (O \cup \tau(n)) \\ \dagger, & x \in n\triangleleft \setminus O \\ S(x), & \text{otherwise.} \end{cases}$$

Definition 3 of state transitions above focuses on the tokens of edges similarly to token-game semantics: $\bullet$ and $\dagger$ are propagated through the process model. After skipping or executing a node, $\bullet$ tokens get $\top$ and $\dagger$ get $\bot$. This makes it possible to trace how a process model is and was executed. After executing an XOR-split, just one outgoing edge gets $\bullet$, the others are dead and get $\dagger$. After the execution of an OR-split, a non-empty subset of outgoing edges gets $\bullet$ whereas the others get $\dagger$.

Signals also play an important role during execution: Skipped nodes ignore caught signals and throw signals with $\dagger$ tokens to inform communication partners in the best case; non-skippable nodes wait if one signal gets $\bullet$ (if any), and throw signals with $\bullet$ tokens *after* finishing. Thus, there could be non-processed signals and there can be a "signal-based deadlock" if a task is waiting for a signal with $\bullet$ but never gets one.

Node executions and skippings are only possible, if *all* incoming edges have $\bullet$ or $\dagger$ tokens. As a consequence, instead of executing a node immediately as in token-game semantics, the execution in TTS is delayed until tokens are propagated to all the incoming edges of a node. This delayed execution does not have a great impact on the performance of the engine: In sound acyclic process models, each path from a node to an end node is "safe", i.e., there cannot be two edges with a $\bullet$ token (cf. the Path-to-End Theorem in [27]); therefore, one token cannot "overtake" another. Furthermore, propagating $\dagger$ tokens and skipping nodes is fast as no invocation, service call, etc. is performed. Skipped nodes do not have any effect on environments, services, actors, etc. Skipping is mainly used to propagate information through process models (and communication partners) and, therefore, to localize the semantics. In summary, sound acyclic process models are executed as usual just with a $\dagger$ propagation.

As attentive readers may already have recognized, the semantics defined by Definition 3 do not differentiate between the kind of nodes (task, XOR-split, XOR-join, etc.) in deciding which node can be executed or skipped. Soundness and the "waiting semantics" of OR-joins enable it: Each XOR- and AND-join get the right number of tokens in

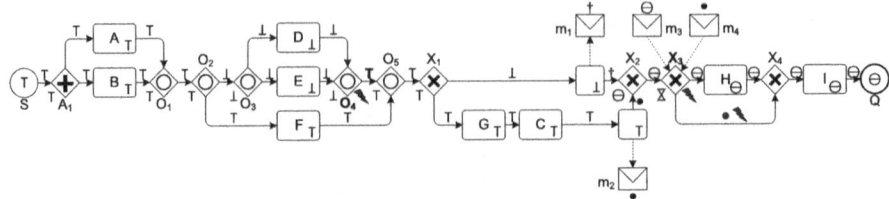

Fig. 4. The workflow graph of Fig. 3 in an *invalid* (TTS) state.

token-game semantics and could be replaced with and handled as OR-joins. Of course, it would be possible to check, e.g., for AND-joins whether all incoming edges have • tokens, or not. However, in sound process models, each AND-join, which gets a • token on one of its incoming edges during execution, will get exactly one • token on each other incoming edge in a reachable state (to be finally executed under guarantee) [14,24]. For this reason, an AND-join could be considered as an OR-join for which either none or all incoming edges will get • tokens. As a consequence, checking if all of its incoming edges have • tokens is unnecessary as a sound process model ensures its correct behavior.

In Fig. 3, the XOR-join X_2 is executable. Prior to this state, the task throwing signal m_1 was skippable and skipped. Other nodes cannot be executed or skipped. Executions and skippings can be traced through the different kinds of ⊤ and ⊥ tokens, e.g., in Fig. 3, nodes $S, A_1, A, B, O_1, O_2, F, O_5, X_1, G, C$, and the task-throwing signal m_2 were executed, whereas nodes O_3, D, E, O_4, and the task throwing signal m_1 were skipped. All nodes can decide locally whether they are executable or skippable: Through the † token propagation, X_2 has only to consider its incoming edges and caught signals. Following the step-wise execution of process models by Definition 3 of state transitions, we can also define *reachability* as in token-games:

Definition 4 (Reachability). In a sound workflow graph $(N, E, \lambda, \Lambda, \tau, \gamma, M)$, the state S' is *directly reachable* from a state S, depicted $S \to S'$, if there is a node $n \in N$ with $S \xrightarrow{n} S'$, $S \xrightarrow{n}_⊠ S'$, or $S \xrightarrow{n}_⊤ S'$. S' is *reachable* from S, $S \to^* S'$, if a sequence $S_1 \to S_2 \to \cdots \to S_{m-1} \to S_m$ of directly reachable states exists with $S_1 = S$, $S_m = S'$, and $m \geq 1$.

The state depicted in the workflow graph shown in Fig. 3 is reachable from the initial state. It can be easily checked that at most only one outgoing edge of each XOR-split has a ⊤ token, that all outgoing edges of A_1 have ⊤ tokens, etc. As a result, reachability can be utilized to define *valid* states as well:

Definition 5 (Valid States). A state S of a sound workflow graph *WFG* is *valid* if it is reachable from an initial state S_0 of *WFG*, $S_0 \to^* S$.

As mentioned before, the state of the workflow graph in Fig. 3 is reachable from the initial state and, thus, valid. On the contrary, Fig. 4 shows the same workflow graph in an *invalid* state for several reasons: (1) XOR-split X_3 has a pending ⊠ token although its incoming edge has a ⊖ token; (2) the edge between X_3 and X_4 has a • token although X_3 has not finished; and, finally, (3) O_4 has a ⊤ token on its outgoing edge although

O_4's incoming edge did never had • tokens. Note that the • token at signal m_4 must *not* invalid as signals may receive *before* the execution reaches it.

TTS enables a runtime and post-execution verification of whether there was an observable manipulation or misaligned execution of a process model. The reason for this possibility are the ⊤ and ⊥ tokens, which highlight (non-) executed paths of a process model. For reasons of space, we explain this verification mechanism on the example of a task. However, the interested reader can derive such rules from Definition 3.

We consider a task $t \in N$ in a state S. If $S(t) = \ominus$, then t's incoming edge and caught signals (if any) have to be $\ominus$, •, or † since t was not executed yet. t's outgoing edge and thrown signals have to be $\ominus$. The case $S(t) \in \{\bullet, †\}$ is not possible by Definition 3 and would be, therefore, invalid. If $S(t) = \mathsf{x}$, then $S(\triangleright n) = \{\top\}$. In this case, if t has caught signals, then one has to be ⊤, the others are $\ominus$, •, †, or ⊥, since they were not processed by the task or arrived *after* t's execution was started. t's outgoing edge and thrown signals have to be $\ominus$. If $S(t) = \top$, then the same rules for t's incoming edge and caught signals are valid as for $S(t) = \mathsf{x}$. However, its outgoing edge and thrown signals are all either • or ⊤. Eventually, if $S(t) = \bot$, then n's incoming edge has to be ⊥ and its caught signals are either $\ominus$, •, or †. t's outgoing edge and thrown signals are all either † or ⊥. Note that in the case of post execution verification, the rules simplify since there are no $\ominus$, •, x, and † tokens anymore (except for signals).

It is possible to check the validity of a TTS state with such derived rules in linear time, as summarized in the following corollary:

Corollary 1. *Checking whether a given state S is valid (and reachable) for a sound workflow graph is possible in the worst case in linear time $O(|N| + 2 \cdot |E| + 2 \cdot |M|)$ by examining each node ($|N|$) once with all its incoming and outgoing edges ($2 \cdot |E|$) as well as caught and thrown signals ($2 \cdot |M|$).* □

TTS does not simplify the deployment of *Progression*. However, it reduces the complexity of its implementation, therefore, strengthens its robustness. Furthermore, it allows for an unsupervised execution of process models as the next section will show, thus, simplifying the deployment of process models.

4 The *Progression* Workflow Management System

Currently available WMS mostly follow an *orchestrated* execution approach, i.e., a supervised system guides the execution by using different services to implement tasks and other flow elements [33]. However, such supervision incurs a bottleneck and is not well scalable [3]. In *choreographies*, different autonomous services interact without any supervision to realize a process model—with the drawback that such an interaction is more difficult to analyze, understand, and control [33]. TTS introduced in Sect. 3 allows for a strictly local execution of a sound acyclic process model and, therefore, enables a more simple realization and monitoring of choreographies. This was realized in our architecture of the *Progression* WMS, which is discussed in the following whereas implementation-related details are explained in the next Sect. 5.

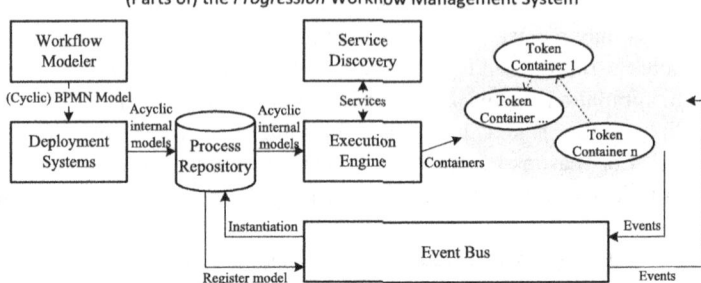

Fig. 5. Parts of the *Progression* workflow management system with local process execution.

Progression was designed as a lightweight WMS following the software architecture illustrated in Fig. 5. This architecture complements partly the platform proposed by Flechsig et al. [15] and adheres partly to the proposals for a virtual machine for business process models by Prinz [22]. In the following, we will explain the main principles of *Progression* on this architecture and how the theoretical foundations of Sect. 2 as well as TTS of Sect. 3 have impacts on the architecture.

As usual, *Progression* covers a BPMN *workflow modeler*, which enables users to create or modify process models. To ensure that the models accurately reflect real-world processes, the modeler must allow for the creation of cyclic models. A further crucial functionality of the modeler is an immediate soundness feedback system with detailed diagnostic information and explanations in case of any soundness violations. As discussed in Sect. 2, Prinz et al. [27], Favre and Völzer [14], and Prinz and Amme [24] build the foundation of such a soundness verifier.

Once a process model has been modeled or modified, a *deployment system* transfers it into an executable format. During this process, loop decomposition (as discussed in Sect. 2) is applied to break down any cyclic model into acyclic ones using interactions via events. The internal resulting format of acyclic process models need not to orientate on BPMN. It can, e. g., use Definition 1 of acyclic workflow graphs as foundation. Instead of storing cyclic models into a *process repository*, the *Progression* architecture stores the acyclic models. However, we strongly recommend storing the original BPMN visual model along with the acyclic models for state visualization and debugging purposes.

Process models in the process repository are instantiated when they need to be executed. There are two situations in which an instantiation is required: 1) when the user (or a bot) or engine specifies a process model to be executed, or 2) when an event thrown from a process instance matches the start node of another model. The latter is realized by registering each process model at an *event bus*, which handles and redirects events during execution. An *execution engine* prepares the execution of a process model. For reasons of simplicity, all tasks in the process model are represented as *services*. Of course, there can be engine-local services for simple tasks such as replacing values or executing quick "atomic" scripts. For complex services, the *Progression* architecture contains a *service discovery*, which assigns a specific service instance to a task when a process model is instantiated.

Once a process model is scheduled for execution, the execution engine assigns a unique identifier to it. The engine builds a *token container* for each node, flow, and exchanged event of the process model. These token containers are *Progression*-standardized services, realizing data (de-)serialization, data transfer, and token management. A container accepts the tokens of TTS as defined in Definition 2, i.e., $\ominus$, $\bullet$, $\dagger$, $\mathbf{x}$, $\top$, and $\bot$. Furthermore, the container is assigned the unique identifier of the process instance. Each token container is registered at the *event bus* that broadcasts local token changes to the token containers, which require them. Resulting of the acyclicity and soundness of the executed process models, each container is used (in case of a task and in case of an execution) just for a single call of a service. Note that this increases the security and transparency of process execution as invalid states can be identified locally and quickly by Definition 3 of state transitions, Corollary 1, and the explanations in Sect. 3. In addition, services cannot be accessed and executed without a token container, which in turn can only be invoked or skipped once. In this way, any duplicate execution of a service within the same instance is blocked by the engine with an alert.

When the token containers are created, they are deployed as a choreography, i.e., they do not require special monitoring and can be executed independently because they make local decisions based on received information of the event bus. This approach enables a high degree of concurrency without the execution engine becoming a bottleneck and, at the same moment, a high security without monitoring each service. Of course, it is of benefit to log every time a token container receives or sends an event, along with the token type and a unique identifier for the instance. This allows for precise process model mining out of the log when process models got lost.

5 Implementation and Discussion

In the last section, we have introduced the desired requirements of the *Progression* architecture. This section will discuss the current state of its implementation.

As mentioned in the introduction, *Progression* evolved in the *Coast* information system for creating questionnaires, surveys, and reports [25]. For this reason, some parts of the implementation (such as the persistence layer and the user interface) are used from *Coast*. These parts are not part of *Progression* being available at https://github.com/guybrushPrince/progression under GPL-3.0 license. *Coast* is implemented in PHP 8 and so is the current *Progression* implementation. For the future, we have planned to also provide an open source Java implementation.

We have implemented a *workflow modeler* (cf. Fig. 5) based on the JavaScript BPMN modeler provided by *bpmn.io* [6]. This modeler was extended with a JavaScript implementation of a soundness verification based on Prinz et al. [27] and Prinz and Amme [24]. This verification is performed in each modification step of the process model and annotates the model with detailed diagnostic information. We have planned to share this implementation as well in the future.

The *deployment system* (cf. Fig. 5) is currently part of the *workflow modeler* as it has to perform a normalization and loop decomposition of the process model as part of

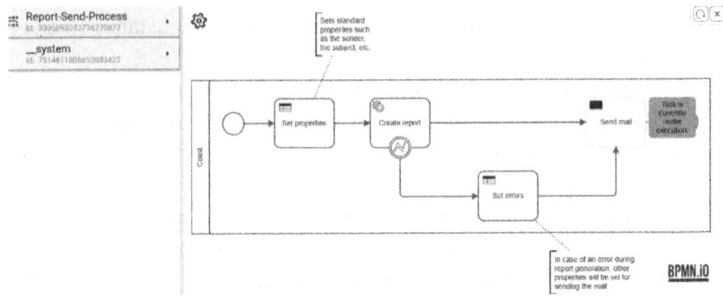

Fig. 6. A "screenshot" of a process model instance in the *Coast Progression* implementation. The process model generates a report with survey data and sends that report or an error message as e-mail. *Progression* illustrates the tokens of TTS as different colors (e. g., dark green artifacts inform about ⊤ tokens whereas dark red artifacts inform about ⊥ tokens).

the verification algorithm. Furthermore, the *bpmn.io* BPMN modeler has a very good parser of BPMN models that is reused. The modeler loads a BPMN diagram and sends the decomposed acyclic sound internal process models on request to the *Progression* server. The internal process model is available in the form of PHP classes on GitHub.

The *process repository* as part of the *Progression* architecture is solved as MySQL database, which stores the process models (with the BPMN diagrams), process instances, token containers, data, etc. The available implementation at GitHub uses an abstract persistence layer. *Coast* provides its own currently proprietary implementation of this layer, which currently has to be replaced for test purposes if necessary.

The *execution engine* is realized as a PHP class with methods for instantiating and canceling process models as well as getting information about stored process models and running instances. Since this *Progression* instance runs in the closed environment of *Coast*, the *service discovery* is relatively simple as specialized tasks in the *workflow modeler* already specify the services to use. The engine also creates the token containers illustrated in Fig. 5. It uses three different token containers for the three kinds of stateful model elements: One for flows (called *Token* in the implementation), one for *signals* (called *Incident*), and one for *nodes* (called *LocalState*).

The drawback of PHP is its missing ability to perform concurrent actions. Other programming languages and environments such as Java are more powerful in this direction. As a consequence, the current PHP implementation follows a "clock tick"—an asynchronous Unix *Cron* job invokes the engine in an interval of some individual seconds whereas such an invocation is called a *tick* in the following. Instead of sending events to an *event bus*, changing tokens affects the database (i. e., the objects of *Token*s, *Incident*s, and *LocalState*s). In each tick, the engine loads all objects of *LocalState* having a ⊠ token. Subsequently, it checks if any execution of the tasks behind those *LocalState*s has been finished (these asynchronous tasks are executed by a separate *Cron* job). Then, each finished local state propagates ● or † tokens through connected *Token*s and *Incident*s. Note that gateways and other nodes with non-computational intensive tasks are executed directly. For this reason, skipping nodes is also performed directly and † tokens are propagated quickly through the models.

Besides the architecture as illustrated in Fig. 5, the *Coast*-related *Progression* implementation uses the web user interface capabilities of *Coast*. This allows to investigate already defined process models, process instances under execution, and already executed instances. Figure 6 shows a "screenshot" of a simple process model currently under execution. Clicking on a flow or node gives information about the data that was transferred or used. Furthermore, task-specific forms are generated and presented to the users belonging to the tasks. Process models can be started by users or are started by the engine as they are "invoked" by other process instances. Soundness and loop decomposition guarantee that there are no deadlocks if the provided data is correct.

Overall, the implementation of *Progression* comprises around 5k lines of code including comments and empty lines and can, therefore, also be interpreted as "lightweight" with respect to much longer code lines of existing WMS. Since the execution engine of *Progression* relies on an acyclic internal model, each BPMN model has to be transformed as described in Sect. 4. Currently, this transformation does not cover all elements of BPMN but will be extended in the future.

6 Conclusion

The creation and execution of business process models on modern workflow management systems should be done by experts. However, for small and medium-sized companies as well as for training or teaching purposes, such systems can be a big hurdle. For this reason, this paper introduced a new workflow management system called *Progression*, which is part of the *Coast* information system [25] for creating questionnaires, surveys, and reports. *Progression* has a lightweight architecture and builds up on research outcomes such as soundness checking with detailed diagnostic information, a most liberal semantics of converging inclusive gateways ("OR-joins"), and loop decomposition allowing the execution of only acyclic models interacting with signals. In utilizing these research results, a new *Track-and-Trace Semantics* was introduced. This allows for local, distributed, and unsupervised execution of process models. Furthermore, the security during execution is increased as services and tasks cannot be invoked without an execution context. Once a process model is executed, the signals, flows, and nodes that have been actively and passively followed can be read from the process instance and, therefore, enable detailed event logs to optimize process models. Besides a new execution semantics, this paper explained the architecture of *Progression* from modeling a process model until a process instance is running on the engine. This architecture was further investigated in terms of its available implementation in PHP.

The business process management community profits from this new lightweight workflow management system as it can be extended for its own use cases and it can be integrated more easily in the tool landscape of small and medium-sized companies.

In the future, we want to extend the implementation of *Progression*, e.g., with more standardized tasks and the removal of any dependence on the *Coast* system. However, the main focus will be a version of *Progression* in Java. This shall utilize an event bus as stated in the architecture being not possible to implement in the PHP-version of *Progression* because of PHP's missing support of concurrency.

References

1. van der Aalst, W.M.P.: Verification of workflow nets. In: Azéma, P., Balbo, G. (eds.) Application and Theory of Petri Nets 1997, 18th International Conference, ICATPN '97, Toulouse, France, June 23-27, 1997, Proceedings. LNCS, vol. 1248, pp. 407–426. Springer (1997). https://doi.org/10.1007/3-540-63139-9_48
2. Alfresco: Activiti (2025). https://www.activiti.org/, Business Process Modeling and Automation System
3. Barker, A., Weissman, J.B., van Hemert, J.I.: Reducing data transfer in service-oriented architectures: the circulate approach. IEEE Trans. Serv. Comput. **5**(3), 437–449 (2012)
4. Bizagi: Bizagi Modeler (2025). https://www.bizagi.com/de/plattform/modeler, Business Process Modeling and Automation System
5. Bonitasoft: Bonita Platform (2025). https://www.bonitasoft.com/bonita-platform, Business Process Modeling and Automation System
6. Camunda Services GmbH: bpmn.io (2025). https://bpmn.io/, Business Process Modeling and Simulation System
7. Camunda Services GmbH: Camunda BPM (2025). https://camunda.com/, Business Process Management System
8. Corradini, F., Muzi, C., Re, B., Rossi, L., Tiezzi, F.: BPMN 2.0 OR-join semantics: global and local characterisation. Inf. Syst. **105**, 101934 (2022). https://doi.org/10.1016/j.is.2021.101934
9. van Dongen, B.F., Mendling, J., van der Aalst, W.M.P.: Structural patterns for soundness of business process models. In: Tenth IEEE International Enterprise Distributed Object Computing Conference (EDOC 2006), 16-20 October 2006, Hong Kong, China, pp. 116–128. IEEE Computer Society (2006). https://doi.org/10.1109/EDOC.2006.56
10. Dumas, M., Rosa, M.L., Mendling, J., Reijers, H.A.: Fundamentals of Business Process Management, Second Edition. Springer (2018)
11. Fahland, D., Favre, C., Koehler, J., Lohmann, N., Völzer, H., Wolf, K.: Analysis on demand: instantaneous soundness checking of industrial business process models. Data Knowl. Eng. **70**(5), 448–466 (2011) (2024)
12. Fahland, D., Völzer, H.: Dynamic skipping and blocking, dead path elimination for cyclic workflows, and a local semantics for inclusive gateways. Inf. Syst. **78**, 126–143 (2018)
13. Favre, C., Fahland, D., Völzer, H.: The relationship between workflow graphs and free-choice workflow nets. Inf. Syst. **47**, 197–219 (2015)
14. Favre, C., Völzer, H.: Symbolic execution of acyclic workflow graphs. In: Hull et al. [17], pp. 260–275. https://doi.org/10.1007/978-3-642-15618-2_19
15. Flechsig, C., Völker, M., Egger, C., Weske, M.: Towards an integrated platform for business process management systems and robotic process automation. In: Marrella, A., et al. (eds.) Business Process Management: Blockchain, Robotic Process Automation, and Central and Eastern Europe Forum - BPM 2022 Blockchain, RPA, and CEE Forum, Münster, Germany, September 11-16, 2022, Proceedings. Lecture Notes in Business Information Processing, vol. 459, pp. 138–153. Springer (2022)
16. Flowable AG: Flowable (2025). https://www.flowable.com/, Business Process Modeling and Automation System
17. Hull, R., Mendling, J., Tai, S. (eds.): Business Process Management - 8th International Conference, BPM 2010, Hoboken, NJ, USA, September 13-16, 2010. Proceedings, vol. 6336, Lecture Notes in Computer Science. Springer (2010)
18. KIE: jBPM (2025). https://www.jbpm.org/, Business Process Modeling and Automation System

19. Kindler, E.: On the semantics of EPCs: resolving the vicious circle. Data Knowl. Eng. **56**(1), 23–40 (2006). https://doi.org/10.1016/j.datak.2005.02.005
20. OASIS: Web Services Business Process Execution Language Version 2.0 (2007). http://docs.oasis-open.org/wsbpel/2.0/OS/wsbpel-v2.0-OS.pdf, standard
21. Object Management Group (OMG): Business Process Model and Notation (BPMN) Version 2.0. formal/2011-01-03. http://www.omg.org/spec/BPMN/2.0 (2011). standard
22. Prinz, T.M.: Proposals for a virtual machine for business processes. In: Heinze, T.S., Prinz, T.M. (eds.) Proceedings of the 7th Central European Workshop on Services and their Composition, ZEUS 2015, Jena, Germany, February 19-20, 2015. CEUR Workshop Proceedings, vol. 1360, pp. 10–17. CEUR-WS.org (2015)
23. Prinz, T.M., Amme, W.: A complete and the most liberal semantics for converging OR gateways in sound processes. Complex Syst. Inform. Model. Q. **4**, 32–49 (2015)
24. Prinz, T.M., Amme, W.: Control-flow-based methods to support the development of sound workflows. Complex Syst. Inform. Model. Q. **27**, 1–44 (2021)
25. Prinz, T.M., Apel, S., Bernhardt, R., Plötner, J., Vetterlein, A.: Model-centric and Phase-spanning software architecture for surveys - report on the tool coast and lessons learned. Int. J. Adv. Softw. **12**(1 & 2), 152–165 (2019). iSSN 1942-2628
26. Prinz, T.M., Choi, Y., Ha, N.L.: Understanding and decomposing control-flow loops in business process models. In: Ciccio, C.D., Dijkman, R.M., del-Río-Ortega, A., Rinderle-Ma, S. (eds.) Business Process Management - 20th International Conference, BPM 2022, Münster, Germany, September 11-16, 2022, Proceedings. Lecture Notes in Computer Science, vol. 13420, pp. 307–323. Springer (2022)
27. Prinz, T.M., Choi, Y., Ha, N.L.: Soundness unknotted: an efficient soundness checking algorithm for arbitrary cyclic process models by loosening loops. Inf. Syst. **128**, 102476 (2025). https://doi.org/10.1016/J.IS.2024.102476
28. Prinz, T.M., Ha, N.L., Choi, Y.: Transformation of cyclic process models with inclusive gateways to be executable on state-of-the-art engines. In: Filipe, J., Smialek, M., Brodsky, A., Hammoudi, S. (eds.) Proceedings of the 27th International Conference on Enterprise Information Systems, ICEIS 2025, Porto, Portugal, April 4-6, 2025, Volume 2, pp. 280–291. SCITEPRESS (2025). https://doi.org/10.5220/0013386400003929
29. Prinz, T.M., Spieß, N., Amme, W.: A first step towards a compiler for business processes. In: Cohen, A. (ed.) Compiler Construction - 23rd International Conference, CC 2014, Held as Part of the European Joint Conferences on Theory and Practice of Software, ETAPS 2014, Grenoble, France, April 5-13, 2014. Proceedings. Lecture Notes in Computer Science, vol. 8409, pp. 238–243. Springer (2014)
30. Prinz, T.M., Welsch, T., Ha, N.L.: Recognizing relationships: detecting the 4C Spectrum in $O(P^2 + T^2)$ for acyclic sound process models. In: Borbinha, J., Sales, T.P., Silva, M.M.D., Proper, H.A., Schnellmann, M. (eds.) Enterprise Design, Operations, and Computing - 28th International Conference, EDOC 2024, Vienna, Austria, September 10-13, 2024, Revised Selected Papers. Lecture Notes in Computer Science, vol. 15409, pp. 281–299. Springer (2024). https://doi.org/10.1007/978-3-031-78338-8_15
31. Sadiq, W., Orlowska, M.E.: Analyzing process models using graph reduction techniques. Inf. Syst. **25**(2), 117–134 (2000). https://doi.org/10.1016/S0306-4379(00)00012-0
32. SAP Signavio: SAP Signavio Process Manager (2025). https://www.signavio.com/, Business Process Modeling and Automation System
33. Valderas, P., Torres, V., Serral, E.: Towards an interdisciplinary development of IoT-enhanced business processes. Bus. Inf. Syst. Eng. **65**(1), 25–48 (2023)
34. Verbeek, H.M.W., Basten, T., van der Aalst, W.M.P.: Diagnosing workflow processes using Woflan. Comput. J. **44**(4), 246–279 (2001)
35. Völzer, H.: A new semantics for the inclusive converging gateway in safe processes. In: Hull et al. [17], pp. 294–309

Predicting Newcomer Capabilities and Performance in Process Execution

Roy Jing Yang[1,2](✉), Chun Ouyang[1,2], and Remco Dijkman[1,3]

[1] School of Information Systems, Queensland University
of Technology, Brisbane, Australia
roy.j.yang@qut.edu.au
[2] Centre for Data Science, Queensland University
of Technology, Brisbane, Australia
[3] Eindhoven University of Technology (TU/e),
Eindhoven, Netherlands

Abstract. Companies are constantly hiring new employees. To efficiently allocate these newcomers to tasks, we need to predict which tasks they can do and how well they will perform on these tasks. However, making such predictions for newcomers is challenging, particularly at the early stage of onboarding, due to the limited availability of observational data on their past experience. In this research, we explore the problem of predicting newcomer capabilities and performance in the context of process execution and propose a solution to address the challenge. The proposed approach uses data augmentation from historical event data, guided by organizational model mining, and generates predictions for the tasks that newcomers may perform and the time it will take them to perform those tasks. Experiments based on several real-life event logs showed that the proposed approach achieves accurate predictions for newcomers given limited data availability.

Keywords: newcomer · process mining ·
predictive process monitoring · organizational model · onboarding

1 Introduction

Large companies frequently appoint new employees or change the roles of employees in the organization. As a result, there frequently are "newcomer" resources deployed into the execution of business processes. For event logs from practice that we studied, this amounted to approximately five per month. When new employees are appointed, managers often need to get an idea of the tasks that a newcomer will perform and how well the newcomer will perform on those tasks. This knowledge is useful for managers and supervisors to support newcomers during their onboarding and adjust their roles and work assignments [2,8]. It can also be used to help automated workflow systems in efficient allocation of tasks to employees, e.g., assisting simulation when extra, new resources are deployed to model the impact on a process [5].

While techniques from the area of process mining exist to make predictions about what might happen for a process, including case outcome, case remaining time, next event, and next activity duration [4], existing techniques are not fully up to the task of handling novel data values in a prediction task [10]. For newcomers—novel data points regarding the resource perspective—the key challenge lies in the limited amount of observed process execution data needed for modeling newcomer behavior and experience at the early stage of their onboarding. However, this early stage is often when good predictive information is most valuable for managers [2].

How may we make predictions about newcomers in process execution with limited event data? In this paper, we present an approach to address the question. The approach consists of techniques for predicting both (i) the tasks that newcomers are likely to perform and (ii) the time it will take them to perform those tasks. Our approach is based on organizational model mining [13,16], which discovers from event logs a model that represents the grouping of resources with similar capabilities of executing tasks in different process execution contexts. When a newcomer can be linked with existing resources in an organizational model, historical events related to those resources may be used to augment the limited data for the newcomer and be used to make (more) accurate predictions about the capabilities and likely performance of the newcomer.

We implemented an open-source prototype of the approach and evaluated it through experiments on two real-life datasets. Results show that our approach is effective for both prediction tasks.

The remainder of the paper is structured as follows. Section 2 reviews related work on predicting resource behavior and task timing. Section 3 covers preliminary definitions. Section 4 details our proposed approach, and Sect. 5 evaluates it. Finally, Sect. 6 presents the conclusion and future work.

2 Related Work

This paper proposes techniques in two areas in process mining. On the one hand, the idea of linking newcomers to similar resources is related to several existing topics on the organizational perspective of process mining, including the discovery of organizational models (e.g., [16]), business role discovery (e.g., [3]), and resource social network mining (e.g., [18]). Research on these topics considers measuring the similarity between human resources based on their performed activities and cases, and clustering similar ones into groups to describe the organizational structures in a process context.

On the other hand, prediction about newcomers is concerned with resource performance prediction using event logs. Predictive Process Monitoring (PPM) [4] is an emerging area in process mining, where methods are developed to address a number of predictive tasks ranged from case outcome to next activity duration (e.g., [6,14,15]). Resource information contributes important features for developing PPM methods, which include the direct use of resource ID (label) as a feature (as in Reference [14]) and the more systematic modeling of resource experience features [7] using resource historical data. While the prediction of resource performance for the next activity is not commonly addressed as a specific task, it can be viewed as the prediction of next activity duration given that the performing resource is known.

The key challenge in making predictions for a newcomer is the lack of sufficient observational data. This is related to the challenges in PPM of handling unseen data (see e.g., [9,12]) and, more broadly, updating predictive models (see e.g., [10,11]). Newcomers are a source of unseen/novel data for PPM from the resource perspective and may manifest as unseen categorical values (if resource ID is used) or potentially inaccurate resource experience features due to the lack of historical events.

The core idea of our approach presented in this paper is the augmentation of newcomer data based on newcomer's similarity with existing resources. As such, our research contributes to the literature of process mining from the organizational perspective, by showing how discovered organizational models may be used beyond a descriptive purpose. Moreover, we also contribute to the literature of PPM by providing a solution for handling new resources at runtime.

3 Preliminaries

Let $\mathcal{E}$ be the universe of event identifiers, $\mathcal{U}_{Att}$ the universe of possible attribute names, and $\mathcal{U}_{Val}$ the universe of possible attribute values. $\mathcal{C} \subseteq \mathcal{U}_{Val}$, $\mathcal{A} \subseteq \mathcal{U}_{Val}$, $\mathcal{T} \subseteq \mathcal{U}_{Val}$, and $\mathcal{R} \subseteq \mathcal{U}_{Val}$ denote the universes of case identifiers, activity names, timestamps, and resource identifiers, respectively.

Process execution data is recorded by process-aware information systems and stored in the form of event logs (Definition 1). An event log consists of a set of uniquely identifiable events and a set of event attribute names. Each event records some event attribute values corresponding to the attribute names. In this work, we consider event logs recording at least the case identifier, activity name, timestamp, and resource identifier.

Definition 1 (Event Log). *An event log is a tuple $L = (E, Att, \pi)$ with $E \subseteq \mathcal{E}$, $E \neq \emptyset$, $Att \subseteq \mathcal{U}_{Att}$, and $\pi \colon E \to (Att \not\to \mathcal{U}_{Val})$. An event $e \in E$ has attributes $dom(\pi(e))$. For an attribute $x \in dom(\pi(e))$, $\pi_x(e) = \pi(e)(x)$ is the attribute value of x for event e.*

We consider event logs with at least four event attributes, i.e., {case, act, time, res} $\subseteq dom(\pi(e))$. For any $e \in E$, $\pi_{case}(e) \in \mathcal{C}$ is the case to which e belongs; $\pi_{act}(e) \in \mathcal{A}$ is the activity e refers to; $\pi_{time}(e) \in \mathcal{T}$ is the time at which e occurred; and $\pi_{res}(e) \in \mathcal{R}$ is the resource that originated e.

An organizational model describes the grouping of resources with similar characteristics in process execution and their capabilities as members of resource groups (Definition 2). The notion of *execution context* is used to describe resource capabilities [16]. An execution context is a tuple $co = ($ *case_type, activity_type, time_type*$)$, where the type names together represent a type of activity instance executed in some type of case during a certain time period. For example, ("VIP", "contact customer", "Afternoon") is an execution context for an insurance claim handling process, with case type "VIP" covering all cases from high-value customers, activity type "contact customer" covering all process activities related to the job of phoning or emailing customers, and time type "Afternoon" covering process activities executed at afternoon time. A set of execution contexts can be applied to classify events originated by resources with similar characteristics and therefore describe their capabilities.

Definition 2 (Organizational Model, adapted based on Ref. [16]). $\mathcal{R}$ *is the universe of resource identifiers. Let* $\mathcal{CO}$ *denote the set of execution contexts. Given a set of events* $E \subseteq \mathcal{E}$, *an execution context* $co \in \mathcal{CO}$ *can be associated to (and therefore select) events with specific attribute values.* $[E]_{co}$ *denotes the set of events associated with co.*

An organizational model is a tuple $OM = (RG, mem, cap)$ *where* RG *is a set of resource groups, mem*: $RG \rightarrow \mathcal{P}(\mathcal{R})$ *maps each resource group onto its members, and cap*: $RG \rightarrow \mathcal{P}(\mathcal{CO})$ *maps each resource group onto its possible execution contexts (i.e., activities under specific execution contexts that can be performed by resources).*

4 Approach

This section introduces our approach to predicting newcomer performance. Specifically, it supports two types of prediction tasks. First, we consider the task of predicting the "full" capabilities of newcomers in process execution, i.e., types of work that newcomers are expected to perform besides the ones that have been observed in the process so far. Second, we consider the task of predicting newcomer's performance in ongoing cases, i.e., activity duration time when it is known that a newcomer will be executing the next activity.

Figure 1 presents the proposed approach. The prediction tasks are illustrated in blue and green, respectively. The starting point of the approach is an event log recording observations of all resources in the process so far. We consider two subsets of the input event log: A *recent log*, which consists of a small number of events executed by resources known as newcomers; and a *historical log*, which consists of events that were executed by resources other than the newcomers, i.e., existing resources, and that occurred before those in the recent log.

Subsequently, three steps are performed, namely *Discover organizational model*, *Identify similar resources*, and *Augment event data*, generating inputs for the predictive tasks. The augmented event data can be used for the first task of predicting the newcomer's capabilities. To conduct the second prediction task, a predictive model is trained utilizing the historical log along with the similar resource information and the augmented event data for all newcomers. The trained predictive model can then be applied to generate predictions about activity duration when performed by newcomers in ongoing cases. The remainder of this section discusses these steps and tasks. Note that the log completeness assumption needs to hold to enable the proposed approach.

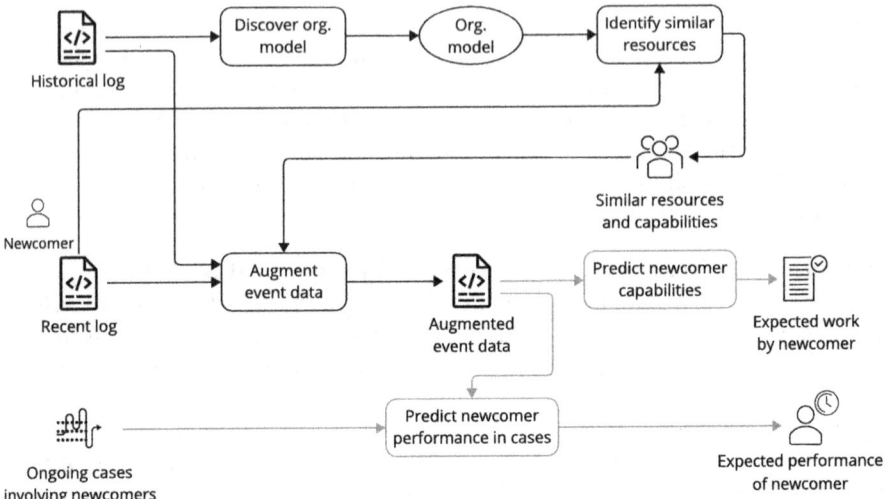

Fig. 1. An overview of the proposed approach to predicting newcomer performance. Blue and green correspond to the tasks of predicting newcomer capabilities and predicting newcomer's performance in ongoing cases, respectively. (Color figure online)

4.1 Discovering Organizational Model

The discovery of an organizational model from the historical log follows the general approach in the *OrdinoR* framework [16], which consists of three key steps: (i) Learning execution contexts, which results in a set of execution contexts (CO) that will be used to describe resource capabilities; (ii) Discovering resource groups, which includes the selection of a method for characterizing resource features (featurization function δ) and a resource similarity measure (ω). Note that the featurization function needs to normalize the volume of work across resources [1] so that resources with the same capabilities are not differentiated by the number of events, e.g., full-time vs. part-time employees in the same position; (iii) Profiling resource groups, which results in a discovered organizational model that captures the grouping and capabilities of resources according to the log.

Definition 3 (Organizational Model Discovery). *Given an event log L, an organizational model OM can be discovered, along with: (i) $CO \subseteq \mathcal{CO}$, a set of execution contexts; (ii) $\delta \colon \mathcal{R} \times L \times CO \to \mathbb{R}^m$, a featurization function that characterizes resources by transforming their relevant event data into m numeric values[1] ($m \in \mathbb{Z}, m \geq 1$); and (iii) $\omega \colon \mathbb{R}^m \times \mathbb{R}^m \to \mathbb{R}_{\geq 0}$, a resource similarity measure selected to quantify the extent of resource similarity in $\mathbb{R}^m$.*

Note that the configuration of organizational model discovery may be varied and lead to different organizational models as outputs. The model-log fitness and

---

[1] It is possible to use other data types to describe resource features, e.g., categorical type. Here, we consider only numeric values for simplicity.

precision measures in the *OrdinoR* framework provide a reference to discover an "optimal" model with respect to the historical log, along with the specific intermediate artifacts (CO, δ, ω). For details, readers are referred to the work in organizational model mining [16,17].

4.2 Identifying Similar Resources

The discovered organizational model groups existing resources in the historical log by similarity. To identify those that are similar to a given newcomer, we need to compare the newcomer to the resource groups. This is achieved by first characterizing the features of resource groups in the discovered organizational model. For a resource group in the model, its feature values are computed as consisting of the aggregated values of the features of its member resources (Definition 4).

Definition 4 (Resource Group Features). *Let L_H be a historical log, $OM = (RG, mem, cap)$ an organizational model discovered from L_H, and $g \in RG$ a resource group in the model, the set of features characterizing the member resources in g is $\{\delta(r, L_H, CO) \mid r \in mem(g)\}$. Group g can be characterized by the centroid aggregating the features of member resources, denoted as $\mathbf{c}_g = \frac{1}{|mem(g)|} \sum_{r \in mem(g)} \delta(r, L_H, CO)$.*

Then, using the same resource featurization function (δ), the feature values of the newcomer are computed, based on the newcomer's events in the recent log (Definition 5). As such, the newcomer and all the resource groups in the model are now represented as feature vectors in the same vector space specified during the discovery of the organizational model. Finally, the similarity between the newcomer and each of the resource groups can be computed using the same resource similarity measure (ω) as in model discovery. The most similar resources are therefore the members of the top-k group(s) most similar to the newcomer. Note that more than one most similar resource group may be used ($k > 1$), resulting in a larger set of similar resources identified. The choice of a larger k depends on domain knowledge about the newcomer and the existing resources.

Definition 5 (Resources Similar to Newcomer). *Given a newcomer $n \in \mathcal{R}$, a recent log L_R recording its originating events, and a discovered organizational model OM, resources in OM similar to n can be identified by selecting the top-k resource groups that are similar to n, i.e.,*

$$\widetilde{G}_n = \{g \in RG \mid |\{x \in RG \mid \omega(\delta(n, L_R, CO), \mathbf{c}_x) > \omega(\delta(n, L_R, CO), \mathbf{c}_g)\}| < k\},$$

where $k \in \mathbb{Z}, k \geq 1$ is a configurable parameter. Note that in the case where more than one group has the same similarity with the newcomer, more than k resource groups will be selected. Resources identified as similar to the newcomer are therefore all members in the top-k selected resource groups, i.e., $\widetilde{R}_n = \bigcup_{g \in \widetilde{G}_n} mem(g)$.

4.3 Augmenting Event Data

As mentioned, the key challenge in making predictions for newcomers is the lack of observational data about their past experience, i.e., the limited number of events recorded in the recent log. Our solution is to augment the event data for newcomers. The key idea is to use the identified similar resources and their capabilities to guide the selection of events from the historical log—these events were not executed by the newcomers, but are most closely relevant to the possible behavior of them. The selected historical events are then combined with the actual observations, creating a set of augmented event data.

Given a newcomer and its similar resources, an intuitive solution for augmentation is to simply use all events (Definition 6) in the historical log that were executed by the similar resources.

Definition 6 (Using All Events by Similar Resources). *Given a historical log* $L_H = (E_H, Att, \pi)$, *an organizational model OM discovered from* L_H, *a newcomer* $n \in \mathcal{R}$, *and* $L_R = (E_R, Att, \pi)$ *a recent log, the set of augmented event data with respect to newcomer* n *is*

$$E_n^+ = \{\, e \in E_H \mid \pi_{res}(e) \in \widetilde{R}_n \,\} \cup \{\, e \in E_R \mid \pi_{res}(e) = n \,\}.$$

The selection of all such events maximizes the possibility to cover all likely behavior of a newcomer when making predictions. However, it may increase the chance of considering excessive behavior, especially when many similar resources are considered, and may therefore cause imprecise predictions when using the augmented event data. Also, from the perspective of onboarding employees, a newcomer is often gradually exposed to different tasks and would not perform a broad range of work as existing employees in the same position do, at least at the early stage of onboarding.

A sensible solution that mitigates the risk of using excessive events is to choose only events that closely resemble the observations of newcomers so far, i.e., events in the recent log. Here, given that events are mapped onto execution contexts to describe resource capabilities (Definition 2), we may define event distance with regard to their corresponding execution contexts (Definition 7). Concretely, the distance between any pair of events, given their execution contexts, computes the proportion of types shared between the two execution contexts.

Definition 7 (Event Distance within Execution Contexts). *Let* $CO \subseteq \mathcal{CO}$ *be a set of execution contexts.* $d_{CO} \colon \mathcal{E} \times \mathcal{E} \to [0,1]$ *is a distance measure between events within a set of execution contexts CO. Concretely, for any* $e \in \mathcal{E}$ *and* $e' \in \mathcal{E}$, d_{CO} *can be defined as the normalized hamming distance between the two execution contexts respectively associated with* e *and* e', *i.e., the proportion of shared types between the two execution contexts (see Eqn. (2) in Reference [17]).*

Events selected from the historical log can be limited to only those within the nearest distance from events in the recent log (Definition 8) by specifying a distance threshold (θ). When the normalized hamming distance is used as in Reference [17], we have $\theta \in \{0, 1/3, 2/3, 1\}$, meaning that an event from the historical log is selected only if its corresponding execution context is exactly the same ($\theta = 0$), shares at least two types ($\theta = 1/3$), or shares at least one type ($\theta = 2/3$) with regard to an event in the recent log. It follows that with $\theta = 1$, this selection yields the same result as the use of all events by the similar resources (Definition 6).

Definition 8 (Selecting Events from Nearest Execution Contexts). *Given a historical log $L_H = (E_H, Att, \pi)$, an organizational model OM discovered from L_H, a newcomer $n \in \mathcal{R}$, and $L_R = (E_R, Att, \pi)$ a recent log, the set of augmented event data with respect to newcomer n is*

$$E_n^+ = \{\, e \in E_H \mid \pi_{res}(e) \in \widetilde{R}_n \wedge \exists_{e' \in E_R}(\pi_{res}(e') = n \wedge d_{CO}(e, e') \leq \theta)\,\} \\ \cup \{\, e \in E_R \mid \pi_{res}(e) = n \,\},$$

where $\theta \in (0, 1]$ is a configurable threshold that determines the nearest execution contexts to select events from.

4.4 Predicting Newcomer Capabilities

We use execution contexts to characterize the different types of work that resources are capable of performing in process execution. To predict the capabilities of a newcomer, we combine observations of the newcomer (i.e., execution contexts where it performed so far) and its similar resources (i.e., execution contexts where the similar resources performed). Note that all such observations are readily captured in the corresponding augmented event data for the newcomer. Formally, let $N \subseteq \mathcal{R}$ be a set of newcomers and $L_R = (E_R, Att, \pi)$ a recent log. Predictions of newcomer capabilities is a mapping $f_{work} \colon \mathcal{R} \to \mathcal{P}(\mathcal{CO})$, such that

$$f_{work}(n) = \{\, co \in \mathcal{CO} \mid [E_n^+]_{co} \neq \varnothing \,\},$$

where E_n^+ is the set of augmented event data with respect to n (Sect. 4.3).

4.5 Predicting Newcomer Performance in Ongoing Cases

We consider this task as the prediction of next activity duration time in a running case, assuming that (i) the next activity name and (ii) the next resource (who may be a newcomer) to perform the activity are both known. This can be viewed as making predictions once a task is allocated to a newcomer during process execution. Built on the general predictive process monitoring (PPM) framework [15], we propose a solution that utilizes *augmented* resource experience features [7]. Readers are referred to Reference [15] for the introduction of PPM and Reference [7] for the resource experience framework used in PPM.

Figure 2 provides an overview of our solution regarding how resource features are computed at both the training (offline) and testing (online) phases of PPM to enable dealing with predictions for newcomers. At the training phase, a predictive model is built using historical cases in an event log. Note that here

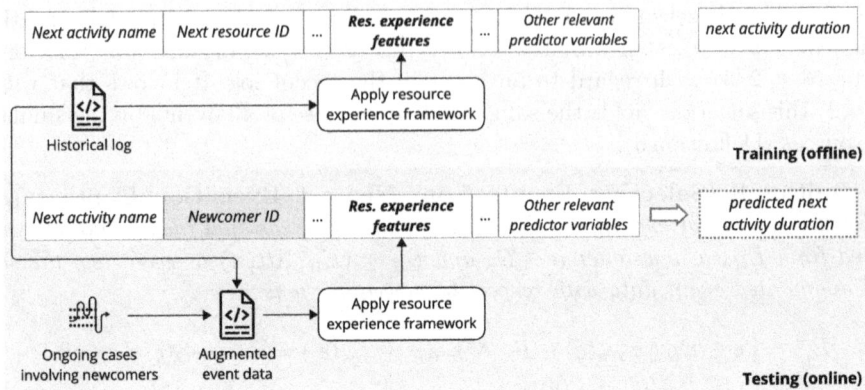

Fig. 2. An illustration of the proposed solution for predicting next activity duration, assuming that the activity is known to be performed by a newcomer. We apply the resource experience framework [7].

only the historical log is used as input, following the same assumption (as in Sect. 4) that the recent log contains limited events of the newcomers. Resource experience features [7], describing resources by aspects including specialization and generalization, are derived for the existing resources and used in feature encoding. At the testing phase, the trained predictive model is applied to make predictions of activity duration time given the next activity and resource. For existing resources, this application is trivial—whether at the training phase or the testing phase, the resource experience features of an existing resource can be computed using all its historical events. However, for a newcomer, such a history is limited to its recent log and is thus scarce. Our solution addresses this issue by using the augmented event data to compute resource experience features for newcomers.

The procedure of creating augmented event data is already discussed in the previous sections. Here, we discuss two issues related to the computation of resource experience features when the input is a set of augmented event data instead of the recent log.

The first issue is concerned with how to use events sourced from a historical log. We consider them as events recording activity instances that the newcomer *would do* in process execution. Therefore, elementary measures in the resource experience framework [7] related to activities and activity instances ("number of work items" and "number of unique tasks"), can be computed and be combined systematically to compute the resource experience features. The use of case identifiers in the augmented event data require special attention, since more than one similar resource related to the newcomer may have performed in the same case in the historical log. In this situation, events executed by the multiple similar resources should be viewed as relating to more than one case identifiers according to the number of similar resources appeared in the original case.

This brings the second issue, concerned with the choice of resource experience features. Most of the events in the augmented event data for a newcomer are sourced from the historical log and were performed by potentially multiple resources similar to the newcomer. As such, those events may not represent the *volume of work* that would normally be performed by a newcomer. Therefore, we choose only resource experience features that are invariant to volume of work. Frequency-related features are therefore excluded. Also, since we considered only resource execution of activity instances but not the interactions among resources, we do not choose resource experience features concerned with their involvement in handover of work. As a result, the following resource experience features are used: sp_work_item_case, sp_curr_case, and sp_curr_task, concerned with resource specialization; gen_task, gen_case, concerned with resource generalization. The definitions of these features can be found in Reference [7].

Our solution is decoupled from the PPM workflow. Note that the discovery of organizational models is required only at the offline phase, which is independent of the (re)training of the predictive models in PPM. Once the resource experience framework [7] is applied, the set of resource features used in the PPM workflow remain static. At the testing phase, augmented event data is dynamically created for given ongoing cases with newcomers, following the steps of identification of similar resources and augmentation of event data. Therefore, the proposed solution incurs only two types of costs in its application: (i) The deployment of the resource experience framework [7], which may require training a more complex while effective model to incorporate the resource experience features; and (ii) the computational cost to generate an augmented event data on-the-fly.

5 Experiment Evaluation

We developed an open-source prototype as implementation of the proposed approach and conducted experiments using publicly available, real-life event logs to evaluate the effectiveness of the approach. The prototype and experiment details are made open in an online repository: https://doi.org/10.5281/zenodo.15048773.

5.1 Experiment Datasets and Design

The experiment datasets consist of several event logs from the Business Process Intelligence Challenges (BPIC), available through 4TU Research Data center[2]. Since these event logs are not accompanied by organizational information that identifies any newcomer resources, we used time-based splitting to create a simulated setting, where resources who appeared later in process execution are assumed as newcomers. Table 1 shows a summary of the datasets.

For each group of split logs created from the original event log, we used the train set as the historical log. Then, for each newcomer resource, we generated

---

[2] 4TU.ResearchData: https://data.4tu.nl/.

Table 1. Key statistics of the experiment datasets (Res: resources, Exec. Ctx.: execution contexts; NC: newcomers).

Dataset	Train-Test split point	Train set				Test set	
		#Events	#Cases	#Res	#Learned Exec. Ctx.	#NC	#NC events
BPIC12	2012-01-01	61842	5899	57	42	6	3219
BPIC15-1	2014-01-01	31866	744	17	189	2	2168
BPIC15-2	2014-01-01	26166	503	9	164	2	888
BPIC15-5	2014-01-01	35993	731	15	148	2	1151

multiple recent logs. For the task of predicting newcomer capabilities, we created recent logs with different sizes of 10/20/50/100 events, respectively. For the task of predicting next activity duration by newcomers, the recent log for a newcomer is generated "on-the-fly"—when an event in a case prefix is originated by a newcomer, all the previous events by this newcomer constitute the recent log at that point to simulate predictions made at runtime. Predictions can then be generated following the proposed approach. Note that the remaining events, i.e., the subset of the latest events occurred after the recent log, are only used to extract the "ground-truth" information for an evaluation purpose, including: (i) the work that was observed as performed by the newcomers, and (ii) the known next activity duration. We applied existing metrics from the data mining and machine learning literature to quantitatively evaluate the effectiveness of our approach.

5.2 Predicting Newcomer Capabilities

The output of this task is a set of predicted execution contexts for each newcomer. Each execution context, defined by the type names, can be viewed as a label in a *multi-label learning* problem. Therefore, we selected typical metrics *accuracy, precision*, and *recall* [19] for evaluation. We varied two key parameters in our approach: the number of selected resource groups (k in Definition 5) and the threshold for determining nearest events (θ in Definition 8).

Figure 3 summarizes the evaluation results, taking the mean metric values over newcomers in each log. For the parameter configuration of the proposed approach, we observe that a larger k yields lower precision but higher recall, and that a smaller θ yields higher precision but lower recall. Both are expected, since enlarging k (the number of selected groups) results in more similar resources and their capabilities being selected for a newcomer, and reducing θ imposes a stricter selection and results in the opposite. In the meantime, we observe that, in most of the cases, our approach has an increase in recall as the size of the recent log grows (from 10 to 100) while maintains precision at a consistent level. Note that a large number of execution contexts in these event logs (Table 1) means many labels in multi-label learning and tends to yield lower accuracy compared to dataset with fewer labels [19]. Therefore, we conclude that the prediction accuracy is satisfactory. This is evidenced by the lower prediction accuracy observed in the BPIC15 logs, which have more execution contexts derived (see Table 1). Moreover, prediction accuracy in general exhibits an increasing trend as the size of the recent log grows, which is as expected.

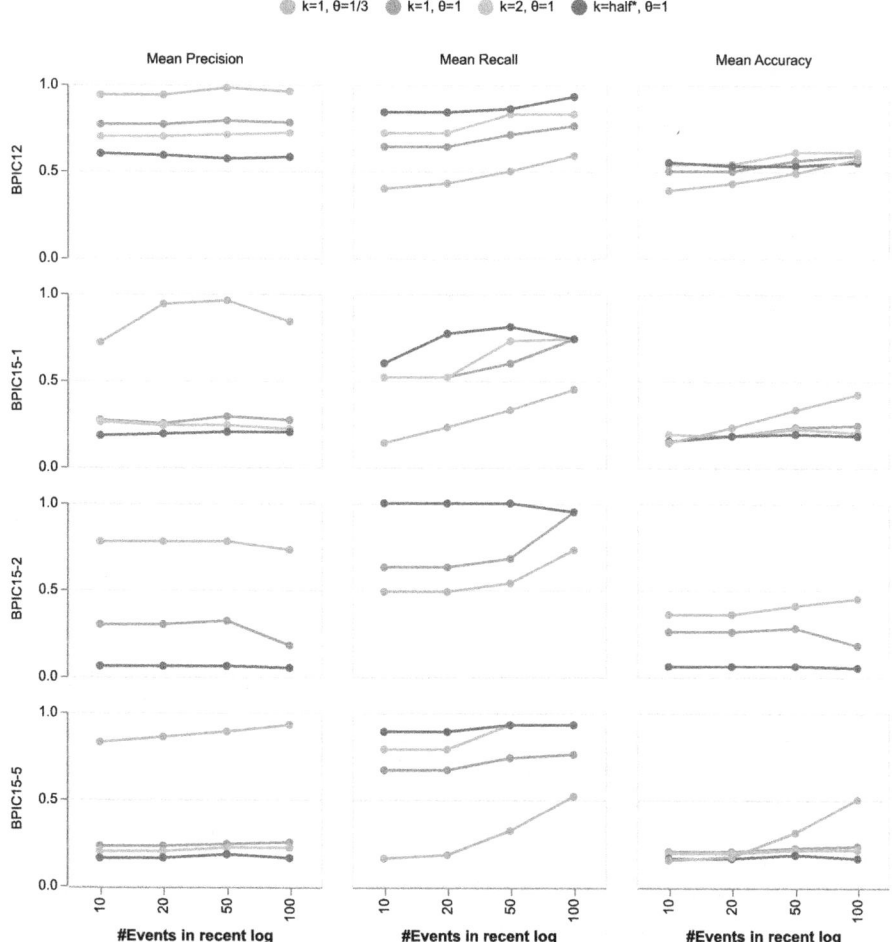

Fig. 3. Comparison between varied parameters k and θ of the proposed approach when generating predictions for newcomer capabilities. `k=half*`: setting k to half the number of resource groups in the organizational model.

It is worthwhile noting some counterexamples, such as the one in BPIC15-2. For the configuration of $k = 1, \theta = 1$, there is a noticeable drop in precision (green line) when the recent log enlarges from 50 to 100 events. Figure 4 illustrates the cause. Newcomer "22445896" is identified to be similar to existing resources in Group 2 when its recent log has 50 or fewer events. But as more events were recorded, it can be observed that the newcomer started to focus more on other work (new types of cases denoted by CT.3 and CT.4) and its capabilities became more related to those of Group 1 instead. Upon examination, we found that Group 1 consists of more resources than Group 2. Therefore, this change in the identified similar resource group yields the same impact as that from the increase of k, i.e., reducing precision and increasing recall.

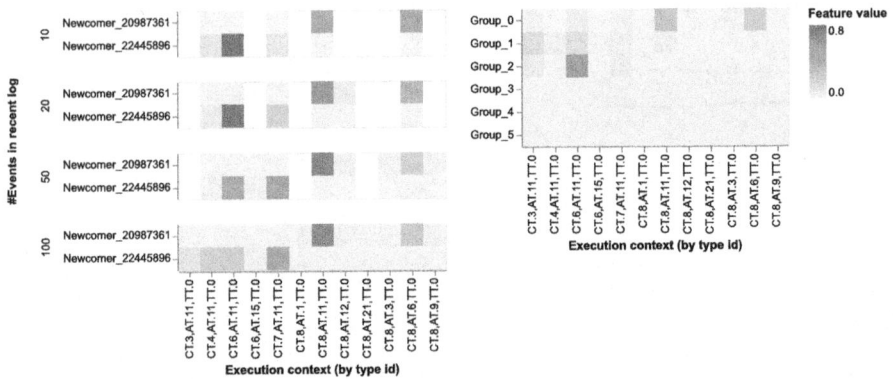

Fig. 4. BPIC15-2: The drop in precision is due to the changed result of identifying similar resources for newcomer "22445896" as its recent log enlarges. The heatmaps visualize the change of newcomer resource feature values corresponding with the most dominant execution contexts (left), compared to the resource group features in the organizational model (right).

The effect of various parameter values is aligned with our expectation. In applications of our approach, we suggest the use of a "validation set" split from the observational data (recent logs of newcomers) at hand to choose a suitable configuration based on the metrics. Depending on the weight assigned to precision and recall, one may decide on whether to start from using low or high values for k and θ. However, if the overall accuracy is valued as most important, it may be best to start tuning from low values, i.e., $k = 1, \theta = 1$, particularly when many execution contexts are derived from the event log in the step of discovering an organizational model.

5.3 Predicting Newcomer Performance in Ongoing Cases

For this task, we selected only the BPIC12 log. The BPIC15 logs do not record the start time of process activities required for deriving activity duration time to be used as the prediction target. In the experiments, we did not consider process-specific conditions that may limit the length of prefixes at which newcomers may take up activities. Therefore, zero bucketing was used. We applied last-state encoding for the next activity name and the next resource ID, and aggregation encoding to capture the activities performed so far in prefixes. In addition, we included the time elapsed from case start (with last-state encoding) as a feature to further characterize the progress of cases. For the predictive algorithm, we applied XGBoost with hyperparameters tuned with grid search. The configuration follows that reported in Reference [15]. Details of the experiment setting can be found in the aforementioned online repository.

Mean Absolute Error (MAE) was used as the evaluation metric. We compared our solution to a baseline method that does not explicitly consider if the next resource is a newcomer. That is, the baseline method applies the resource experience framework and uses only the recent log events to compute resource experience features for a newcomer.

Table 2 summarizes the evaluation results. By design, our solution is not different from the baseline when an existing resource is the next resource. Our first observation is that the baseline method performs worse when the next resource is a newcomer. Compared to when the next resource is an existing one (not a newcomer), prediction error for newcomers (MAE) increases from 0.23 to 0.346 (50% ↑). This is likely due to the limited events in the recent log. Then, compared to the baseline, our solution achieves better results when the next resource is a newcomer with an MAE of 0.333 compared to 0.346 (4% ↓).

Table 2. Evaluation of the proposed solution and the baseline method by test error measured in MAE.

#Cases	MAE (hours): Next Res is not NC		MAE (hours): Next Res is NC		
	#Prefixes	Baseline/Ours	#Prefixes	Baseline	Ours
6813	5997	0.230	1748	0.346	0.333

We further investigate the prediction error by relating it to each newcomer individually and the size of the recent log available for the newcomer at the point of prediction.

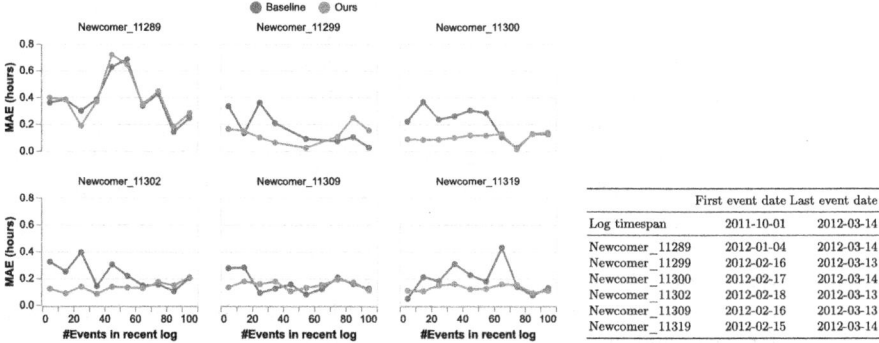

Fig. 5. Left: Testing error for predictions per newcomer in BPIC12. Right: The BPIC12 log timespan, and the first and last event times of the newcomers.

Figure 5 (left) shows the results, with the recent log size varying between (0, 100]. In general, we can see that the prediction error of our method is in general lower than that of the baseline. The difference between the two method becomes smaller as the number of events increase in the recent log. This trend is as expected. The augmented event data used in our solution is, at best, an approximation of the newcomers' "experience" based on resource similarity. Therefore, as more events are observed for newcomers, observational data should start to characterize the newcomers' true experience with more accuracy. Specifically, we highlight the case of newcomer "11829", for which we do not observe much testing error difference between our method and the baseline, as in the case of other newcomers. This may be attributed to the inherent difference between "11829" and other newcomers, or the purpose of deploying them into process execution, e.g., specialists to resolve bottlenecks or the use of additional resources. This is reflected in their different deployment times. As shown in Fig. 5 (right), "11829" had its first event logged in January 2012 while all others had the first events logged around the same time in mid-February 2012. Note that further process performance and resource analysis are required to determine the underlying reason, which we leave as out of the scope of this paper.

6 Conclusion and Future Work

Making accurate predictions for newcomers in a process is valuable for improving the management of business processes and providing support to the new employees as they fit into the organization. In this paper, we proposed a solution to address the problem of predicting the capabilities and performance of newcomers in a process mining context. The proposed approach is underpinned by organizational model mining and predictive process monitoring and can generate accurate predictions as shown in the experiments.

Our work contributes to resource management in BPM and opens several opportunities for future work. First, it would be interesting to investigate the relation between discovered organizational models and newcomer prediction accuracy. For that, model evaluation measures in the *OrdinoR* framework [16] may be applied. Second, data other than historical logs may be sourced and integrated, e.g., employee profiles from HR systems. It would be thus necessary to ensure that data privacy concerns are addressed properly. Moreover, other predictive tasks about newcomers may be explored to provide useful insights on newcomers during their onboarding process, for example, predicting their interactions with existing resources and their impact on case performance.

Acknowledgments. The authors would like to thank Arthur H.M. ter Hofstede for his advice and comments. The research leading to this paper is partly supported by the Dutch Growth Fund under the NEXTGEN Hightech project.

References

1. Van der Aalst, W.M.P., Reijers, H.A., Song, M.: Discovering social networks from event logs. Comput. Support. Coop. Work **14**(6), 549–593 (2005)
2. Bauer, T.N., Erdogan, B., Ellis, A.M., Truxillo, D.M., Brady, G.M., Bodner, T.: New Horizons for newcomer organizational socialization: a review, meta-analysis, and future research directions. J. Manag. **51**(1), 344–382 (2025)
3. Burattin, A., Sperduti, A., Veluscek, M.: Business models enhancement through discovery of roles. In: IEEE Symposium on Computational Intelligence and Data Mining, CIDM 2013, Singapore, 16-19 April, 2013, pp. 103–110. IEEE (2013)
4. Ceravolo, P., Comuzzi, M., De Weerdt, J., Di Francescomarino, C., Maggi, F.M.: Predictive process monitoring: concepts, challenges, and future research directions. Process Sci. **1**(1), 1–22 (2024)
5. Fehrer, T., Fischer, D.A., Leemans, S.J.J., Röglinger, M., Wynn, M.T.: An assisted approach to business process redesign. Decis. Support Syst. **156**, 113749 (2022)
6. Gunnarsson, B.R., Broucke, S.V., De Weerdt, J.: A direct data aware LSTM neural network architecture for complete remaining trace and runtime prediction. IEEE Trans. Serv. Comput. **16**(4), 2330–2342 (2023)
7. Kim, J., Comuzzi, M., Dumas, M., Maggi, F.M., Teinemaa, I.: Encoding resource experience for predictive process monitoring. Decis. Support Syst. **153**(113669), 113669 (2022)
8. Lee, A.S.: Supervisors' roles for newcomer adjustment: review of supervisors' impact on newcomer organizational socialization outcomes. Eur. J. Train. Dev. **48**(5/6), 521–539 (2024)
9. Mangat, A.S., Rinderle-Ma, S.: Next-activity prediction for non-stationary processes with unseen data variability. In: Lecture Notes in Computer Science, pp. 145–161. Lecture Notes in Computer Science, Springer International Publishing, Cham (2022)
10. Rinderle-Ma, S., Winter, K., Benzin, J.V.: Predictive compliance monitoring in process-aware information systems: state of the art, functionalities, research directions. Inf. Syst. **115**(102210), 102210 (2023)
11. Rizzi, W., Di Francescomarino, C., Ghidini, C., Maggi, F.M.: How do I update my model? On the resilience of Predictive Process Monitoring models to change. Knowl. Inf. Syst. **64**(5), 1385–1416 (2022)
12. Roider, J., Wang, W., Zanca, D., Matzner, M., Eskofier, B.M.: Predictions in Predictive Process Monitoring with Previously Unseen Categorical Values. In: De Smedt, J., Soffer, P. (eds.) Process Mining Workshops (2024)
13. Song, M., van der Aalst, W.M.P.: Towards comprehensive support for organizational mining. Decis. Support Syst. **46**(1), 300–317 (2008)
14. Teinemaa, I., Dumas, M., Rosa, M.L., Maggi, F.M.: Outcome-oriented predictive process monitoring: review and benchmark. ACM Trans. Knowl. Discov. Data **13**(2) (2019)
15. Verenich, I., Dumas, M., La Rosa, M., Maggi, F.M., Teinemaa, I.: Survey and cross-benchmark comparison of remaining time prediction methods in business process monitoring. ACM Trans. Intell. Syst. Technol. **10**(4), 1–34 (2019)
16. Yang, J., Ouyang, C., van der Aalst, W.M.P., ter Hofstede, A.H.M., Yu, Y.: OrdinoR: a framework for discovering, evaluating, and analyzing organizational models using event logs. Decis. Support Syst. **158**, 113771 (2022)
17. Yang, J., Ouyang, C., ter Hofstede, A.H.M., van der Aalst, W.M.P.: No time to dice: learning execution contexts from event logs for resource-oriented process

mining. In: Business Process Management - 20th International Conference, BPM 2022, Münster, Germany, September 11-16, 2022, Proceedings, pp. 163–180. Lecture Notes in Computer Science, Springer (2022)
18. Ye, J.H., Li, Z.W., Yi, K., Al-Ahmari, A.: Mining resource community and resource role network from event logs. IEEE Access **6**, 77685–77694 (2018)
19. Zhang, M.L., Zhou, Z.H.: A review on multi-label learning algorithms. IEEE Trans. Knowl. Data Eng. **26**(8), 1819–1837 (2014)

Management

A Case for Public Process Documentation: Robodebt an Automated Decision Making System

Adam Banham(✉), Azumah Mamudu, and Rehan Syed

Queensland University of Technology, Brisbane, Australia
a.banham@qut.edu.au

Abstract. Governments worldwide are embracing the evolution of digital economies and automation for efficiency gains to overcome scarce resourcing. However, achieving efficiencies without burdening the government or community they serve is not trivial. We explore a case study following the creation of an automated system to recover debts from welfare recipients who supposedly misrepresented their income, colloquially known as 'Robodebt'. Business process management was extensively used in the project's design and implementation; however, the owners and agents struggled to achieve their desired outcomes. On the flip side, the recipients of the automated system faced erroneous debts and a faceless system that ignored their existence. The system would ultimately fail; ruled illegal due to its process assumptions and, never came close to the goal of $4.772 billion in savings. Our analysis synthesises the project's design choices, deriving how the process mindset failed the owners and agents. We posit that public process documentation is required to prevent similar failures in the future.

Keywords: Australia · Public Service · Business Process Improvement · Business Process Management · Robodebt

1 Introduction

Public sector organisations around the world are constantly faced with demands for improving and providing efficient services by adopting digital solutions [15, 17]. Business process management (BPM) is a field that provides a wide range of tools and techniques for process owners to strategically identify improvements and quantify their impact *before* they are implemented. Integrating Information Technology (IT) with Business Process Improvement (BPI) is seen as a viable solution to enhance transparency and tackle the persistent problem of resource scarcity in public sector organisations [13].

Government departments provide social benefits and many "services" to their communities and, like corporate companies, face the need to improve processes to efficiently use their limited resources. We explore how BPM was used in an Australian government department enacting and implementing a BPI project.

We investigate the implementation of a specific budget measure (Scheme) from the Australian government, enacted into policy by the Department of Social Services (DSS), and operationalised by the Department of Human Services (DHS). The budget measure aimed to return **$4.772 billion** to the Commonwealth of Australia by raising debts on welfare recipients, who had supposedly incorrectly reported earned income while receiving social welfare. The Scheme cost **$971.391 million** to implement, administer, and ultimately wind back. However, it returned $*406.196 million* in actual savings, resulting in a net loss [14].

Relying on publicly available evidence, we conduct an exploratory case study, ascertaining how the Scheme failed to achieve desired outcomes. Specifically, we address the question: *How did a process mindset influence the operationalisation of the Scheme?* We use three terms throughout this paper. Firstly, "[the] *Scheme*" refers to the continual operationalisation of the budget measure, 'Strengthening the Integrity of Welfare Payments', or its more colloquially known term of 'Robodebt'. We reason that using the term 'Robodebt' may form a biased view of the Scheme. Secondly, "*recipient/s*" refers to an individual/s targeted by the Scheme's mechanisms [14, pp.viii-ix]. We use *recipient* over "customer" as no value was derived for these individuals; instead, only business value was derived for the Australian government. Finally, we use "*owner/s*" to refer to employees of DSS and "*agent/s*" to refer to employees of DHS.

We analyse evidence procured by the Royal Commission into the Scheme [14]and provide links to the evidence that supports our conclusions and reasoning[1]. For example, the first referenced piece of evidence is from a previous DHS CEO[2]:

> "*There appears to have been assumptions underlying [the Scheme] that everyone was computer literate and that debt was widespread. This is contrary to the knowledge of frontline workers and external agencies. Nothing was checked before hand*", Sue Vardon AO

Our analysis focuses on how the owners and agents operationalised a process to handle discrepancies between a recipient's claims and their annual taxation information. During normal operations, agents would perform compliance activities on historical welfare payments and use governmental powers to procure information. These efforts would result in a discrepancy being placed on a recipient, which may result in a debt being raised against them. However, their processes at the time meant that they could only handle the most likely or the most at-risk discrepancies. The introduction of the budget measure meant that owners and agents had responsibilities to increase processing capacity, by a factor of 50, to ensure that all discrepancies were processed within the year they were raised. The resultant was a purposefully designed system to automate the

---

[1] The paper references evidence from the commission in the following manner: rc-[document-id].

[2] rc-ANO.9999.0001.0056.

processing of discrepancies into debts, based on several dubious assumptions. For recipients, it was a nightmarish experience that surprised them when the Scheme extracted wealth from them. One recalls[3]:

> "I have never had a good understanding of why the debt was raised against me or what had caused the debt to exist...To this day, I feel worried when I check the mail out of a fear that [the Scheme] will have written to me about further debts or raising the amount that I owe them", Recipient

Notably, our discussion is not an attack on stakeholders handling the thankless job of facilitating public service in Australia. From a process perspective, typical owners have many options on hand to consider when *choosing* what processes should be improved or *what* new processes to implement [4]. It is reasonable to assume that typical owners have at least five options: fully commit to change, a short term change, a mid-term change, a long-term change, or ignore the process, moving on to bigger issues [10]. In contrast, public service *owners* have one: faithfully implement the elected officials goals into policy, and then the *agents* need to operationlise that policy [14, pp.xxiii][4]. They never stood a chance of changing anything, unless the elected voices said so [14, pp.iii]. These forces can unwillingly combine to confuse/mislead excelling public service in Australia, the Scheme is one such anomaly to *learn from*.

One might question whether owners and agents were process-minded, but it is clear that processes were thoughtfully modelled (in BPMN); for example, the detailed requirements for the initial version of the scheme[5]. Thus, we apply a process lens to the Scheme's designs to show that simple solutions exist [14]. Our findings present how public process documentation might have aided in clarifying the behavioural changes and their impact on the performance of processes. Lastly, we synthesise how aforementioned constrained process change mindset affected the owners and agents. For instance, it can lead to an echo-chamber, where they start shouting " *We know boats*"[6] as reasoning for their good work.

We argue that BPM affords opportunities to increase transparency in the operationalisation of public policy. As learning from the scheme's finale, it may not be until the public can question the process changes in detail, or ask for a debt to be repaid with *interest* [14, pp. 288,297-298,316-317] in Federal Court (as without interest it can be easily side-stepped), that an anomaly can be halted. We posit that by having publicly available process documentation that accurately reflects process changes when automated delegation is adopted will not only prevent future similar schemes but also may increase the likelihood of recipient's participation when they stand to gain nothing except disillusionment.

We structure our discussion as follows: Sect. 2 present existing discussions, Sect. 3 presents our methodology, Sect. 4 describes the unfolding of the scheme, Sect. 5 contains our synthesise, Sect. 6 articulates our insights, Sect. 7 concludes.

---

[3] rc-FBU.9999.0001.0002_R, p 10, para 49
[4] rc-RBD.9999.0001.0216.
[5] rc-CTH.3023.0004.8451_R, pages 70 –76, 110, 111, 168.
[6] Transcript, Scott Britton, 23 February 2023 [p 3686: lines 38-46].

2 Related Work

Discussions of the Scheme have spanned several contexts, such as managerial governance [9], automated decision-making [19], public service [18], social justice [5] and legal administration [30]. Previous work that most extensively looked at the findings of the Royal Commission was presented by Clarke, Michael, and Abbas [9]. Using a socio-technical framing, the authors present an in-depth chronological timeline based on several public commentaries on the Scheme. While their timeline touched on the process lens, our discussion distils this aspect in greater detail. Work by Braithwaite [5] positions the strategic goals of the Scheme and its design choice within the context of nudging, a reward philosophy of nudging people down the "correct" path to achieving a desired outcome. Rinta-Kahila et al. [19] discussed how algorithmic decision-making (ADM) is influencing governments, but to be successful, implementations must overcome several unique challenges as the scheme demonstrates. Finally, Nikidehaghani, Andrew, Cortese [17] considered another perspective on the Scheme, whether basing accountability on algorithms affects the uni-directionality of the relationship and reinforces instrumentation power. The authors [17] argue that the Scheme should be considered in the broader context of digital economies and the growing reliance on data-driven decision-making. Their work showed and discussed how the lived experiences demonstrated that the Scheme was indifferent to the circumstances or difficulties of proving that the debt was incorrect. Notably, prior studies have *hardly* investigated the Scheme from a process lens.

Process Mindset. BPM adopts a socio-technical approach to improve operational excellence [6,10,28]. Within the BPM toolkit are a range of techniques and approaches, some more qualitatively focused [10], and others, like process mining [28], are more quantitative. Nonetheless, a key characteristic of BPM is a *process mindset*, which starts by viewing organisations as interconnected, end-to-end business processes [20]. But the goal of those embracing a process mindset is not perfection or to be "*done*"; instead the goal is to achieve a constant state of change [3,6]. vom Brocke et al. [6] describes the mindset as the study of socio-technical processes over time, which consist of a coherent series of changes involving actions between actors and technologies on various levels. The process mindset is framed by the BPM lifecycle [10] to institutionalise continuous improvement and iterative change towards operational excellence.

Embracing the process mindset at the operational level is not enough; leadership has to support longer and continuous improvement projects [16]. Even when the mindset is embraced at the operational and managerial levels, a naive approach can lead to fail-fix cycles of growing complexity [1]. Notably, in terms of flexibility within the mindset, tensions can occur between the differences in strategic goals and operational expertise [2]. In the public domain, these tensions are further stressed by the need to provide efficient services while meeting public expectations and their diverse interests [23]. Thus, it is important to consider the role that a *process mindset* played in the development of the Scheme.

Theoretical Lens: Technology Discontinuity. Technology implementations require thoughtful planning and execution strategies to yield the required results. For technology to be successful, its acceptance and appropriate use is a critical element [8,27]. For the Scheme, the tension between the design of the system and processes was influenced by different pressures that emerged from a variety of stakeholders. This tension requires exploring and understanding the key factors that influence the objectives, performance expectations, political norms, and design variables that result in the discontinuation of a technology.

Therefore, we explored discontinuity and technology adoption theories as a theoretical lens to analyse the case [24]. The technology discontinuation theory offers two aspects to understand how technology use can be discontinued by stakeholders in a social system [26]. The *Competence-destroying* practices explains how new processes requiring new skills and technologies are fundamentally different from existing practices. Whereas, *Competence-enhancing* practices explain the improvements to existing methods that do not render current skills obsolete but incrementally improve technology [26].

These discontinuities often arise from competing and competitive changes that give birth to new initiatives and innovative technologies. Moreover, introducing new technology can lead to organisational change and resistance, as stakeholders must abandon old practices. The goal is to replace old technologies with new concepts, processes, and systems. Old technologies, initially innovative, can become obstacles due to their familiarity, experience, and sustained long-term practices [29]. Using the technology discontinuity lens will help explain the tensions that arose as stakeholders abandoned the Scheme and reverted to old processes and systems discussed in this case study.

3 Methodology

This section presents our methodological approach. Typical to the case study methodology [31], this study provides an in-depth analysis of the chronology and key occurrences of the scheme from a process lens. Relying on publicly available secondary evidence (see Table 1), we conduct a document analysis to ascertain how the Scheme failed despite owners and agents having a process mindset. Publicly available qualitative data are a rich source for investigating and theorising about socio-technical systems [11], providing a diverse coverage that would otherwise have been impossible to collect first-hand.

3.1 Case Description

The case was sanctioned by elected officials of the Commonwealth of Australia and operationalised by the owners and agents. The owners had oversight responsibility of the Scheme while the agents was responsible for its implementation. The goal was to generate revenue, supporting a surplus in the governmental budget, estimated at AUD $4.772 billion.

These estimates were based on the prior success of a manual process that reconciled discrepancies between a recipient's welfare remittances and their earned income. Owing to a growing backlog of unprocessed discrepancies, which was a major justification for the implementation of the Scheme. The response to the growing backlog was to introduce and adopt automation, increasing processing capacity. Therefore, an automated decision-making system was used to mass-produce debt invoices for outstanding discrepancies. This change from a manual to an automated system is the focus of our investigation. The Scheme was mostly driven by management (both owners and agents) in a top-down manner, hunting for the savings to become reality. The key change was that agents were not going to use their powers to produce evidence to support debts; instead, responsibility was shifted to recipients to "disprove" a presumptuous debt.

Table 1. Data sources considered in our analysis.

Document	Year	Source Type	# of items	Content
Royal Commission [14]	2023	Primary	1	1052 pp.
Deloitte report	2023	Primary	1	76 pp.
Procured Docs	2021-23	Secondary	10,933	$\geq$10,933 pp.
Hearings	2022-23	Secondary	46 days	4,998 pp.
Whiteford report	2023	Tertiary	1	114 pp.
Podger report	2023	Tertiary	1	42 pp.
Public Submissions	2022-23	Tertiary	293	$\geq$293 pp.
Unpub. PwC report	2017	Tertiary	1	106 pp.
Budget Reports (DHS)				
2014-15	2015	Secondary	1	325 pp.
2015-16	2016	Secondary	1	316 pp.
2016-17	2017	Secondary	1	322 pp.
2017-18	2018	Secondary	1	352 pp.
2018-19	2019	Secondary	1	392 pp.
Senate reports				
2017	2017	Tertiary	1	155 pp.
2022	2022	Tertiary	1	62 pp.
Ombudsman reports				
2017	2017	Primary	1	113 pp.
2019	2019	Secondary	1	45 pp.
Legal proceedings				
Masterton	2019	Tertiary	3	43 pp.
Amato	2019	Tertiary	2	12 pp.
Prygodicz	2019	Tertiary	4	258 pp.
2015 advice	2015	Tertiary	1	6 pp.
2017 advice	2017	Tertiary	1	2 pp.
Solicitor-General	2019	Tertiary	1	47 pp.
		total:	11,298	$\geq$ 20,084 pp.

3.2 Data Extraction and Analysis

Table 1 describes the included data sources. Qualitative data was grouped into primary, secondary and tertiary sources. Primary sources were used as our basis for synthesising insights. Secondary sources provided specifications on our synthesis, such as process performance, culture insights, and management mindsets. Tertiary sources provided additional perspectives on the Scheme but were not explored in-depth. In total 20,084 pages of reports and supplementary documents were inductively searched.

Following the Gioia et al. [12] approach, we inductively extracted initial insights and considered how emanating themes could be viewed from a theoretical perspective. First, two authors extracted stakeholder narratives of key occurrences of the Scheme's implementation from 2016 to 2020 from the Primary sources (see Table 1). These first-level codes described the chronology of events and challenges encountered during the Scheme's operationalisation. Where reference was made to secondary or tertiary data sources, relevant sections were extracted. Secondly, three authors reviewed the extracted first-level codes to derive key themes around the operationalisation of the Scheme. This began with comparing first-level codes and reviewing emanating themes to ensure proper alignment and parsimony. Thirdly, we sought to identify a theoretical lens to best describe and explain how the Scheme failed despite the owners and agents having a process mindset. A dominant feature in our data was that the Scheme's failure resulted from stakeholder tensions regarding maintaining legacy practices and implementing a process innovation. We selected the theory of technology discontinuity [25] to explain the discontinuities in the Scheme.

4 Chronology of the Scheme

This section outlines the public reports that became available during the duration of the Scheme and at high-level describes the unfolding of the Scheme.

One of the first pieces of public information about the Scheme came from the then-incoming minister, who proclaimed that they were going to be a *"strong welfare cop"*[7] and that the relationship between ministers, owners, and agents needs to be corrected[8]. It is within this culture of *welfare cops* that the initial idea was born, grown and evolved into a budget measure, *Strengthening the Integrity of Welfare Payments*, in the 2015-2016 Budget[9].

It would not be until the first online refinement of the Scheme went live that further public information would be released. The first iteration, the online compliance intervention system (OCI), went live in July of 2016[10], and by December of 2016 major criticism of the Scheme was being presented by media outlets[11]. Criticism was voiced as incorrect debts being mass produced by the Scheme. The outrage from the public in the media and a large number of complaints meant that during this time Ombudsman's office contacted the agents to request information for an upcoming inquiry report[12]. In 2017, several public reports and

---

[7] rc-RBD.9999.0002.0002.
[8] rc-RBD.9999.0001.0216, p.4
[9] rc-DSS.5124.0001.1240_R, 2015-2016 Budget Paper 2: Budget Measures, p.132
[10] rc-KHA.9999.0001.0001_2_R2, para. 71
[11] rc-CTH.3000.0023.6773.
[12] rc-CTH.1000.0006.8741_R.

inquiries were published. Responding to growing criticism of the Scheme, owners proclaimed the following in a media release [14, p.156][13]:

> "I think this [Scheme] is about as reasonable a process as you could possibly derive...It really is an incredibly reasonable process...It's very significant. Four billion dollars over four years is evidently a very significant amount of money. That is helping us get back into surplus", 3 January 2017, Mr Porter

These media statements by the minister did little to address the more serious claims of the public outrage being covered in the media. In April of 2017, the Ombudsman would release its findings about the Scheme[14], containing real incidents surrounding the scheme over 113 pages, including eight recommendations. Notably, the report is only 30 pages and the remaining are left to appendices which somewhat defend the Scheme and design choices. However, the findings of the commission was that wording used within the report where handpicked by the owners and agents, as they were"...*given a great opportunity to* ***effectively co-write*** *the report...*"[15]. Nonetheless, the report sheds light on the machinations of the scheme, albeit in the best possible light. In June of 2017, a parliamentary committee presented findings on the Scheme [21], outlining (emphasis added):

> "It was made clear to the committee during the course of this inquiry that the evidence consistently demonstrated a key flaw in the [Scheme], a flaw which filtered throughout the [Scheme]: a fundamental lack of **procedural fairness**... This lack of procedural fairness disempowered people, causing emotional trauma, stress and shame... What also became clear through the inquiry is that the department has a fundamental conflict of interest – **the harder it is for people to navigate this system and prove their correct income data, the more money the department recoups**...", Commonwealth of Australia 2017

This senate inquiry also presents 21 recommendations, starting off with the recommendation that the OCI program should be put on hold until all procedural fairness flaws are addressed. Notably, the dissenting deport [21, pp.115-122] relied on influenced the Ombudsman report to dismiss the findings.

Throughout 2018, the Scheme rolled on with little to no pushback from its owners and agents [14, p.259]. However, the year did produce new public information in terms of Administrative Appeals Tribunal (AAT) decisions on recipients claims against the Scheme. In the previous year, a growing number of decisions from the AAT had set aside decisions surrounding the Scheme [14, pp.239,240]. In September of 2017, a decision was made by the AAT on a previous decision in March of 2017, setting aside a debt based on income averaging for want of an evidentiary basis [14, p.268]. Following on in April of 2018, an article by an AAT member (Professor Carney AO) is published [7], and opens with:

---

[13] rc-CPO.9999.0001.0009.
[14] rc-CTH.3044.0003.7539.
[15] rc-CTH.3007.0004.4949_R.

A Case for Public Process Documentation   335

> "...the so-called 'practical onus' to establish a debt and its size continues to remain with [the Scheme]; the failure of a person to 'disprove' the possibility of a debt is not a legal foundation for a debt... when confronted with suggestions of having an overpayment, often from up to seven years ago, the least literate, least powerful, and most vulnerable alleged debtors will simply throw up their hands...", Professor Carney AO

The response to the article was to state that the process was reasonable[16] and the esteemed law professor was mistaken[17] [14, p.271] in saying that the onus of proof had been transferred to the recipient [14, p.269].

In 2019, two major legal cases occurred in the Federal Court of Australia[18][19]. The first case, *Masterton vs DHS*(see footnote 18), would be sidestepped by the owners by simply reducing the debt to zero and repaying the recipient. In the second case, *Amato vs the Commonwealth of Australia*[20], a difference was that the recipient had already been garnished from their tax return to recover part of their debt [14, p.298]. However, for this case, the extinguishment of the debt did not stop the proceedings as their claim for interest had not been handled[21].

In parallel, the agents received legal advice noting that if the case would proceed, it would be unfavourable for the Scheme[22][23]. The opinion would later conclude that recipients had no duty to act on the discrepancy notices, and in the event that they didn't respond, agents should have used their compulsory power to make inquiries[24]. The effect of this opinion made clear that the Scheme in all iterations had been unlawfully raising debts against recipients [14, p.305]. It wouldn't be until 18th November 2019 that the Scheme would stop using the averaging approach, almost two months after the opinion was received.

By May of 2020, recipients would be refunded for all repayments made on debts raised using averaging[25]. The Scheme was closed altogether on 30 June 2020. The commission clarifies the importance of the court cases as [14, p.317]:

> "[They] succeeded in exposing the illegality of [the Scheme] where other possible forms of check on the scheme – the AAT, the Commonwealth Ombudsman, the sound advice of some lawyers – did not or could not", Royal Commission into the Robodebt Scheme

At the end of 2019, the Senate referred an inquiry into the Scheme to the Senate Community Affairs References Committee [22]. In May 2022, the inquiry would publish its final report [22] with one recommendation:

---

[16] rc-CTH.3007.0008.5900.
[17] rc-CTH.4750.0003.3228_R, "A former member of the AAT - what a lofty authority"
[18] rc-VLA.9999.0001.0001.
[19] rc-VLA.9999.0001.0075_R.
[20] rc-VLA.9999.0001.0075_R.
[21] rc-CTH.3004.0014.3119_R.
[22] rc-CTH.3007.0011.6407_R.
[23] rc-CTH.3007.0011.6408_R, para.A2
[24] rc-CTH.2013.0012.5070_R, Answer 3, Answer 4
[25] rc-DSS.5015.0001.0048_R.

> "...that the Commonwealth Government establishes a Royal Commission into the [Scheme]", Commonwealth of Australia 2022

The *Royal Commission into the Robodebt Scheme* [14] exercised powers forcing entities to produce evidence, producing over 958,000 documents. However, not all documents produced were made publicly available. Those that were, often had some redaction applied. In total 10,933 documents were made public through the commission[26]. The commission held four separate hearing blocks between October 2022 and March 2023, calling 115 witnesses to discuss the Scheme; video and transcripts of these hearings are publicly available[27].

5 The Scheme

This section synthesises the development of the Scheme across two stages. We first consider the manual process leading to the generation of a debt for a recipient, followed by the first iteration of the automated version for the Scheme. We synthesise process documentation for these phases and use document analysis to inductively derive key problems faced by the Scheme. These problems are positioned into pulling and pushing forces over tensions for the Scheme [24,25].

5.1 Manual Process

This section outlines the manual processing of discrepancies between fuzzy data [14, ch.16.5.3] derived from taxation information and reported income to the agents. The manual process for handling discrepancies is shown in Fig. 1. This BPMN model was derived from procured evidence from Dr Elea Wurth, Deloitte[28] and further synthesised using the commission's findings [14]. The process model describes five phases that could occur while handling a discrepancy for a recipient: *initial outreach*, optionally an *extended outreach*, *confirmation*, optionally *third party collection*, and finally *entitlement assessment*.

In the *initial outreach* phase, a compliance officer would generate a notice from the intervention case data and post/email the notice to the recipient. The officer may attempt to contact the recipient before sending a notice if the recipient meets vulnerability criteria. If the recipient makes contact, the process moves to *confirmation*. Otherwise, the process moves to *extended outreach*. During the *extended outreach* phase, the officer makes several attempts to contact the recipient, waiting up to 35 days across several races. If contact is made, the process moves to *confirmation*; otherwise, it moves to *third party collection*.

The *confirmation* phase is a communication between agents and the recipient to discuss information that may influence the assessment of the discrepancy. A period of waiting may occur in this phase while a race between documents

---

[26] robodebt.royalcommission.gov.au/document-library.
[27] robodebt.royalcommission.gov.au/hearings.
[28] rc-RBD.9999.0001.0485.

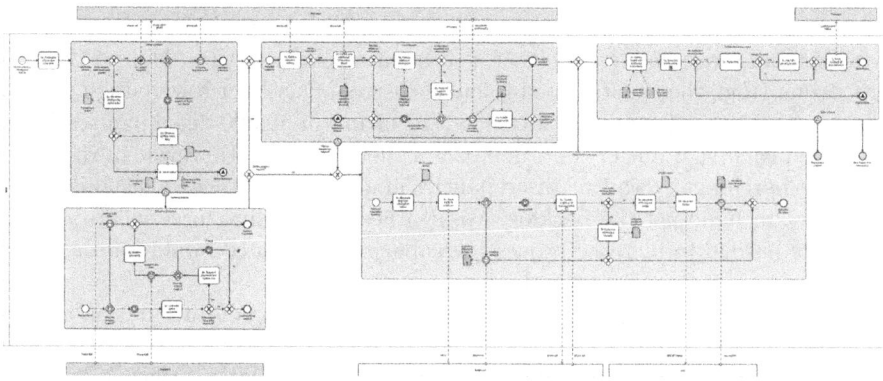

Fig. 1. The BPMN model for the process before the implementation of the robodebt. Green expanded sub-processes highlight the non-optional phases, gold sub-processes highlight optional phases. (Color figure online)

being provided by the recipient resolves. If the recipient is unable to provide acceptable documentation for the discrepancy, the process moves to *third party collection*. *Third party collection* consists of a series of races between collecting information from entities about the recipient and a long waiting period for a response. Otherwise, it moves to *entitlement assessment*, where the discrepancy is considered in light of additional information to determine if a debt should be raised.

Problems. The nature of the problems faced by the implementation of the scheme under this design was somewhat unorthodox in comparison to a typical case in business process management. The first is that the process does not have an arrival rate of process instances waiting to be enacted, instead large batches of instances are added to the backlog after receiving updated fuzzy profiles for recipients. For example, by October 2014, there were 866,857 unique recipients with a discrepancy and a total of 1,080,028 discrepancies[29][30] [14, pp.39] in the backlog from the taxation matching in previous years. This aspect is unusual as process instances are typically received in the form of say, 5 instances a day or once a week, meaning analysts focus on improving process time to be faster than the arrival rate. However, in this case, the large batching of awaiting discrepancies made manual processing significantly inefficient in handling the backlog. We synthesise these issues into a pulling force, *automation* over the tension for transparency in later iterations.

The second problem was the processing time was on average 186 days per discrepancy[31], mainly due to several phases, *initial outreach*, *extended outreach* and *third party collection*, due to races with long deadlines with external entities.

---

[29] rc-CTH.3000.0001.8417.
[30] rc-CTH.3000.0001.8680_R.
[31] rc-CTH.3023.0002.0503.

These waiting periods can be seen in Fig. 1 when we have a race between a deadline and a response from either a recipient or a third party. Due to the long processing time, the agents opted to enact discrepancies which were high risk, or when a large discrepancy was identified, resulting in 20,000 discrepancies being processed each year out of the 1,080,028 in the backlog[32]. However, in this case, the large batching of awaiting discrepancies made manual processing significantly inefficient in handling the backlog. We synthesise a pulling force where *owners & agents* needed to increase processing capacity. This force would interact with the tensions of transparency and generating debts.

The third problem was the assessment and processing of information related to a discrepancy required an expert agent to apply complex business rules. This problem meant the process could only be scaled up by hiring a larger resource pool. Estimates were that over 1,000 new employees would need to be hired and trained over the coming three years to handle incoming backlogs[33]. Ultimately, this ended up happening[34] at great cost. We synthesise a pulling force, *Expertise*, related to this problem for the scheme over the tension of generating debt.

These problems are not unique to the scheme and, in fact, are common problems faced by any business, regardless of domain. Scaling up process capacity is not a trivial task for any analyst, but is ideal for a process mindset to handle.

5.2 The Online Compliance Intervention Process

This section outlines the first implementation of the Scheme for processing discrepancies between income reported by recipients and taxation income identified by an internal assessment by agents, known as the "Online Compliance Intervention" process (OCI). The OCI process was *officially* launched after completing a pilot program[35], a manual mirroring program and a staging program.

In Fig. 2 shows our synthesis of the OCI process for the (staging) version that was "launched" at the end of September 2016 and operated unchanged until February 2017. Once again, the Deloitte process maps were used as a base[36] model and further information from the commission was used to annotate our process model [14, Chapters 5-6][37][38][39]. We synthesised six phases: *Issue Notice, Intervention, Assessment,* and optionally *Update, Disagreement,* and *Override*.

The process starts with *initial notice*, where a compliance officer issues a notice to a recipient and raises an intervention activity in the OCI system, moving to *intervention*. Importantly, no communication with the recipient occurs as the notice is sent to the last known contact address for the recipient. The intervention starts a race between its resolution and a deadline of 21 days.

---

[32] rc-CTH.3001.0030.3987_R.
[33] rc-CTH.3053.0044.0425_R.
[34] rc-CTH.3004.0009.0076.
[35] rc-CTH.9999.0001.0033_R, para.1.5,2.1,3.1
[36] rc-RBD.9999.0001.0486.
[37] rc-CTH.3715.0002.6783_R, frequencies
[38] rc-CTH.3715.0002.4820, guide
[39] rc-RBD.9999.0001.0451_R.

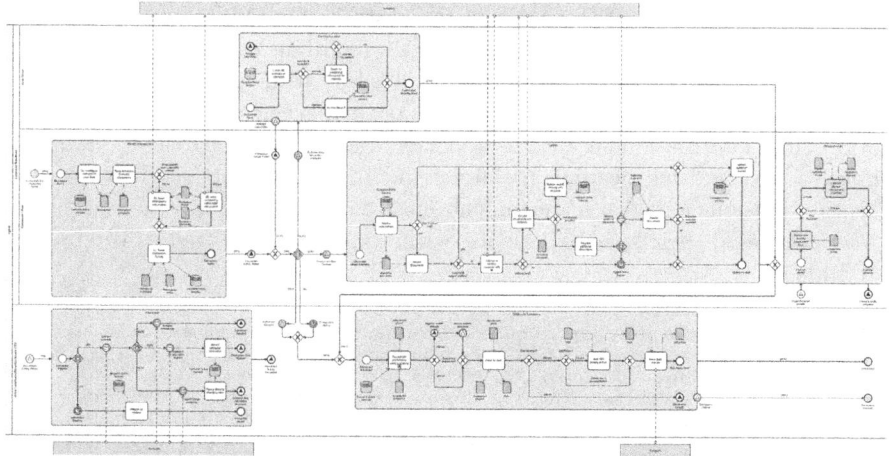

Fig. 2. The BPMN model for the process after OCI implementation, initiating the robodebt scheme.

The *intervention* phase is a race between the deadline and the recipient interacting with the system to provide further information about the discrepancy. If the recipient interacts, they have three options: accept the information, supply further information, or deny the employer information. The latter two require the recipient to provide documents through the system to prove their disagreements.

If the recipient supplies further information, the process moves to the *update* phase where the information is processed in a somewhat similar fashion to the *confirmation* phase in Sect. 5.1, after which moves to *assessment*. If the recipient denies the employer, the process moves to the *disagreement* phase where a case officer processes the claim, either reissuing the intervention or moving to *assessment*. In all other cases, the process moves to *assessment*. The *assessment* phase consists of an automated evaluation, which uses the averaging approach in 94.9% of cases where a debt may be raised. However, many IT issues[40][41] were present in the system at the time, meaning that a manual override may occur to complete the evaluation. After which, a raised debt usually includes an additional 10% fee and a debt advice notice is sent to the recipient. Otherwise, if no debt is raised, no further interaction with the recipient occurs.

Problems. In response to the lengthy processing time of the scheme in Sect. 5.1, owners and agents sought ways to reduce the needed actions before a debt was raised. Their conclusion was to simply do *less work* and at a strategic level, change their business model to place the full onus of "correcting" the discrepancy on the recipient[42]. Going as far as creating a script that compliance officer should

---

[40] rc-RBD.9999.0001.0486.
[41] rc-CTH.3018.0020.8802.
[42] rc-CTH.3715.0001.4283_R.

follow[43], which reinforced that no additional work should be undertaken *unless* explicitly asked by the recipient. This strict handling of behaviour created a tension over *whose* expertise should be emphasised while generating debts. The change in operation was in the face of many experienced compliance officers noting that information in the discrepancy was prone to erroneous data and duplication of income, leading to larger and incorrect debts(see footnote 43). Creating tensions between management and frontline operators, one compliance officer expressed their lived experience at the time as(see footnote 43)(emphasis not added):

> "*I know that not all our [recipients] have the capacity to engage in a meaningful discussion about what their circumstances were and what they have provided to [agents]...There will be [recipients] who will repay debts that they should never have had...we are being asked to ignore evidence that no debt exists and to 'collude' in raising a debt when none should exist.* **That is, we are being asked to commit a fraudulent act**",
> an agent with 30 years of expertise, Jan 2016

We synthesise several forces from this problem: a pushing force, *Compliance Officers*; a pulling force, *Strategic Decisions*; and a pulling force, *Expertise*; these forces melded over a tension on generating debts.

The change of operation and its impact on the behaviour of the process is also quite pronounced in terms of communication to the recipient in Fig. 2. The change in onus on the recipient can be seen in the fact that 70% of recipients were not interacting with the system at all. Management pushed for the dubious assumption that these recipients had simply accepted the discrepancy, not the more likely case that they were unaware of the letter or misunderstood the hidden meanings behind them [14, ch.5.2]. This decision created tensions over the surrounding culture handling the Scheme and over these one-sided assumptions chosen to suit the mass generation of debt. If our modelling truly reflects the scheme of the time, 94.9% of instances never interacted with a compliance officer, in contrast with the nearly 0% in the pre-scheme process. Unsurprisingly, the reduced interaction meant processing time was now 22 days[44], compared to 186 days previously. The short-term thinking, coupled with the dubious assumption of acceptance, generated short-term value but fed into long-term issues.

One might jump to the conclusion that the implementation of the scheme was unthoughtful. We argue against this notion of unthoughtfulness based on the following information. Before launching the OCI, several implementation phases were conducted, and in each of these phases, they considered several metrics for "success"[45]. These are not the acts of the unthoughtful, but did they rush over many problems? Absolutely [14, p.136]. Several forces were permeating from the top-level (i.e. politicians) down to the owners and their culture issues, which led to hyper-toxic culture [14, p.243], and from agents needing to show their excellence for career progression [14, chp.23]. These exogenous forces [18] had

---

[43] rc-CTH.3001.0035.3159_R.
[44] rc-CTH.3023.0002.8902_R.
[45] rc-CTH.3715.0002.6783_R.

only one place to go under the constrained process mindset, *forward towards the goal*; as such, these forces often turned heads away from underlying problems. We synthesise these exogenous forces into a pulling force for the *Ministers* influences over public servants, adding to the tension for generating debts.

6 Discussion

This study synthesises secondary evidence to understand how the process mindset failed the owners and agents of a process improvement initiative for a Scheme within the Australian public service. Using publicly available data, we elucidate insights into the prior and post-states of the Scheme. We also identify the tensions and push and pull factors between actors of the Scheme.

Our conceptual model (Fig. 3) identifies transparency in operational behaviour and the generation of debts as the key tension areas.

The first tension was over the **transparent operationalisation** of the Scheme. The pushing away forces consist of recipients and independent bodies, such as the senate inquires [22] or AAT decisions [30], which questioned how these debts came to be and why clear errors were not addressed. This tension is crystallised in lived experiences describing the sudden appearance of debts in their lives, as 70% did not respond to the so-called "call-to-action" letters. The pulling forces continued on with their debt collection regardless of these experiences, and the owners and agents saw their non-response as an informed decision. In result, the overall acceptance was heavily impacted, the Scheme was viewed negatively by the general public, causing the agents and owners to defend their choices. They defended their operationalisation by influencing the ombudsman report in 2017 to clear them of any wrong doing.

The second tension was over the **automated generation of debts**. Concerns were raised on the validity of material evidence, change in business model and legality of the averaging approach used to generate debts at a mass scale. Ministers influenced owners and agents to increase processing capability to crack down on welfare frauds. The owners and agents insisted that averaging was a standard calculation measure for debts. This was one of the many strategic decisions to reduce the burden of the business model. Relying on their domain expertise, they justified the legal and business ramifications of the Scheme.

The management of the agents and owners used the tensions to twist outcomes towards generating perceived value about the Scheme in media and to presumptuously generate debts. Recall that the Scheme's goal was to generate debt from discrepancies; inducing a lot of short-term thinking without understanding the growing long-term consequences. From our analysis, it is clear that they forced themselves into a constrained process mindset of finding success wherever possible. As they were locked into faithfully implementing and operationalising policy without any recourse to make the decision to stop the scheme.

We posit that a constrained process mindset, albeit agents and owners could do little about this, could quickly be replaced. When process improvement projects are placed under such a constrained mindset, the only lever to stop

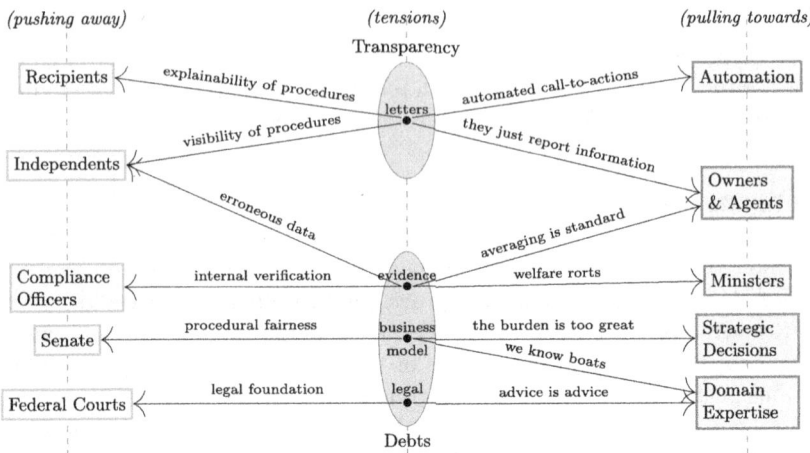

Fig. 3. The transparency and debt generation tensions within the Scheme.

the project tends to be from the outside. Agents and owners must create ways for reality to pull that lever with some amount of authority. Otherwise, the hunt for "savings" will be unstoppable and affect another 526,000 Australians[46] [14]. Public process documentation, as demonstrated in Sect. 5, may have enlightened reality in ways that could not be misdirected or dismissed in a public discourse.

7 Conclusion

Scaling processing capacity despite scarce resources will continue to be a painpoint for public services. Our exploratory case study unfolds the Scheme to investigate how owners and agents failed to achieve desired improvement outcomes despite having a process mindset. We presented a chronology of the Scheme and its related implementation challenges, based on publicly available data. Our findings identify two tensions influencing the adoption and continuation of the scheme: a tension over its transparency and another on the generation of debt. These tensions were significant in the Scheme's failure. We posit that the constrained process mindset created a myopic vision to continue the project, whereby tensions were resolved in favour of the continuation, feeding into longterm problems. To ensure process owners do not fall into constrained process mindsets, meaningful transparency into process documentation must be introduced to prevent similar failures.

Our study holds some limitations. It relies on sources not collected for our purpose of investigation. Our findings are bounded by the scope, bias and limitations of the data, such as personal views and interview codes. We acknowledge that we are limited by selection and researcher bias in the code selection and

---

[46] rc-DSS.9999.0001.0051_R.

data analysis. In future work we will expand the scope to cover all iterations of the Scheme and further improve our initial framework.

References

1. Adams, N., Augusto, A., Davern, M., Rosa, M.L.: Why do banks find business process compliance so challenging? An Australian perspective. In: Business Process Management Forum. Springer, Cham (2022)
2. Baiyere, A., Salmela, H., and, T.T.: Digital transformation and the new logics of business process management. EJIS **29**(3) (2020)
3. Baygi, R.M., Introna, L.D., Hultin, L.: Everthing flows: studying continuous sociotechnical transformation in a fluid and dynamic digital world. MIS Q **45** (2021)
4. Beerepoot, I., et al.: The biggest business process management problems to solve before we die. Comput. Ind. **146**, 103837 (2023)
5. Braithwaite, V.: Beyond the bubble that is robodebt: how governments that lose integrity threaten democracy. Aust. J. Soc. Issues **55**(3) (2020)
6. vom Brocke, J., et al.: Process science: the interdisciplinary study of socio-technical change. Process Sci. **1**(1) (2024)
7. Carney, T.: The new digital future for welfare: debts without legal proofs or moral authority? UNSW Law J. Forum (1) (2018)
8. Christensen, C.M., Overdorf, M.: Meeting the challenge of disruptive change. Harvard business review **78**(2) (2000)
9. Clarke, R., Michael, K., Abbas, R.: Robodebt: a socio-technical case study of public sector information systems failure. AJIS **28** (2024)
10. Dumas, M., Rosa, M.L., Mendling, J., Reijers, H.A.: Fundamentals of Business Process Management, Second Edition. Springer (2018)
11. Ghazawneh, A., Henfridsson, O.: Balancing platform control and external contribution in third-party development: the boundary resources model. Inf. Syst. J. **23**(2) (2013)
12. Gioia, D.A., Corley, K.G., Hamilton, A.L.: Seeking qualitative rigor in inductive research: Notes on the Gioia methodology. Organ. Res. Methods **16**(1) (2013)
13. Gulledge, T.R., Sommer, R.A.: Business process management: public sector implications. Bus. Process. Manag. J. **8**(4) (2002)
14. Holmes, C., Greggery, J., Scott, A., Berry, R., Marsh, S., Freeburn, D.: Royal Commission into the Robodebt Scheme. Commonwealth of Australia (2023). https://robodebt.royalcommission.gov.au/publications/report
15. Jurisch, M., Ikas, C., Palka, W., Wolf, P., Krcmar, H.: A review of success factors and challenges of public sector BPR implementations. In: HICSS-45. IEEE Computer Society (2012)
16. Malewska, K., Roszyk-Kowalska, G., Chomicki, M.: The impact of leadership on business performance. the role of process performance. In: Business Process Management Forum. Springer (2023)
17. Nikidehaghani, M., Andrew, J., Cortese, C.: Algorithmic accountability: Robodebt and the making of welfare cheats. Acc. Auditing Accountability J. **36**(2), 677–711 (2023)
18. Podger, A., Kettl, D.F.: How much damage can a politicized public service do? Lessons from australia. Public Adm. Rev. **84**(1) (2024)
19. Rinta-Kahila, T., Someh, I., Gillespie, N., Indulska, M., Gregor, S.: Algorithmic decision-making and system destructiveness: a case of automatic debt recovery. Eur. J. Inf. Syst. **31**(3) (2022)

20. Rosemann, M., Brocke, J.v., Van Looy, A., Santoro, F.: Business process management in the age of AI–three essential drifts. Inf. Syst. e-Bus. Manag. (2024)
21. Senate Community Affairs Committee: Design, scope, cost-benefit analysis, contracts awarded and implementation associated with the Better Management of the Social Welfare System initiative. Commonwealth of Australia (2017)
22. Senate Standing Committees On Community Affairs: Centrelink's compliance program – Accountability and justice: Why we need a Royal Commission into Robodebt. Commonwealth of Australia (2022)
23. Syed, R., Bandara, W., French, E., Stewart, G.: Getting it right! Critical success factors of BPM in the public sector: a systematic literature review. Australas. J. Inf. Syst. **22** (2018)
24. Syed, R., Leemans, S.J.J., Eden, R., Buijs, J.A.C.M.: Process mining adoption - a technology continuity versus discontinuity perspective. In: Fahland, D., Ghidini, C., Becker, J., Dumas, M. (eds.) Business Process Management Forum-2020. Lecture Notes in Business Information Processing, vol. 392, pp. 229–245. Springer (2020)
25. Tushman, M.L., Anderson, P.: Technological discontinuities and organizational environments. Adm. Sci. Q. **31**(Sep 86) (1986)
26. Tushman, M.L., Anderson, P.: Technological discontinuities and organizational environments. Routledge (2018)
27. Tushman, M.L., Murman, J.P.: Dominant designs, technology cycles, and organization outcomes. Acad. Manag. Proc. **1**, A1–A33 (1998)
28. van der Aalst, W.M.: Process Mining Data Science in Action. Springer (2016)
29. Venkatesh, V., Thong, J.Y.L., Xu, X.: Unified theory of acceptance and use of technology: a synthesis and the road ahead. J. Assoc. Inf. Syst. **17**(5) (2016)
30. Whiteford, P.: Debt by design: the anatomy of a social policy fiasco – or was it something worse? Aust. J. Public Adm. **80**(2) (2021)
31. Yin, R.K., Campbell, D.T.: Case study research and applications: design and methods. SAGE, sixth edition. edn. (2018)

Affective Business Process Design

Thomas Grisold[1](✉) and Michael Rosemann[2]

[1] Vienna University of Economics and Business, Vienna, Austria
thomas.grisold@wu.ac.at
[2] Queensland University of Technology, Brisbane, Australia
m.rosemann@qut.edu.au

Abstract. Business Process Management and design have been mainly concerned with transactional efficiency and effectiveness of workflows reflecting its grounding in scientific management, engineering and computer science. Affective, human-centered design goals have been largely absent, making BPM an under-designed discipline constrained in its ambition and limited in its impact. Addressing this shortfall and motivating a new research agenda, this paper proposes 'affective business process design' to explicitly consider how human actors experience emotions, feelings and moods during process interactions. Making affect a primary concern of business process design counterbalances the established, rational and analytical focus of this stage in the process lifecycle. The contribution of this paper is two-fold. First, we conceptualize affective business process design and outline its foundations and premises. Second, we discuss process design options to realize affective business processes. We conclude by outlining future research avenues to advance, operationalize and integrate the broader agenda of 'affective BPM'.

Keywords: Affect · human-centered design · emotion · customer experience · design options

1 Introduction

Business Process Management (BPM) emphasizes efficiency and effectiveness in workflows [e.g., 1]. While the experiences of customers—both internal and external—are recognized as crucial, they are typically not explicitly considered in business process design. They are only implicitly assumed when a process achieves its targeted operational performance [2]. For instance, a customer's experience is regarded as high when the process has an optimal throughput time, or when it leads to reliable outcomes [3].

This prevailing view overlooks that a customer's experience goes beyond workflow-related KPIs. Customer experiences do not only involve rational assessments (e.g., net promoter score, willingness to proceed), but also affective reactions, typically taking shape without explicit intention and on subliminal levels. A key finding in modern psychology is that affective states are key drivers of perception, interpretation and—more or less rational—decision-making [e.g., 4, 5]. And while there are process-related approaches that capture emotional states of customers (and employees) along a business process (e.g., customer journey mapping, human-centered design), these lack the rigor and specificity commonly associated with BPM methods. They have not made it into standard process lifecycle models or methodologies and technologies (e.g., Lean Six Sigma, RPA). This absence of a systematic approach to capturing customer affect during the process design stage is a shortfall of contemporary BPM, especially in light of calls for customer-first approaches to business process (re-)design [e.g., 6].

The purpose of this paper is to establish affective business process redesign as a new lens complementing the established body of cumulative knowledge around effectiveness and efficiency [7, 8]. Affective business process design refers to the design of business processes under consideration of customers' experiences, including emotions, feelings and moods. Thus, this paper is driven by the following research question: *How to integrate affect into business process design?* We propose ways in which organizations can design touchpoints to make customers feel affectively catered to, appreciated, and empowered. These design options bring the affective impact of business processes closer to the BPM community, also by using a pattern-like approach which is a common form of conceptualization in the field [e.g., 9].

We proceed as follows. In Sect. 2 we cover the related body of knowledge dedicated to customer experiences in BPM and adjacent domains. We conceptualize affective business process design in Sect. 3 and differentiate dimensions of affect. The core contribution of our paper is made in Sect. 4 where we present design options for affective BPM along the core stages of a business process. In Sect. 5 we discuss our research before presenting a brief summary, limitations and future research options in Sect. 6.

2 Background

2.1 The (Missing) Role of Customer Experience in Business Process Design

BPM as a field is concerned with designing and managing efficient and effective workflows [1, 8]. Efficiency in business processes refers to aims such as reduced costs and time, while effectiveness implies striving for better quality and reliability [3]. A plethora of works emphasizes the role of customer experience or customer value [e.g., 10], but there is little awareness of what this implies in the context of common BPM approaches (e.g., process (re)design, process modelling/mining). Considering the history and dominant logics of the field [1], two observations are important in this regard.

From a *historical* point of view, BPM is rooted in scientific management. Well-established concepts, such as waste reduction, cycle-time reduction or process standardization illustrate how the field is biased towards increasing the efficiency and effectiveness through appropriate business process design [1, 11]. Certainly, customer experience is considered important [12]—but it is backgrounded in favor of other, workflow-related key performance indicators. For instance, in the widely used book by Dumas et al. [2], the role of customer experience is stressed in several places, but there is no specific guidance on how it can be designed. Consider, for example, the common and well-established identification of 'non-value add' activities in a business process. Such activities are identified by asking 'Would the customer pay for it?', thereby making a customer more of a rational voice rather than a process stakeholder with affective expectations and experiences. Similarly, the process modeling standard BPMN can capture customers as a separate entity (lane) and identify customer touchpoints. However, it does not dedicate symbols or semantics towards their experiences [e.g., 13].

With an eye on the established *conventions* in BPM research, the field has a bias towards the perspective of organizations and companies which own and operate business processes. So far, the primary focus of BPM research, and business process design in particular, has been to offer a prescriptive and analytical focus on behalf of organizations providing business processes [2, 3]. Commonly, positive experiences of customers are a secondary, implicit concern and seen as welcome by-products of streamlined workflows. For instance, it is assumed that customers will be satisfied with a business process when it is designed to be faster [3], cheaper [14] or more innovative [15]. In short, the implicit consideration of customer experiences in the core BPM literature poses limitations. However, designing affect-aware customer experiences is essential for BPM as a discipline in the spirit of process science claiming to take a comprehensive view on all stakeholders [16].

2.2 Customer Experiences in Adjacent Research Fields

Customer experience has played an integral role in other business and management-related fields. Product management, innovation management, sales management and service management, for instance, are widely known for integrating and fostering customer experience in the design of services and products [e.g., 17]. Consider how *design thinking* shifts full attention to customers, trying to get to "the root of what is actually, truly, going on for them" [18]. Customer experience advances as an organization's central ambition which is realized through a "multi-function endeavor, including and integrating many parts of the business" [19], thereby involving increasingly specialized customer experience designers.

A common outcome of design thinking is a *customer journey map*, a graphical visualization of the end-to-end experiences of a customer. At so called moments-of-truth (i.e., 'micro moments'), a customer journey captures pain points and—in extended versions—experiences and expectations. This motivates organizations to prioritize action taking at these micro moments to improve the overall process experience for a customer. Going beyond touchpoint design and exploring '*optimal patterns of a customer journey*', De Freitas [15] defines evolving experience patterns along a customer journey (e.g., negative, deteriorating; positive, improving). By capturing sentiment scores and foregrounding most memorable process moments, organizations can make experiences an explicit design variable. This also requires an experience feedback loop, for example by tracking customers up/downvoting an embedded chatbot's response.

Service blueprinting is another popular approach, predominantly in domains that take a service-based view of the firm [16]. A service blueprint emphasizes a customer viewpoint with a special attention towards the demarcation created by the line-of-visibility. A service blueprint helps to improve those process parts that the customer does not experience (back-stage processes) with a view on efficiency and standardization, while directing attention to those actions in the line-of-visibility (front-stage processes) that are noted by the customer and require personalization and attentiveness.

Without doubt, established perspectives—such as design thinking, service blueprinting and others—are valuable, but they tend to foreground specific activities or events (such as when customers make use of a service). But BPM, and business process design, are concerned with entire sequences of activities and events that together constitute products or services [2]. Here, customer experience needs to be designed and implemented such that it is considered along the business process while a customer undergoes a set of steps. In addition to that, customer experience is commonly seen to be heavily relying on affective states, emotions and feelings, triggered by events, such as surprise or delight [20]. BPM, however, stresses a rationale, analytical point of view where affect is only implicitly considered. The main reason for this is the primacy of the organization, i.e. the 'vendor of the process,' as opposed to an at least equal consideration of the customer. This is also evidenced by the fact that the term Customer Process Management [21] never could come even close to the significance of Business Process Management. In light of these considerations, the question is how a business process design approach could look like that emphasizes customer experience, particularly on the grounds of what a customer feels? This is what we discuss in the following.

3 Conceptualizing Affective Business Process Design

3.1 Defining Affective Business Process Design

'Affect' is a psychological term that broadly refers to a human's "experience of feeling or emotion" [22]. The term bears two important implications in the context of this paper. First, it subsumes emotions, feelings and moods, and thus refers to a wide spectrum of possible "affective states," ranging from negative to positive ones [22]. Affect occurs as some form of reaction, meaning that it is caused by a certain perception, impression or interpretation of an event. Second, affect is seen as one important component of the human mind that is connected to, yet distinct from, rational cognition. According to modern psychology, humans have limited rational control over affect, while vice versa, affect strongly influences decision-making and sense-making [4, 5].

Building on findings from psychology, we define affective business process design as *the design of business processes under consideration of customers' experiences*. Affective business process design is human-centered in that it highlights the role of individual information-processing [23]. The focus of affective BPM lies in the experience-aware design and management of activities and events in business processes and how they influence the emotions, feelings and moods of customers. Different from other psychology-inspired BPM research, such as on business process modelling [24], affective business process design focuses on emotions, feelings and moods in business process interactions. This implies an extension beyond the analytical focus of BPM and business process design [1, 8] which typically emphasize outcome-related key process performance indicators, such as throughput time or cost-to-serve.

3.2 Clarifying the Role of Affect in BPM

We draw on an influential and widely used categorization of affective states: Russell's Valence Arousal Model [e.g., 25, 26]. Accordingly, affective states are distinguished along two dimensions of valence (low/high) and arousal (low/high). This results in five broad affect categories [25]:

- *High-Arousal-High-Valence*: positive and involved states, such as pleasure, joy, excitement.
- *High-Arousal-Low-Valence*: negative and involved states, such as anger, anxiety, fear.
- *Low-Arousal-Low-Valence*: negative and uninvolved states, such as boredom, being asleep.
- *Low-Arousal-High-Valence*: positive and uninvolved states, such as being calm, relaxed, peaceful.
- *Neutral*: affective states not located in the previous dimensions.

Table 1 summarizes the dimensions.

Table 1. Different dimensions of affect according to Russell's Valence Arousal Model.

	Positive Valence		**Negative Valence**
High Arousal	Excited, Delighted, Happy	**Neutral**	Tense, Angry, Frustrated
Low Arousal	Content, Relaxed, Calm		Depressed, Bored, Tired

From the viewpoint of BPM, an affect can be desired or undesired. For instance, business processes may be designed to promote a neutral affective state (e.g., conducting a risk appetite statement with a financial advisor), or affect with positive valence, either with high arousal (e.g., visiting a Disney theme park is connected with excitement) or low arousal (e.g., being a relaxed customer in a Spa). In general, organizations avoid affect with negative valence.

Business processes can be designed to have an affective impact by creating, amplifying or changing affect. Affect, unlike cost or time, however, is a dependent variable that so far has not led to the conscious design of its causal independent design variables. For example, when a customer experiences negative affect during a complaint handling process, it might be desirable to induce positive affect, for the well-being of the customer, but also the well-being of the process (so it leads to completion).

Certainly, which affects are desired or perceived by the customer, depends on various factors. First, it depends on the business process, its purpose, its customers, their expectations and the overall context of the process execution. For instance, a claims process in an insurance company should lead to different affects than purchasing limited-edition sneakers in a hip fashion store. Furthermore, affect is related to the organization and how it aims to present itself to the customer. Established, tradition-based financial institutions, for instance, seek to make the customer feel trusted and well taken care of in a discrete environment, while younger FinTechs want to appear simple, convenient and proactively engaging (e.g., via gamified solutions and stimulating messages). Affect in business process design is also related to the sectorial and cultural context in which a business process operates. For example, facial recognition within a well-regulated departure process at an international airport might lead to far less trust concerns and related negative valence than the same facial recognition solution at the checkout process of a grocery store. Finally, national cultures may favor higher or lower arousal. For instance, a sales process in Japan may emphasize long-term relationships (consider, e.g., how the term "nemawashi" highlights the need to build a trustful foundation between business parties). As a result, a sales process in the Japanese setting might be frontloaded with activities seeking a positive valence whereas a neutral valence may be considered sufficient in other cultures.

4 Design Options to Realize Affective BPM

Affective business process design requires awareness for and adequate responses to the affective states of a customer, proactively creating desired emotional states and flows. Given that affective business process design carries implications beyond those of traditional design interests, the key question we address is: *How to integrate affect into business process design?*

We answer this question following a *meta-design approach* [e.g., 27]. To this end, we do not develop actual design artefacts (e.g., methods, frameworks, patterns), design principles or even a design theory as it is common to Design Science Research [28]. Rather—and as a very first step towards affective business process design—we provide *design options*, i.e. higher order (meta) design guidance for designers. These design options do not come as implemented process elements (artefacts) and do not have the richness of a design principle (e.g., an aim, context, mechanism, reasoning, examples), but they can inform business process design activities. We derived these design options from secondary data and known business processes, and positioned them along different stages in a business process from the viewpoint of a customer (see Fig. 1).

The 'pre-process'-stage implies that affect can be anticipated before a customer gets involved in a business process. The 'process start'-stage suggests that affect can be initiated when the customer joins the process. The 'during process'-stage specifies how affect can be upheld, sensed and steered as the customer interacts with process steps. The 'process end'-stage describes how affect can be emphasized before a business process terminates. Finally, the 'post-process'-stage implies that customers are reminded of affective states after the process has been terminated. Importantly, the stages—and the respective design options—are not mutually exclusive. Depending on the process and the broader context in which the process is embedded, one or more design options can be used. In the following, we describe affective process design options along these process stages including examples sourced from secondary data.

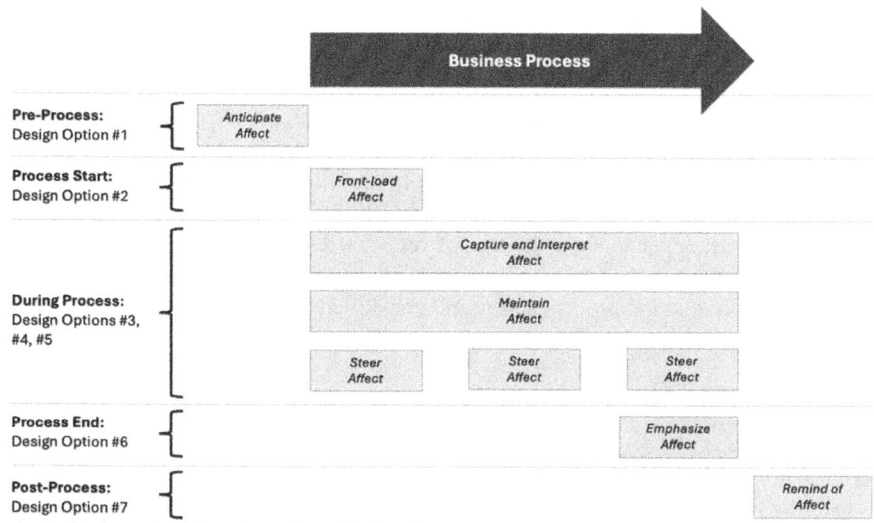

Fig. 1. Design options for affective business process design in different stages of a business process (indicated through the green arrow). (Color figure online)

4.1 Pre-process

Design Option #1: Anticipate Affect
Before customers interact with a business process, they may be in a distinct affective state. Such states can be desired from an organization's point of view, such as a customer who is in a positive affective state with high arousal as they call a travel agency to finally book the vacation they have longed for in a while. Affective states can also be negative and potentially with high arousal, such as when a customer contacts the customer service line of their mobile phone provider to interrogate about unexpected roaming costs. Customers might also be apprehensive knowing that the process will lead to either positive or negative valence, but they are uncertain which of these will materialize (e.g., when receiving a call with a medical diagnosis). In all of these cases, it is crucial for an organization to anticipate a customer's affective state before the process starts, such that desired states can be amplified and undesired ones can be mitigated. Hence, the design option *anticipate affect* refers to the organizational capability to foresee a customer's affective state and take appropriate actions.

Organizations have a track record in transactional preparedness as they specify in their process models what type of start events lead to what type of organizational actions [2]. Anticipating affective states, however, implies that organizations find and interpret relevant data that mirror the affective states of a customer. This can be done in different ways. For instance, a company may anticipate a customer's affective state based on records of previous interactions, and other customer profile-related data. Consider, for instance, how Telstra, Australia's biggest telecommunication company, uses a generative AI solution for what is called a 'one sentence summary' as a sensitizing additional step in their in-bound call center process [29]. Triggered by an incoming call, a genAI solution consolidates information from Telstra's CRM system, social media or previous call documents, and creates a one sentence summary not only about the context of the calling customers, but also about the likely affective state of the customer allowing the call center agent to prepare for it [29].

Other means to anticipate a customer's affective state include probability-based or population-based inferences. For example, when a certain geographical region faces electricity black outs, call center agents might be informed about the negative affective state of the customer before they answer a call [30]. Importantly, anticipating affective states is not only dependent on the use of relevant data sources, but it also implies that those who work in the process are provided with affect-related information in an actionable way before the process starts, and are skilled to interact with the customer in appropriate ways.

4.2 Process Start

Design Option #2: Front-Load Affect
A customer can be affectively influenced when the process starts. Again, this can occur in two ways. On the one hand, when customers start a process with negative valence, attempts to initiate positive affect right away can turn subsequent process interactions into more positive experiences. On the other hand, when customers start the process with a neutral or even positive valence, their experience might be amplified further. This leads to an emotional credit, i.e. a net-positive affectual state, thereby creating a positive bias of the customer. Hence, the design option *front-load affect* is concerned with considering and influencing affect at the starting point of a process.

An example for front-loading affect is Disney and its selective practice to open their theme parks a few minutes before the official opening hour. This leads to reduced waiting time for their often impatient customers who cannot wait to experience the joy and fun of Disney's processes on offer. Another example are surprise upgrades at the start of a process, which are used, for instance, by hotels and airlines. Providing customers with unexpected benefits has proven to be a well-working strategy to not only strengthen relationships between organizations and customers, but also to nudge customers to experience subsequent process steps in a more positive way [31].

Applying this design option requires organizations to have a comprehensive knowledge of their customers' mindsets and schemas when they enter the process as well as the costs of creating the affect (e.g., available underutilized capacity). Also, it is crucial to decide whether affect should be influenced in every instance of a business process, in which case it might become expected by customers over time [32]. Alternatively, this design option can be used by means of probability-based or situational allocations where, for instance, only a certain share of business processes will be started off with front-loaded affect.

4.3 During Process

Design Option #3: Capture and Interpret Affect
As customers interact with business processes, their affect might change, both in intended and unintended ways. For instance, unexpected and undesired experiences may cause

customers to develop affect with negative valence, such as when they become annoyed during long waiting times in customer support processes. To understand affect in customers at given points in time in a business process, it is important that affect is continuously monitored and addressed. The third design option thus suggests *capturing and interpreting affect* throughout the business process.

Since business process interactions leave behind an increasing share of digital data [16], a variety of sources can be used to assess the affective states of customers, such as e-mail communication or biometric data. Consider how call centers in the US use AI-based tools to scan and monitor a customer's voice to infer information about their affective state. In customer complaint processes, for example, such approaches can reveal if a customer expresses signs of frustration, to which the call center agent can respond in appropriate ways [33]. Similarly, AI-supported video recording can be used to analyze affect-related facial expressions, for example, during tele health processes or comedy club visits [34]. Other sources to capture affect refer to written communication, such as e-mails, customer reviews or incident tickets, which can be scanned for hints of affective states through text mining and topic modelling tools [35]. This also includes the conversational interactions a customer has with a chatbot or the semantics of the words they are using in recorded phone calls.

When capturing and interpreting affect, it is important that organizations have a clear understanding for what kinds of affect they are looking for. For instance, travel booking processes might not only aim to create affect with positive valence, but they may specifically promote high arousal (e.g., being delighted) over low arousal (e.g., being only content). Furthermore, it is important to make sure that meaningful data points are considered and measured at relevant points during process interactions. Finally, attempts to capture and interpret affect need to be aligned with privacy regulations [36]. Depending on the country, for instance, any initiatives involving digital trace data require explicit customer consent and other legal considerations (e.g., GDPR-related issues in the European Union).

Design Option #4: Maintain Affect
When customers are interacting with a business process, it can be important to maintain an affective state throughout. For instance, in onboarding processes, it is key that new employees develop and maintain a positive impression of the company. In terms of affect, this might involve the maintenance of positive valence such that employees become and stay eager to work for the organization. In other processes, it might be more important to ensure that customers do not develop affect with negative valence, such as in a customer complaint process; or that customers are not confronted with positive valence when the process is concerned with serious topics (e.g., hospital diagnosis) Therefore, this design option suggests organizations should *maintain affect* throughout a process execution.

At the core of this design option is that affect is considered and addressed along a sequence of events and activities. Accordingly, process design implies that any interaction point with the customer is considered in terms of the customer's affect at each point, and how this affect can be addressed. In processes in which the customers' affective states in subsequent activities are not aligned, affective tension can arise. Consider departure processes at airports. A security check is an intrusive process which can lead to stress; inconvenience can result from delays resulting from extra checks, and furthermore, passengers dislike the feeling of being screened [37]. By design, this is followed by a shopping experience which competes for passengers' dwell time and relies on them having a positive affect where they are willing to explore and purchase offers. In response, airports have started to redesign the security experience and created a more homely process environment consisting of a soft carpet, indoor plants and background music to soften the emotional tensions of this process part [see e.g., 38].

Design Option #5: Steer Affect
So far, the assumption has been that affect in business processes is homogenous in the sense that it remains largely the same. But in some processes, affective states can—or should—change throughout a customer's interactions. This can be the case, for instance, when unexpected events occur during the process. Consider how the luxury hotel chain Ritz Carlton strategically applies actions to steer and redirect the affective states of guests during process interactions. For example, when the guest goes through the check-in process and faces unexpected problems (such as the room is not properly cleaned), hotel staff may respond through surprising gestures, such as providing complementary dinner or sending personalized handwritten notes with small gifts [39]. In fact, Ritz Carlton has ensured that steering of customer affects' is not a scripted process, but done by empowered staff. For example, a concierge has a monthly budget to be spent on creating affect with positive valence among their customers [40].

This design option requires alignment of the affective states created by the process provider and those desired by the process consumer. If this is not the case, they are either over-affective or under-affective. An example for an over-affective process is Walmart as it exported its retail process to Germany in 1997 [e.g., 41]. In the nearly 100 stores Walmart acquired, management followed the established US process practice of having a greeter at the shop entry and providing an over-supply of sales assistants who frequently and proactively offered their help to arriving customers. German customers, however, did not appreciate this assistance and preferred to conduct their shopping journey on their own. They did not value the smiling welcome and had no appetite for an increased arousal as part of their retail experience. They demanded a transactional, and not an affective process. A similar experience was made with 'KLM Surprise,' where the Dutch airline welcomed its passengers with a personalized gift (e.g., a baseball cap) after inferring the passenger's individual travel motivation from their personal social media profiles. The passengers reacted with negative—and not the intended positive—affect as they perceived the airline's actions as privacy invading [42].

In contrast, an under-affective process exists when a customer is looking for an additional emotional experience in a process which remains unfulfilled. A typical class of business processes here are personal service processes in the areas of care, beauty and healthcare. Though the customer is charged for a transactional service (e.g., haircut, physiotherapy), the one-on-one experience with the individual service provider can offer additional affective experiences. Consider how the Dutch retailer Jumbo has converted an under-affective, transactional process into one with positive valence by rolling out slow checkout lines ('Kletskassa') across 100s of their supermarkets. Contrary to the typical process desire of acceleration, the retailer uses such an affect-aware process design to fight loneliness among its customer base [43].

4.4 Process End

Design Option #6: Emphasize Affect

Business processes may terminate by prompting the customer to take a decision or response in a certain way. Consider, for instance, a car sales process. At the end of the process, the customer may ideally be in an affective state with positive valence and high arousal, such as by being excited about the prospect of purchasing and owning a particular car. In such a case, it might make sense to emphasize the desired affective state before the process is terminated (e.g., by putting a ribbon on the car or taking a picture of the customer with the new car). Similarly, after a long and cumbersome legal process, a law firm may emphasize a positive affect for their client when the court decision is at their favor. Thus, this design option suggests *emphasizing affect* at the end of the business process.

Consider, for instance, how Disney deploys a variety of actions to emphasize positive affect when customers leave their theme parks. Fireworks and parades are timed as concluding activities for the day, intended to provide visitors with strong and lasting impressions of their stay. Similarly, staff members are trained to deliver warm and personalized interactions as visitors leave the park, thereby reinforcing positive affect with high arousal (e.g., joy or excitement) [44].

From a psychological point of view, this design option leverages the "peak end rule" effect whereby individuals disproportionately remember emotions at the peak and the very end of experiences [45]. Organizations seeking to apply this design option should make sure that they are able to consider and respond to affective states of individuals in personalized ways (e.g., when a customer is tired, overly enthusiastic interactions might lead to negative affect after all).

4.5 Post-Process

Design Option #7: Remind of Affect

In some cases, it can be important to recall affective states after the business process has been finished. For instance, business processes that rely on positive affect may follow up with the customer within a certain period of time. This can apply to positive valence with low arousal, such as when a customer has been subject to an exclusive spa treatment where they should remain in a sustained relaxed state. This can also apply to high arousal affect, such as when customers have purchased a sports car and the vendor makes sure that the customer is still excited about their decision. Hence, the design option suggests *reminding a customer of affect*.

Typical examples where companies reinforce affect in the post-process stage are located in exclusive and high-priced sectors. For instance, it is common practice for Michelin stars-decorated fine-dining restaurants to contact guests after their visit. During personalized phone calls, restaurant managers make sure that the stay was pleasant and reinforce customers' memories of delicious dishes [44]. Similarly, Facebook brings up memories from a few years ago to recreate affects.

Organizations aiming to apply this design option should be careful about the frequency and timing in which they remind customers of a certain affect. For example, when a fine-dining restaurant contacts the guest weeks after their visit, the guest might not specifically recall the dishes or even might find the call obtrusive. Furthermore, it is important that employees who are involved in the reinforcement of affect have emotional intelligence and connect with the customer in adequate ways.

5 Discussion

Business processes can have a significant affective impact on their customers. This, however, is typically not an explicit and primary concern of BPM in general, and business process design in particular. As a result, most customer-facing business processes are under-designed, i.e. they are inadequate for the impact they have. This motivated this paper and its proposed design options for affect-aware business processes. By postulating a new process design goal, this research has implications for BPM practices, creating future research opportunities.

5.1 The Role of Affective Business Process Design in the Age of AI

Affective business process design can play an increasingly important role for organizations in light of the growing ubiquity of advanced AI-based systems in business processes [46]. AI-based systems can be used for all kinds of business process-related tasks, typically with an interest to achieve productivity gains [47]. Along these lines, Rosemann et al. [46] argue that the age of AI enables three major drifts whereby business processes become more conversational, autonomous, and sophisticated.

However, as much as AI presents an evolving set of opportunities to advance efficiency and effectiveness in business processes [48], more and more tasks will be delegated away from humans, such as employees in organizations [49]. Consider, for instance, how chatbots take over a variety of tasks in service support where they respond to human inquiries, interact with them, and provide advice and recommendations [50]. Research shows that one side-effect of these developments is that authentic experience, spontaneous surprise, and meaningful, human-based interaction can fade into the background [e.g., 51]. Against this backdrop, affective business process design provides a specific viewpoint on how customer experience can be taken seriously and how affect can be implemented while more tasks are executed by AI-based systems.

5.2 From Affective Business Process Design to Affective BPM

A central goal of BPM is to guide and inform organizations on how to design and manage business processes [52]. The design options presented here provide guidance for the affect-aware (re-)design of a process and with this extend existing exploitative and explorative design patterns [53] with a new set of design options. However, elevating affective business process design as the broader agenda of 'affective BPM', it is clear that considering affect has impact on most methods and tools along the BPM lifecycle [2]. This already starts with the definition of the *goal of a process* which also should include desired affect. It continues with the *identification of the most important business processes* of a BPM initiative. Competing on affect might be an option for organizations who otherwise would not be competitive in terms of sheer transactional process performance.

Competing on affect may also be crucial for contexts where organizations produce similar products and need other means to stand out from their competition. Designing for affect can then create distinct value propositions as many of the cases above showed, and as a result might motivate to prioritize otherwise overlooked processes. Like all primary concerns, affect also deserves to be captured in *as-is* and *to-be process models*. The *process execution* requires ensuring that affect-building or mitigating actions are adequately embedded and their impact needs to be monitored via affect-specific metrics. Finally, affect-aware process mining can enable organizations to extract customer affect from logfiles, but for this requires an extension of the data types commonly collected.

5.3 Future Research Directions

Affective BPM is in its infancy. Future research is needed to further mature the related skillsets, toolsets and datasets. In what follows, we exemplify how future research could extend our understanding of affective BPM.

- *Skills and Capabilities for Affective Business Process Design*: Considering affect in business process design requires organizations to be able to plan, design and manage such initiatives in appropriate ways. This entails the context-adequate utilization of the design options presented in this paper. One research direction could be to investigate how established business process design techniques can be adopted for affective BPM, or what new capabilities are required [3, 15]
- *Affect-aware Process Modelling:* Process modelling is a core element of BPM [2]. BPMN, for instance, has evolved as the standard modelling language that provides modelers with a wide set of notations to denote workflow-related cues, such as information exchanges or time-related information. To embrace affect in business process design, one could study how established notations in BPMN models can be extended to inform an actor, for example, that they should convey a certain affect.
- *Affect-based Key Performance Indicators (KPIs):* Affect needs to be operationalized and measured in order to be considered in business process design. While established KPIs measure efficiency and effectiveness, such as cycle time or throughput rate. Arguably, managing and measuring affect will require other KPIs, such as affect satisfaction rate (overall affective response) or net affective value (difference between frequency of positive and negative affects). Future research could develop KPIs for affective business process design systematically.
- *Affective BPM for External and Internal Customers:* In this paper, we have focused on external customers, that is, customers who make use of services of organizations or buy their products. Importantly, however, the term 'customer' in the BPM field can refer to external as well as internal process participants [2]. For instance, organizations may have interests to create affective experiences for employees. Future research could study affective BPM initiatives with other types of customers.
- *Considering Affective BPM along the entire BPM Lifecycle*: While this paper focused on design options only, the concept of affective BPM can be studied along the entire BPM lifecycle [2]. To this end, future studies can see, for instance, how affect can be discovered along with process-relevant information. Text-mining tools, for example, can search through e-mail correspondences that occurred before, during or

after process performances, revealing how, when and why affect can or should be considered.

6 Conclusion

In this paper, we presented affective business process design as a new lens for BPM research and practice. Shifting attention to the affective states of customers, this view complements the established focus on efficiency and effectiveness of business processes. The design options we presented in this paper suggest how affective business process design can be realized. Future research is needed to fully embrace the idea of affective BPM as a holistic effort and to build related theories rigorously based on comprehensive primary data.

As this paper represents a first step towards affective business process design, it inevitably comes with limitations. First, we have relied on selective secondary data only. Second, a semi-formalization of affective business process design is missing so far; the design options can be integrated with existing BPM methods more systematically. Third, the design options are not as comprehensive and normative as design principles; they are a first step towards building research-informed design knowledge on affective business process design.

References

1. Baiyere, A., Salmela, H., Tapanainen, T.: Digital transformation and the new logics of business process management. Eur. J. Inf. Syst. **29**(3), 238–259 (2020)
2. Dumas, M., et al.: Fundamentals of Business Process Management, 2nd edn. Springer, Heidelberg (2018). https://doi.org/10.1007/978-3-662-56509-4
3. Reijers, H.A., Limam Mansar, S.: Best practices in business process redesign: an overview and qualitative evaluation of successful redesign heuristics. Omega **33**(4), 283–306 (2005)
4. Pessoa, L.: On the relationship between emotion and cognition. Nat. Rev. Neurosci. **9**(2), 148–158 (2008)
5. Edwards, K.: The interplay of affect and cognition in attitude formation and change. J. Pers. Soc. Psychol. **59**(2), 202 (1990)
6. Kreuzer, T., Röglinger, M., Rupprecht, L.: Customer-centric prioritization of process improvement projects. Decis. Support. Syst. **133**, 113286 (2020)
7. Wessel, L., et al.: Unpacking the difference between digital transformation and IT-enabled organizational transformation. J. Assoc. Inf. Syst. **22**(1), 6 (2021)
8. Mendling, J., Pentland, B.T., Recker, J.: Building a complementary agenda for business process management and digital innovation. Eur. J. Inf. Syst. **29**(3), 208–219 (2020)
9. Brambilla, M., Fraternali, P., Vaca, C.: BPMN and design patterns for engineering social BPM solutions. In: Daniel, F., Barkaoui, K., Dustdar, S. (eds.) BPM 2011, Part I. LNBIP, vol. 99, pp. 219–230. Springer, Heidelberg (2012). https://doi.org/10.1007/978-3-642-28108-2_22
10. Davenport, T.H.: Process Innovation: Reengineering Work Through Information Technology. Harvard Business Press (1993)
11. Recker, J.: Suggestions for the next wave of BPM research: strengthening the theoretical core and exploring the protective belt. J. Inf. Technol. Theory Appl. **15**(2), 5–20 (2014)
12. Kettinger, W.J., Teng, J.T.C., Guha, S.: Business process change: a study of methodologies, techniques, and tools. MIS Q. **21**(1), 55–98 (1997)

13. Chinosi, M., Trombetta, A.: BPMN: an introduction to the standard. Comput. Stand. Interfaces **34**(1), 124–134 (2012)
14. Harmon, P.: Business Process Change: A Business Process Management Guide for Managers and Process Professionals. Morgan Kaufmann (2019)
15. Gross, S., et al.: The business process design space for exploring process redesign alternatives. Bus. Process. Manag. J. **27**(8), 25–56 (2021)
16. vom Brocke, J., et al.: Process science: the interdisciplinary study of socio-technical change. Process Sci. **1**(1), 1 (2024)
17. Homburg, C., Jozić, D., Kuehnl, C.: Customer experience management: toward implementing an evolving marketing concept. J. Acad. Mark. Sci. **45**, 377–401 (2017)
18. Critchley, S.: Using design thinking methods to improve customer experiences. Forbes (2022)
19. Clark, K., Smith, R.: Unleashing the power of design thinking. Des. Manag. Rev. **19**(3), 8–15 (2008)
20. Palmer, A.: Customer experience management: a critical review of an emerging idea. J. Serv. Mark. **24**(3), 196–208 (2010)
21. Surbakti, F.P.S.: Customer process management: a systematic literature review. Eng. Manag. Res. **4**(2), 1 (2015)
22. American Psychology Association, T.: 'Affect'. APA Dictionary of Psychology. APA: Washington D.C. (2018)
23. Felin, T., Foss, N.J., Ployhart, R.E.: The microfoundations movement in strategy and organization theory. Acad. Manag. Ann. **9**(1), 575–632 (2015)
24. Djurica, D., et al.: Effective presentation of ontological overlap of multiple conceptual models. Decis. Support. Syst. **187**, 114327 (2024)
25. Cittadini, R., et al.: Affective state estimation based on Russell's model and physiological measurements. Sci. Rep. **13**(1), 9786 (2023)
26. Posner, J., Russell, J.A., Peterson, B.S.: The circumplex model of affect: an integrative approach to affective neuroscience, cognitive development, and psychopathology. Dev. Psychopathol. **17**(3), 715–734 (2005)
27. Fischer, G., Giaccardi, E.: Meta-design: a framework for the future of end-user development. End user development, pp. 427–457 (2006)
28. Gregor, S., Hevner, A.R.: Positioning and presenting design science research for maximum impact. MIS Q., 337–355 (2013)
29. Starc, A.: Telstra Expands Use of GenAi for Frontline Teams. CRN Australia (2024)
30. Pardede, A., et al.: Implementation of data mining to classify the consumer's complaints of electricity usage based on consumer's locations using clustering method. J. Phys. Conf. Ser. (2019). IOP Publishing
31. Lindgreen, A., Vanhamme, J.: To surprise or not to surprise your customers: the use of surprise as a marketing tool. J. Cust. Behav. **2**(2), 219–242 (2003)
32. Gyung Kim, M., Mattila, A.S.: Does a surprise strategy need words? The effect of explanations for a surprise strategy on customer delight and expectations. J. Serv. Mark. **27**(5), 361–370 (2013)
33. Simonite, T.: This call may be monitored for tone and emotion. WIRED (2018)
34. Wakefield, J.: Comedy club charges per laugh with facial recognition. BBC Technology 2014. https://www.bbc.com/news/technology-29551380. Accessed 11 Apr 2020
35. Acheampong, F.A., Wenyu, C., Nunoo-Mensah, H.: Text-based emotion detection: advances, challenges, and opportunities. Eng. Rep. **2**(7), e12189 (2020)
36. Grisold, T., et al.: Digital surveillance in organizations. Bus. Inf. Syst. Eng. **66**, 401–410 (2024)
37. Malvini Redden, S.: How lines organize compulsory Interaction, emotion management, and "emotional taxes" the implications of passenger emotion and expression in airport security lines. Manag. Commun. Q. **27**(1), 121–149 (2013)

38. Hasanzade, M.P., van Oel, C.J., Pazhouhanfar, M.: Passengers' preferences for architectural design characteristics in the design of airport terminals. Archit. Eng. Des. Manag. **19**(6), 586–601 (2023)
39. Michelli, J.A.: The New Gold Standard: 5 Leadership Principles for Creating a Legendary Customer Experience Courtesy of the Ritz-Carlton Hotel Company. (No Title). McGraw-Hill, New York (2008)
40. Hall, J.M., Johnson, M.E.: When should a process be art, not science? Harv. Bus. Rev. **87**(3), 58–65 (2009)
41. Hunt, I., Watts, A., Bryant, S.K.: Walmart's international expansion: successes and miscalculations. J. Bus. Strategy **39**(2), 22–29 (2018)
42. O'Neill, M.: KLM stalks passengers through social media & buys them gifts. AdWeek (2011)
43. Rosemann, M., Ostern, N., Voss, M., Bandara, W.: Benevolent business processes - design guidelines beyond transactional value. In: Di Francescomarino, C., Burattin, A., Janiesch, C., Sadiq, S. (eds) BPM 2023. LNCS, vol. 14159, pp. 447–464. Springer, Cham (2023). https://doi.org/10.1007/978-3-031-41620-0_26
44. Pine, B.J., Gilmore, J.H.: The Experience Economy. Harvard Business Press (2011)
45. Fredrickson, B.L., Kahneman, D.: Duration neglect in retrospective evaluations of affective episodes. J. Pers. Soc. Psychol. **65**(1), 45 (1993)
46. Rosemann, M., et al.: Business process management in the age of AI – three essential drifts. Inf. Syst. e-Bus. Manag. (2024, forthcoming)
47. Grisold, T., et al.: BPM is dead, long live BPM!"–an interview with tom davenport. Bus. Inf. Syst. Eng., p. 1–4 (2024)
48. Davenport, T.H., Ronanki, R.: Artificial intelligence for the real world. Harv. Bus. Rev. **96**(1), 108–116 (2018)
49. Baird, A., Maruping, L.M.: The next generation of research on IS use: a theoretical framework of delegation to and from agentic IS artifacts. MIS Q. **45**(1), 315–341 (2021)
50. Suhaili, S.M., Salim, N., Jambli, M.N.: Service chatbots: a systematic review. Expert Syst. Appl. **184**, 115461 (2021)
51. Puntoni, S., et al.: Consumers and artificial intelligence: an experiential perspective. J. Mark. **85**(1), 131–151 (2021)
52. Mendling, J., et al.: Pluralism and pragmatism in the information systems field: the case of research on business processes and organizational routines. Data Base Adv. Inf. Syst. **52**(2), 127–140 (2021)
53. Rosemann, M.: Explorative process design patterns. In: Fahland, D., Ghidini, C., Becker, J., Dumas, M. (eds.) BPM 2020. LNCS, vol. 12168, pp. 349–367. Springer, Cham (2020). https://doi.org/10.1007/978-3-030-58666-9_20

Process Autonomization: Rethinking Business Process Management

Christian Janiesch[1,2(✉)], Marek Kowalkiewicz[2], and Michael Rosemann[2]

[1] TU Dortmund University, Otto-Hahn-Str. 12, 44227 Dortmund, Germany
`christian.janiesch@tu-dortmund.de`
[2] Centre for Future Enterprise, Queensland University of Technology, Brisbane, Australia
`{marek.kowalkiewicz,m.rosemann}@qut.edu`

Abstract. Process automation has been a cornerstone of business process management, enabling organizations to streamline operations through predefined rules and procedures. However, a new paradigm is emerging: process autonomization. It has the potential to disrupt the core of business process management and the methods and tools that have characterized it for the last century. This paper introduces and conceptualizes process autonomization as distinct from traditional automation, characterized by the key capabilities of artificial autonomy: independence, neglect tolerance, and indeterminism. We demonstrate how autonomization transcends predefined machine behaviors to handle tasks that have traditionally required human involvement. Further, we identify three critical aspects for process autonomization: explicit goal specification, macro-level resource allocation, and appropriate constraint definition. We discuss them across the process lifecycle stages of build time, run time, and change time. Our analysis also reveals that process autonomization introduces new reliability and explainability caveats. We contribute to theory by providing a conceptual framework for understanding process autonomization and offer practical insights for organizations considering the transition from automated to autonomized processes.

Keywords: Process Automation · Process Autonomization · Artificial Intelligence · Artificial Autonomy · Goal Modeling · Resource Modeling · Declarative Modeling

1 Introduction

The evolution of automated systems has profoundly shaped the landscape of modern business, tracing its roots back to early industrial innovations like the assembly line and advancing through the digital revolution with enterprise software and robotics. Automated processes have been the foundation for the industrial age and enabled streamlined workflows, massive reduction in error rates, and consistent outcomes, driving productivity, and cost savings [1, 2].

While automated processes continue to deliver value, a new paradigm is emerging: process autonomization. Unlike automated processes, autonomous processes make

© The Author(s), under exclusive license to Springer Nature Switzerland AG 2026
A. Senderovich et al. (Eds.): BPM 2025, LNBIP 564, pp. 361–377, 2026.
https://doi.org/10.1007/978-3-032-02929-4_21

complex decisions and execute dynamically adapting to changing conditions without predetermination in their execution logic or in the outcome [3]. Whereas automated processes require input on *how* they have to do something (e.g., steps required to assemble a car or how to conduct a payroll process), autonomized processes need input in terms of *what* to do (e.g., identify cancerous cells, clear the floor of a warehouse). In short: An automated process does not need to know what to do (as it does not care), an autonomized process does not need to know how to do it (as it can work it out by itself).

This distinction marks a significant departure from traditional process automation and its focus on prescribed execution guided by process models and the post-execution mining of processes in the form of event files. In process autonomization, the control flow is no longer the primary area of interest for business process management (BPM).

To further exemplify the difference between process automation and process autonomization, consider two travelers planning their vacation through different systems:

> *Alice uses a traditional automated booking platform, while Bob interacts with an autonomized travel planning system. Both have a desire for a 'perfect family vacation in Europe'.*

> *Alice's automated system prompts her via a series of transactions through a predetermined sequence: selecting dates, choosing from preset destination categories (beach, city) and specifying a budget. The system then matches these parameters against its database, presenting a list of pre-packaged itineraries that match her inputs. When Alice mentions her children's interests in history, the system can only respond by filtering for destinations tagged as "historical" in its database. Throughout the process, Alice needs to be the orchestrator of this experience requiring transactional literacy.*

> *In contrast, Bob does not have to have any form of transactional booking process skills. His interaction with the autonomized system begins with his natural expression of preferences: "I want to plan a three-week trip that will keep my teenagers engaged with history while giving us all a chance to relax." The system engages in a dialogue, learning that Bob's family enjoys interactive experiences, that his eldest is studying World War II in school, and that his youngest gets easily overwhelmed by crowds. Drawing from patterns observed across thousands of family vacations, the system crafts a personalized itinerary that combines lesser-known historical sites in northern France.*

This distinction exemplifies the shift from rigid, rule-based automation to adaptive, context-aware autonomization. While Alice's system executes predefined workflows and their variants according to its transactional logic, Bob's system demonstrates independence through contextual decision-making and execution as well as self-efficacy through its ability to respond and act according to conversational input.

The evolution towards process autonomization represents a fundamental transformation in BPM. While traditional automation relies on handcrafted, deterministic responses, autonomized processes have an indeterministic capacity for dynamic adaptation and cognition-based decision making. This shift holds the potential to revolutionize industries, and BPM, as organizations can deploy processes that learn from experience, handle ambiguous and underspecified situations, self-optimize decisions, address change, and continually improve performance – moving beyond the constraints of predefined rules and outcomes to achieve continuous performance improvement.

Real-world examples of this transition are already evident in various industries. Stock trading processes automatically react to contextual changes and human guidance (e.g., risk appetite) to make decisions in milliseconds today and advanced autonomous trading capabilities are clearly emerging in the future.[1] Amazon's warehouse management systems already showcase the evolution from traditional automation to autonomous operations, using artificial intelligence (AI) robots that adapt to changing inventory and order patterns.[2] Singapore's Smart City initiatives employ autonomous systems that dynamically adjust to real-time conditions, such as traffic patterns and public safety needs, showcasing autonomization at a city-wide scale.[3]

In light of this context, the purpose of this paper is to *identify the defining themes and dimensions of process autonomization* and discuss the impact it will have on the BPM discipline. Thereby, we answer to the call by Grisold et al. [4] to conceptualize accounts of potential futures of BPM while discarding superseded practices to advance.

Our paper is structured accordingly: First, we explore the background of automation and artificial autonomy from a socio-technical perspective. After a presentation of our research method of conceptual research, we conceptualize process autonomization. Last, we discuss its impact on the BPM discipline before we close with a summary.

2 From Industrial Automation to Artificial Autonomy

Automation, or predeterminations embodied in machines, can be defined as the use of technology to perform tasks with minimal human intervention referring to the pre-established rules, algorithms, and programming that dictate how these machines operate [2, 5]. This involves the integration of various technologies such as physical machines, control systems such as BPM systems, and decision-making software such as business rule management systems to enable automated process execution. The essence of automation lies in its ability to increase efficiency, accuracy, and consistency in performing repetitive, complex, or dangerous tasks while reducing the need for direct human oversight [2].

In the domain of BPM, automation has played a crucial role in optimizing workflows and enhancing operational efficiency. By automating processes such as sales order management, payroll processing, and producing goods, organizations significantly improve their processes, reduce or even eliminate human error, and allocate resources more effectively [1]. Automation software, such as BPM systems and robotic process automation (RPA), enable businesses to model, analyze, and improve their processes by embedding predefined rules and logic into their operations [1].

---

[1] https://medium.com/@adelstein/ai-revolution-in-the-stock-market-how-automated-trading-is-transforming-global-exchanges-and-77eab4152e9a, visited 2025–06-11.

[2] https://www.exotec.com/insights/how-amazon-robotics-has-changed-the-landscape-of-fulfilment/, visited 2025–06-11.

[3] https://www.thalesgroup.com/en/worldwide-digital-identity-and-security/iot/magazine/singapore-worlds-smartest-city, visited 2025–06-11.

In contrast, *artificial autonomy* can be defined as a *quality of systems enabling them to act independently based on their inherent potential* to evaluate options and choose a course of action [5, 6]. Artificial autonomy emphasizes the ability to operate with a degree of self-governance, allowing a system to adapt to changing circumstances and to make choices that align with predefined goals or preferences.

In the BPM community, first steps have been undertaken towards autonomous systems by introducing the concept of AI-augmented BPM systems [7]. Such technical systems are intended to be characterized by their autonomy in decision-making, adaptability to changing environments, self-improvement through predictive maintenance, explainability for user trust, and conversational engagement with human agents. However, the current concept does not extend to socio-technical aspects but remains on the capability level of AI for process automation technology.

Summarizing, while automation is concerned with executing predetermined tasks according to fixed rules (early process binding), artificial autonomy incorporates learning and adaptability. As a result, it allows for complex interactions and problem-solving capabilities tailored to the situational context of a process (late process binding).

3 Research Method

Inherently, process autonomization is a concept that embraces the dynamics and change catalyzed by current technological advances. Grisold et al. [4] have elaborated on the process of managing dynamics in and around business processes. As a structure for future developments, they propose the dimensions of closed- and open-ended futures as well as of operational and contextual level.

The closed-ended perspective suggests that businesses can predict future outcomes by extrapolating from historical data, allowing for focused scenario management. In contrast, the open-ended perspective highlights the uncertainty and complexity of future events in turbulent environments, necessitating the consideration of multiple scenarios with unpredictable impacts. The operational level emphasizes the detailed analysis and design of control-flow, activities, resources, and performance measures within actual business processes. The broader contextual level considers the overarching conditions influencing these processes and highlights the importance of context in process design and analysis, allowing for the development of general patterns or guidelines and BPM capabilities that shape how dynamics are managed within organizations.

Using these dimensions, they have devised four approaches to manage dynamics in and around business processes. Here, "Approach IV" addresses the management of unknown dynamics in *open-ended futures* by emphasizing the importance of broader *contextual factors* and acknowledging uncertainty across various dimensions. It encourages organizations to envision narratives about possible futures while unlearning outdated practices to thrive less predictable environments. Our research problem conforms to this approach as process autonomization is a prime example of such an open-ended development that is at this stage only characterized by its contextual factors rather than operational concerns.

To approach the complex and – in itself – underspecified concept of process autonomization, we followed a conceptual research approach. Conceptual research in general plays a crucial role in advancing academic discourse and theoretical development across various fields by taking a macro perspective [8]. By focusing on the exploration and refinement of ideas, frameworks, and models, conceptual research allows scholars to challenge existing paradigms, identifies gaps in knowledge, and proposes innovative approaches that may not yet be empirically tested. Conceptual research fosters critical thinking and encourages interdisciplinary dialogue, as it often draws upon diverse perspectives to address complex issues [9, 10].

More specifically, we followed the approach of *creative synthesis* [11, 12] with the goal to build a *nascent theory* [13] that requires a blend of existing knowledge and conceptual creativity. In particular, we reviewed the literature on automation, (artificial) autonomy, and BPM in a hermeneutic fashion [14] to identify, combine and innovate beyond the current state-of-the-art of (intelligent) process automation. This was followed by conceptual synthesizing discussions among the authors on a regular basis as well as gathering insights from practice that eventually enabled specific exemplar ideas as a shared way of understanding. These were refined iteratively and eventually, the concepts described in this paper emerged. See Fig. 1 for an overview of the process.

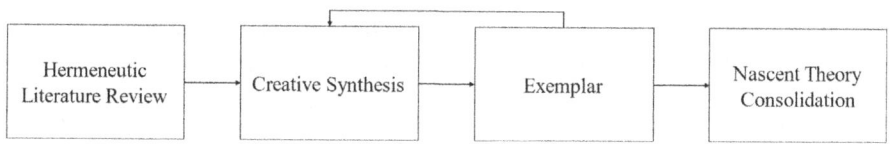

Fig. 1. Creative Synthesis Process

4 A Socio-Technical Conceptualization of Artificial Autonomy in Process Autonomization

Conceptualizing machine decision-making or self-governance as artificial autonomy is similar to the conceptualization of complex mental or cognitive machine performance as AI [15, 16]. Both, AI and artificial autonomy, are relative concepts that describe the technological capabilities of a synthetic system to mimic or even excel beyond human capabilities. Both concepts allude to novel and not yet well-understood technology while systems based on outmoded technology are typically not called intelligent or autonomous but automated [17].

Autonomous systems as well as artificial autonomy have been the subject of computer science research on multi-agent systems since the 1990s [18, 19]. Here, the focus was foremost on enabling systems to solve problems that are hard (or even impossible) to approach for a single system. Then, technology has not been on par with human capabilities for many cognitive tasks as it is now and it was rarely discussed in a sociotechnical setting. Hence, a (re-)conceptualization of artificial autonomy becomes necessary to guide the potential of autonomization (and limit the proliferation of narratives pertaining seemingly autonomous systems).

Yet, similar to intelligence, autonomy is difficult to define. In this context, we refer to artificial autonomy as self-dependence, which has been granted to an entity by an overseeing authority that itself still retains ultimate authority over that entity. It entails that in process autonomization, the process has the right and ability to make its own decisions without outside interference other than being constrained to an operating region [20] analog to concepts such a work envelope [21] in robotics or sandbox [22] in software engineering. This may involve restrictions in process initialization as well as in execution (e.g., compliance requirements) and termination (e.g., result expectations). Consequently, we do not speak about artificial autonomy as a theoretical or idealistic construct since such autonomy, as the case of complete independence from context, is neither a reasonable nor a feasible socio-technical criterion of autonomy. Rather, we consider autonomy as a relative concept guided by reason, distinct from notions of complete self-sufficiency (autarky) or the absence of rules (anarchy) [23, 24].

Based on this understanding, we systematize artificial autonomy using the three capabilities of *indeterminism*, *(framed) independence*, and *neglect tolerance* (as perceived from the user's perspective).

Indeterminism. While indeterminism is only arguably a 'capability' of artificial autonomy, it is a quality and – so to speak – a consequence of its agentic nature entailing "the power to act without being acted upon" [25, 26]. It must be accepted and cannot be changed as it is innate to the concept of agents. Indeterminism sets artificial autonomy apart from the deterministic nature of traditional automation and enables adaptability and flexibility within a reasonable frame. Indeterminism does not imply unpredictability, yet process autonomization operates in a dynamic environment where decisions are made in real-time, often influenced by a variety of contextual factors and evolving data. This inherent flexibility allows for adaptive responses to unforeseen circumstances, enabling the system to navigate complexities that would be impossible to handcraft for deterministic models, where outcomes are largely predictable and follow predefined rules.

(Framed) Independence. Independence refers to an entity's ability to determine and pursue its own actions [25–27]. In the context of process autonomization, this independence needs to be constrained or framed [7, 20]. Artificial autonomy is a kind of independence which has been granted by an overseeing authority (i.e., delegated independence) that itself retains ultimate control [20]. In process autonomization. It is not desirable to achieve strong artificial autonomy resembling human autonomy. Instead, the concept of framed independence constrains autonomous behavior by defining boundaries of permissible autonomous action [7]. Process autonomization must exhibit the capability to make independent, goal-aligned decisions and to execute action accordingly. While the former requires self-learned decision models and own decision-making, the latter involves not only the operational ability to execute but also the resources to do so.

Neglect Tolerance. Neglect tolerance constitutes the capacity of a synthetic entity to endure periods of inattention, oversight, or absence of active maintenance without experiencing significant degradation in performance, functionality, or well-being [28, 29]. In process autonomization, it entails that the process can be left unattended for extended amounts of time. Neglect tolerance is similar to independence in so far as it is limited to some operations and some time-frame that is reasonable considering its goal. Thus, autonomous processes must exhibit – to some degree – the capability to adapt and to self-maintain. While the former requires sensing and updating, the latter involves self-monitoring, self-repair, and resource management. Neglect tolerance measures the resilience with which and grace period during which an autonomous process can operate without active oversight or intervention (by humans or other synthetic agents).

In summary, independence is the ability to make decisions and to execute within a set of constraints, while neglect tolerance affords self-sufficiency and resilience when responding to incomplete, conflicting, novel or complex tasks over an extended amount of time. In addition, indeterminism represents an inherent quality of artificial autonomy's nature that affords unscripted flexibility and adaptability. All of them are the significant capabilities of process autonomization that distinguish the concept from traditional process automation.

5 The Concept of Process Autonomization

5.1 Overview

In the following, we present our conceptualization of process autonomization. It comprises the three themes of *artificial autonomy*, *aspects*, and *stages* each with several dimensions. Figure 2 shows these different themes.

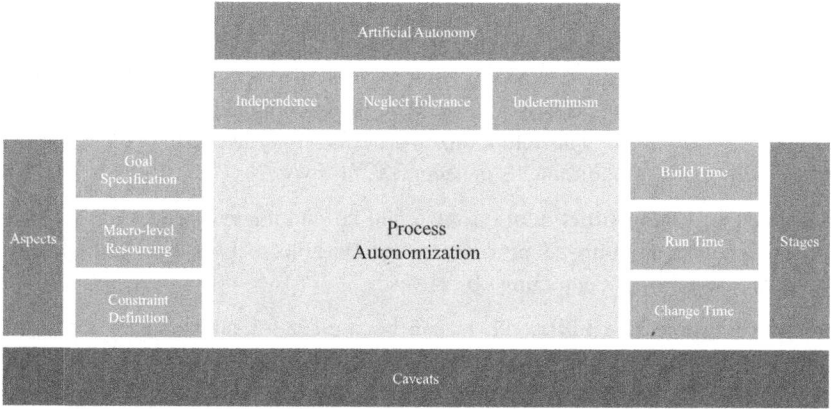

Fig. 2. Themes and Dimensions of Autonomization

We have already discussed the theme of artificial autonomy in the preceding section detailing the capabilities of independence, neglect tolerance, and indeterminism as its subdimensions. The theme of artificial autonomy considers process autonomization from a theoretical and at times philosophical viewpoint. On the more operational side, we regard three aspects of process autonomization as essential in contrast to traditional process automation: the explicit specification of goals rather than tasks, macro-level resourcing instead of micro-level resourcing, and the appropriate definition of constraints rather than control flows. Lastly, we find that each stage of the BPM lifecycle, build time, run time, and change time, focusses on different aspects of autonomization and, thus, should be regarded distinctively.

5.2 Aspects in Process Autonomization

To fully appreciate the transformative potential of process autonomization, it is essential to highlight those aspects that distinguish it from traditional process automation: that is a *goal* the process needs to fulfill, *resources* that are allocated for the entirety of the autonomous process, and the presence of hard and soft *constraints*. None of these three aspects is a new requirement as such for BPM as previous research has dealt with goal-driven business processes, resource allocation, and declarative modeling allowing constraint specifications in the past. However, process autonomization requires a substantial expansion of research in all of these three aspects at the same time for reasons we will detail in the following.

5.2.1 Goal Specification

Goals define the intentions behind processes, tasks, and decisions. They offer design rationales and guide improvements [30]. Though previous business process modeling research explored goal-oriented approaches [e.g., 31, 32], goals have not made it as primary concerns into today's process definitions. Instead, the focus is on *how* the process is executed (functional requirements) as opposed to the *why* the process exists in the first place and *what* the process needs to achieve (non-functional requirements). Thus, the intentional aspect of the process tends to be missing in academic and professional process specifications.

However, the capabilities of artificial autonomy as themes of process autonomization necessitate an explicit goal-setting as it is this goal that guides the situational composition of the process. Thus, whereas an explicit goal has been optional for automated processes, it is a mandatory attribute of an autonomized process. Thus, an overhaul of goal-oriented BPM satisfying novel requirements is needed as follows:

Functional and Non-functional Goals. Unlike automated processes with fixed functional goals, autonomized processes must be able to prioritize variable, multi-dimensional, sometimes competing objectives.

Goal Specification Flexibility. Goals can be pre-set for all instances (attribute of the process model) or defined individually per process (attribute of the process instance). These goals can evolve during the process execution, requiring the process to dynamically adjust its behavior.

Contextual Goals. In addition to the process-specific goals, an autonomized process must be able to interpret and configure itself to overarching internal goals (e.g., the strategic goal of cost leadership or customer intimacy) and external goals (e.g., an ESG requirement for a minimized carbon footprint).

5.2.2 Macro-level Resourcing

As business processes evolve from automation to autonomization, resource management must shift from static allocation on a task level to dynamic resourcing on a process level. Traditional resourcing in process automation requires the predefined (micro-level) allocation of resources (e.g., physical, digital, human resources) to each task to be executed.

Here, the process is a white box and the process designer needs to ensure that each step on its own has what is needed for executability [33, 34]. In contrast, the process itself is a black box when it comes to process autonomization. Here, resources are assigned at process-level as the individual tasks are unknown and thus beyond the possibility of having their own resource allocation. As a result, an autonomized process has resources assigned as a consolidated asset from which its goal- and context-sensitively consumes along the emerging process. Unlike automated processes, this could also entail (meta) financial resources that are at the disposal of the process and which can be invested into the resourcing as required (e.g., purchasing an API-access or synthetic agents). This shows that not only the autonomized process itself is the result of late binding, the same is also true for the resources involved.

Thus, the following are unique requirements of process autonomization and its embedded need for macro-level resourcing.

Goal-Aligned Resource Allocation. In automated systems, resources are assigned based on fixed rules, such as always choosing the least expensive supplier. In contrast, autonomized processes resources are allocated based on the goal valid for the respective process. For example, a process might look different based on the individual, situational risk appetite of the process user. Autonomized processes may even adjust resource allocation continuously in response to changes of the goal or context.

Meta Resources. Autonomized processes might have allocated meta resources, that is resources that allow processes to secure further resources (such as funds). As autonomized processes have an inherent decision capability, they can invest these meta resources to procure those resources needed for the execution of the required tasks (e.g., hire human or synthetic agents). This dynamic, allocation of meta resources is in sharp contrast to the static, specific resource allocation that exists in automated processes.

Late Binding of Resources. Traditional BPM hardcodes resource assignments early in the process. In contrast, autonomized processes embrace late resource binding, where resource selection occurs as late as possible to maximize flexibility as well as goal and context alignment.

5.2.3 Constraint Definition

Due to the indeterministic nature of process automation and the fact that artificial autonomy is a relational notion, constraint specification becomes a critical element in ensuring aligned execution with the overseeing authority. Unlike traditional automated systems, where constraints are predefined and static [35], autonomized processes may require dynamic, multi-level, and context-sensitive constraints that evolve based on process goals, environmental conditions, and user preferences.

Explicit vs. Implicit Constraints. Automated processes are implicitly constrained by the span of processes specified in the process model. Autonomized processes require constraints to be explicitly defined and dynamically adjustable, allowing for flexibility in execution.

Soft vs. Hard Constraints. Hard constraints must always be met (e.g., compliance with regulations). Soft constraints are context-sensitive (e.g., constraints depending on the process stakeholders involved such customer preferences). Autonomized systems use soft constraints to balance competing objectives dynamically. Further, autonomized processes may allow late constraint binding, meaning constraints can be defined or modified during execution.

Conflict Resolution and Constraint Trade-Offs. Autonomous processes must be able to resolve conflicts when multiple constraints compete, otherwise they would deadlock.

5.3 Stages in Process Autonomization

In the following, we show how these three distinct aspects of process autonomization play out across its various dimensions in the three process stages of build time, run time, and change time.

See Table 1 for a summary of these dimensions across the three stages. The table also differentiates between what used to be agreed upon principles in traditional business process automation and what are novel propositions of process autonomization.

5.3.1 Build Time

In process automation, build time is dedicated to developing a *prescriptive* process model that captures *how* a sequence of tasks, the control flow, is to be executed, and what type of data and resources are available to complete this process. This process model is an *imperative* artifact, which is used by the process execution engine to deduce the correct process. The more fine-granular the process is specified, the less the human agents involved have the freedom to specify themselves how to conduct the process at hand. Synthetic agents (e.g., enterprise systems) involved tend to be transactional and serve to manifest the process. Such systems help improving the process' cost effectiveness, quality, reliable and time-to-execute. The outcome of build time in process automation is a fully specified executable process model satisfying design criteria covering performance and conformance potentially covering all known variants of the process.

The primary concern at build time for process autonomization is not the control flow, but the *goal* of the process and the definition of *what* needs to be done. Specifying this goal is *directive* for the system that is used to support this process. The control flow as such is *underspecified*. Thus, the act of determining the control flow is part of the run time, not the build time, when it comes to process autonomization. Human designers need to ensure that the process is sufficiently framed, that is any (legal, technical, ethical) *constraints* need to be captured, otherwise the system can execute processes as they fit within this set of constraints. This means, it is a *declarative* approach to build time. Finally, the user needs to specify the data sources available, not just for the initial execution but over the lifetime of the processes as these data are used to *induce* the process to be executed.

Table 1. Build Time, Run Time, and Change Time in Automation and Autonomization.

Stage	Dimension	Automation	Autonomization
Build Time	Type	*Prescriptive*: users specify a process model, which serves as a template for the process execution	*Directive*: users specify aims of the process in a goal model
	Paradigm	*Imperative*: users specify what needs to be done	*Declarative*: users specify what not to do (constraints)
	Object	*How* it needs to be done	*What* needs to be done
	Input	*Human* input pre-execution	Human input pre-execution and *continuous input* during execution
	Level of detail	*Complete specification* including variants	*Underspecified* process model (if at all) based on *inherent capabilities*
Run Time	Reference point	*Process* model compliant execution	*Goal-* and *constraint*-compliant execution
	Level of allocation	Use of *micro*-allocated resources and data as they are associated with each task	Use of *macro*-allocated resources and data as they are associated with the entire process
	Context	Consideration of *immediate* process context (process and organization)	Consideration of *extended* process *context* (incl. sector and economy)
	Degree of predetermination	*Predetermined* execution of sequence of tasks	*Probabilistic* execution of sequence of tasks
	Escalation	*Early binding* of escalation procedure	*Late binding* of escalation procedure
Change Time	Frequency	*Periodic*	Periodic and *ongoing*
	Authority	*User* endorsed	*System* endorsed
	Trigger	*Discrete* (e.g., defined bottleneck, quality issue, process drift): defined moments of process change	Discrete (defined moments of change) plus *continuous* as new data emerges and adjusts the process
	Support	*Process* mining	Process and *goal* mining

5.3.2 Run Time

Run time is the phase in process automation during which processes are executed *in compliance* to their *predetermined* design specification with *resource allocation at task level.* Ensuring that this indeed occurs and processes comply with their process models has motivated a substantial amount of related research in the area of process mining. The control flow constitutes the core in process automation and only *immediate contextual information* – as it can be captured in the profile of available resources, the characteristics of the case (token) that is executed, and any type of explicit event information – can be considered by the process engine. Any exceptions can be only handled by *predefined* escalation routines or embedded process agility as the entire process run time is based on the paradigm of *early binding* (of the process execution to the defined process model). This guarantees predictable reliability, the core value proposition of process automation.

Process autonomization is not free of compliance requirements, but here the focus is on adherence to the specified process *goal and constraints*. Otherwise, run time in process autonomization follows a much more liberated paradigm. Resources are *not prespecified* and the contextual information considered can be as *broad* as it is accessible via large language models. Based on a *late process binding* paradigm, an autonomized process can be impacted by any type of relevant information. As a result, the process execution is no longer predetermined but *probabilistic*.

5.3.3 Change Time

Change time refers to the phase in process automation where adjustments to processes are made based on predetermined intervals. In traditional process automation, changes are typically *periodic* and occur at set frequencies, often triggered by *discrete* events such as identified bottlenecks, quality issues, or signs of process drift. The authority to implement these changes usually rests with *users* who endorse modifications based on their observations and experiences. Support for these changes is primarily provided through *process mining* techniques, which analyze historical data to identify areas for improvement and ensure compliance with the established process model.

In contrast, change time within process autonomization operates on a more dynamic and *ongoing* basis. Changes can be both periodic and continuous, allowing for real-time adjustments as new data emerges. This approach recognizes that relevant information can arise at any moment, prompting necessary adaptations to the processes being executed. Authority in this context shifts from user endorsement to *system* endorsement, where the autonomized system itself identifies when changes are needed based on its analysis of current conditions and *goals*. The support mechanisms also evolve from traditional process mining to include both process mining and goal mining, enabling a more holistic understanding of how processes can be optimized in alignment with overarching objectives. This shift towards *continuous adaptation* fosters an environment where processes remain agile and responsive to changing circumstances.

5.4 Caveats

The novel themes and dimensions of process autonomization including the framing of artificial autonomy do not come with without caveats and, thus, demands for control and oversight. In the following, we highlight two caveats, reliability and transparency, viz. explainability.

5.4.1 Reliability Caveats

The indeterministic nature of process autonomization means that the process outcomes are potentially less uniform. As these systems operate based on advanced algorithms, their ability to make real-time decisions introduces a level of unpredictability that can challenge traditional notions of reliability [36]. Users expect these systems not only to perform tasks autonomously but also to do so consistently and accurately. To meet these expectations, it is essential that process analysts prioritize rigorous testing to validate the robustness of decision-making processes. Additionally, implementing comprehensive monitoring can help identify anomalies or unexpected behaviors promptly.

For example, when a travel agency transitions from Alice's automated system to Bob's autonomous system, it must become comfortable with greater variety in travel recommendations and quality control. While Alice's system produces consistent, predictable itineraries based on fixed rules, Bob's system generates individualized suggestions likening to the personal travel planner. These autonomous recommendations could challenge established travel planning assumptions and should not be delivered to a client without oversight.

5.4.2 Transparency and Explainability Caveats

Process autonomization potentially comes with compromised interpretability. This is the case when deep learning is deployed as these analytical models make decisions by processing multiple layers of advanced algorithms. While this computational complexity enables more nuanced and contextual decision-making, it is practically impossible to trace decision steps from input to output as the decision model constitutes a de facto black box [17]. It can even make it challenging to provide clear, straightforward explanations for specific outcomes.

Consider how explainability differs between Alice's and Bob's travel planning experiences. When Alice's automated system recommends a particular flight routing, the logic is transparent and rule-based: it selected these flights because they matched her specified criteria. However, Bob's autonomous system makes more sophisticated recommendations that consider and interrelate multiple variables: e.g., historical booking patterns, seasonal tourism flows, and cultural event schedules. While this complexity enables more nuanced and personalized travel planning, it makes it more challenging to explain exactly why the system suggested one particular itinerary over another.

This explainability challenge becomes particularly important when autonomous processes make critical decisions as explanations are not a simple remedy for responsible AI adoption, and they can be misused and cause harm [37]. Be that as it may, some transparency in how these systems function is crucial for building user trust, as it allows stakeholders to assess some rationale behind decisions made by the system.

6 Discussion

Our research is by no means a call to autonomize any process. When considering whether to autonomize an existing process, organizations must carefully weigh the benefits against the risks of disrupting proven practices. While the enhanced capabilities of process autonomization could unlock new opportunities and enable more personalized service, it fundamentally changes how a business operates. Staff would need to shift from executing pre-determined procedures to monitoring and guiding autonomous decisions. Customer expectations would need to be recalibrated – they would gain more personalized, but less predictable services and might need to accept longer interaction times as the system engages in more detailed preference discovery. That is, the potential benefits of process autonomization must be weighed against the risks of disrupting established, well-functioning operations and the investments required in retraining staff and reshaping customer expectations. Amazon's transition from traditional automated warehouse systems to more autonomous robots illustrates this consideration. While their automated systems were already highly efficient, the shift to autonomous robots required careful implementation to ensure that the benefits of adaptability outweighed the risks of disrupting established processes. The more a business process requires consistency of outcomes the less autonomous processes would be advisable.

Further, the shift from automated to autonomized processes necessitates a fundamental rethinking of BPM, particularly in how goals, resources, and constraints are specified and managed in contrast to control flow and rules using process models and decision tables. Traditional BPM approaches prioritize efficiency, often embedding functional goals and constraints implicitly into process models. However, autonomized processes require explicit goals and constraints as they must dynamically adjust based on evolving priorities, environmental changes, and user preferences. This necessitates using analytical models that can balance multiple competing objectives and the BPM community does not yet have the tools to do so.

In addition, process autonomization introduces significant socio-technical challenges. Unlike traditional automated systems that operate within rigidly defined parameters, autonomized processes interact dynamically with human decision-makers, organizational policies, and societal expectations. This interplay requires a governance framework that ensures transparency, accountability, and ethical compliance in decision-making. For instance, when autonomized processes allocate resources or optimize processes, they must consider not only operational efficiency but also responsible process design including criteria such as fairness, inclusivity, and trustworthiness. Additionally, organizations must establish mechanisms for human oversight and intervention, ensuring that automated decisions align with strategic priorities and ethical standards. Successful autonomization depends on an integrated approach that balances technical efficiency with human-centered values.

Lastly, process autonomization is not an end in itself and we are already experiencing the interaction of multiple synthetic agents with each other. Correspondingly, we posit that while task autonomization with limited scope and goal, simple resourcing, and strict constraints is already permeating, process autonomization will be a substantial change as it will make the core focus of common BPM – the control flow – not the primary focus anymore. Looking ahead, process autonomization has the potential to be a core foundation of autonomous departments and enterprises with first glimpses already here (e.g., decentralized autonomous organizations based on blockchains). This may ultimately evolve into autonomous ecosystems and markets.

7 Conclusion

The evolution from traditional process automation to process autonomization marks a pivotal shift in the landscape of BPM as the primary areas of concern are transitioning from elaborated specification and monitoring of process execution via process models and process mining to adaptive design specifications comprising goals, macro-level resources, and constraints. As illustrated through the contrasting experiences of Alice and Bob, autonomized processes can not only enhance and personalize the user experience by allowing for a natural conversational instead of a transactional interaction but also empower organizations to enable mass service with the batch size of one.

We established the defining capabilities of artificial autonomy with independence in decision management and task execution, neglect tolerance towards self-management, and indeterminism enabling agency. Further, we introduced process autonomization based on the aspects of goal definition, macro-level resourcing, and constraint specification. We explored them across the three stages of build time, run time, and change time. By embracing process autonomization, businesses can go beyond the restrictions of static process models and navigate ambiguity more effectively because they can *autonomize* business processes that could not be automated before due to the cognitive effort required to orchestrate and execute them.

The potential applications across various sectors, as evidenced by real-world examples like Singapore's Smart City initiatives and Amazon's warehouse management systems, highlight the transformative impact that autonomized processes can have on operational efficiency and customer satisfaction across various sectors. However, as seen with hallucinations faced by systems like ChatGPT, this transition also brings new caveats in terms of reliability and explainability. Organizations must carefully weigh these factors when considering the move from automation to autonomization.

An early investigation into an entirely new phenomenon like process autonomization has to come with a number of limitations. First, this paper is grounded in conceptual research. Complementary, valuable theoretical and empirical research is still outstanding. Second, there is no (semi-)formalization of process autonomization in this paper which limits its integration into existing BPM methods and techniques, or a more formalized comparison with process automation. Finally, a detailed discussion of the boundary conditions of process autonomization has not been presented yet. This includes a discussion of the types of processes that are suitable to process autonomization.

Conceptual research artifacts such as ours can serve as a foundation for future empirical studies by providing a structured understanding of phenomena, thereby guiding researchers in formulating hypotheses and designing methodologies. In an era where rapid technological advancements and societal changes continuously reshape our understanding of various domains, conceptual research is essential for generating new insights that can inform both practice and policy while paving the way for more rigorous empirical investigations as intended by Grisold et al.'s Approach IV. Thus, as we look ahead, it is crucial for scholars and practitioners alike to further explore the implications of this paradigm shift on business practices and strategies. Understanding how to integrate autonomized systems into existing workflows will be essential for organizations aiming to maintain a competitive edge in an increasingly automated world.

References

1. Dumas, M., La Rosa, M., Mendling, J., Reijers, H.A.: Fundamentals of Business Process Management. Springer, Berlin (2018). https://doi.org/10.1007/978-3-662-56509-4
2. Nof, S.Y. (ed.): Springer Handbook of Automation. Springer, Berlin (2009). https://doi.org/10.1007/978-3-540-78831-7
3. Castelfranchi, C., Falcone, R.: Founding autonomy: the dialectics between (social) environment and agent's architecture and powers. In: Nickles, M., Rovatsos, M., Weiss, G. (eds.) AUTONOMY 2003. LNCS, vol. 2969, pp. 40–54. Springer, Heidelberg (2004). https://doi.org/10.1007/978-3-540-25928-2_4
4. Grisold, T., Janiesch, C., Röglinger, M., Wynn, M.T.: Managing dynamics in and around business processes. Bus. Inf. Syst. Eng. **66**, 533–540 (2024)
5. Chiodo, S.: Human autonomy, technological automation (and reverse). AI Soc. **37**, 39–48 (2022)
6. Ballou, K.A.: A concept analysis of autonomy. J. Prof. Nurs. **14**, 102–110 (1998)
7. Dumas, M., et al.: AI-augmented business process management systems: a research manifesto. ACM Trans. Manag. Inf. Syst. **14**, 1–19 (2023)
8. MacInnis, D.: Where have all the paper gone? Reflections on the decline of conceptual articles. ACR News (2004)
9. Whetten, D.A.: What constitutes a theoretical contribution? Acad. Manag. Rev. **14**, 490–495 (1989)
10. Gilson, L.L., Goldberg, C.B.: Editors' comment: so, what is a conceptual paper? Group Org. Manag. **40**, 127–130 (2015)
11. Harvey, S.: Creative synthesis: exploring the process of extraordinary group creativity. Acad. Manag. Rev. **39**, 324–343 (2014)
12. Heinonen, K., Gruen, T.: Elevating conceptual research: insights, approaches, and support. AMS Rev. **14**, 1–6 (2024)
13. Gregor, S., Hevner, A.R.: Positioning and presenting design science research for maximum impact. MIS Q. **37**, 337–355 (2013)
14. Boell, S.K., Cecez-Kecmanovic, D.: A hermeneutic approach for conducting literature reviews and literature searches. Commun. Assoc. Inf. Syst. **34** (2014)
15. Turing, A.M.: I.—Computing machinery and intelligence. Mind **LIX**, 433–460 (1950)
16. Russell, S.J., Norvig, P.: Artificial Intelligence: A Modern Approach, 4th edn. Pearson, Hoboken (2021)
17. Wanner, J., Herm, L.-V., Heinrich, K., Janiesch, C.: The effect of transparency and trust on intelligent system acceptance: evidence from a user-based study. Electron. Mark. **32**, 2079–2102 (2022)
18. Henderson, T.C., Dalton, P.: Z-infinity: a framework for autonomous agent specification and analysis. Technical report No. UUCS-90-018, University of Utah (1990)
19. Shoham, Y., Leyton-Brown, K.: Multiagent Systems. Cambridge University Press, Cambridge (2009)
20. Antsaklis, P.J.: Setting the stage: some autonomous thoughts on autonomy. In: Proceedings of the IEEE ISIC/CIRA/ISAS Joint Conference, Gaithersburg, MD, pp. 520–521. IEEE (1998)
21. Lehto, M.R., Buck, J.: Introduction to Human Factors and Ergonomics for Engineers. CRC Press, Boca Raton (2007)
22. Stephens, M.: Sandbox. In: Jajodia, S., Samarati, P., Yung, M. (eds.) Encyclopedia of Cryptography, Security and Privacy, pp. 2158–2162. Springer, Cham (2024). https://doi.org/10.1007/978-3-030-71522-9_791
23. Beavers, G., Hexmoor, H.: Types and limits of agent autonomy. In: Nickles, M., Rovatsos, M., Weiss, G. (eds.) AUTONOMY 2003. LNCS, vol. 2969, pp. 95–102. Springer, Heidelberg (2004). https://doi.org/10.1007/978-3-540-25928-2_8

24. May, T.: The concept of autonomy. Am. Philos. Q. **31**, 133–144 (1994)
25. Taylor, R.: Determinism and the theory of agency In: Hook, S. (ed.) Determinism and Freedom in the Age of Modern Science, pp. 224–230. Collier Books, New York (1958)
26. Williams, C.: Indeterminism and the theory of agency. Philos. Phenomenol. Res. **45**, 111–119 (1984)
27. Weigand, H., Dignum, V.: I am autonomous, you are autonomous. In: Nickles, M., Rovatsos, M., Weiss, G. (eds.) AUTONOMY 2003. LNCS, vol. 2969, pp. 227–236. Springer, Heidelberg (2004). https://doi.org/10.1007/978-3-540-25928-2_18
28. Crandall, J.W., Goodrich, M.A., Olsen, D.R., Nielsen, C.W.: Validating human-robot interaction schemes in multitasking environments. IEEE Trans. Syst. Man Cybern. Part A Syst. Hum. Part A Syst. Hum. **35**, 438–449 (2005)
29. Schultz, A.C., Goodrich, M.A.: Human-robot Interaction: a survey. Found. Trends Hum. Comput. Interact. **1**, 203–275 (2007)
30. Covrigaru, A.A., Lindsay, R.K.: Deterministic autonomous systems. AI Mag. **12**, 110–117 (1991)
31. Kueng, P., Kawalek, P.: Goal-based business process models: creation and evaluation. Bus. Process. Manag. J. **3**, 17–38 (1997)
32. Greenwood, D., Rimassa, G.: Autonomic goal-oriented business process management. In: Proceedings of the 3rd International Conference on Autonomic and Autonomous Systems, Athens, pp. 1–6. IEEE (2007)
33. Kumar, A., van der Aalst, W.M.P., Verbeek, E.M.W.: Dynamic work distribution in workflow management systems: how to balance quality and performance. J. Manag. Inf. Syst. **18**, 157–193 (2015)
34. Huang, Z., van der Aalst, W.M.P., Lu, X., Duan, H.: Reinforcement learning based resource allocation in business process management. Data Knowl. Eng. **70**, 127–145 (2011)
35. Xu, J., Liu, C., Zhao, X., Yongchareon, S., Ding, Z.: Resource management for business process scheduling in the presence of availability constraints. ACM Trans. Manag. Inf. Syst. **7**, 1–26 (2016)
36. European Commission: Regulation (EU) 2024/1689 of the European Parliament and of the Council of 13 June 2024 (2024). https://eur-lex.europa.eu/legal-content/EN/TXT/HTML/?uri=OJ:L_202401689
37. Martens, D., et al.: Beware of "explanations" of AI. arXiv:2504.06791 (2025)

Automation to Agitation: Unveiling RPA-Induced Technostress

Ishadi Mirispelakotuwa[✉], Rehan Syed, and Moe T. Wynn

Queensland University of Technology, Brisbane, Australia
ishadi.mirispelakotuwa@hdr.qut.edu.au, {r.syed, m.wynn}@qut.edu.au

Abstract. Robotic Process Automation (RPA) is a task-level process automation technology that enables hybridisation where employees and bots collaboratively execute tasks. RPA improves process efficiency and employee productivity, allowing them to concentrate on value-added tasks. Although RPA offers many organisational benefits, its increased adoption presents new challenges, such as technostress—a psychological strain employees experience when interacting with information systems (IS). Building on seminal technostress research, this study explores the technostress induced by RPA through a single case study. A conceptual model depicting attributes of technostress creators, dependencies, and outcomes is developed via qualitative data analysis. The findings indicated that RPA reduces techno-invasion compared to other IS while it contributes to techno-complexity, techno-uncertainty, techno-overload, and techno-insecurity. Enhancing the explanation of hybridisation, the findings revealed that techno-overload presents in hybridised processes by design and in fully automated processes that become hybridised due to exceptions handled by employees. This study contributes to the technostress literature by discovering dependencies between techno-complexity and techno-overload, as well as techno-uncertainty and techno-overload in the RPA context. The conceptual model will help managers to take countermeasures to prevent psychological and behavioural outcomes associated with RPA-induced technostress by identifying their sources.

Keywords: Robotic Process Automation · Technostress · Case Study · Conceptual Model

1 Introduction

Robotic Process Automation (RPA) was promised to bring a healthy work-life balance and a carefree working environment where humans can enjoy working on innovative tasks [13,26]. RPA is a distinct form of task-level process automation that enables hybridisation by replacing or sharing rule-based and routine tasks with bots [13,28]. RPA also changes how employees interact with processes by introducing bots as "digital colleagues" who invoke a sense of anthropomorphism [9]. The use of such automation technologies with anthropomorphic qualities has led to early signs of employees experiencing psychological and behavioural issues in the work environment [17]. Thus, just as with any other technology, symptoms have begun to emerge in the RPA context, indicating that not all promises are fulfilled.

This paper explores the issues that arise when employees interact with RPA from the human stress point of view. It is defined as ***technostress***–a psychological strain that employees experience when interacting with information systems (IS) [29,35]. Although existing literature hints at the symptoms of technostress in the RPA context [5,23,27], studies have not primarily focused on examining the impact of RPA on technostress and its outcomes. This demands us to explore; *"how does RPA impact technostress?"*

Building on seminal technostress research [29,35], this study is an early attempt to explore RPA-induced technostress through an exploratory case study with a global apparel manufacturing company. Through thematic analysis of qualitative data, a conceptual model is developed depicting the attributes of technostress creators (i.e., the factors that create technostress [35]), dependencies, and outcomes. The attributes of technostress creators and dependencies are specific to the RPA context and have not been discovered in previous studies.

The paper unfolds as follows. Section 2 includes the literature review. Section 3 presents the methodology. Section 4 presents the case background. Section 5 synthesises case findings. Section 6 presents a discussion of the findings, contributions, and limitations. Section 7 concludes the paper with future work.

2 Literature Review

Lazarus and Folkman's [16] transactional model of stress (TMS) is a widely adopted theoretical framework for studying technostress. According to TMS, stress arises from the interaction between an individual and their environment, particularly when environmental demands exceed the individual's ability to cope. When TMS is applied to human interactions with IS, technostress is identified as a type of psychological strain that arises from the use of IS [29,35].

The relationship between technostress and IS artifacts has been extensively discussed in previous studies. Early research explored factors that contribute to technostress, known as technostress creators [29,35]. The framework by Tarafdar et al. [35] is widely recognised for identifying five sub-dimensions of technostress referred to as the'Technostress Creator Inventory': techno-complexity, techno-uncertainty, techno-invasion, techno-overload, and techno-insecurity.

- **Techno-complexity** refers to the stress that arises when employees feel that IS requires significant effort to understand and learn.
- **Techno-uncertainty** is the stress associated with the constant evolution and change of IS.
- **Techno-invasion** describes the stress employees feel when the work demands associated with IS invade their personal time.
- **Techno-overload** refers to the stress experienced when employees are overwhelmed by IS, forcing them to work more and faster.

– **Techno-insecurity** is the threat of replacement that employees feel due to more tech-savvy colleagues or new technologies [35].

A recent study by Nastjuk et al. [25] argued that, when exploring the impact of technostress creators in IS, it is critical to consider the context in which technostress manifests and its outcomes. There are two contexts: organisational and private. Research in the organisational context focuses on IS such as enterprise resource planning (ERP) systems [1,15] that aim to increase efficiency [35] and performance by serving a utilitarian purpose [25]. Research in the private context is often associated with social networking services [34], private smartphone usage [18], and IS in vehicles [24]. Given the inherent link between job security concerns and techno-insecurity, techno-insecurity remains relevant only to the organisational context [24]. There are two types of outcomes: psychological and behavioural [25]. Psychological outcomes indicate one's conscious state of mind [25] and are linked to exhaustion, burnout, role overload, job dissatisfaction, and role conflict [29,35,36]. Behavioural outcomes reflect one's conscious or unconscious engagement of actions in a particular context [25].

There is limited evidence on technostress related to RPA [5,23,27]. Parsley et al. [27] examined technostress in the accounting profession due to the rapid proliferation of technologies, including RPA. Studies [5,23,27] also highlighted how IS, including RPA, contribute to technostress due to the fear of job loss. While not directly addressing technostress, concerns about potential job loss are frequently discussed in the RPA literature [12,26,33]. As RPA takes over repetitive tasks [40], employees are required to shift towards more creative and analytical tasks [13]. While this transition presents growth opportunities, it has introduced job security concerns [26,32]. Job security concerns associated with RPA were linked to the outcome - knowledge hiding [21]. Changing job roles was linked to signs of employee frustration [26]. When employees are given additional responsibilities to manually execute tasks in case of a bot failure [13,26], it has increased workload pressure and frustration [26]. Consequently, employees have sought more fulfilling employment opportunities [26]. The uncertainty due to repeated bot failures has led to a reduction in overall trust in RPA [33]. However, existing studies are limited to investigating technostress associated with different job roles (e.g., [27]) or highlighting its symptoms [26]. Thus, no study has primarily focused on examining the impact of RPA on technostress and its outcomes.

3 Methodology

A single exploratory case study was conducted following the guidelines of Yin [41]. Technostress in RPA is not thoroughly studied, specially in comparison to ERP [1] and other general IT use [4]. Thus, an exploratory case study was conducted to help bridge this gap by investigating how RPA impacts technostress.

Data, method, and investigator triangulations were used to ensure the credibility of qualitative data [10]. Data triangulation was achieved by gathering data from various stakeholder groups, including RPA users, technical staff, and managerial staff, through 8 semi-structured interviews as shown in Table 1.

Table 1. Details of the case study participants.

Code	Role	Stakeholder group	Interview duration
IN1	RPA Project Manager	Managerial Staff	1 h
IN2	Clerical Officer	RPA User	45 min
IN3	Software Engineer	Technical Staff	34 min
IN4	RPA Project Lead	Managerial Staff	43 min
IN5	Director	RPA User/ Managerial Staff	1 h
IN6	Head of Department	RPA User	51 min
IN7	Senior Clerical Officer	RPA User	58 min
IN8	Clerical Officer	RPA User	58 min

To support method triangulation, in addition to interviews, secondary data were gathered from various documents, including project presentations and process design documents (PDD). PDDs were primarily used to observe task allocation among bots and employees to gain an understanding of participants' backgrounds. Project presentations were used to understand the various elements of the governance framework.

Case data was analysed using the thematic analysis method of Clarke and Braun [6]. Step 1 - familiarised with the data by repeatedly reading the transcripts of each interview. Step 2 - conducted the coding of case study data by applying open, axial, and theoretical coding [31] using the NVivo software. To analyse data, a hybrid approach combining inductive and deductive approaches was used [37]. First, data was deductively analysed using Nastjuk's framework [25]. This framework was selected because (1) it allows investigating technostress associated with IS, taking context and outcomes into consideration, and (2) it facilitates studying technostress in the organisational context related to IS, like RPA that is used for business purposes over private use [40]. Second, new open codes (52 codes) were inductively identified. Step 3 - developed and identified themes. At this stage, attributes of each technostress creator were identified as new themes. Step 4 - conducted axial coding to recategorise the attributes (8 attributes as discussed in Sect. 5) under each technostress creator. Step 5 - finalised the themes and relationships by identifying the outcomes linked to each technostress creator and the dependencies between technostress creators. Step 6 - completed the write-up [6]. As the analysis evolved, sense-making annotations were used to document key insights and areas for further investigation. The coding, analysis, and interpretation phases included collaboration with a second and third coder to maintain rigour. A coding rulebook following Bandara et al. [3] was created to ensure consistent terminology among the research team, promoting transparency and inter-coder reliability in the coding process [7]. The codebook was iteratively developed and updated in multiple iterations as new themes emerged.

4 Case Background

This section provides an overview of the case organisation - MAS Holdings and its RPA journey. MAS Holdings was established in 1987 in Sri Lanka. It is valued at US$2 billion and is the largest apparel manufacturing conglomerate in Southeast Asia [20]. MAS designs, develops, and produces intimate apparel, sportswear, activewear, and swimwear. It has a network of 53 manufacturing facilities across 16 countries and employs over 99,000 employees [38]. Since 2000, SAP has served as the ERP system for MAS, supporting its vertically integrated business model. As a participant in the United Nations Global Compact program, MAS aims to support employee work-life balance and enhance process efficiency by using automation technologies like RPA [39]. Its RPA journey began in 2017. Currently, it has about 1120 users. MAS uses the RPA tool, UiPath, to develop bots. RPA was initially introduced as a proof-of-concept project in the labour order placement process. Their initial implementation strategy focused on achieving quick wins through small-scale task automation. For instance, in the bill of material creation process, the data entry tasks of the sub-process - purchase order creation were first automated. With evolving RPA capabilities, MAS scaled from quick wins to more complex automation in end-to-end processes. Thus, MAS has both hybridised and fully automated processes. Subsequently, MAS has increased from 7 to 35 unattended bots. They operate in queues across 150 processes in areas like human resources, finance, merchandising, and logistics. At MAS, RPA governance is managed by a centralised centre of excellence (CoE) team of 15. RPA governance policies outline that in a hybridised process, task ownership of an automated task is assigned to the employee. Thus, when a bot fails, the assigned employees must execute the process manually to ensure process continuity.

While MAS has a mature RPA initiative of 5+ years, it faces challenges due to increasing complaints by RPA users. RPA users have reported difficulties related to increasing responsibilities and workloads with bots, especially when bots fail due to errors or exceptions. This situation has been apparent in both hybridised and fully automated processes. RPA users have exhibited negative emotions such as frustration and anxiety, along with behaviours such as a lack of trust in bots and hiding knowledge, as the adoption of RPA increases. Accordingly, the MAS case study sets a rich foundation for exploring technostress in the RPA context.

5 Findings

This section presents the key findings of the case study. All findings are supported by relevant interview quotes. Figure 1 depicts the technostress creators linked to RPA, highlighting their attributes, corresponding outcomes, and dependencies.

5.1 Factors Impacting Technostress Creators

Techno-complexity refers to the stress that arises when employees feel that IS like RPA requires considerable effort and time to understand and learn [35]. Techno-complexity exists in the RPA context. A participant highlighted, *"there are many complaints from employees because they don't understand bots"* (IN8). This statement highlights the symptoms of techno-complexity as the employees find it challenging to understand bot operations. Three RPA-related factors were identified to induce techno-complexity: RPA awareness, task visibility, and RPA-IT integration. A lack of RPA awareness leads to techno-complexity. A participant mentioned, *"If those interns didn't have prior exposure to a similar set-up [RPA environment], it would be like 'Greek' [an expression of confusion in the Sri Lankan culture] for them"* (IN8). This statement demonstrates that employees found the learning process overwhelming in the absence of prior exposure to RPA. In contrast, employees with prior RPA experience showed a positive attitude, as they found it easy to understand and use. As per a participant, *"I was able to understand because I was there from the beginning. I engaged with xxx's team to build the bots. I can identify if the bot fails due to a technical or business exception"* (IN8). This statement demonstrates that employees' participation in the development phase has allowed them to improve their understanding of the broader RPA environment. Task visibility contributes to techno-complexity. As per a participant, *"You have to wait until bot does its work and send the email. We are blindfolded until then. We didn't know how to explain it when something happened"* (IN2). This statement explains that due to a lack of visibility in bot operations, employees found it challenging to explain or troubleshoot when problems occur. RPA-IT integration contributes to techno-complexity. As per a participant, *"These are automation done on top of other applications. There are changes in those applications that users don't see all the time... It makes it difficult for them to understand the changes later"* (IN1). This statement highlights that any changes made to the other systems integrated with RPA may not be immediately visible to employees. As a result, when changes occurred, employees struggled to understand them due to being unaware of prior changes (Fig. 1).

Techno-uncertainty refers to the stress in employees feeling that IS, like RPA, is constantly evolving and changing [35]. Techno-uncertainty exists in the RPA context. As per a participant, *"When we start doing, the entire process or parts have changed because of errors. Now, the bot requires different inputs. Users must enter the data into a specific template. If not bot would reject. So, what gets lost is the knowledge to trigger the bot. Now they panic and come to us"* (IN1). This statement emphasises that, over time, employees forgot how to accurately trigger a bot due to the frequent modifications made to the bots to correct errors. When employees faced issues, they reacted with panic because they were unable to adapt to the evolving changes in how the bots function. Two factors were identified as causing techno-uncertainty: the frequency of process changes and the frequency of errors or exceptions. Continuous process changes make it difficult for employees to maintain a solid RPA knowledge base. As per a participant, *"When someone makes a request to make changes to that process,* *sometimes it impacts the process I do. My process does not continue. The bot stops suddenly. We had frequent problems like that [business exceptions]"* (IN8). This statement explains that changes in other processes can trigger uncertainty due to bot failures in the current process. The second factor is the frequency of errors/exceptions. As per a participant, *"Bot kept failing. We did many changes to it because of that change in customs policy. I can't even remember those now"* (IN2). This statement demonstrates the difficulty employees had in recalling the knowledge of multiple bot modifications caused by a policy change (i.e., business exception). Therefore, relentless changes due to errors/ exceptions lead to techno-uncertainty in the RPA context (Fig. 1).

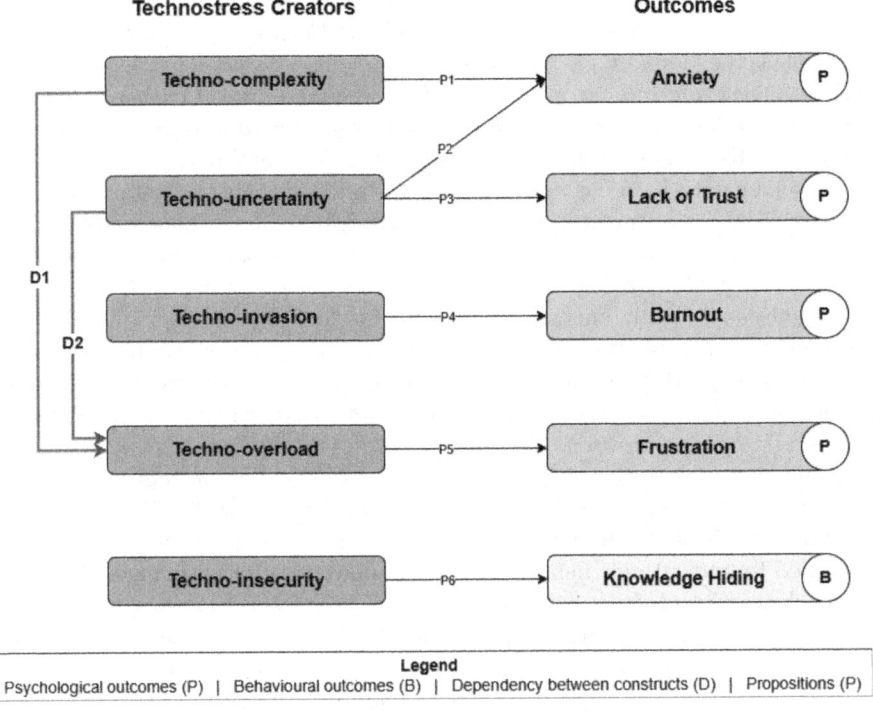

Fig. 1. An RPA-induced technostress conceptual model

Techno-Invasion refers to the stress employees feel when the work demands of IS like RPA invade personal time [35]. Techno-invasion was reduced in the case context as RPA has helped to improve work-life balance. As per a participant, *"We have unattended bots that work at night and weekends...When it comes to people's work-life balance, RPA has helped very much. Now we don't have to work overtime, and we do only 9–5"* (IN6). This statement highlights that RPA can avoid employees' constant connectivity to work. Unattended bots that operated independently outside of regular work hours handled tasks that would otherwise require employees. While bots managed after-hour tasks, employees could maintain a standard 9–5 schedule. Hence, RPA has improved employees' work-life balance, reducing techno-invasion. RPA-IT integration was identified as a factor that reduces techno-invasion. A participant mentioned, *"Everyday we get that email from the [internal] system at 5.00 pm. We have automated that task, and now employees are happy because they don't have to do overtime or monitor it"* (IN6). The participant explains that before RPA, employees had to work overtime due to receiving a regular email from an internal system at 5:00 PM that required action afterwards. This situation reduced their personal time leading to burnout. However, the integration between RPA and the internal IS has increased employee satisfaction and decreased techno-invasion (Fig. 1).

Techno-overload refers to the stress that employees feel when overwhelmed by IS like RPA as they are forced to work more and faster [35]. Techno-overload exists in the RPA context. A participant mentioned, *"I just send my data set and the bot is asking me to fill all these master data in an Excel sheet...they might get annoyed and think, this is just a booking, I might create this manually, rather than going through all that hassle to get the master data"* (IN1). This statement highlights that RPA demanded employees to work with multiple steps and applications to create inputs to trigger a bot. This has resulted in techno-overload as employees felt 'annoyed' as they had to complete several steps for a simple task. There are five factors that contribute to techno-overload in the RPA context. The first factor is the frequency of errors/ exceptions. A participant mentioned, *"We had frequent problems [bot failures]. To meet critical deadlines, we had to always do parts or all of the processes manually... Some steps have changed...Sometimes, the steps were confusing...it took us some time to understand"* (IN8). This statement reflects a hybridised process by design, where employees are responsible for executing the process in case of a bot failure. The steps have eventually changed due to errors/ exceptions. The participant highlights how they had to invest additional time and effort to understand the new steps and figure out how to manually execute them when a bot fails. The second factor is the frequency of process changes. A participant mentioned, *"We get constant tweaks and updates to our process...we have seen these issues. A new variation or requirement would come...it will have additional steps that we need to do to extract data. We had to always keep up with these [frequent process changes]"* (IN7). This statement highlights that process changes contribute to extra work by adding steps to extract data from a bot. The third factor is the frequency of bot modifications. *"I had to work really hard when something was not right. Sometimes, I had to work until 11 or 12 at night and work with xxx's team [development team]... It wasn't the best experience"* (IN8). This statement emphasises how employees with process expertise were needed to collaborate with the development team when there were frequent bot modifications. Consequently, employee frustration has increased due to increasing workloads. The fourth factor is the level of hybridisation. A participant stated, *""The users would have the knowledge, but they wouldn't have enough capacity to accom-*

modate the data load...We used to have 10 people. Now that we've automated this process, only two people are doing that, and the rest is done by bots. The bot went down... They[employees] can't create all the purchase orders within the time frame, even if they do overtime" (IN1). This statement highlights a hybrid process designed to reduce the number of employees, sharing tasks with bots. In such processes, when a bot fails, the remaining employees struggle to meet the additional data processing demands within a given time. Techno-overload was observed in fully automated processes. As per a participant, "It's all automated, but they call me if they can't figure out why the bot went down...it's not even related to my work now" (IN8). This statement highlights how a fully automated process that became hybridised due to exceptions causes techno-overload. In such processes, employees with process expertise are pulled from other processes to handle exceptions. As a result, employees are frustrated due to tasks beyond their current scope. Fifth, task ownership imposed by RPA governance was identified as another factor. A participant stated, "We keep that responsibility and accountability with the user...they might feel stressed because everything is being automated, they still are accountable for that" (IN1). This statement indicates that strict RPA governance policies regarding task ownership can create a sense of stress to consistently meet work demands, causing techno-overload (Fig. 1).

Techno-insecurity refers to the threat of replacement that employees feel due to more tech-savvy co-workers or IS like RPA [35]. Techno-insecurity exists in the RPA context. A participant stated, "The employees have a fear of job loss. So, the users are resistant to automation" (IN3). This statement reflects employee reluctance towards RPA due to job security concerns. RPA awareness was identified as a factor that impacts techno-insecurity. A participant mentioned, "I think it's [resistance to RPA] happening because of the lack of knowledge of bots. We try to remind the user the automation is not to replace them, but to help them to grow" (IN3). This statement highlights that the resistance to RPA primarily stems from a lack of understanding regarding its purpose and capabilities. Employees have mistakenly believed that bots are meant to replace rather than assist them. In contrast, employees with high RPA awareness felt more secure despite the presence of RPA. A participant mentioned,"*there are several automation projects I've worked with...Even if RPA is implemented, it is not like we are separated from the process. It is more like another person working with us*" (IN8). This statement highlights that employees with high RPA awareness felt more secure in their job roles as they were aware of the RPA capabilities. RPA governance was identified as a factor that mitigates techno-insecurity. At MAS, strategies were implemented to reduce techno-insecurity. As per a participant, "We initially conducted workshops to tell the user the automation is not to replace them but to help them to grow" (IN3). This approach was taken to shift their mindset from the fear of job loss due to RPA towards opportunities for career growth. Another strategy was assigning task ownership to employees. A participant stated, "We always keep the accountability and responsibility with the user" (IN1). This statement emphasises that assigning the accountability and responsibility of an automated task promotes mandatory human intervention. As a result, employees feel secure in their job roles (Fig. 1).

5.2 Dependencies Between Technostress Creators

Techno-Complexity impacts **techno-overload** in the RPA context. As per a participant, *"I didn't know RPA. I have to spend hours trying to figure out if the issue is with the SAP or bot"* (IN8). This statement highlights that employees lacking RPA knowledge spent more time and effort detecting issues during bot failures. The complexity of handling multiple systems has increased employees' stress, as they must identify issues outside their expertise (D1 in Fig. 1).

Techno-Uncertainty impacts **techno-overload** in the RPA context. As per a participant, *"even if we do a certain part by a bot, we always have to follow up. We don't know when the bot goes down. We don't have time to do other tasks. Suddenly, we get error emails to quickly fix"* (IN8). This statement emphasises that in a hybridised process by design, employees were often disrupted from their primary tasks by sudden error emails, which they needed to resolve quickly. As a bot can fail unexpectedly, employees were assigned continuous monitoring duties. Thus, their workload has increased. The participant also stated that if a bot fails in a fully automated process, they are called upon to identify exceptions. *"they call me if they can't figure out why the bot went down...it's not even related to my work now"* (IN8). This statement demonstrates a fully automated process that became hybrid due to exceptions. Employee frustration was evident in these processes due to due to additional out-of-scope tasks (D2 in Fig. 1).

5.3 Organisational Context: Outcomes of Technostress Creators

RPA-induced **techno-complexity** contributes to the psychological outcome: **anxiety**. As per a participant, *"on top of the complexity of SAP, if additional things from RPA add up, it will be very difficult for new employees to understand. They are constantly on edge"* (IN6). This statement highlights that the complexity of integrating RPA with SAP creates a steep learning curve, causing anxiety among employees who lack prior RPA exposure (P1 in Fig. 1).

RPA-induced **techno-uncertainty** contributes to the psychological outcome: **anxiety**. A participant mentioned, *"After RPA, all we did was 'let it take over the task and run behind it' because we don't know when it fails. It's more like we always wait for the next issue to come"* (IN8). This statement reflects how RPA shifts an employee's responsibilities to constantly monitor bots instead of reducing workload. The phrase "run behind it" suggests that without using the saved time for more valuable tasks, employees continuously chase after bot failures, uncertain about when and why they occur. This uncertainty associated with bots has created anxiety among employees (P2 in Fig. 1).

RPA-induced **techno-uncertainty** contributes to the psychological outcome: **lack of trust**. A participant mentioned, *"We don't trust the bot to rely in the next 10 min because we don't know what's next and when it goes down"* (IN6).

This statement indicates that employees struggle to trust the bot due to uncertainty about when it might fail. The lack of trust leads employees to depend less on RPA bots for task execution. Accordingly, techno-uncertainty contributes to employees' lack of trust in RPA bots (P3 in Fig. 1).

RPA reduces **techno-invasion** and the the psychological outcome: **burnout**. As per a participant, *"we have created bots to improve work-life balance... Employees come to us to automate tasks that require them to work after 5.00 pm like those that need their input if they get an email...we had so many complaints before. We have focused on those tasks for RPA"* (IN1). This statement highlights how prior to RPA, employees experienced techno-invasion associated with other IS that required them to work after hours. Employees were experiencing burnout due to excessive overtime work which led them to request automation. After integrating RPA with other IS, employees have been able to reduce using their personal time for after-hour work. As a result, it has improved employees' work-life balance, reducing techno-invasion and burnout (P4 in Fig. 1).

RPA-induced **techno-overload** contributes to the psychological outcome: **frustration**. As per a participant, *"It's more like we did more after a bot fails...I feel like I'm constantly playing catch-up, doing usual tasks while correcting issues I never knew"* [IN8]. This statement highlights the signs of employees' frustration due to constant monitoring and taking corrective actions for bot failures. When bots fail, employees face additional work demands to maintain process continuity. Consequently, there is growing frustration among employees due to working more hours than before instead of enjoying the benefits of RPA (P5 in Fig. 1).

RPA-induced **techno-insecurity** contributes to the behavioural outcome: **knowledge hiding**. A participant stated, *"new employees might fear job loss, and they are not very good with RPA. There was a lack of support in the final stages [of the development phase]"* (IN3). This statement highlights that the new employees may not have enough experience with RPA. A lack of RPA awareness has created a fear of job loss among them, hindering full participation in bot development. The same participant explained, *"Some people do not share everything because they think that knowledge is valuable for their next role"* (IN3). This statement demonstrates how employees considered their knowledge as a valuable asset that made them indispensable in their current role or secure in their next role. Consequently, employees intentionally withhold knowledge due to job security concerns that arise from implementing RPA (P6 in Fig. 1).

6 Discussion

This section summarises the creators, dependencies, and outcomes of RPA-induced technostress.

RPA causes techno-complexity. As identified in the literature, our findings confirmed that characteristics of RPA tools that limit task visibility in bot operations [2,26] cause techno-complexity. Our findings discovered a link between RPA awareness and employee anxiety. We identified that employees who lack sufficient RPA awareness feel anxious as they struggle to understand bot operations when integrated with other systems. These findings align with Tarafdar et al. [36] that highlight that techno-complexity arises when employees take a long time to understand and use IS or feel they lack knowledge of IS.

RPA causes techno-uncertainty. Our findings indicated that techno-uncertainty arises from frequent bot failures caused by process changes, errors, or exceptions. Frequent bot failures are identified in the literature as a limitation of RPA [22,26]. RPA executes tasks exactly following its choreography [40]. Any deviation from the designed path limits the capabilities, causing bot failures [13]. Hence, RPA is associated with a high level of techno-uncertainty. In line with [33]'s argument, we found a link between techno-uncertainty and employees' trust in RPA. When bots do not perform as expected, throwing frequent errors, it diminishes employees' trust by failing to meet their expectations. Our findings are consistent with the Tarafdar et al. [36] where techno-uncertainty can manifest as employees' negative emotions stemming from a loss of control due to constant changes in software, hardware, or networks.

RPA causes techno-overload. RPA is intended to replace repetitive and rule-based tasks with bots, allowing employees to save time for valuable tasks [13,40]. Our findings challenge this notion by indicating that even after bots take over specific tasks, employees are assigned additional responsibilities to monitor and correct bot failures, which distracts them from primary tasks. Employees are frustrated because the expectations set by RPA have not been met. These observations correspond with the literature that highlights changing work habits to adapt to new technologies, time pressure due to increasing workloads, and complexity-triggered negative emotions [36]. Furthermore, the findings discovered that techno-overload is present in hybridised processes. However, there is ambiguity with the concept of hybridisation in the RPA context. As per Ruiz et al. [30], the hybrid RPA means the vertical segmentation of activities in a process between humans and bots. The literature on RPA conveys a similar idea [26,40]. However, we uncovered two types of hybridisation: (1) hybridised by design: a process that is originally designed to allocate tasks between bots and employees, and (2) hybridised due to exceptions: a fully automated process that is hybridised because employees intervene to handle exceptions. Accordingly, we enhance the existing explanation. Within a spectrum ranging from 'not automated' to 'fully automated', all processes except for the 'not automated' category are hybridised. Thus, as the number of errors/ exceptions increases, hybridisation also increases due to increased human intervention. However, both types of hybridisation led to employee frustration due to additional work demands.

RPA causes techno-insecurity. In the literature, techno-insecurity refers to the threat of replacement that employees feel due to more tech-savvy co-workers or IS [35]. Similarly, in the RPA literature, fear of job loss due to the introduction of RPA is highlighted [12,13]. Therefore, techno-insecurity in the RPA context can be interpreted as the fear of job loss. In agreement with the literature [12,26], our findings discovered a link between a lack of awareness of RPA and fear of job loss. Many employees mistakenly believe they can be replaced due to RPA's ability to mimic their behaviour [26]. Supporting the argument of Mirispelakotuwa et al. [21], the findings indicated that job loss concerns lead to knowledge-hiding behaviours. These findings align with the existing technostress literature, where employees limit sharing IS knowledge with others due to fear of being replaced [36].

RPA reduces techno-invasion. Early research on technostress in IS indicates that technologies for business use, such as ERP contribute to techno-invasion [1,15]. Our findings emphasise that RPA differentiates itself from other IS as it reduces techno-invasion by sharing work responsibilities that require employees to spend their personal time. Consequently, RPA improves employee work-life balance as highlighted in RPA literature [12,26,40].

As per the literature, *RPA governance* involves the practice of development, management, and supervision of bots, employees, and automated processes [14, 26,32]. Similarly, our findings indicate that specific RPA governance practices, such as raising awareness of RPA and assigning task ownership to employees, can be introduced to mitigate techno-insecurity. In contrast, the same policies on the rigidity of task ownership can cause techno-overload. However, the influence of RPA governance on other technostress creators needs to be further explored.

Overall, out of the five technostress creators in Nastjuk's framework [25], RPA was identified to cause techno-complexity, techno-uncertainty, techno-overload, and techno-insecurity. However, RPA reduces techno-invasion.

6.1 Contributions

The study contributes to the body of knowledge in several ways. First, RPA differentiates itself from other IS as it reduces techno-invasion. Technostress research suggests that IS causes techno-invasion [1,15,36]. However, we found that RPA has a positive impact on techno-invasion because it acts as a digital colleague [32], taking over process tasks that typically require employees' personal time to manage excessive workloads. The overall impact leads to an improved work-life balance in organisations adopting RPA. Second, our findings emphasise that RPA contributes to techno-complexity, techno-uncertainty, techno-overload, and techno-insecurity. Certain limitations of RPA, like frequent bot failures, and its characteristics, like the ability to mimic human behaviour and integrate with other IS, cause techno-complexity, techno-uncertainty, and techno-overload. Particularly, the ability to mimic human behaviour causes techno-insecurity due to employees' fear of being replaced. Third, unlike previous technostress research (e.g., [29,36]), we demonstrate how technostress creators relate to one another. Our findings discovered two dependencies between techno-complexity and techno-overload, as well as techno-uncertainty and techno-overload in the RPA context. These dependencies can determine how a technostress creator impacts another and the direction of the relationship. Several studies have examined the outcomes of technostress [35,36]. However, mitigating the outcomes of technostress requires awareness of its related technostress creators in different technological contexts. For instance, to mitigate the impacts of techno-overload, techno-uncertainty surrounding RPA due to frequent process changes, errors, and exceptions should be reduced. These findings inform the investigation of technostress associated with other process automation technologies. Fourth, we highlight how technostress is impacted by process hybridisation in the RPA context by enhancing the current explanation. In addition to hybridisation by design that is explained in RPA literature [30,40], we found that hybridisation due to exceptions increases the techno-overload due to additional workloads.

There are several practical contributions. First, managers should be aware that RPA contributes to significant psychological and behavioural outcomes. While behavioural outcomes are commonly tracked through a series of actions [25], such as knowledge hiding behaviour due to job security concerns, psychological outcomes can develop unnoticed (e.g., frustration). Hence, it is crucial to identify relevant indicators and take measures to prevent the psychological health risks associated with the use of RPA. Second, the attributes in the conceptual model will help RPA stakeholders identify the relevant sources of technostress creators and develop specific countermeasures. For instance, identifying processes with a high rate of exceptions may help managers recognise the symptoms of techno-overload among employees. Additionally, the RPA developers can enhance the design of workflows by reducing the chances of known business and technical exceptions. Third, managers can use the model to understand how a technostress creator impacts another through dependencies. This will help in managing the related technostress creator to mitigate the targeted technostress creator.

6.2 Limitations

There are several limitations. First, the study is based on a single case study with a limited number of participants. Although the case study was conducted with a company that has a long-standing RPA adoption, the findings may be unique to the case context. Thus, the manifestation of technostress creators and their outcomes may differ across organisations adopting RPA based on their specific circumstances. Furthermore, the limited number of participants may have constrained the depth of insights obtained. Second, in addition to the identified attributes of technostress creators, there may be other attributes specific to RPA that were not revealed through a single case study. Third, while the current case study provides rich insights into dependencies among technostress creators in the RPA context, it does not uncover all the dependencies.

7 Conclusions and Future Research

This study explores technostress in the RPA context through a single revelatory case study. Based on the results, an RPA-induced technostress conceptual model was developed, depicting the attributes of technostress creators, dependencies, and outcomes. The findings differentiate RPA from other IS as it reduces techno-invasion. However, RPA contributes to techno-complexity, techno-uncertainty, techno-overload, and techno-insecurity. Enhancing the explanation of hybridisation, the findings revealed that techno-overload was present both in hybridised processes by design and due to exceptions. This study contributes to the technostress literature by identifying dependencies between techno-complexity and techno-overload, as well as techno-uncertainty and techno-overload, in the RPA context, which was not previously discovered. The conceptual model will help managers take measures to prevent psychological and behavioural outcomes associated with RPA-induced technostress creators.

The current RPA-induced technostress model is developed based on rich insights from participants in a global apparel manufacturing company. In future studies, researchers should focus on conducting multiple case studies to enrich the existing insights on dependencies and attributes relevant to technostress creators in the RPA context. Additionally, a cross-case analysis can be conducted to construct a theoretical framework for explaining RPA-induced technostress.

References

1. Agrawal, K., Tarafdar, M., Vaidya, S.: Monitoring, surveillance and technostress-an enterprise application case. In: AMCIS, pp. 1–10. AIS, New Orleans (2018)
2. Asatiani, A., Penttinen, E., Rinta-Kahila, T., Salovaara, A.: Implementation of automation as distributed cognition in knowledge work organizations: six recommendations for managers. In: ICIS, AIS, Munich (2019)
3. Bandara, W., Syed, R.: The role of a protocol in a systematic literature review. J. Decis. Syst. **33**(4), 583–600 (2024)
4. Brod, C.: Technostress: The Human Cost of the Computer Revolution, 1st edn. Addison-Wesley (1984)
5. Cieslak, V., Valor, C.: Moving beyond conventional resistance and resistors: an integrative review of employee resistance to digital transformation. Cogent Bus. Manag. **12**(1), 1–32 (2024)
6. Clarke, V., Braun, V.: Thematic analysis. J. Posit. Psychol. **12**(3), 297–298 (2017)
7. DeCuir-Gunby, J.T., Marshall, P.L., McCulloch, A.W.: Developing and using a codebook for the analysis of interview data: an example from a professional development research project. Field Methods **23**(2), 136–155 (2011)
8. Eulerich, M., Waddoups, N., Wagener, M., Wood, D.A.: The dark side of robotic process automation (RPA): understanding risks and challenges with RPA. Account. Horiz. **38**(2), 143–152 (2024)
9. Fettke, P., Czarnecki, C.: Robotic Process Automation: Management, Technology, Applications, 1st edn. De Gruyter, Berlin, Germany (2021)
10. Flick, U.: An Introduction to Qualitative Research, 7th edn. SAGE, Los Angeles (2023)
11. Fortune Business Insights. https://www.fortunebusinessinsights.com/robotic-process-automation-rpa-market-102042. Accessed 15 Feb 2025
12. Hartikainen, E., Hotti, V., Tukiainen, M.: Improving software robot maintenance in large-scale environments–is center of excellence a solution? IEEE Access **10**, 96760–96773 (2022)
13. Hofmann, P., Samp, C., Urbach, N.: Robotic process automation. Electron. Mark. **30**(1), 99–106 (2020)
14. Kedziora, D., Penttinen, E.: Governance models for robotic process automation: the case of Nordea Bank. J. Inf. Technol. Teach. Cases **11**(1), 20–29 (2020)

15. Khanzada, M., Khan, S., Alam, F., Kamal, M.: Investigating the relationship between ERP-related technostress and employee performance. Pak. Bus. Rev. **26**, 149–174 (2024)
16. Lazarus, R.S., Folkman, S.: Stress, Appraisal, and Coping, 1st edn. Springer, New York (1984)
17. Linkedin. https://www.linkedin.com/pulse/lucy-effect-when-automation-creates-work-how-smarter-systems-agoston-boole/. Accessed 17 March 2025
18. Lee, Y., Chang, C., Lin, Y., Cheng, Z.: The dark side of smartphone usage: psychological traits, compulsive behavior and technostress. Comput. Hum. Behav. **31**, 373–383 (2014)
19. Lin, Y., Yu, Z.: An integrated bibliometric analysis and systematic review modelling students' technostress in higher education. Behav. Inf. Technol. **44**(4), 631–655 (2024)
20. MAS. https://www.masholdings.com/media-centre/mas-recognised-as-one-of-most-innovative-companies-2/. Accessed 15 Feb 2025
21. Mirispelakotuwa, I., Syed, R., Wynn, M.T.: Is RPA causing process knowledge loss? Insights from RPA experts. In: Köpke, J., et al. (eds.) BPM: Blockchain, Robotic Process Automation and Educators Forum, LNBIP, vol. 491, pp. 73–88. Springer, Cham (2023)
22. Modliński, A., Kedziora, D., Jiménez Ramírez, A., del-Río-Ortega, A.: Rolling back to manual work: an exploratory research on robotic process re-manualization. In: Marrella, A., et al. (eds.) BPM: Blockchain, Robotic Process Automation, and Central and Eastern Europe Forum. LNBIP, vol. 459, pp. 154–169. Springer, Cham (2022)
23. Murphy, L.: The productivity dilemma: examining the truth behind automation's impact on employment, and the mediating role of augmentation. Int. J. Organ. Anal. **33**(3), 622–644 (2025)
24. Nastjuk, I., Kolbe, L. M.: On the duality of stress in information systems research: the case of electric vehicles. In: ICIS, pp. 1–22. Fort Worth, USA (2015)
25. Nastjuk, I., Trang, S., Grummeck-Braamt, J.V., Adam, M.T.P., Tarafdar, M.: Integrating and synthesising technostress research: a meta-analysis on technostress creators, outcomes, and IS usage contexts. Eur. J. Inf. Syst. **33**(3), 361–382 (2023)
26. Oshri, I., Plugge, A.: What do you see in your bot? Lessons from KAS Bank. In: Oshri, I., Kotlarsky, J., Willcocks, L.P. (eds.) Global Sourcing 2019. LNBIP, vol. 410, pp. 145–161. Springer, Cham (2020). https://doi.org/10.1007/978-3-030-66834-1_9
27. Parsley, S., Wieck, M.R., Compton, S.G.: Technostress and the accounting profession: certified public accountant. CPA J. **92**(11), 72–75 (2022)
28. Plattfaut, R., Borghoff, V.: Robotic process automation: a literature-based research agenda. J. Inf. Syst. **36**(2), 173–191 (2022)
29. Ragu-Nathan, T.S., Tarafdar, M., Ragu-Nathan, B.S., Tu, Q.: The consequences of technostress for end users in organizations: conceptual development and empirical validation. Inf. Syst. Res. **19**(4), 417–433 (2008)
30. Ruiz, R.C., Ramírez, A.J., Cuaresma, M.J.E., Enríquez, J.G.: Hybridising humans and robots: an RPA horizon envisaged from the trenches. Comput. Ind. **138**, 1–17 (2022)
31. Saldana, J.: The Coding Manual for Qualitative Researchers, 4th edn. SAGE, Los Angeles (2021)
32. Syed, R., et al.: Robotic process automation: contemporary themes and challenges. Comput. Ind. **115** (2020)

33. Syed, R., Wynn, M.T.: How to trust a Bot: an RPA user perspective. In: Asatiani, A., et al. (eds.) BPM 2020. LNBIP, vol. 393, pp. 147–160. Springer, Cham (2020). https://doi.org/10.1007/978-3-030-58779-6_10
34. Tarafdar, M., Maier, C., Laumer, S., Weitzel, T.: Explaining the link between technostress and technology addiction for social networking sites: a study of distraction as a coping behavior. Inf. Syst. J. **30**(1), 96–124 (2020)
35. Tarafdar, M., Tu, Q., Ragu-Nathan, B.S., Ragu-Nathan, T.S.: The impact of technostress on role stress and productivity. J. Manag. Inf. Syst. **24**(1), 301–328 (2007)
36. Tarafdar, M., Tu, Q., Ragu-Nathan, T.S., Ragu-Nathan, B.S.: Crossing to the dark side: examining creators, outcomes, and inhibitors of technostress. Commun. ACM **54**(9), 113–120 (2011)
37. Thompson, J.: A guide to abductive thematic analysis. Qual. Rep. **27**(5), 1410–1421 (2022)
38. UiPath. https://www.uipath.com/resources/automation-case-studies/mas-holdings-manufacturing-rpa. Accessed 15 Feb 2025
39. UN. https://unglobalcompact.org/what-isgc/participants/6398-MAS-Holdings-Pvt-Ltd-. Accessed 15 Feb 2025
40. Van der Aalst, W.M.P.: Hybrid intelligence: to automate or not to automate, that is the question. Int. J. Inf. Syst. Proj. Manag. **9**(2), 5–20 (2021)
41. Yin, R.K.: Case Study Research Design and Methods, 5th edn. SAGE, Los Angeles (2014)

FairPM: A Taxonomy of Bias and Interventions in Process Mining

Kate Revoredo[1](✉) , Saimir Bala[1,2] , and Flavia Santoro[3]

[1] Humboldt-Universität zu Berlin, Berlin, Germany
{kate.revoredo,saimir.bala}@hu-berlin.de
[2] SAP Signavio, Berlin, Germany
[3] Inteli, Sao Paulo, Brazil
flavia@inteli.edu.br

Abstract. As organizations increasingly rely on data-driven methods to support decision-making, ensuring fairness in their processes becomes critical. Fairness in responsible process mining involves preventing unfair outcomes and recognizing potential biases that may arise in the different stages of process mining initiative. Acting fairly entails treating individuals equitably, irrespective of inherent or acquired characteristics such as gender, race, or disability, while ensuring compliance with legal and organizational fairness standards. While fairness in process mining has been explored in prior research, there remains a lack of conceptualization to identify, understand, and address fairness issues. To bridge this gap, we propose **FairPM**, a taxonomy that conceptualizes biases in process mining and the corresponding interventions to mitigate them. Our approach builds on theory adaptation as research method. It integrates an adaptation of biases and interventions from prior machine learning research into process mining. We illustrate the applicability of **FairPM** through three scenarios, demonstrating its relevance for both academia and industry. This research contributes to the growing field of fair process mining by providing a structured conceptualization that enables researchers and practitioners to diagnose biases and implement fairness interventions, ensuring equitable and unbiased process mining outcomes.

Keywords: Fairness · Process mining · Bias · Debiasing interventions

1 Introduction

Process mining [2] is a data-driven discipline that analyzes event logs recorded by information systems to derive knowledge about how business processes are executed in practice. This knowledge is leveraged to improve existing processes, and support predictions and decision-making. Process mining outcomes may influence organizational actions and resource allocation, making it critical to consider potential fairness concerns. Fairness in responsible process mining involves preventing unfair outcomes and recognizing biases that may arise in a process mining initiative [17]. Bias can be defined as a systematic deviation of an estimated parameter from its true value [6], which may manifest in data, algorithms, or interpretation. While not all biases are inherently negative [25], understanding their impact is crucial when fairness is a goal. Acting fairly implies striving to avoid discrimination, especially in data-driven decisions, by preventing the misuse of data based on protected characteristics, such as gender, sexual orientation, religion, race, or disability, as defined by law or organizational policy. At the same time, fairness may require treating individuals differently to account for structural inequalities, acknowledging that treating everyone exactly the same can also lead to unfair outcomes. For example, Article 22(4) of the European General Data Protection Regulation (GDPR[1]) prohibits decision making based on protected data as defined in Article 9 of the regulation.

Although fairness in process mining has been discussed for a while [1,17,24] and different technical approaches have been proposed [15,19,22,26,31], there is still a lack of conceptualization to support researchers and practitioners to systematically identify, understand, and address fairness issues in process mining. To fill this gap, this paper aims to answer the following research question: *What are biases in process mining and what are possible interventions to mitigate them?*.

We propose FairPM, a taxonomy of 16 biases in process mining and corresponding 12 interventions to mitigate them. We adopt *theory adaptation* [10] as a research method, adapting the intervention taxonomy from [3] to the context of process mining by considering the types of bias that need to be addressed. To define the types bias, we adapt the conceptualization proposed by Mehrabi et al. [18] to the specifics of process mining. We demonstrate the applicability of our taxonomy using three scenarios. This research contributes to academia by expanding the body of knowledge on fairness in process mining. For industry, it provides a structured approach for guiding practitioners to analyze their processes, thereby supporting equitable and unbiased process mining outcomes.

This paper is structured as follows. Section 2 elaborates on the concept of fairness, describes three scenarios where biases negatively influence fairness at different stages of the process mining pipeline, and summarizes contributions from related work. Section 3 describes our theory adaptation method. Section 4 presents our FairPM taxonomy of biases and interventions in process mining. Section 5 how interventions can be used address the fairness issues in the three scenarios. Section 6 concludes the paper.

2 Background

2.1 Fairness

Fairness is a core ethical principle, but its interpretation varies across disciplines. Philosophers and theorists have approached fairness from different perspectives, often linking it to justice, equality, and moral responsibility. We highlight four

---

[1] https://gdpr-info.eu/.

perspectives demonstrating its nuanced nature and the need for contextual adaptation. Rawls [27] defines fairness as a principle of justice, proposing the "veil of ignorance" to ensure impartiality. He argues for equal basic liberties and structured inequalities benefiting the least advantage, prioritizing equity over mere equality. Sen [30] critiques Rawls' institutional focus, emphasizing real-life outcomes. His "capability approach" shifts attention from abstract rights to individuals' actual opportunities, framing fairness as dynamic and context-dependent. Yet, Rescher [28] challenges egalitarian views, arguing fairness should align distributions with valid claims rather than strict equality. He differentiates between resource ownership types, emphasizing objective criteria over subjective preferences. Furthermore, Pazner [21] critiques idealized fairness theories, highlighting real-world limitations. He warns against rigid applications that may hinder innovation or market efficiency, advocating for fairness as a flexible guideline integrated with economic and social realities.

In algorithmic decision-making field, fairness is a complex concept with no universal definition [29]. In distributive justice, fairness ensures outcomes are based on relevant attributes while avoiding discrimination [4]. Different definitions lead to varying approaches: Dwork et al. [5] emphasize consistency in decisions for similar individuals, while Joseph et al. [11] advocate for meritocratic fairness, prioritizing better-qualified individuals. Liu et al. [16] introduced calibrated fairness, aligning resource distribution with merit, which was most favored in empirical studies. Public attitudes toward fairness can vary, particularly concerning sensitive attributes like race. Saxena et al. [29] found that individualists favored fairness definitions acknowledging historical injustices when the disadvantaged individual was more qualified. This highlights the dynamic, context-dependent nature of fairness, suggesting rigid definitions may not fully address ethical concerns in algorithmic decision-making.

In machine learning, fairness is defined as the absence of bias [3,18] towards an individual or group based on sensitive[2] features such as gender, religion or race. Bias can propagate through a feedback loop between data, algorithms, and human decision-making [18]. This cycle can be broken down into three key stages: data to algorithm, algorithm to user, and user to data. In the data to algorithm stage, *historical* and *measurement* biases, among others (see Fig. 1 for all types of bias and [18] for a detail on them), embedded in training data can influence how mining algorithms learn patterns, leading to biased outcomes. In the algorithm to user stage, the learned artifact itself may contain biases due to *aggregation*, *sampling*, or *evaluation* decisions, influencing human decision-making. In the user to data stage, humans interacting with the system can reintroduce or reinforce biases, which then propagate back into future training data, completing the feedback loop. Intervention methods target different stages [3]: *pre-processing intervention* removes bias before training, *in-processing intervention* adjusts the learning process, and *post-processing intervention* corrects the learned artifact.

---

[2] Sensitive variables refer to attributes like race or gender that are directly tied to fairness concerns. Proxy variables are correlated features (e.g., zip code, department) that can unintentionally reveal or substitute for these sensitive attributes.

These techniques aim to enhance fairness and reliability in machine learning. In this paper, we will conceptualize fairness in process mining based on the conceptualization of fairness in machine learning.

2.2 Problem Description

To illustrate the types of bias, we consider the fictitious case study of *TrustBank*, a financial institution that used process mining to refine its credit scoring and risk management practices. By analyzing vast amounts of data, including customer transaction histories and loan repayment behaviors, the bank sought to improve accuracy, minimize financial risk, and ensure compliance with regulations. However, during this analysis, it discovered that certain demographic groups were statistically more likely to default on loans.

Scenario 1: Data to Algorithm. TrustBank's dataset contains *historical biases* from past loan approvals and *representation biases* due to flawed demographic data collection (e.g., non uniform *sampling*). For instance, if the bank historically denied loans to certain racial or socioeconomic groups, the dataset will reflect this trend. Assume predictive process monitoring techniques are trained on such data. They will associate high risk with these groups, even if they are creditworthy. Another issue arises from *measurement bias*. Suppose TrustBank estimates income using ZIP codes. Since neighborhoods are often segregated by income and historical policies (e.g., redlining), individuals from lower-income areas may receive lower credit scores, even if they have stable jobs and strong financial habits. ZIP codes are used as a proxy for financial stability. Furthermore, there is an *aggregation bias*, where the individual's creditworthiness is judged upon group creditworthiness rather than their own.
Problem: The dataset does not reflect true creditworthiness but rather past prejudices, leading to unfair predictions for marginalized groups.

Scenario 2: Algorithm to User. TrustBank employs a process mining algorithm to support the improvement of the process aiming at better scoring and risk assessment. However, it introduces *aggregation bias* by generalizing risk across broad demographic groups. Even if the data is unbiased, *design bias* is introduced due to design choices. A process discovery algorithm based on directly-follows graphs is used. However, this algorithm also filters and ranks the traces by frequency. As a rule of thumb, it creates a model that can roughly explain 80% of the traces. Therefore, less frequent traces related to cases by minority groups, are left out and not explained by the model. This in turn, leads to problems in conformance checking results, when the discovered model is used as a reference. Then, traces from minority groups are flagged as non-conformant and interpreted as unacceptable deviations, giving them a low credit scoring. Moreover, *user interaction bias* can occur when the user interacts with the results of process mining algorithms. For example, common process discovery techniques visualize a directly-follows graph or rank the discovered variants by

the most frequent behavior. When the user starts to analyze process behavior, variants of the process that rarely occur but are equally important have less chances getting the user's attention.
Problem: The process mining algorithm generalizes risk unfairly across diverse groups, leading to inaccurate and biased credit scoring.

Scenario 3: User to Data. Even if TrustBank had an unbiased credit-scoring algorithm, human users can still introduce *user interaction bias* through subjective decision-making. A loan officer reviewing applications may have implicit biases and reject applicants based on demographic factors, overriding the conformance checking fair results. Repeated user interactions may lead to *emergent bias* when these decisions get stored back into the event logs. Over time, the data from user interactions end up in the process history tracing biased user behavior, reinforcing discriminatory trends rather than correcting them. If bank employees consistently deny loans to applicants from specific backgrounds, the future traces in the event log lead the algorithms to prioritize similar decisions, reinforcing the bias of the analyst even if the algorithm was initially unbiased.
Problem: Users' actions and feedback loops can amplify biases, even when an algorithm is designed to be neutral.

2.3 Related Work

Fairness in process mining is an emerging concern and several studies have explored the implications of bias in process analytics and proposed methods to mitigate discrimination. These works highlight both the opportunities and challenges of ensuring fairness in predictive and process-oriented AI models. Fairness has been approached by the process mining community through two main research streams: one defining fairness within the context of process mining, and the other proposing fair process mining solutions.

The first research stream frames fairness as one of the FACT challenges (Fairness, Accuracy, Confidentiality, and Transparency) [1]. Van der Aalst [1] identifies fairness concerns in process mining tasks such as event data management, process discovery, conformance checking, and performance analysis. However, the study only briefly illustrates fairness issues with an example and does not systematically explore causes of discrimination or mitigation strategies. Mannhardt [17] builds upon this by applying machine learning fairness definitions to process mining, discussing algorithmic fairness challenges through examples. Expanding this perspective, Pohl et al. [24] investigate discrimination in process mining, focusing on unfair treatment of cases and resources. The authors categorize fairness concerns into three areas: discrimination in processes, where biased historical practices affect outcomes; discrimination in conclusions, where unfair patterns emerge in analysis results; and the impact of process mining techniques on fairness, where algorithms may reinforce bias. Their work highlights the need for domain-specific fairness notions rather than direct application of machine learning fairness principles.

The second research stream leverages machine learning techniques to mitigate bias in process mining. Predictive process monitoring [7] has been the main focus, with three proposed approaches addressing the prediction of binary outcomes [22,26,31] and one approach covering both binary outcome prediction and numeric prediction [15]. Process discovery was tackled by one approach [19]. Qafari et al. [26] construct a fair classifier by relabeling the leaves of a decision tree learned from situations extracted from event logs and transformed into a tabular dataset, aiming to reduce the level of discrimination. Da Silva et al. [31] explore an intervention method implemented[3] in the machine learning field for each of the three stages of intervention. The pre-processing intervention is addressed through a *Reweighing technique*, which adjusts training instance weights to ensure equal representation of protected and non-protected groups. *Adversarial Debiasing* serves as an in-processing method to reduce bias during model training. Finally, post-processing intervention applies *Equalized Odds (Calibration)*, which ensures fairness by equalizing true positive and false positive rates across protected groups. Peeperkorn et al. [22] address group fairness using the independence criterion, exploring demographic parity and threshold-independent, distribution-based fairness metrics. They introduce a loss function that combines binary cross-entropy with a fairness-based loss, enabling trade-offs between predictive accuracy and fairness. De Leoni et al. [15] apply adversarial learning to train predictive models that are independent of a sensitive variable, thereby reducing bias in decision-making. Finally, Muskan et al. [19] propose a fairness-aware process discovery approach. Their method uses a genetic algorithm to learn a process model by partitioning the event log into two sublogs based on a sensitive variable. To ensure fairness, they introduce two equalizing log fitness measures that balance process fitness across the sublogs. The results demonstrate that this approach supports uncovering unfairness in business processes.

Despite increasing interest, the process mining field still lacks a comprehensive fairness framework tailored to its domain. Existing studies highlight fairness challenges and propose algorithmic solutions, but a holistic approach (i.e., integrating perspectives across the process lifecycle actors, technical layers, and organizational practices) remains underdeveloped.

This paper addresses this gap by proposing a structured fairness taxonomy for process mining, identifying key fairness risks (i.e., bias), and outlining mitigation strategies to support equitable and unbiased process mining outcomes.

3 Research Method

The research method adopted is the theory adaptation according to [10]. Jaakkola [10] proposes four distinct approaches to guide the development of conceptual papers: (i) Theory Synthesis: integrates previously unconnected concepts to offer a new or enhanced understanding of a phenomenon; (ii) Theory Adaptation: modifies existing theories using insights from other theories through

---

[3] AiFairness 360 (AIF360) library: https://ai-fairness-360.org/.

a process called problematization; (iii) Typology: classifies conceptual variants to provide a more precise and nuanced perspective. and (iv) Model: focuses on building a theoretical framework that predicts relationships. Jaakkola [10] also emphasizes that, like empirical studies, conceptual papers must be grounded in a clear research design. Researchers should justify their choice of theories and explicitly define the role each theory plays in their analysis, ensuring a logical and complete chain of evidence is crucial for constructing robust arguments. The research design elements in conceptual papers should be: (i) choice of theories and concepts used to generate novel insights, (ii) choice of theories and concepts analyzed, (iii) perspective; level(s) of analysis/aggregation, (iv) key concepts to be analyzed/explained or used to analyze/explain, (v) translation of target phenomenon in conceptual language; definitions of key concepts, and (vi) approach to integrating concepts; quality of argumentation.

Theory adaptation fits our purpose: we revisit and apply the concept of fairness from machine learning to business process management (BPM), specifically process mining. We adopt a broad understanding of theory that includes frameworks, models, taxonomies, and related scientific constructs [9]. The methodological steps are described below.

First, a comprehensive analysis of existing frameworks and models that theorize fairness and fairness in process mining were examined in Sect. 2.1 and Sect. 2.3, respectively (elements (i) and (ii)). Based on our research question and the data-driven nature of process mining, we chose the machine learning framework described in [3,18] as the basis to our proposal. As *domain theory* (i.e., a particular set of knowledge on a substantive topic area situated in a field or domain) we used the existing knowledge about process mining and fairness in process mining. As method theory (i.e., a meta-level conceptual system for studying the substantive issue(s) of the domain theory at hand) we used the conceptualization of bias presented in [18] and the intervention taxonomy proposed in [3] from machine learning.

Second, once we had selected the domain theory and method theories, we performed the theory adaptation (elements (iii), (iv) and (v)) using these theories as follows. In the theory adaptation, we looked for commonalities and incompleteness between the domain theory and the method theories. Here, we confront the understanding of biases and intervention of biases in process mining and in machine learning. Our analysis was guided by the questions: how can fairness in process mining account for bias in data, algorithms, and human decision-making? And how these biases can be addressed in the pre-processing, in-processing and post-processing stages of intervention. Section 4 describes the taxonomy result of the theory adaptation. Finally, we provide a discussion of the taxonomy based on the scenarios described in Sect. 2.2 (element (vi)).

4 FairPM Taxonomy

In this section, we describe the FairPM taxonomy, which is derived as an adaptation of two theories: a taxonomy of interventions and a conceptualization of bias from machine learning. The taxonomy described eleven methods categorized into pre-processing, in-processing or post-processing interventions. We include one additional intervention for post-processing in process mining called Visualization. The conceptualization of bias described 19 biases. We adapt them into 16 biases relevant for process mining. Figure 1 depicts `FairPM` and the following subsections detail the adaptation. For each intervention method, we indicate the *fairness objective*, the *bias(es)* that it is addressing, if it was already considered in the process mining literature (*Synopsis*) and some reflections on how it can be used to process mining (*Commentary*).

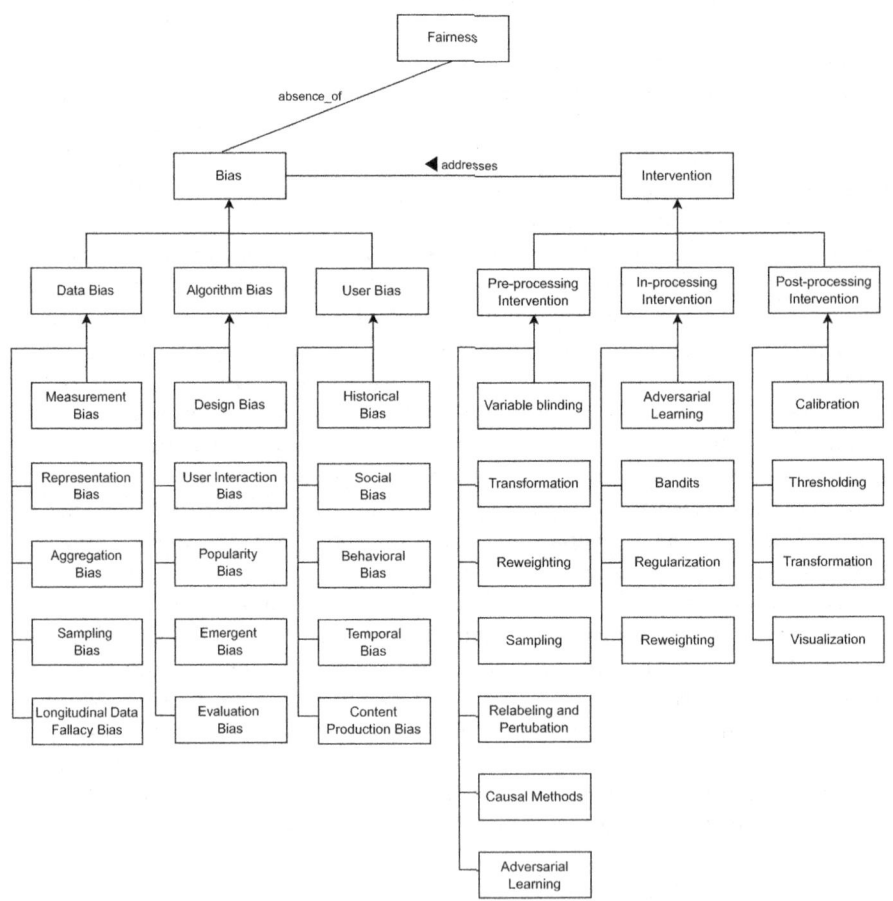

Fig. 1. FairPM taxonomy

4.1 Variable Blinding

Fairness objective: to make the artifact immune to sensitive variables.
Bias addressed: Measurement bias, emergent bias, historical bias.
Synopsis: In [26], a blinding method was used so the classifier was learned without considering the sensitive variable (i.e., omission technique).
Commentary: A process model is, for example, gender blind if its variants are independent of gender (i.e., the possible control-flows are not differentiable by gender). One possible technique is to omit the sensitive variable from the data, preventing the algorithm from using it. In this regards, process discovery techniques can be considered as already applying a blinding solution given that only the sequence of activities are used to learn a process model (i.e., no sensitive variable is used to guide the learning). Another possible solution is to build an artifact that it is immune to the sensitive variable and its possible proxy variables. For process discovery techniques, given that the learned process model represents how the process is running in reality, make the process model immune to a sensitive variable and its proxy variables means make explicit the discrimination that it is happening. For instance, the gender discrimination is implicit in the data and the variants of the process model are indeed reflecting discriminatory behavior (e.g., certain checks during the evaluation of a loan application are only done for women). An enhanced process model indicating that 10% of the process instances have a certain outcome when a choice is made, it is not making explicit that this behavior happen only for one of the genders. Implicitly the choice in this process is done based on the sensitive variable. In this scenario the process model can be used to provide awareness of the discrimination. Highlighting in the enhanced model that a particular behavior only happens for a certain value of the sensitive variable or its proxy variables can support improvement of the process towards a fair process.

4.2 Transformation

Fairness objective: Learn a new fair representation of the data while still preserving the fidelity of the mining task.
Bias addressed: Measurement bias, social bias, behavioral bias, content production bias, emergent bias, historical bias.
Synopsis: In [26], a transformation method was applied as a post-processing intervention. This approach used dynamic programming and rounding to select an approximately optimal set of leaves from the decision tree learned for relabeling. The relabeling adjusts the probability distribution of the output, aiming to balance accuracy and fairness.
Commentary: Techniques in this group are relevant to process mining methods that rely on attributes beyond control-flow, such as predictive process monitoring. For instance, numeric attributes that serve as proxies for sensitive variables could be transformed to preserve only the order while removing the actual values.

4.3 Reweighing

Fairness objective: Change the impact of instances on the artifact during learning to promote fair handling of sensitive variables and groups.

Bias addressed: Representation bias, design bias, popularity bias, evaluation bias, social bias, emergent bias.

Synopsis: In [31] an exploratory study was conducted to evaluate the impact of intervention techniques on the task of predicting a binary outcome using Random Forest. As pre-processing intervention technique, a reweighing method was used.

Commentary: Unlike transformation, relabeling, and perturbation approaches that alter the data, reweighing assigns weights to instances of the training data while leaving the data itself unchanged. Process discovery techniques could benefit from this group by increasing the relevance of variants associated with underrepresented groups, allowing these variants to appear in the final process model. Such weights could also be applied in enhancement techniques, which, when combined with filtering methods, would prevent the underrepresentation of certain groups from being retained in the final model. In the context of predictive process monitoring, techniques from this group help make the predictive model more balanced, thereby promoting fairness in its outputs. Additionally, this method can be used as an in-processing intervention, where algorithms learn the weights for instances and use them during the learning task.

4.4 Sampling

Fairness objective: Create samples for the training where the bias is eliminated. Also, to identify groups of data that are in disadvantage.

Bias addressed: Aggregation bias, longitudinal data fallacy, temporal bias, representation bias, emergent bias.

Synopsis: In [19], the event log is split into two subgroups based on a sensitive variable. The approach requires the prior identification of this variable. One subgroup contains the majority of cases where unfairness is not observed, while the other subgroup contains the minority cases where unfairness issues are present.

Commentary: A challenge for this method is how to sample [12] or create groups when the sensitive variable is not known. Causal methods can support in pre-processing the event log to identify sensitive and proxy variables. Additionally, the technique in [19] used a single sensitive variable at a time. A key challenge is how to sample and create groups when considering multiple sensitive variables simultaneously. Furthermore, in [19], a single process model was created using both subgroups. An alternative consideration would be to discover two separate process models. So far, only process discovery has been addressed by this method. For process mining, conformance checking would also benefit from the use of subgroups to indicate the alignment of both event logs with the existing process model. In predictive process monitoring, the different subgroups can be used to train specific predictors. When applying such a method, one must be aware of the potential inclusion of other biases, such as representation bias.

4.5 Relabeling and Perturbation

Fairness objective: Modify the training data such that underprivileged and privileged instances are treated similarly and/or explore the effects of such modifications on artifact fairness.
Bias addressed: Representation Bias, sampling Bias, social bias, behavioral bias, historical bias.
Synopsis: Not yet considered by process mining approaches.
Commentary: The techniques in this group are a subset of the Transformation techniques. They focus on changing the labels of instances (relabeling) or the values of attributes (perturbation) when a model is to be learned through supervised learning. They modify the data to achieve equilibrium between the target variable and the sensitive variable. Given their focus on supervised learning, these techniques are particularly suited for predictive process monitoring.

4.6 Causal Methods

Fairness objective: Identify potentially useful relationships between sensitive and non-sensitive variables to provide insights for fairness-related methodological decisions.
Bias addressed: Measurement bias, social bias, behavioral bias, temporal bias, content production bias.
Synopsis: Not yet considered by process mining approaches.
Commentary: The techniques in this group focus on identifying dependencies between the sensitive variables and other variables in the data. The outcomes of these techniques can guide the choice of how to address the bias. For example, if the sensitive variables and any other identified dependent variables are to be excluded from the data in the pre-processing phase, or if a post-processing technique will be used. This set of techniques can also provide transparency regarding how decisions were made, which groups are most (un)fairly treated, and help differentiate the types of bias exhibited.

4.7 Adversarial Learning

Fairness objective: Tutor an artifact to be fairer by providing in-training feedback or modifying the training data to promote immunity to one or more sensitive variables.
Bias addressed: Design bias, popularity bias, evaluation bias.
Synopsis: Adversarial Learning [8] is a technique proposed initially to improve the prediction of a neural network. For that it uses two neural networks with competing goals to force the first network to mislead the second network. In process mining it was used as in-processing method by [15] and the exploratory study performed in [31].
Commentary: In process mining, this method has been used only for predictor learning but could also be used to support transformation techniques during pre-processing.

4.8 Bandits

Fairness objective: Instrument fair online decision making with little or no training data.
Bias addressed: Design bias, popularity bias, evaluation bias.
Synopsis: Not yet considered by process mining approaches.
Commentary: This in-processing set of techniques analysis potential unfairness to individuals or groups during learning and promote a fair distribution of the possible actions to be taken.

4.9 Regularization

Fairness objective: Extend the artifact's evaluation function such that it penalizes unfair behavior.
Bias addressed: Popularity bias, evaluation bias.
Synopsis: In [19], two quality measures were proposed by leveraging alignment costs between the event log and the process model being learned, a technique typically used to calculate log fitness. These quality measures aim to equalize the log fitness scores for both sublogs. In [22], fairness in binary outcome process prediction was addressed. The model evaluation function combined a standard supervised learning loss function, such as binary cross-entropy, with an integral probability metric, such as the Wasserstein distance, which quantifies the difference between the two prediction distributions. These components were combined using a weighted sum, where a hyperparameter controls the trade-off between maximizing predictive accuracy and ensuring fairness. To evaluate the fairness of the learned model, the study used the average propensity difference and the difference in the proportion of positive predictions between the two groups, alongside standard binary classification metrics such as accuracy, F1 score, and AUC.
Commentary: The technique proposed in [19] was applied to process discovery, while the approach in [22] was used for predictive process monitoring. However, both methods could be adapted for all process mining tasks. In general, techniques developed using this approach can be applied across different process mining tasks. The key idea is to extend standard evaluation functions, such as fitness in process discovery and conformance checking or accuracy in predictive process monitoring to incorporate penalties for unfair behavior.

4.10 Calibration

Fairness objective: To adjust the probability outputs of a model such that the portion of predicted positive outcomes matches that of positive examples across (or within) all groups in the dataset.
Bias addressed: Design bias, popularity bias, evaluation bias.
Synopsis: The approach described in Sect. 4.2 (i.e., the method proposed in [26]) also incorporated a calibration technique. In [31], the labels of a Random Forest model were adjusted using linear programming. This calibration method

was referred to as Equalized Odds in the paper.
Commentary: In process discovery, calibration can be used, for example, to ensure that a certain activity is executed in the same proportion across different values of the sensitive variable. In predictive process monitoring, calibration guarantees that the output of the learned model occurs in the same proportion as observed for the values of the sensitive variable.

4.11 Thresholding

Fairness objective: Consider fairness metrics in the setting of thresholds applied to the artifact learned.
Bias addressed: Design bias, popularity bias, evaluation bias.
Synopsis: In [22], fairness in binary outcome prediction was assessed by measuring the difference in the proportion of positive predictions between the two groups. This measure relies on an indicator function that equals 1 if the predicted value exceeds a predefined threshold and 0 otherwise. Similarly, in [26], a threshold was used to define the acceptable level of discrimination in the final decision tree. The relabeling of the tree's leaves was performed to ensure that discrimination remained below this threshold.
Commentary: Thresholding is a post-processing approach based on the idea that discriminatory decisions often occur near decision-making boundaries due to decision-makers' biases and that humans tend to apply threshold rules when making decisions. Approaches in this group focus on learning threshold values to be used in the post-processing phase to assess whether the final artifact meets fairness criteria. While predictive process monitoring has been the primary focus of this method, other process mining tasks could also benefit from its application. For instance, thresholding could be used to limit the extent of discrimination in a learned process model.

4.12 Visualization

Additionally, we incorporated the *Visualization* method as a post-processing intervention.
Fairness objective: Adjust visualization of the artifact to single out unfair representation.
Bias addressed: User Interaction Bias
Synopsis: Although it was not the primary focus of the paper, [19] highlighted unfair behavior in the visualization of the learned process model, allowing users to gain awareness of unfairness in the process.
Commentary: The user interpretation may lead to unfairness due to the way the outcome is displayed, even when the data and algorithm are fair. This method leverages graphical elements [13], filters, activity and path highlighting, and allows users to adjust thresholds, layouts, colors, and alternative visualizations to identify and mitigate unfair representation. It can be explored by all the process mining techniques.

5 Discussion

This section discusses potential interventions to reduce bias in the scenarios introduced in Sect. 2.2, followed by practical implications and limitations.

5.1 Scenario 1: Data to Algorithm

This scenario was affected by historical, representation, measurement and aggregation biases. Such biases happen in the data-to-algorithm stage and must be addressed by interventions in the pre-processing phase. Let us use the introduced interventions to tackle each bias in turn. A historical bias regarding previous loans denied to certain racial or socioeconomic groups was present in the dataset used by TrustBank. This bias can be addressed via a number of interventions. Variable blinding can be used by removing the sensitive variables representing the racial or special socio-economic group. In this way, the process prediction algorithm would not consider these attributes to discriminate the relative cases. When variable blinding is not possible because the algorithm requires such input, other interventions can be used. Transformation is an alternative intervention that can operate on the dependent variable. For example, we can map individuals to an input space that is independent of the specific protected subgroup. Another option is the relabeling of outcome based on the sensitive variable so that the cases have an equal chance of acceptance across all groups.

The aforementioned interventions can also be used to reduce the representation bias, by balancing the shares of outcomes (i.e., deny loan, accept loan) across all groups. A further intervention to reduce representation bias is by reweighing. For instance, if only a few cases from minority groups are present in the event log, we can assign them a higher weight (e.g., we can repeat the number of such traces in the event log). Measurement bias, where ZIP codes were used as a proxy for financial stability, can be addressed through causal methods by identifying relationships between variables. Consequently, a decision can be made to rule out proxies (e.g., through blinding). Finally, the aggregation bias can be addressed by sampling (i.e., creating a sublog that fairly represents the relations among the super and subgroups) and subgroup analysis, consequently creating event logs for each sensitive group [19].

5.2 Scenario 2: Algorithm to User

By relying on the algorithms for the analysis of the financial scores of their customers, TrustBank also relies on their internal bias. In the context of this scenario, as aggregation bias follows from using a one-model-fits-all solution, that does not distinguish among the peculiarities of the different groups. To address this aggregation issue, knowing how the process mining algorithm works, input data can be modified to achieve a fair outcome. For instance, the event log can be split by group of interest, following a subgroup analysis pre-processing intervention. Knowing the decisions made by the algorithm, reweighing is also an option, giving sufficient weight (e.g., frequency) to the lowly represented cases.

As well, in-processing interventions can be used to address design bias. Design biases in this case emerge from the ranking, filtering and decision making that incorporated in the algorithm. Adversarial learning can be used with a two-fold goal, to discover accurate traces and to discover a fair distribution of all the traces relative to the sensitive groups. Bandits can also be used. A fairness-aware bandit can be used to explore alternative decisions (rather than denying loan to low-income applicants), ensuring that all applicants are treated fairly.

Post-processing techniques can be used to address biases, too. Calibration can be used on the model or variants, by increasing the occurrence of certain activities or variants associated to the underrepresented groups. Thresholding is also an intervention, using an indicator function that assigns a threshold to the various groups. This can counter filtering or ranking effects introduced by the algorithmic bias.

5.3 Scenario 3: User to Data

In the context of this scenario, user introduced bias deriving from their interaction with the process mining tool can be tackled at different levels. A visualization intervention can be used in order to nudge users to not take biased decision by highlighting potential fairness pitfalls when interacting with process mining results. Another kind of bias affecting this scenario is emergent bias. To address this bias, process mining algorithms should take into account how outcome of the different applications, especially from the sensitive group, evolves over time. If there is a positive trend towards rejecting specific groups (e.g., based on ZIP code, race, gender, etc.) then, one of the pre-processing interventions should be used to either transform, sample or blind the corresponding variables.

Summary. Regarding our research question *What are biases in process mining and what are possible interventions to mitigate them?*, we provide the `FairPM` taxonomy with 16 biases and 12 interventions. This taxonomy distills knowledge from the current state of the art solutions for fair process mining. The above scenarios show how our proposed taxonomy helps the practitioners in two ways: i) to identify the type of bias and ii) to obtain the relative interventions.

5.4 Practical Implications and Limitations

Organizations can apply the taxonomy in a phased manner: first by assessing potential bias points in their process pipeline using diagnostic tools, and then by selecting targeted interventions that address the identified issues. Not all detected bias may be relevant for a particular organization, and not all interventions are universally applicable; their relevance depends on the type of process, business goals, existing organizational awareness of bias, and resource constraints. Our taxonomy is as a guide for structured reflection and context-sensitive action.

This study is subject to several limitations. The taxonomy is primarily conceptual and grounded in literature synthesis, which may limit its completeness.

While we acknowledge the limitations of using a fictional case, it was designed based on recurring patterns observed in actual organizational practices and interviews. The structure mirrors documented bias challenges in domains hiring, renting lending and hospital [23].

6 Conclusion

This paper examines fairness challenges in process mining by adapting machine learning fairness principles to its unique context. We identify key biases across data collection, algorithmic processing, and user interactions, mapping them to known computational fairness issues. To address these biases, we propose a fairness taxonomy for process mining, conceptualizing 16 bias and 12 interventions . These techniques adjust data, learning algorithms, or model outputs to mitigate bias. We discuss the use of the taxonomy in three scenarios.

Fairness in process mining remains an evolving field. From the analysis of the state-of-the-art using FairPM, we observed that three intervention methods were not yet considered. Future research should refine domain-specific fairness metrics [24], evaluate fairness interventions on real-world event logs, develop automated fairness tools, and balance fairness with accuracy and efficiency. Embedding fairness at all process mining stages enhances transparency, accountability, and equitable decision-making, aligning organizations with ethical and regulatory standards. Furthermore, as future work we will evaluate the use of the FairPM taxonomy on real-world cases and conduct interviews to investigate on their use in practice by organizations [14, 20].

Acknowledgments. Kate Revoredo is funded by the Berliner Chancengleichheitsprogramm (BCP) as part of the DiGiTal Graduate Program. Saimir Bala is supported by the Einstein Foundation Berlin under grant EPP-2019-524.

References

1. van der Aalst, W.M.P.: Responsible data science: using event data in a people friendly manner. In: ICEIS (Revised Selected Papers). LNiBIP, vol. 291, pp. 3–28. Springer (2016)
2. van der Aalst, W.: Process Mining: Data Science in Action. Springer (2016)
3. Caton, S., Haas, C.: Fairness in machine learning: a survey. ACM Comput. Surv. **56**(7), 166:1-166:38 (2024)
4. Chouldechova, A.: Fair prediction with disparate impact: a study of bias in recidivism prediction instruments. Big Data **5**(2), 153–163 (2017)
5. Dwork, C., Hardt, M., Pitassi, T., Reingold, O., Zemel, R.S.: Fairness through awareness. In: ITCS, pp. 214–226. ACM (2012)
6. Feuerriegel, S., Dolata, M., Schwabe, G.: Fair AI. Bus. Inf. Syst. Eng. **62**(4), 379–384 (2020)
7. Francescomarino, C.D., Ghidini, C.: Predictive process monitoring. In: Process Mining Handbook. LNiBIP, vol. 448, pp. 320–346. Springer (2022)

8. Goodfellow, I.J., et al.: Generative adversarial networks. Commun. ACM **63**(11), 139–144 (2020)
9. Gregor, S., Benbasat, I.: Explanations from intelligent systems: theoretical foundations and implications for practice. MIS Q. **23**(4), 497–530 (1999)
10. Jaakkola, E.: Designing conceptual articles: four approaches. AMS Rev. **10**(1), 18–26 (2020)
11. Joseph, M., Kearns, M.J., Morgenstern, J., Roth, A.: Fairness in learning: classic and contextual bandits. In: NIPS, pp. 325–333 (2016)
12. Kabierski, M., Richter, M., Weidlich, M.: Quantifying and relating the completeness and diversity of process representations using species estimation. Inf. Syst. **130**, 102512 (2025)
13. Kaur, H., Mendling, J., Kampik, T., Rubensson, C.: Towards timeline-based layout for process mining. In: PoEM. LNiBIP, vol. 538, pp. 192–206. Springer (2024)
14. Kundisch, D., et al.: An update for taxonomy designers. Bus. Inf. Syst. Eng. **64**(4), 421–439 (2022)
15. de Leoni, M., Padella, A.: Achieving fairness in predictive process analytics via adversarial learning. In: CoopIS. LNCS, vol. 15506, pp. 346–354. Springer (2024)
16. Liu, Y., Radanovic, G., Dimitrakakis, C., Mandal, D., Parkes, D.C.: Calibrated fairness in bandits. arXiv preprint arXiv:1707.01875 (2017)
17. Mannhardt, F.: Responsible process mining. In: Process Mining Handbook. LNiBIP, vol. 448, pp. 373–401. Springer (2022)
18. Mehrabi, N., Morstatter, F., Saxena, N., Lerman, K., Galstyan, A.: A survey on bias and fairness in machine learning. ACM Comput. Surv. **54**(6), 115:1-115:35 (2022)
19. Muskan, Mannhardt, F., van Dongen, B.F.: Extending genetic process discovery to reveal unfairness in processes. In: ICPM Workshops. LNiBIP, vol. 533, pp. 751–763. Springer (2024)
20. Nickerson, R.C., Varshney, U., Muntermann, J.: A method for taxonomy development and its application in information systems. Eur. J. Inf. Syst. **22**(3), 336–359 (2013)
21. Pazner, E.A.: Pitfalls in the theory of fairness. J. Econ. Theory **14**(2), 458–466 (1977)
22. Peeperkorn, J., Vos, S.D.: Achieving group fairness through independence in predictive process monitoring. arXiv preprint arXiv:2412.04914 (2024)
23. Pohl, T., Berti, A., Qafari, M.S., van der Aalst, W.M.P.: A collection of simulated event logs for fairness assessment in process mining. In: BPM (Demos / Resources Forum). CEUR Workshop Proceedings, vol. 3469, pp. 87–91. CEUR-WS.org (2023)
24. Pohl, T., Qafari, M.S., van der Aalst, W.M.P.: Discrimination-aware process mining: a discussion. In: ICPM Workshops. LNiBIP, vol. 468, pp. 101–113. Springer (2022)
25. Pot, M., Kieusseyan, N., Prainsack, B.: Not all biases are bad: equitable and inequitable biases in machine learning and radiology. Insights Imag. **12**(1), 1–10 (2021). https://doi.org/10.1186/s13244-020-00955-7
26. Qafari, M.S., van der Aalst, W.: Fairness-aware process mining. In: Panetto, H., Debruyne, C., Hepp, M., Lewis, D., Ardagna, C.A., Meersman, R. (eds.) OTM 2019. LNCS, vol. 11877, pp. 182–192. Springer, Cham (2019). https://doi.org/10.1007/978-3-030-33246-4_11
27. Rawls, J.: A Theory of Justice: Original Edition. Harvard University Press (1971)
28. Rescher, N.: Fairness: Theory & Practice of Distributive Justice. Transaction Publishers (2002)

29. Saxena, N.A., Huang, K., DeFilippis, E., Radanovic, G., Parkes, D.C., Liu, Y.: How do fairness definitions fare?: Examining public attitudes towards algorithmic definitions of fairness. In: AIES, pp. 99–106. ACM (2019)
30. Sen, A.: The Idea of Justice. Harvard University Press, Cambridge, MA (2011)
31. da Silva, M.C., Fantinato, M., Peres, S.M.: Towards fairness-aware predictive process monitoring: evaluating bias mitigation techniques. In: CoopIS. LNCS, vol. 15506, pp. 150–166. Springer (2024)

Author Index

A
Aa, Han van der 128
Adan, Ivo 256

B
Bala, Saimir 395
Balaktsis, Christos 147
Banham, Adam 327
Barbaro, Luca 3
Bukhsh, Zaharah 256
Burattin, Andrea 23

C
Caballero-Villalobos, Juanita 23
Choi, Yongsun 291
Comuzzi, Marco 147

D
David, Yuval 40
de Leoni, Massimiliano 274
De Moor, Jakob 165
De Smedt, Johannes 165
De Weerdt, Jochen 165
Di Ciccio, Claudio 3, 238
Dijkman, Remco 256, 308

E
Elyasi, Keyvan Amiri 128

F
Farkas, Martin 58
Fournier, Fabiana 40
Frazzetto, Paolo 274

G
Gaaloul, Walid 221
García-Pérez, Álvaro 221

Gounaris, Anastasios 147
Grisold, Thomas 345
Grohs, Michael 183

H
Henry, Tiphaine 221

J
Janiesch, Christian 361

K
Kabierski, Martin 75
Kirchdorfer, Lukas 128
Kocsis, Imre 58
Kowalkiewicz, Marek 361
Kryston, Michele 238
Kuzmanoski, Aleksandar 92

L
Lee, Deoksang 204
Leemans, Sander J. J. 92
Lemaire, Victor 221
Limonad, Lior 40
López, Hugo A. 23

M
Maggi, Fabrizio Maria 147
Mamudu, Azumah 327
Marangone, Edoardo 238
Marcelletti, Alessandro 238
Mavroudopoulos, Ioannis 147
Middelhuis, Jeroen 256
Mirispelakotuwa, Ishadi 378
Montali, Marco 3

N
Navarin, Nicolò 274

O
Ouyang, Chun 308
Özdemir, Konrad 128

P
Padella, Alessandro 274
Péter, Bertalan Zoltán 58
Prinz, Thomas M. 291

R
Rebmann, Adrian 183
Rehse, Jana-Rebecca 75, 183
Revoredo, Kate 395
Rinderle-Ma, Stefanie 110
Rosemann, Michael 345, 361

S
Santoro, Flavia 395
Schumann, Felix 110
Skarbovsky, Inna 40
Song, Minseok 204

Stuckenschmidt, Heiner 128
Syed, Rehan 327, 378

T
Tucci-Piergiovanni, Sara 221

V
van der Aalst, Wil M. P. 204
van der Heijden, G. Wessel 110
van der Werf, Jan Martijn E. M. 75
van Detten, Jan Niklas 92
Varricchione, Giovanni 3
Vetterlein, Anja 291

W
Weytjens, Hans 165
Wynn, Moe T. 378

Y
Yang, Roy Jing 308

Made in the USA
Monee, IL
03 May 2026